Mobile and Blended Learning Innovations for Improved Learning Outcomes

David Parsons
The Mind Lab by Unitec, New Zealand

A volume in the Advances in Mobile and Distance Learning (AMDL) Book Series

Information Science REFERENCE
An Imprint of IGI Global

Published in the United States of America by
Information Science Reference (an imprint of IGI Global)
701 E. Chocolate Avenue
Hershey PA, USA 17033
Tel: 717-533-8845
Fax: 717-533-8661
E-mail: cust@igi-global.com
Web site: http://www.igi-global.com

Library of Congress Cataloging-in-Publication Data

Names: Parsons, David, 1959 October 13- editor.
Title: Mobile and blended learning innovations for improved learning outcomes / David Parsons, editor.
Description: Hershey, PA : Information Science Reference, [2016] | Includes bibliographical references and index.
Identifiers: LCCN 2016006155| ISBN 9781522503590 (hardcover) | ISBN 9781522503606 (ebook)
Subjects: LCSH: Mobile communication systems in education. | Blended learning--Technological innovations.
Classification: LCC LB1044.84 .M58 2016 | DDC 371.33--dc23 LC record available at https://lccn.loc.gov/2016006155

This book is published in the IGI Global book series Advances in Mobile and Distance Learning (AMDL) (ISSN: 2327-1892; eISSN: 2327-1906)

British Cataloguing in Publication Data
A Cataloguing in Publication record for this book is available from the British Library.

All work contributed to this book is new, previously-unpublished material. The views expressed in this book are those of the authors, but not necessarily of the publisher.

For electronic access to this publication, please contact: eresources@igi-global.com.

Advances in Mobile and Distance Learning (AMDL) Book Series

Patricia Ordóñez de Pablos
Universidad de Oviedo, Spain

ISSN: 2327-1892
EISSN: 2327-1906

Mission

Private and public institutions have made great strides in the fields of mobile and distance learning in recent years, providing greater learning opportunities outside of a traditional classroom setting. While the online learning revolution has allowed for greater learning opportunities, it has also presented numerous challenges for students and educators alike. As research advances, online educational settings can continue to develop and advance the technologies available for learners of all ages.

The **Advances in Mobile and Distance Learning** (AMDL) Book Series publishes research encompassing a variety of topics related to all facets of mobile and distance learning. This series aims to be an essential resource for the timeliest research to help advance the development of new educational technologies and pedagogy for use in online classrooms.

Coverage

- Managing Sustainable Learning
- Economics of Distance and M-Learning
- Distance Learning
- Pedagogy and Design Methodology
- Educational Apps
- Online Class Management
- Course Design
- Ethical Considerations
- Student Achievement and Satisfaction
- Ubiquitous and Pervasive Learning

IGI Global is currently accepting manuscripts for publication within this series. To submit a proposal for a volume in this series, please contact our Acquisition Editors at Acquisitions@igi-global.com or visit: http://www.igi-global.com/publish/.

The Advances in Mobile and Distance Learning (AMDL) Book Series (ISSN 2327-1892) is published by IGI Global, 701 E. Chocolate Avenue, Hershey, PA 17033-1240, USA, www.igi-global.com. This series is composed of titles available for purchase individually; each title is edited to be contextually exclusive from any other title within the series. For pricing and ordering information please visit http://www.igi-global.com/book-series/advances-mobile-distance-learning/37162. Postmaster: Send all address changes to above address.

Titles in this Series

For a list of additional titles in this series, please visit: www.igi-global.com

Integrating Touch-Enabled and Mobile Devices into Contemporary Mathematics Education
Maria Meletiou-Mavrotheris (European University Cyprus, Cyprus) Katerina Mavrou (European University Cyprus, Cyprus) and Efi Paparistodemou (Cyprus Pedagogical Institute, Cyprus)
Information Science Reference • copyright 2015 • 341pp • H/C (ISBN: 9781466687141) • US $190.00 (our price)

Identification, Evaluation, and Perceptions of Distance Education Experts
Gulsun Eby (Anadolu University, Turkey) and T. Volkan Yuzer (Anadolu University, Turkey)
Information Science Reference • copyright 2015 • 354pp • H/C (ISBN: 9781466681194) • US $195.00 (our price)

Assessing the Role of Mobile Technologies and Distance Learning in Higher Education
Patricia Ordóñez de Pablos (University of Oviedo, Spain) Robert D. Tennyson (University of Minnesota, USA) and Miltiadis D. Lytras (American College of Greece, Greece)
Information Science Reference • copyright 2015 • 377pp • H/C (ISBN: 9781466673168) • US $195.00 (our price)

Critical Examinations of Distance Education Transformation across Disciplines
Abigail G. Scheg (Western Governors University, USA)
Information Science Reference • copyright 2015 • 333pp • H/C (ISBN: 9781466665552) • US $185.00 (our price)

Promoting Active Learning through the Integration of Mobile and Ubiquitous Technologies
Jared Keengwe (University of North Dakota, USA)
Information Science Reference • copyright 2015 • 331pp • H/C (ISBN: 9781466663435) • US $195.00 (our price)

Artificial Intelligence Applications in Distance Education
Utku Kose (Usak University, Turkey) and Durmus Koc (Usak University, Turkey)
Information Science Reference • copyright 2015 • 329pp • H/C (ISBN: 9781466662766) • US $195.00 (our price)

Handbook of Research on Emerging Priorities and Trends in Distance Education Communication, Pedagogy, and Technology
T. Volkan Yuzer (Anadolu University, Turkey) and Gulsun Eby (Anadolu University, Turkey)
Information Science Reference • copyright 2014 • 480pp • H/C (ISBN: 9781466651623) • US $315.00 (our price)

Practical Applications and Experiences in K-20 Blended Learning Environments
Lydia Kyei-Blankson (Illinois State University, USA) and Esther Ntuli (Idaho State University, USA)
Information Science Reference • copyright 2014 • 519pp • H/C (ISBN: 9781466649125) • US $175.00 (our price)

List of Reviewers

Sohaib Ahmed, *Bahria University, Pakistan*
Gökçe Akçayır, *Gazi University, Turkey*
Laurent Antonczak, *Auckland University of Technology, New Zealand*
Shamsul Arrieya Ariffin, *Sultan Idris Education University, Malaysia*
Onur Ceran, *Gazi University, Turkey*
Thomas Cochrane, *Auckland University of Technology, New Zealand*
Neil Davie, *South Westphalia University of Applied Sciences, Germany*
Sonal Dekhane, *Georgia Gwinnett College, USA*
Francisco Delgado-Cepeda, *Tecnologico de Monterrey, Mexico*
Carrie Demmans Epp, *University of Pittsburgh, USA*
Giovanni Ferreira de Farias, *Athabasca University, Canada*
Virendra Gawande, *College of Applied Sciences – Sur, Oman*
Serçin Karataş, *Gazi University, Turkey*
Cynthia Johnson, *Georgia Gwinnett College, USA*
Noorminshah Lahad, *Universiti Technology Malaysia, Malaysia*
Katja Lehmann, *Kassel University, Germany*
Jan Marco Leimeister, *Kassel University, Germany*
Kathryn MacCallum, *Eastern Institute of Technology, New Zealand*
Marva Mirabolghasemi, *Universiti Technology Malaysia, Malaysia*
Camilius Sanga, *Sokoine University of Agriculture, Tanzania*
Matthias Söllner, *Kassel University, Germany*
Michel Plaisent, *University of Quebec Montreal, Canada*
A. Prabaharan, *Nehru Memorial College, India*
Mai Yin Tsoi, *Georgia Gwinnett College, USA*
Ülkü Ülker, *Gazi University, Turkey*

Table of Contents

Section 2
Mobile Learning

Section 3
Mobile and Blended Learning

Detailed Table of Contents

The Community of Inquiry (CoI) model has been used for exploring various aspects of blended learning courses using technology such as Course Management Systems (CMSs). However, there is a lack of research literature evaluating CoI in environments where students use an educational social network in addition to a CMS and face to face teaching and learning. This study investigates a learning experience in blended learning that combines the usage of a CMS, Edmodo, which is an educational Social Network Site (SNS) and face to face teaching and learning. The results have shown a significant relationship and prediction of overall learning experience in relation to teaching, cognitive and social presence in blended learning. Moreover, social presence is the most significant factor to predict the overall learning experience of students.

Higher Education in developing countries like Oman offers several unique circumstances that suggest curriculum delivery using Blended Learning is a rational choice. Research has been conducted to identify the factors affecting blended learning adoption and to determine whether the teaching and learning process is amenable to its adoption at HEIs in Oman. Based on the findings a model, Blended Learning Acceptance Model (BLAM), has been developed. It was inferred from the findings that there is a positive relationship among the demographic factors and behavioral Intention i.e. blended-learning adoption. In addition, the teaching and learning styles also had significant influence on adoption of blended learning. BLAM was primarily intended to be used at HEIs in Oman, but may also be used in other developing countries as a reference for the adoption of blended learning.

through the Internet. The report demonstrates the effectiveness of using such educational resources to improve pedagogical results by applying the TPACK (Technological Pedagogical Content Knowledge) model to measure the impact of MRE by STEM teachers.

With the advent of mobile technology, students and academics are using mobile devices, such as mobile phones, for personal daily activities. Due to the heavy overseas 'Mat Salleh' content from English speaking countries, such as the U.S.A. and the United Kingdom this contributes to the lack of content. In fact, there is a lack of local content for local cultural subjects. Furthermore, there are limited qualitative studies that take into consideration the potential of students in the Malaysian universities to make use of mobile phones as tools to enhance mobile learning or mLearning, particularly for studying local cultural subjects. Interviews and focus group discussions were conducted for this largely qualitative study. Thematic analysis was then conducted on the data collected. The main themes that emerged from this study comprise multimedia usage and software for daily use. Thus, this study contributes to understanding by revealing future potential, including suggestions for the use of the multimedia functions of the mobile phones for better learning outcomes.

This chapter explores the potential of mobile learning for creativity in formal and informal contexts of learning with a focus on media production and self-expression. In doing so it attempts to move beyond binary views around the nature of creativity and the role of technologies for creative learning. It presents the literature on how creativity and its relationship with technologies have been conceptualized, especially in education, and provides the theoretical underpinnings that supported the study. In particular, it refers to the Vygotskyan perspective of creativity as a transformative (social) process of culture and the self, and looks at digital technologies' affordances to reflect on their potential for learning. It describes three projects addressing young people and entailing the creation of digital artefacts through mobile devices. It highlights learners' and teachers' perspectives and shows how mobile devices can serve as cultural resources that young people may use for meaning making and transforming themselves.

Preface

Using technology in teaching and learning is mostly an exercise in appropriation. Most of the technology tools we use in the classroom are consumer devices, not pedagogical devices. We might, for example, use a digital tool for a learning activity, but the way that it is used is open to interpretation both by the teacher and the student. The same device may be used for many other purposes during the day, some educational, some not. Indeed a learner may be constantly task switching between multiple activities, moving in and out of learning and virtual social contexts (Rosen, 2010.) Increasingly, activities become harder to classify as learning, as opportunities for learning become more informal and opportunistic. Online resources and portable digital devices give us access to information anywhere, any time. This gives us increasing control over where and when we choose to learn, and how we choose to separate or integrate learning and other activities. Thus there is a blurring of boundaries between what is learning and what is something else, and what the roles of digital tools might be in this blending of learning and increasingly mobile lives.

This is not to say that there are no technology tools that have a specific educational intent. Areas where we see them being targeted at learning tend to be in the interface between hardware and software, for example Makey Makey (Beginner's Mind Collective & Shaw, 2012), Arduino (Banzi, 2011) and Lego Mindstorms (Ferrari & Ferrari, 2002.) Such tools provide opportunities for social constructivism and constructionist learning (Papert & Harel, 1991). Indeed this is a well-established theme in educational ICT that includes past initiatives such as the BBC Micro (Fitzpatrick, 2011) and the Logo Turtle (Solomon, 1978). These types of activities exemplify the blend between the use of technology and the social, collaborative learning styles of the contemporary classroom.

The plethora of devices and learning modalities that are now available to us tend to blur the boundaries between different types of technology enhanced learning. This has long been the case for mobile and blended learning, where the use of mobile devices has frequently been integrated into blended learning scenarios. This overlap becomes even more common now that the definition of what is, or is not, a mobile device has more to do with how a particular tool is used than its form factor. We can see this in the study of affordances, where the definition of mobile learning is focused more on types of learning activity than it is on which device the learner might have. Nevertheless, there are some distinctive characteristics of both mobile and blended learning that make it possible to differentiate most research as being primarily in one camp or the other.

Regardless of how we classify these investigations, the overarching question that we address in this volume is what types of mobile and blended learning innovations might lead to improved learning outcomes? The International Journal of Mobile and Blended Learning has been published since 2009, exploring these themes from a multitude of perspectives, so what new frontiers remain to be addressed in this field of research? In the following sections, we explore the key characteristics of both blended and mobile learning. Through these analyses, it is clear that both types of learning encompass complex interactions between technologies, pedagogies, contexts and individuals that provide an almost limitless set of opportunities for innovative work that explores new ways to teach and learn.

DEFINING BLENDED LEARNING

From simple beginnings at the BBC in 2000, when free online courses were supplemented with web based resources (Parsons, 2011), we now see multiple interpretations of what blended learning might mean. One view is that we can categorise blended learning practice into a range of specific types; the rotation model, a classroom centric approach where students move between online and face to face learning (this model includes the flipped classroom), the flex model, where some learning takes place onsite but is primarily delivered online, the a la carte model where online courses are mixed with other face to face courses, and the enriched virtual model where the online learning complements onsite learning (Clayton Christensen Institute, 2015). Such distinctions are somewhat arbitrary but may help in planning the mechanics of blended learning delivery. In essence, behind these models, there is a long continuum from blended learning that is predominantly face to face with some online support, through to learning that is predominantly online with some face to face support. It has been suggested that blended learning is where 30%-70% of the instruction is delivered online but, like the various models, this guideline should not be considered as any kind of rule (Blended Learning Toolkit, n.d.) As Garrison & Vaughan (2008) point out, the balance between face to face and online learning should be a thoughtful fusion, not just an arbitrary division. In addition there is a distinction between learning that takes place online, regardless of location, and online learning that is intended to take place in a specific situation, such as within a formal on-site class. Thus within the scope of blended learning there are two relationships; one between online and face to face learning, and another between on-site and distance learning, which will both vary depending on the context. Figure 1 represents these relationships as a quadrant diagram. One important feature of the diagram is that it indicates (as dark grey areas) combinations of factors that are not blended learning, namely learning that is all online, learning that is all face to face, and learning that is all distance. However it is possible to blend learning to take place exclusively on site, provided that some online learning is involved. Whilst this diagram does not necessarily imply any value judgment, it would seem likely that a blended learning programme that is taking full advantage of the opportunities offered by online learning would fall into either of the two right hand quadrants.

DEFINING MOBILE LEARNING

In the early days of mobile learning, finding a definition for it was perhaps simpler than it is today, since the devices themselves (early mobile phones and personal digital assistants) were easily distinguished from less portable devices such as desktop computers and early generation, heavy laptops with limited connectivity options. Voice, text and data connectivity through cellular networks was often an advantage in the days when WiFi connections were few and far between (indeed it still is in many situations.) As devices and networks have converged and there is a blurring of boundaries between what is, or is not, a mobile device, what is clear is that mobile learning is defined more by its affordances than by a par-

Figure 1. Quadrant diagram of the dimensions of blended learning

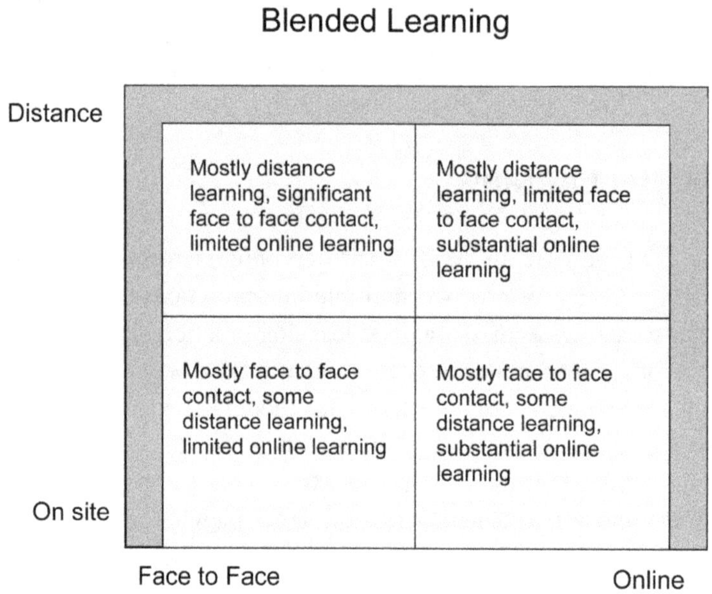

ticular piece of hardware or software. The most obvious affordance of mobility is the level of portability (Naismith et al., 2004.) Whether I can use a device in my hand that I carry my pocket and is immediately available, or whether I have to carry it in my bag or my backpack and start it up when required, really makes a difference to the simplistic concept of anytime anywhere learning. Nevertheless if we dig a little deeper, we can identify a range of other affordances of mobile learning, including gathering data (Orr, 2010), communicating (Liang et al., 2005), interacting with the interface (Lai et al., 2007), contextual, active learning (So, Kim & Looi, 2008) interacting with an outdoor environment (Tan & So, 2015) and interacting with other devices (Parsons, Thomas & Wishart, 2016). Thus even in a world of device convergence, mobile learning as an activity still has unique characteristics.

If there are two dimensions of relationships in blended learning, are there similar types of relationship in mobile learning that will also help us to understand this field of research? A review of the literature makes it clear that there are some core concepts that can be put into a similar quadrant view. Along one dimension is the difference between formal and informal learning, since one of the main benefits of learning on a mobile device is that it is personal, and its use is managed by the owner, not by a third party instructor (Clough, Jones, McAndrew & Scanlon, 2009). Nevertheless mobile devices can be, indeed are, used to great effect within formal learning environments. Another important concept that relates to mobile learning is that of context. It is worth noting that the annual conference of the International Association for Mobile Learning (IAmLearn) is called the World Conference on Mobile and Contextual Learning. Thus context is seen as being equally as important as mobility. This is because the mobility of a device enables it to be brought into many different learning contexts, where features such as location awareness, sensors and ubiquitous technologies can be leveraged to provide rich learning experiences. Figure 2 represents the relationships between formality and context in mobile learning as a quadrant diagram.

Unlike Figure 1, there are no areas of the quadrant that are 'out of bounds' for mobile learning. For example formal, context free learning with a mobile device is still mobile learning. Again, like Figure 1, this diagram does not necessarily imply any value judgment, but it would seem likely that a mobile

Figure 2. Quadrant diagram of the dimensions of mobile learning

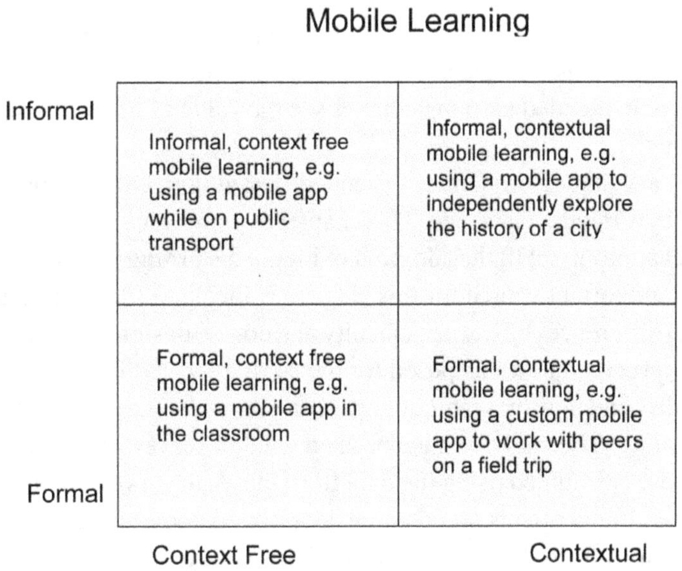

learning activity that takes full advantage of mobile affordances (including communicating for collaboration) would fall into the lower right hand quadrant.

Although the quadrant relationships outlined here for blended and mobile learning have different concerns there is, in fact, some overlap between them. For example context is important in both cases. In blended learning, the key component of context is engagement within the classroom, whereas in mobile learning perhaps the most important concept of context is that which engages the learner outside the classroom. We might also perhaps recognise that distance learning can in some ways be aligned with informal learning. However there is an important difference here, in that blended learning programs often provide formal structures, content and processes for the distance component, whereas when we talk about informal learning with mobile devices there is no expectation that this learning is being driven by educators, but rather that it is being driven by the individual.

This book has been arranged into sections based on the primary concerns of the authors. The first section includes those chapters that address the topic of blended learning. This is followed by a section of chapters that focus primarily on mobile learning. The third section includes chapters that address both mobile and blended learning in an integrated manner. Finally, there is a section of review papers that reflect on previous research into mobile or blended learning.

BLENDED LEARNING CHAPTERS

The first chapter in the blended learning section is "Evaluating Learning Experience through Educational Social Network Support in Blended Learning" by Marva Mirabolghasemi and Noorminshah Lahad. This chapter explores Garrison and Vaughan's (2008) Community of Inquiry (CoI) model in an environment where an educational social network (Edmodo) was used in conjunction with a course management system (i.e. a learning management system such as Moodle) and face to face teaching and learning. The authors investigate the three components of Garrison, Anderson, and Archer's (2000) CoI model, namely teaching presence, cognitive presence and social presence in blended learning. While the results of the study confirm the importance of all three types of presence, they particularly emphasise the importance of social presence in the overall learning experience of students. This may be because participants who engage in blended learning courses see the value of opportunities for social interaction to support their learning.

This is followed by a chapter from Oman, "Blended Learning in Oman: Confronting the Challenges of Higher Education" by Virendra Gawande. This chapter explores the development of a Blended Learning Acceptance Model appropriate to the adoption of blended learning in Oman, based on an e-learning acceptance model. An important source for this chapter is Picciano (2009) because of the reference to cultural and individual differences of learners, faculty and course designers in the application of multiple blended approaches. Research questions posed for the study reported in the chapter looked at teaching styles and learning styles, and how these might impact on user acceptance and behavioural intention with regard to blended learning. The study was undertaken using a survey of 841 students and 30 teachers. There were 18 hypotheses examined using the data from these surveys. Although learning and teaching styles appeared to have limited effects on potential adoption, demographic factors were found to play a role. Interpreting the results from this study is not straightforward, but it represents an important part of the process of developing blended learning policies in a specific national and cultural context.

Continuing the theme of investigating blended learning innovations in specific national and cultural contexts, the third chapter in this section is "Is Flipped Classroom a Tendency or a Fad? The Point of View of Future Teachers in the Philippines" by Michel Plaisent, Filomena Dayagbil, Angeline Pogoy and Prosper Bernard. This chapter tackles the popular blended learning approach of the flipped classroom, and analyses its role in the Philippines. The authors cite Mazur (2013) in identifying the following four aspects of the flipped classroom approach; constructivist pedagogy, inquiry, student-centered learning and active learning. Whilst recognising the potential of the flipped classroom, the chapter also investigates the potential negative aspects including equity of access to digital tools and internet connections required for flipped classroom activities to be undertaken. The survey reported in the study investigated pre-service teachers' knowledge and attitudes towards various aspects of the flipped classroom. The survey questions were structured around the four pillars of FLIP; flexible environment, learning culture, intentional content and professional educator (Flipped Learning Network, 2014). The authors gathered 153 valid responses for analysis. Although the pre-service teachers demonstrated enthusiasm for the flipped classroom, they lacked knowledge of the approach, and the authors caution that other studies do not necessarily show benefits. There are further reservations about how such an approach would work within the Filipino context. Thus the chapter concludes that further investigation is required to ensure that any move towards the flipped classroom in the Philippines recognises both evidence from past practice and the specific national and cultural context.

The last chapter in the blended learning section of the book is "Learning about Blended Learning through Postgraduate Student Experiences in Australia and Singapore: An Exploratory Study in Guidance and Counselling" by Margaret Carter, Paul Pagliano, Cecily Knight and Donna Goldie The study discussed in this chapter investigated seventy three postgraduate students' experiences with blended learning in guidance and counselling programs in one Australian university with an offshore branch campus in Singapore. In their exposition of blended learning, the authors reference both the CoI model and the seven principles of good practice in undergraduate education (Chickering & Gamson, 1987.) The study described in this chapter examines students' perceptions of their experiences with the blended guidance and counselling programs taught across international university campuses. There were two online surveys, followed by a focus group that further explored issues raised by the two surveys. The study aimed to find out about the experiences of students in their blended guidance and counselling programs and how these experiences might be improved. The research took over three years and involved 73 postgraduate students. Eight themes were reported within the data: blend as a multidimensional concept; communication and connectedness; arranged collegiality; digital literacy; information communication technologies; discipline knowledge; authentic support; and reflecting about action. The chapter concludes with a number of recommendations and actions that the study has revealed as important in improving the delivery of blended learning in this context.

MOBILE LEARNING CHAPTERS

The opening chapter in the mobile learning section is "Improving Teaching Practice in Early Childhood Supported by Mobile Technology" by Kathryn MacCallum and Heather Bell. This chapter discusses an ethnographic case study investigating mobile learning in an early childhood centre (kindergarten) where mobile devices are being used to actively engage children in the learning environment and support teaching inquiry. The study provides insights into how mobile technology can be embedded into early childhood teaching. Using a qualitative approach, the research involved informal interviews and direct observation of how the teachers at the kindergarten were interacting with the children mediated by the mobile devices. Data was collected through field notes and recordings (photos, audio and video.) In addition, semi-structured interviews were conducted with the teachers and copies of children's artefacts produced using the devices were collected. The research was framed within the Teaching as Inquiry Cycle (Timperley, Wilson, Barrar & Fung, 2007). This process is guided by evidence from learner and teacher practice, what learners and teachers will do differently, and the impact or shift in learning resulting from different experiences and actions. The study showed that the smart devices had become an embedded part of the kindergarten teaching philosophy as a ubiquitous tool to capture and evaluate children's learning stories and providing teachers with an opportunity to collaboratively reflect on and support the children's needs.

The next chapter is "Increasing Learner Interaction in Large-Scale Lectures by Using a Mobile Learning Application" by Katja Lehmann, Matthias Söllner and Jan Marco Leimeister. In this chapter, the authors address the problem of increasing numbers of students in large lectures, leading to decreased interaction and feedback. They present the design, use and evaluation of a mobile learning application aimed at increasing interaction in large-scale lectures. The study applied insights from interaction theory and a focus group workshop. Three types of interaction were identified; Learner-Learner-Interaction, Learner-Lecturer-Interaction and Learner-Content-Interaction. Within these types of interaction, seven important requirements were identified from the literature; learners should be creating learning material collaboratively, discussing amongst each other, receiving and providing feedback, have the possibility to ask questions regarding unclear points, be given content specific assignments and get content specific assignments to discuss. From the subsequent focus group, four priority requirements were further developed: Even in mass settings learners should be able to provide feedback to the lecturer, they should be able to ask questions anonymously, they should create and share true/false-statements about learning content, and lecturers should provide questions to be answered and discussed by the students. For these requirements a mobile application was developed using web components and suitable libraries to support mobile access. After the application was tested in lectures, a survey of 49 students was used to assess its effectiveness. The results showed that three of the four activating functions in the mobile learning application contributed to interaction in large-scale lectures. While further work is necessary, the system provides a firm foundation for development of an effective tool for increasing interaction in large lecture classes.

The next chapter comes from Brazil, "A Remote Mobile Experiment in Brazilian Public Basic Education" by Giovanni Farias, Priscila Nicolete, Juarez Silva, Marta Cristiano, Simone Bilessimo and Silvio Luz Filho. This chapter explores the use of mobile devices to perform remote experiments in STEM (Science, Technology, Engineering and Mathematics) subjects through the Internet. Using such tools, teachers and students can use remote laboratories, equipped with real experiments, to put in practice theoretical concepts learned in class. The work described in this chapter explores these issues in the Brazilian context where a pilot project aims to explore the use of Mobile Remote Experimentation (MRE) by teachers and students of public high schools in Brazil, using the Remote Experimentation Laboratory (RExLab). The overall project within which this study took place involved 287 teachers and 3,201 students across Physics, Biology, and Mathematics. However this particular study focuses on the subject of Physics in one of the partner schools. Providing remote access to experimental labs is particularly important in Brazil, where only 8% of public schools in the country have suitable science laboratories, and less than half of public schools offer computer labs, and these labs only have, on average, eight computers. Using data gathered from surveys of teachers, the authors used TPACK (Technological Pedagogical Content Knowledge) as a framework for their study (Mishra & Koehler, 2006.) The data obtained in this phase was the precursor to training activities for teachers, including blended learning courses, workshops and lectures to empower and motivate teachers to use and integrate the technology. When the training and content design phases were completed, each teacher received access to the remote labs for teaching purposes, twenty-one tablet computers to distribute among their students in their classroom, technical advice from RExLab members and support from high-school peers, who receive scholarships provided by the project. Although the project was in its early stages when reported, initial teacher feedback suggested that both teacher satisfaction and student performance had benefitted.

"Investigating the Daily Use of Mobile Phones for Local Cultural Subjects in the Malaysian Universities Context" by Shamsul Ariffin looks at the potential for students in Malaysian universities to make use of mobile phones, particularly for studying local cultural subjects such as local history, local culture and national heritage, Malay wood craft, cooking, batik, and education action research. The qualitative study described in the chapter was based on a series of Interviews and focus group discussions. These involved 15 focus groups with 148 students and 15 academics participating from two Malaysian universities. Thematic analysis was then conducted on the data collected using NVivo. The main themes emerging from this study were communication, multimedia usage, and software and hardware for daily use. One problem faced in Malaysia, in common with other non-English speaking countries, is a lack of local cultural content for mobile learning. The chapter highlights that local cultural material is not popular among content developers and that 'Mat Salleh' content (a colloquial Malay word referring to English speaking content) is more widely available via mobile phones. The participants were selected from courses in areas related to the local Malay culture, and were questioned both about their general use of mobile devices and how they used them for teaching and learning. The author concludes that, although Malaysian university academics and students could potentially use the multimedia functions of mobile phones to address the challenge of the lack of local content, there was limited awareness of the potential for mobile learning in this context. This leads to the recommendation that, since some students already use multimedia functions on their mobile phones, they could reduce the lack of local content by themselves becoming content producers.

The final chapter in the mobile learning section is "Create, Transform, and Share: Empowering Creativity and Self-expression through Mobile Learning" by Maria Ranieri and Isabella Bruni. This chapter explores the potential of mobile learning for creativity in formal and informal contexts of learning with a focus on media production and self-expression. It describes three projects addressing young people and entailing the creation of digital artefacts through mobile devices. The first of these projects addressed podcasting in the classroom, and involved learners both using and creating podcasts. The second project involved geotagging audio content in and out of school to create audio guides during a school trip. The final project discusses mobile storytelling in a project carried out with a number of young people at risk of social exclusion. The authors note how 'collaborative creativity' took place in these projects, prompted by a series of learning inputs aimed at supporting the transformation of common and familiar media practices into creative self-expression, innovative media use and creative self-representation. The chapter highlights learners' and teachers' perspectives and shows how mobile devices can serve as cultural resources that young people use for meaning making and transforming themselves.

MOBILE AND BLENDED LEARNING CHAPTERS

The section of chapters that integrate aspects of both mobile and blended learning opens with "Journalism and Law 2.0: Collaborative Curriculum Redesign" by Thomas Cochrane, Helen Sissons, Danielle Mulrennan and Vernon Rive. This chapter describes the development of a lecturer Community Of Practice (COP) integrating mobile social media, beginning with journalism staff, and later broadened to include law. The study illustrates how a COP can support pedagogical transformation across different learning contexts. The authors begin this chapter by stating that the initial impetus for their project was a growing realization that traditional journalism was in crisis, in the face of increasing use of social media channels. A participatory action research approach was used for the study, supported by a framework with four stages: the establishment of a supporting community of practice of lecturers, exploring new pedagogies afforded by mobile social media, the implementation of student-generated content and student-generated contexts via mobile social media, and the development of reflective teaching practice enabled by peer-reviewed publication of the project outcomes. The research focused on a series of case studies in journalism and law courses, and how changes had taken place within them during the study. As a result of participating in the community of practice, the lecturers have been empowered to try new approaches to teaching and learning, fostering student-generated content and student-generated contexts. The authors found that nurturing an interdisciplinary community of practice involves significant commitment, but yields creative collaboration in curriculum design, and empowers the exploration of new classroom practices.

The second chapter in this section is "Upgrading a Numerical Methods Course into New Mobile Technologies for Mathematical Education: An Approach Based on Flexibility and Skill Development" by Francisco Delgado-Cepeda. The concern of this chapter is how to continually improve science and engineering education by leveraging emerging educational technologies. The author presents the evolution of educational technology use in a numerical methods course based on the integration of Math and Physics curricula, together with a blended learning approach. The chapter describes the current movement towards a flexible scheme powered by mobile applications. The changes to this course have taken place over several years, and have involved not only the integration of new technologies, such as mobile learning, but changes in how the course is designed, in the balance between blended and face to face evaluation activities. This evolution has made it possible to consider delivering the course in multiple modes, including a purely online option. Part of the change in technology has been to move from a traditional learning management system to a suite of open tools, both mobile and web based. Data analysis in the chapter is based on analytical data available through the various online tools, along with student assessment results. One interesting outcome of moving material into such resources is that there are now formal and informal users of the class resources. While some of the course tools are only available to enrolled students (e.g. a Facebook page) other resources, such as a YouTube channel, are open access, and have attracted many learners who are not enrolled on the course. Another result from the analysis is that online teamwork leads to better student learning than more traditional individual assessment. As the course continues to evolve, building on blended scenarios, online practice and construction of intelligent adaptive tutorials, the author notes that this will require changes in the way that each teacher sees their own role, and each student the way they learn under more self-directed strategies. Thus changes in educational technology are only one aspect of the evolution of course design.

The final chapter in this section is "Mobile Application Development by Students to Support Student Learning" by Sonal Dekhane, Mai Yin Tsoi and Cynthia Johnson. In this chapter the authors discuss how mobile applications can serve two educational purposes; not only to be used by students for learning but also to be created by other students, giving them realistic experience of developing software for end-users. The intention is that software development projects can be used to create a symbiotic, business-like relationship between students enrolled in a software development course and students in other courses (end-users). This chapter discusses the creation of various educational mobile apps, their usability testing results and their effect on student learning. The apps created were in the fields of chemistry and introductory programming. They were written by more senior student studying software development. The authors suggest that the experience of creating these mobile apps for real world users, in teams, gave the students highly relevant experiences that helped to prepare them for future careers in the software industry, as well as providing useful learning tools for other students.

REVIEW CHAPTERS

Review chapters may serve a number of purposes. They may attempt a comprehensive, critical review of the literature, which is a challenging task and would involve analysing many thousands of articles from a wide range of sources. More commonly, they identify a subset of the literature in order to gain insights into a particular publication channel or forum of a research area, or confine themselves to specific subtopics within a research area. One example of analysing a particular publication forum is Wingkvist and Ericsson (2011), which confines its analysis to papers published in multiple proceedings of the World Conference on Mobile and Contextual Learning. In contrast, an example of focusing on a particular subtopic is Viberg and Grönlund (2013) which looked specifically at published research into mobile assisted language learning. Whatever approach literature reviews take, we generally expect that they will give us some insights into areas of research that are well served and those that perhaps warrant further investigation. They may also expose methodological shortcomings that reveal areas where more rigorous work is required. On a simpler level they may just provide us with an easy means of becoming familiar with the background and themes of a given research area, for example an overview of mobile learning (Parsons, 2014).

One of the review chapters in this book analyses the relevant work in a particular publication channel, namely the Web of Science database. This chapter is "A Trend Analysis of Mobile Learning" by Onur Ceran, Serçin Karataş, Ülkü Ülker, Nimet Ünsal Köse, Mustafa Kılıç, Ezgi Tosik Gün, Zeynel Tok and Gökçe Akçayır. The aim of this literature review study is to determine trends in mobile learning research between 2010 and 2015. In this chapter, the authors outline the meaning of trend analysis, which refers to the statistically observable change or general orientation of a general movement. In order for an orientation or a tendency to become a trend, it must comprise a certain period of time, repeat itself during this period, be proven statistically, focus on a specific area, be based on data, and have a certain level of consistency. However, trends report only the facts at the time the research is done, and they do not predict the future. Therefore, it is possible to say that trend studies are repeatable. For this trend study, 221 articles were collected from the Web of Science Core Collection database and examined using content analysis. In this analysis, eight criteria were used, namely; research technique, sampling size, sampling level, learning domain, topical domain, data collecting tool, data analysis method and mobile application development approach. The results suggest that the main tendencies under these categories were experimental method; sample sizes of 31-100 people; higher education students; humanities and social sciences domain; learner outcomes topical domain; mixed data collecting tools; mixed analysis methods; and native mobile application development approach. The authors provide extensive discussion around these results, identifying areas where research has clustered and other areas where useful work might be done.

The other common approach to literature reviews is to identify a subtopic of the research area and analyse only the published work addressing that area. An example of this approach is the chapter "Mobile Learning in Science and Mathematics Teaching: A Systematic Review" by Rosiney Almeida and Carlos Araújo Jr. In this chapter, the authors analyze mobile learning literature addressing formal teaching situations in the field of science and mathematics education. The chapter describes a systematic review of relevant literature, investigating work published between 2005 and 2014. The approach taken was based on work by Kitchenham (2004), who conducted systematic reviews in medicine, and Barcelos (2014), who conducted a systematic review of literature on the relationship between computational thought and mathematics. The chapter also provides an overview of previous literature reviews in the mobile learning field. Following these exemplars, the literature review work was executed in three steps, namely: planning, handling and reporting of results. Relevant publications were identified from the SCIELO, ACM Digital Library and ERIC databases. Based on the findings of this review, the results reveal that since 2009 interest in the academic community for research involving m-learning in science and mathematics teaching has intensified. An emphasis on the growing need for research involving m-learning at undergraduate levels is evident. The study notes positive reported results on the impact of m-learning in science and mathematics teaching, notably in assisting ease of learning, promoting student interest and collaborating with positive, motivating attitudes and encouraging meaningful commitment among students in learning activities.

Our final chapter is "Taking advantage of MOOCs in K-12 Education: A Blended Approach" by Helen Crompton and Samantha Briggs. Unlike the other two review chapters, which focus their attention on aspects of mobile learning, this chapter looks at a particular aspect of blended learning, the potential for

MOOCs to be used in conjunction with K12 (ages 4-19 years) classroom education. In this chapter, the authors explore how the recent popularity of Massive Open Online Courses (MOOCS) might transfer into a blended learning context in K-12 education. This chapter provides a review of the literature of early pioneering work of MOOCs in K-12 within North America. Early MOOC classes are explored and positives and challenges discussed. From their discussion, the authors believe that there is a place for MOOCs in blended learning mode in K-12 education to extend and enhance the existing K-12 curricula.

David Parsons
The Mind Lab by Unitec, New Zealand

REFERENCES

Banzi, M. (2011). *Getting started with Arduino*. Sebastopol, CA: O'Reilly.

Barcelos, T. S. (2014). *Relação entre o pensamento computacional e a Matemática em atividades didáticas de construção de jogos digitais*. (Doctoral Thesis). Universidade Cruzeiro do Sul, São Paulo, Brazil.

Beginner's Mind Collective & Shaw., M. (2012). Makey makey: Improvising tangible and nature-based user interfaces. In *Proceedings of the Sixth International Conference on Tangible, Embedded and Embodied Interaction (TEI '12)*. ACM Press.

Blended Learning Toolkit. (n.d.). *What is blended learning?* Retrieved from https://blended.online.ucf.edu/about/what-is-blended-learning/

Chickering, A.W., & Gamson, Z.F. (1987). Seven principles for good practice in undergraduate education. *American Associates for Higher Education Bulletin*, 3-7.

Clayton Christensen Institute. (2015). *Blended learning definitions*. Retrieved from http://www.christenseninstitute.org/blended-learning-definitions-and-models/

Clough, G., Jones, A., McAndrew, P., & Scanlon, E. (2009). Informal learning evidence in online communities of mobile device enthusiasts. In M. Ally (Ed.), *Mobile learning: Transforming the delivery of education and training*. Athabasca University Press.

Ferrari, M., & Ferrari, G. (2002). *Building robots with Lego Mindstorms*. Rockland, MA: Syngress.

Fitzpatrick, J. (2011). An interview with Steve Furber. *Communications of the ACM*, *54*(5), 34–39. doi:10.1145/1941487.1941501

Flipped Learning Network. (2014). *Definition offlipped learning*. Retrieved from http://flippedlearning.org/site/default.aspx?PageID=1

Garrison, D., & Vaughan, N. (2008). *Blended learning in higher education: Framework, principles and guidelines*. San Francisco, CA: Jossey-Bass.

Kitchenham, B. (2004). *Procedures for performing systematic reviews*. Keele University Technical Report TR/SE-0401.

Lai, C.-H., Yang, J.-C., Chen, F.-C., Ho, C.-W., & Chan, T.-W. (2007). Affordances of mobile technologies for experiential learning: The interplay of technology and pedagogical practices. *Journal of Computer Assisted Learning*, *23*(4), 326–337. doi:10.1111/j.1365-2729.2007.00237.x

Liang, J.-K., Liu, T.-C., Wang, H.-Y., Chang, B., Deng, Y.-C., Yang, J.-C., & Chan, T.-W. et al. (2005). A few design perspectives on one-on-one digital classroom environment. *Journal of Computer Assisted Learning*, *21*(3), 181–189. doi:10.1111/j.1365-2729.2005.00126.x

Mazur, A. D. (2013). *The flipped classroom: Theoretical constructs*. Retrieved from https://ambermazur.files.wordpress.com/2013/06/flipped-classroom-theoretical-constructs-visual-with-explanation.pdf

Mishra, P., & Koehler, M. (2006). Technological pedagogical content knowledge: A framework for teacher knowledge. *Teachers College Record*, *108*(6), 1054. doi:10.1111/j.1467-9620.2006.00684.x

Naismith, L., ... (2004). *Literature review in mobile technologies and learning*. Futurelab Report 11. Retrieved from https://lra.le.ac.uk/bitstream/2381/8132/4/%5b08%5dMobile_Review%5b1%5d.pdf

Orr, G. (2010). A review of literature in mobile learning: Affordances and constraints. In *Proceedings of 6th IEEE International Conference on Wireless, Mobile and Ubiquitous Technologies in Education (WMUTE)* (pp. 107-111). IEEE. doi:10.1109/WMUTE.2010.20

Papert, S., & Harel, I. (1991). *Constructionism*. Norwood, NJ: Ablex Publishing.

Parsons, D. (2011). Foreword. In A. Kitchenham (Ed.), *Blended learning across disciplines: Models for implementation*. Hershey, PA: IGI Global.

Parsons, D. (2014). A mobile learning overview by timeline and mind map. *International Journal of Mobile and Blended Learning*, *6*(4), 1–21. doi:10.4018/ijmbl.2014100101

Parsons, D., Thomas, H., & Wishart, J. (2016). Exploring mobile affordances in the digital classroom. In *Proceedings of IADIS International Conference on Mobile Learning*. IADIS.

Picciano, A. G. (2009). Blending with purpose: The multimodal model. *Journal of the Research Center for Educational Technology*, *5*(1), 4–14.

Rosen, L. (2010). *Rewired: Understanding the iGeneration and the way they learn*. New York, NY: Palgrave Macmillan.

So, H.-J., Kim, I., & Looi, C.-K. (2008). Seamless mobile learning: Possibilities and challenges arising from the Singapore experience. *Educational Technology International*, *9*(2), 97–121.

Solomon, C. J. (1978). Teaching young children to program a logo turtle computer culture. *ACM SIGCUE Outlook*, *12*(3), 20–29. doi:10.1145/964041.964045

Tan, E., & So, H.-J. (2015). Rethinking the impact of activity design on a mobile learning trail: The missing dimension of the physical affordances. *IEEE Learning Technologies*, *8*(1), 98–100. doi:10.1109/TLT.2014.2376951

Timperley, H. S., Wilson, A., Barrar, H., & Fung, I. (2007). Teacher professional learning and development: Best evidence synthesis iteration. Wellington, New Zealand: Ministry of Education; Retrieved from http://www.educationcounts.govt.nz/publications/series/2515/15341

Viberg, O., & Grönlund, A. (2013). Systematising the field of mobile assisted language learning. *International Journal of Mobile and Blended Learning*, *5*(4), 72–90. doi:10.4018/ijmbl.2013100105

Wingkvist, A., & Ericsson, M. (2011). A survey of research methods and purposes in mobile learning. *International Journal of Mobile and Blended Learning*, *3*(1), 1–17. doi:10.4018/jmbl.2011010101

Section 1
Blended Learning

Chapter 1
Evaluating Learning Experience through Educational Social Network Support in Blended Learning

Marva Mirabolghasemi
Universiti Technology Malaysia, Malaysia

Noorminshah A. Iahad
Universiti Technology Malaysia, Malaysia

ABSTRACT

The Community of Inquiry (CoI) model has been used for exploring various aspects of blended learning courses using technology such as Course Management Systems (CMSs). However, there is a lack of research literature evaluating CoI in environments where students use an educational social network in addition to a CMS and face to face teaching and learning. This study investigates a learning experience in blended learning that combines the usage of a CMS, Edmodo, which is an educational Social Network Site (SNS) and face to face teaching and learning. The results have shown a significant relationship and prediction of overall learning experience in relation to teaching, cognitive and social presence in blended learning. Moreover, social presence is the most significant factor to predict the overall learning experience of students.

INTRODUCTION

Many higher education institutions are hoping to enhance student participation. Enhancement of student participation in e-learning is associated with improved learning (Roblyer & Ekhaml, 2000). The technology used to support a blended course may affect the manner and frequency in which faculty and students interact with each other, interact with course materials, and receive and provide feedback (Rubin, Fernandes, & Avgerinou, 2013). Effective blended learning can provide opportunities for students to learn better and faster. Blended course designs support learners and instructors working together in

DOI: 10.4018/978-1-5225-0359-0.ch001

mixed delivery modes, typically technology mediated and face-to-face, to achieve learning outcomes that are pedagogically supported through activities, assignments, and assessments (McGee and Reis, 2012). Haythornthwaite, Kazmer, Robbins, & Shoemaker (2000) suggested that students who fail to interact with other participants in a group feel more isolated compared to participants who interact effectively. Understanding how students participate in blended learning can help us to distinguish how to best engage students.

Although Course Management Systems (CMSs) offer several tools that support social learning, the functionality and scope of these tools can inhibit just-in-time interactions and social connections (Dunlap & Lowenthal, 2010). Jafari et al. (2006) stated that students were interested in CMSs that emulate the kinds of systems they use to interact with socially. A secure social network system in teaching and learning can improve students' social presence, perceived learning, and satisfaction with the instructor (Mirabolghasemi & Iahd. 2013). Özmen & Atıcı (2014) explored the effects of a CMS supported by a SNS on learners' academic achievements in distance education using a survey method. The results indicated that a CMS used in distance education should be supported by a SNS to enhance the academic achievement of learners, and that combining these two platforms could provide more personal interaction, communication, student engagement, social experience and peer-to-peer feedback. Pilli (2014) stated that a CMS and a SNS need to be used together to make sure that individuals can capture the potential benefits of informal learning and collaboration while still providing the formal documentation necessary to meet management and compliance requirements. Since research on usage of social network applications for learning with a CMS in a blended learning environment is still limited, the main objective of this paper is to study how the combination of both technologies influences the overall learning experience.

Garrison, Anderson, & Archer (2000) proposed factors supporting the development of learning experience, namely teaching, cognitive and social presence. The CoI model is a well-cited learning model in education (Daspit and D'Souza, 2012). A critical community of learners is composed of lecturers and students interacting with the specific objectives of constructing, understanding, facilitating, and developing capabilities that inspire further learning (Garrison & Anderson, 2003; Shea et al., 2014). Effective online learning requires the development of a community that provides deep learning. Therefore, it is essential and valuable to create a learning platform where interaction and reflection are sustained, where opinions can be explored and critiqued, and where the process of critical thinking can be modeled. The need to turn from descriptive to predictive quantitative studies using the CoI model that consider the interaction effects of presence and cross disciplinary studies has been identified (Garrison, Cleveland-Innes& Fung, 2010). There are only few previous studies on SNS application in higher education, mostly based on case studies (Jucevičienė& Valinevičienė, 2015).

LITERATURE REVIEW

The literature review for this research study consists mainly of brief descriptions of blended learning, CMSs, SNSs for learning, and the CoI model.

Blended Learning

Blended learning refers to linking e-learning activities to traditional classroom training. E-learning can provide access to both synchronous and asynchronous information and communication (Al-Hunaiyyan

& Al-Sharhan, 2009). In the context of higher education, blended learning consists of the combination of web-based and traditional activities to achieve student learning (Daspit & D'Souza, 2012). Blended instruction has recently received increased usage among private companies and academic institutions that have provided many opportunities associated with place and time (Picciano, 2006).

Currently, many universities and faculties are using a blended learning format. Lim, Morris & Kupritz (2014) stated that the blended delivery format provides clearer instructions to learners and they experienced fewer obstacles and challenges in achieving similar learning levels. Osguthorpe & Graham (2003) defined blended learning as a balance between face to face interaction among individuals and online access to knowledge. This combines various activities such as live e-learning, self-paced learning, and face-to-face classrooms. Driscoll (2002) described that blended learning consists of the following meanings:

- Combining various e-learning technologies to achieve teaching and learning objectives.
- Combining various teaching methods to achieve the best learning outcomes.
- Combining the actual tasks with the teaching techniques.
- Combining any kind of teaching technology with face to face learning.

Course Management Systems

One of the most popular media in blended learning is a CMS such as Moodle. A CMS is an integrated platform that consists of web-based applications to support course management procedures (Severson, 2004). This software package is designed to help instructors to create virtual learning. In spite of the widespread incorporation of CMSs into university teaching and learning, lecturers use the medium primarily for delivery of documents such as lecture notes and course syllabi (Aghili et al., 2014). Many CMSs limit peer conversation to prevent students from copying each other's work. Systems should support peer interaction, not as an opportunity for plagiarism, but as a way to build the kind of academic conversation that resides at the heart of every university (Garrett et al., 2007).

Although CMSs provide learners the freedom to attend to the course they have chosen themselves, they cannot meet the needs of learners for social interactions (Lee & McLoughlin, 2010; Meishar-Tal, Kurtz & Pieterse, 2012). Brady, Holcomb & Smith (2010) stated that CMSs fail to provide the individual participation and network capacity provided by SNSs and tend to be excessively structured. For instance, an instructor using a CMS can post a question on an online discussion board and each learner can send a response. However, the learners could not interact in real-time manner. Therefore, CMSs are not able to provide an adequate level of communication, learner participation, satisfaction and motivation (Meishar-Tal, Kurtz & Pieterse, 2012; Aghili et al., 2014; Özmen & Atıcı, 2014). Naturally, instructors are heading towards SNSs which are becoming an integral part of students' daily life in order to create more efficient and easier learning environments for students (Pilli, 2014).

Social Network Sites for Learning

SNSs can enhance activities such as communication among peers, access to resources, sharing interests, interactions, and collaboration. These characteristics can provide a positive, friendly environment (Selwyn, 2007). SNSs can be considered as complex and large information systems that are composed of the users that operate within the services and tools, using three key main concepts which are; that the content and activities of such sites are provided by users, that users are expected to have active interactions, and that they return to the site at various time intervals (Serajian & Akhgar, 2011).

Lee & McLaughlin (2010) found that SNSs enable instructors and learners to present themselves socially and connect with one another in an online environment while enabling individuals to engage in recurring meaningful experiences with others. SNSs can be used as a new e-learning platform in higher education, and creating personal profiles provides the opportunity for improving learner-centred approaches (Aghili et al., 2014). In this digital environment, existing anytime and anywhere with high connectivity, there is an essential need to expand the concept of learning as a social process supporting learners' personal needs and intentions (McLaughlan, 2007).

SNSs can be used to improve cooperation and solidarity in higher education (Ajjan & Hartshorne, 2008). Similarly, Grant (2008) stated that usage of SNSs in educational environments provides more effective communication between students and teachers, so instructors know their students much better. Moreover, social networking applications are closely related to many aspects of constructivism, and it is claimed that these applications support pedagogical approaches such as active learning, social learning and communities of practice and learning (Ferdig, 2007). In a study conducted by Helou & Rahim (2014) in a Malaysian university context, the majority of respondents clearly indicated that SNSs can be used for group discussions and for discussing assignments as well as to improve interaction between lecturers and classmates

The popularity of SNSs among students has created a concern for privacy in higher education. The existence of SNSs that are focused on education such as Elgg, Ning, and Edmodo provide an opportunity for instructors to minimize these privacy concerns (Brady, Holcomb, & Smith, 2010). However, SNSs such as Elgg and Ning are not offered for free and the cost incurred may lead instructors to look for alternative educational networks such as Edmodo which is a user-friendly and free social network (Iahad, Mirabolghasemi, & Huspi, 2012). Edmodo could provide a secure e-learning environment for students (Tongmak, 2013; Sohaei & Iahad, 2014). Furthermore, it is easy to use in classrooms since its interface is very similar to Facebook (Holland & Muilenburg, 2011). Al-Kathiri (2014) integrated Edmodo into secondary school instruction and studied students' perceptions regarding Edmodo use and its effect on their learning. The experimental group received a six-week daily interaction via Edmodo plus traditional teaching and the control group received traditional teaching only. The results of the post-treatment questionnaire show that students' perceptions towards Edmodo were highly positive.

The CoI Model

The CoI model was developed by Garrison, Anderson, and Archer (2000) as a research tool in blended and online learning. The model is an effective way to explain success in e-learning (Garrison, Anderson, & Archer, 2010). The three interrelated elements – teaching presence, cognitive presence, and social presence - provide a model to guide instructors in their online course design (Weyant, 2013). The effect of each presence in this model should be understood for creating and sustaining a collaborative CoI. Research on the CoI model documents the correlation between the elements of the model and learner perceived learning (Shea et al., 2010) and to online learners' sense of community (Shea, Li, & Pickett, 2006). Figure 1 shows the CoI model.

The first dimension of CoI is teaching presence. To maintain a COI a focused, attentive, and thoughtful teaching presence is needed. The teaching presence is described as process, design and planning of the structure and evaluation aspects of courses. The activities include proposing video/audio mini-lectures,

Figure 1. The CoI model

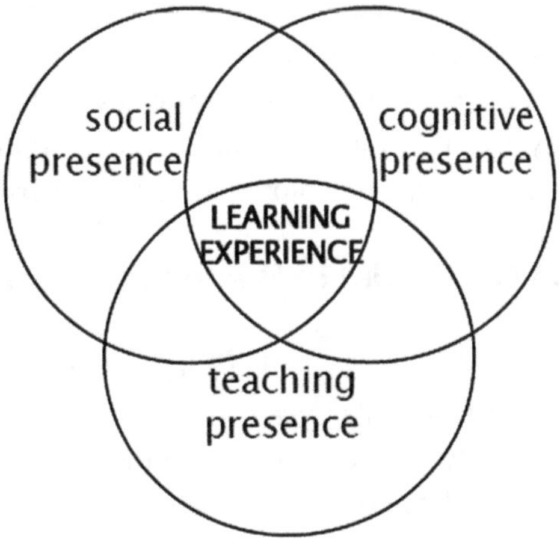

preparing personal insights into the course material, creating lecturer notes into the course site, preparing guidelines on how to use the e-learning platform effectively, and creating a desirable schedule for group activities and individuals. Shea, Li & Pickett (2006) found a link between perceived learning and teaching presence. Previous studies highlighted the prominent role of teaching presence not only in terms of learning consequences but also in the alignment of cognitive and social presence (Garrison & Akyol, 2013).

The second dimension of CoI is cognitive presence, which is defined as the construction, exploration, resolution, and confirmation through reflection and collaboration in the CoI model. The phases are definition of a task or problem, exploration for relevant knowledge/information, integrating ideas, and testing plausible solutions. All of this occurs in an environment of reflection and discourse; analysis and synthesis (Garrison, Cleveland-Innes, & Fung, 2010). With regard to the present study, the integrating ideas emerged during exploration of problem solving by some means of action (Garrison, Anderson, & Archer, 2001). Often, students will continue the exploration phase. As a result, teaching presence is essential to shift the process to more advanced stages of critical thinking. Kozan & Richardson (2014) stated that cognitive presence means being cognitively active, where learners seek the most efficient and effective ways of solving a learning problem, and apply these solutions.

The third dimension of the CoI model is social presence, which describes learners feeling socially and emotionally connected with others (Mirabolghasemi& Iahad, 2013). Short, Williams, & Christie (1976) stated that social presence is fundamental to communication. Social presence is a factor of both the medium in a sequence of interactions and the communicators' perceptions of presence (Gunawardena & Zittle, 1997). Kreijns et al. (2014) argued that social presence should be divided into two constructs which are "the degree of 'realness' of the other in the communication," and the "degree to which social interpersonal relationships are salient." According to Homer, Plass, & Blake (2008), the learning process is more engaging when information is presented in a way that improves social presence. Different researchers have looked for different ways to improve social presence in e-learning. For example, Aragon (2010) described various ways to create social presence. DuVall et al. (2007) explored the use of text messaging to establish social presence. Furthermore, Keil & Johnson (2002) evaluated the use of internet based voice mail to enhance social presence.

It is valuable and necessary to create a CoI where interaction and reflection are sustained, where ideas can be defined and critiqued and where the critical inquiry can be model. The CoI model has been used to explore both blended and online learning in previous research studies. The previous studies illustrate that the medium is one of the prominent factors in achieving higher levels of inquiry (Garrison, Anderson, & Archer, 2001; Meyer, 2004; Vaughan and Garrison, 2005; Stein et al., 2007). Therefore, educational social network support should be explored in blended learning.

Previous Works on using CoI Model in Blended Learning

As online courses are increasing in universities, educators have much to learn regarding how to use technology to enhance effective learning (Perry& Edwards, 2005). Successful e-learning creates a CoI where students interact effectively with each other, while the learning materials and the instructor develop new knowledge. When online courses have a strong CoI, students learn more, participate in discussions, are more satisfied with the learning experience and have greater retention (Garrison & Arbaugh, 2007; Arbaugh, 2008; Boston et al., 2009). The kind of technology utilized for teaching may have different effects on CoI, depending on how students interact with each other, their instructor, and learning materials (Aghili et al., 2014).

In most online instruction, coursework is organized and paced, learning resources are accessed, work is collected and returned, discussion occurs, and feedback is delivered through a CMS. Rubin, Fernandes & Avgerinou (2013) stated that perceived CMS affordances predict social, teaching, and cognitive presence among students. Arbaugh & Benbunan-Fich (2007) found that the amount of student interaction with the system predicts both perceived learning and satisfaction with the online medium, beyond and above the amount of interaction with the instructor and with other learners. Aghili et al. (2014) stated that the CoI model could be applied to SNS platforms for providing collaborative knowledge construction among students.

Few studies have specifically investigated the integration of a SNS and a CMS in higher education (Brady, Holcomb, & Smith, 2010; Iahad, Mirabolghasemi & Huspi, 2012; Özmen& Atıcı, 2014; Rožac et al., 2012; Petrovic et al., 2014; Thoms & Eryilmaz, 2014). Iahad, Mirabolghasemi & Huspi explored the usage of Edmodo in addition to Moodle using the CoI model in blended learning and the results of their study indicated that there was a significant difference between teaching, cognitive and social presence before and after Edmodo intervention. Sinnappan & Zutshi (2011) used the CoI model and found that the pedagogical use of Twitter could facilitate social, teaching and cognitive presences in blended learning, when Twitter was leveraged as a supplement to Blackboard.

METHODOLOGY

This research applied a survey method. The constructs were adopted from existing literature to increase the reliability and validity of the research instrument. The survey instrument of this study has been adapted from Garrison, Anderson, & Archer (2000) and Swan et al. (2008). The questionnaire included questions from the adapted CoI instrument using a five point scale (1 = strongly disagree to 5 = strongly agree). Three experts with blended learning research experience in quantitative methods and scale development have evaluated its content validity. The experts were selected based on their research interest and expertise. Therefore, expert judge approach used to increase the validity of the research. The reliability of the instrument was .953. SPSS version 16.0 was used for data analysis.

Table 1. The demographic information of the sample

Demographic Variable		Sample Composition (N=107)
Gender	Male	49 (46%)
	Female	58 (54%)
Age	18-24 years	107 (100%)
Average time spent on e-learning	Daily	61(57%)
	Several Times Per Week	35(33%)
	Weekly	11(10%)

The respondents were 107 first year undergraduate students in the Faculty of Management and Human Resource Development at Universiti Teknologi Malaysia (UTM) based on the purposive sampling. Table 1 illustrates the sample demographic information. This table shows that most of the participants (57%) use e-learning daily. In contrast, only (10%) stated that they use e-learning weekly. This result indicates that most of the students actively use e-learning to perform their educational duties.

The blended learning consisted of a CMS, a SNS and face to face teaching and learning. The CMS, Moodle, has been used as a managing tool for notes, resources, messages, submitting assignments, and calendar. Meanwhile, Edmodo has been for collaboration activities which are case study analysis, collaborative writing, peer editing, peer help, discussions about course topics, tagging and finding idealistic resources, in the case study. According to Aghili et al (2014), the usage of SNSs for educational purpose especially for first-year students in international universities strengthen the relationship between international and local students and consequently improve their engagement in the course.

The evidence is growing that the CoI model is the appropriate theoretical tool to investigate the relationships between teaching, cognitive, and social presences with learning experience of students in blended learning. The objective of this study was to investigate the role of teaching, cognitive and social presence in relation to students' overall learning experience in blended learning that combines the usage of a CMS and a SNS. The following hypotheses were tested:

1. There is a significant and positive relationship between students' teaching presence and learning experience in blended learning which consisted of a CMS in addition to a SNS and face to face teaching and learning.
2. There is a significant and positive relationship between students' cognitive presence and learning experience in blended learning which consisted of a CMS in addition to a SNS and face to face teaching and learning.
3. There is a significant and positive relationship between students' social presence and learning experience in blended learning which consisted of a CMS in addition to a SNS and face to face teaching and learning.

RESULTS

The relationship and prediction of overall learning experience of respondents in relation to teaching, cognitive and social presence are evaluated in this section. An examination of correlation between teach-

Table 2. The summary of regression analysis

Variables	Standardized Coefficient		Standardized Coefficient	t Value	Sig.
	B	Std. Error	Beta		
Constant	.602	.189		3.177	.002
Teaching presence	.680	.016	.205	42.200	.000
Cognitive presence	.797	.010	.413	82.780	.000
Social presence	.868	.010	.458	88.313	.000

ing, cognitive, social presence and overall learning experience illustrate that overall perceived learning among the respondents yielded a correlation of .890 with overall teaching presence ($p < .05$; $r^2 = .79$) and a correlation of .936 ($p<.05$; $r^2=.87$) with overall cognitive presence. Moreover students' overall learning experience yielded the highest correlation of .945 ($p < .05$; $r^2 = .89$) with students' overall social presence. This indicates a significant relationship between students overall teaching, cognitive, social presence and students overall learning experience.

Multiple regression analysis used for the prediction of overall learning experience of students in relation to teaching, cognitive and social presence. Since learning experience is single criterion (Y), in this approach, we regress all predictors on this criterion. Table 2 shows the summary results of regression analysis.

Because the level of sig. is less than 0.05 teaching, cognitive, and social presence could predict learning experience of respondents in the case study. The level of sig. is less than 0.05 and three presences are prominent in predicting learning experience of respondents which used educational social network in addition to a CMS and face to face teaching and learning. Moreover, the highest coefficient is dedicated to social presence according to regression analysis. A major aspect of social presence is enabling the learners feel like they belong in the online community and have meaningful interactions with others (Kumar& Ritzhaupt, 2014).

DISCUSSION

As there have been various forms of the blended mode further studies needed to identify an appropriate blended mode for the exploration (Szeto, 2015). Previous research has often explored the significance of investigating the application of a CMS such as Moodle in combination with blended learning as well as a SNS as a tool that can contribute significantly to the quality of education (Petrovic et al., 2012). Currently, there is a lack of research tested the CoI model in the environment which students use a CMS in addition to an educational SNS and face to face teaching and learning. Still, little research has specifically investigated the integration of a SNS and a CMS in higher education (Brady, Holcomb, & Smith, 2010; Iahad, Mirabolghasemi & Huspi, 2012; Özmen & Atıcı, 2014; Rožac et al., 2012; Petrovic et al., 2014; Thoms & Eryilmaz, 2014). Thus, the purpose of our study has been to analyze whether or not using Edmodo as an educational tool is more effective than using Moodle alone. Garrison, Cleveland-Innes, & Fung (2010) stated that the three presences do not play equal roles. The results of this study indicated that teaching, cognitive, and social presence are important to predict learning experience of respondents in blended learning. However, social presence is the most significant factor to predict the overall learning experience of students.

Teaching presence plays an essential role in facilitating social presence for fostering student learning and communicative interactions to gain the intended learning outcomes. According to Davis & Roblyer (2005), teaching presence include four clearly different roles (counsellor, assistant, teacher and designer), each calling for the development of specific competences. The nature of a course affects the instructional roles. For instance, a course in education may lend itself more to the cognitive and/or social than to the teaching presence (Szeto, 2015). The respondents of this study were students who used a CMS in addition to a SNS and face to face teaching and learning for a course in education and the findings indicated that the highest coefficients are dedicated to social presence and cognitive presence respectively.

The concept of cognitive presence reflects both creative and thinking processes (Shea et al., 2010). Many studies have highlighted the prominent role of the social dimension in cognitive activity that causes higher-order learning (Armellini & De Stefani, 2015). Kozan & Richardson (2014) claimed that the cognitive presence effect was the most important among the three presences. Despite the different results, developers and course instructors can flexibly improve face-to-face/online instruction or inspire students' inquiry, build in more interactive learning activities to enhance the effects of the three presences (Szeto, 2015). The feature of SNSs is that they provide the opportunity for supporting the social construction of knowledge between and within individuals. Dawley (2009) described a SNS as an environment that supports the literal physical representations of learners' knowledge construction. In the virtual world students have opportunities to represent their knowledge using a variety of formats and the knowledge representation and construction can then be utilized by others to extend learning (Dawley, 2009).

The building of social relationships is foundational for facilitating social presence (Swan, 2003). The results of this study indicated that social presence has the highest effect on students' learning experience in blended learning which consisted of a CMS in addition to a SNS and face to face teaching and learning. Social presence plays a central role in the construction of meaningful cognitive and teaching discourse, higher-order learning and sense-making (Armellini & De Stefani, 2015). This is contingent on students and instructors working in concert rather than a stable pattern related to the passage of time in a course (Shea et al., 2010). Previous researchers explored various ways to support and maintain social presence in blended courses. For example, Keil & Johnson (2002) used Internet based voice mail and DuVall, Powell, Hodge, & Ellis (2007) explored using text. Garrison & Anderson (2003) found that social presence is the facilitator of the learning process. In this research study, a social network in addition to a CMS has shown a very significant level of students' social presence in blended learning. SNSs enable learners and instructors to interact and communicate with each other socially in an online environment and individuals can have a meaningful experience (Wheeler, Yeomans & Wheeler, 2008).

CONCLUSION

The CoI could be an appropriate tool as educators assess the quality and effectiveness of emerging education technologies (McKerlich et al., 2011). Although, SNSs are regularly accessed by students, these sites have not yet been considered as a tool for teaching and learning processes properly (Stanciu, Mihai, & Aleca, 2012). There is a lack of research literature that has tested the CoI model in the environment in which students used a SNS in addition to a CMS in a blended learning environment. Social presence that develops among students is prominent in an online course but can be crucial for improving individuals' connection and retention to the academic institution (Kumar& Ritzhaupt, 2014). The results of this study describe the positive and significant relationship of teaching, cognitive and social presence of students and overall learning experience. Meanwhile, this study shows that social presence is the most significant dimension to predict students learning experience. The reason may be that participants who engage in blended learning courses seek more social presence. Social presence highlights encouraging social interaction as well higher level learning and the underpinnings of critical thinking, and is thought to be of great significance for a CoI (Garrison, Cleveland-Innes, & Fung, 2010).

This study has some limitations. The data collection is from less IT savvy faculty that need more training and in only one university in southern part of Peninsular Malaysia. Moreover, this study used Edmodo as a support of a CMS. For future studies, other SNS platforms can be used to compare students' learning experience with the findings of this research study. Further research should also explore the relationships of the three presences with learning experience across institutions and a larger sample.

The use of CMSs as primary education tools in universities is rapidly increasing (Brady, Holcomb, & Smith, 2010). At the same time, the popularity of SNSs is continuously increasing and they have expanded to a wide population (Boyd & Ellison, 2007). When the popularity of CMSs and SNSs and the educational benefits of SNSs are considered, some educators suggest merging these two platforms to improve the quality of learning (Brady, Holcomb, & Smith, 2010; Iahad, Mirabolghasemi & Huspi, 2012; Özmen& Atıcı, 2014; Rožac et al., 2012; Petrovic et al., 2014; Thoms & Eryilmaz, 2014). Based on the researcher's experience of using a SNS with students, the following guidelines are offered to those who want to use social network intervention:

- **Establish the Clear Purpose for Students:** The purpose of usage of social network in teaching and learning should be defined for students. If students use social networks in irrelevant ways, they will not achieve high quality of learning experience. The researcher in this case study has tried to illustrate different resources and course activities. As a result, when students find the opportunities can help them to improve their learning experiences, they will try to be active participants.
- **Encourage Students to Respond to Other Classmates:** Sharing academic resources and encouraging students to share ideas with other classmates can improve students' knowledge and background. Furthermore, students can be familiar with the ideas of students in same level of understanding. The researcher encourages students to use information and resources that other participants share them so they can improve their background of knowledge.
- **Update Materials and Course Topics:** The materials should be updated regularly. If students are not aware of their responsibilities and specific discussion topics for each session, they may have a misguided discussion and fail to participate as hoped. The researcher defined different types of activities like a case study analysis, relevant materials to course sessions and the lecturer encouraged them to solve and consider these materials weekly.
- **Combining Different Type of Media:** Different types of media, such as audio and video related to course topics, can help students answer questions raised in course activities. Because media in social networks are easy to both use and share, students show a high enthusiasm to share different types of activities and content.
- **Use New Learning Tools for Students:** Encourage students to use more learning applications in a SNS. Different learning tools can offer opportunity to explore and manage their course activities and content related questions.

REFERENCES

Aghili, M., Palaniappan, A. K., Kamali, K., Aghabozorgi, S., & Sardareh, A. S. (2014). Unifying informal and formal learning environments: Educational use of social network sites through implementing community of inquiry framework. *International Journal of e-Education, e-Business, e- Management Learning*, *4*(3), 191–196.

Ajjan, H., & Hartshorne, R. (2008). Investigating faculty decisions to adopt Web 2.0 technologies: Theory and empirical tests. *The Internet and Higher Education*, *11*(2), 71–80. doi:10.1016/j.iheduc.2008.05.002

Al-Hunaiyyan, A., & Al-Sharhan, S. (2009, November). The design of multimedia blended e-learning systems: Cultural considerations. In *3rd International Conference on Signals, Circuits and Systems* (pp. 1-5). IEEE.

Al-Kathiri, F. (2014). Beyond the Classroom Walls: Edmodo in Saudi Secondary School EFL Instruction, Attitudes and Challenges. *English Language Teaching*, *8*(1), 189. doi:10.5539/elt.v8n1p189

Aragon, S. R. (2010). *Creating social presence in online environments. In New Directions for Adult and Continuing Education* (pp. 57–68). San Francisco: Jossey Bass.

Arbaugh, J. B. (2008). Does the community of inquiry framework predict outcomes in online MBA courses? *International Review of Research in Open and Distance Learning*, *9*(2), 1–6.

Arbaugh, J. B., & Benbunan-Fich, R. (2007). The importance of participant interaction in online environments. *Decision Support Systems*, *43*(3), 853–865. doi:10.1016/j.dss.2006.12.013

Armellini, A., & De Stefani, M. (2015). Social presence in the 21st century: An adjustment to the Community of Inquiry framework. *British Journal of Educational Technology*, n/a. doi:10.1111/bjet.12302

Boston, W., Diaz, S. R., Gibson, A. M., Ice, P., Richardson, J., & Swan, K. (2009). An exploration of the relationship between indicators of the Community of Inquiry framework and retention in online programs. *Journal of Asynchronous Learning Networks*, *13*(3), 67–83.

Boyd, D. M., & Ellison, N. B. (2007). Social network sites: Definition, history, and scholarship. *Journal of Computer-Mediated Communication*, *13*(1), 210–230. doi:10.1111/j.1083-6101.2007.00393.x

Brady, K. P., Holcomb, L. B., & Smith, B. V. (2010). The use of alternative social networking sites in higher educational settings: A case study of the e-learning benefits of Ning in education. *Journal of Interactive Online Learning*, *9*(2), 151–170.

Daspit, J. J., & D'Souza, D. E. (2012). Using the community of inquiry framework to introduce wiki environments in blended-learning pedagogies: Evidence from a business capstone course. *Academy of Management Learning & Education*, *11*(4), 666–683. doi:10.5465/amle.2010.0154

Davis, N. E., & Roblyer, M. D. (2005). Preparing teachers for the "Schools that technology built" Evaluation of a program to Train teachers for virtual schooling. *Journal of Research on Technology in Education*, *37*(4), 399–409. doi:10.1080/15391523.2005.10782445

Dawley, L. (2009). Social network knowledge construction: Emerging virtual world pedagogy. *On the Horizon*, *17*(2), 109–121. doi:10.1108/10748120910965494

Driscoll, M. (2002). Blended Learning: Let's get beyond the hype. *E-learning, 1*(4).

Dunlap, J. C., & Lowenthal, P. R. (2010). Investigating Twitter's ability to enhance social presence. In *Annual Meeting of the International Society for Technology in Education* (pp. 1-25). Denver, CO: Academic Press.

DuVall, J. B., Powell, M. R., Hodge, E., & Ellis, M. (2007). Text messaging to improve social presence in online learning. *EDUCAUSE Quarterly*, *30*(3), 24–28.

Ferdig, R. E. (2007). Editorial: Examining social software in teacher education. *Journal of Technology and Teacher Education*, *15*(1), 5–10.

Garrett, N., Thoms, B., Soffer, M., & Ryan, T. (2007). *Extending the Elgg social networking system to enhance the campus conversation*. Pasadena, CA: Second Annual Design Research in Information Systems.

Garrison, D. R., & Akyol, Z. (2013). The community of inquiry theoretical framework. In Handbook of distance education (pp. 104-119). Routledge. doi:10.4324/9780203803738.ch7

Garrison, D. R. & Anderson, T. (2003). *E-Learning in the 21st Century: A Framework for Research and Practice*. Routledge.

Garrison, D. R., Anderson, T., & Archer, W. (2000). Critical inquiry in a text-based environment: Computer conferencing in higher education. *The Internet and Higher Education*, *2*(2), 87–105.

Garrison, D. R., Anderson, T., & Archer, W. (2001). Critical thinking, cognitive presence, and computer conferencing in distance education. *American Journal of Distance Education*, *15*(1), 7–23. doi:10.1080/08923640109527071

Garrison, D. R., Anderson, T., & Archer, W. (2010). The first decade of the community of inquiry framework: A retrospective. *The Internet and Higher Education*, *13*(1), 5–9. doi:10.1016/j.iheduc.2009.10.003

Garrison, D. R., & Arbaugh, J. B. (2007). Researching the community of inquiry framework: Review, issues, and future directions. *The Internet and Higher Education*, *10*(3), 157–172. doi:10.1016/j.iheduc.2007.04.001

Garrison, D. R., Cleveland-Innes, M., & Fung, T. S. (2010). Exploring causal relationships among teaching, cognitive and social presence: Student perceptions of the community of inquiry framework. *The Internet and Higher Education*, *13*(1), 31–36. doi:10.1016/j.iheduc.2009.10.002

Gorsky, P., & Blau, I. (2009). Online Teaching Effectiveness: A Tale of Two Instructors. *International Review of Research in Open and Distance Learning*, *10*(3), 1–27.

Grant, N. (2008, March). On the usage of social networking software technologies in distance learning education. In *Society for Information Technology & Teacher Education International Conference* (Vol. 2008, pp. 3755-3759). Association for the Advancement of Computing in Education (AACE).

Gunawardena, C. N., & Zittle, F. J. (1997). Social presence as a predictor of satisfaction within a computer mediated conferencing environment. *American Journal of Distance Education*, *11*(3), 8–26. doi:10.1080/08923649709526970

Haythornthwaite, C., Kazmer, M. M., & Robins, J. (2000). Community Development among Distance Learners: Temporal and Technological Dimensions. *Journal of Computer-Mediated Communication*, *6*(1). doi:10.1111/j.1083-6101.2000.tb00114.x

Helou, A. M., & Rahim, N. Z. A. (2014). The Influence Of Social Networking Sites on Students' Academic Performance in Malaysia. *International Journal of Electronic Commerce*, *5*(2), 247–254. doi:10.7903/ijecs.1114

Holland, C., & Muilenburg, L. (2011). Supporting student collaboration: Edmodo in the classroom. In *Society for Information Technology & Teacher Education International Conference* (Vol. 1, pp. 3232-3236). Association for the Advancement of Computing in Education (AACE).

Homer, B. D., Plass, J. L., & Blake, L. (2008). The effects of video on cognitive load and social presence in multimedia-learning. *Computers in Human Behavior*, *24*(3), 786–797. doi:10.1016/j.chb.2007.02.009

Iahad, N., Mirabolghasemi, M., & Huspi, S. H. (2012). A blended community of inquiry approach: The usage of social network as a support for course management system. In *International Conference on Computer & Information Science*, *1*, 180–183.

Jafari, A., McGee, P., & Carmean, C. (2006). Managing Courses Defining Learning: What Faculty, Students, and Administrators Want. *EDUCAUSE Review*, *41*(4), 50–70.

Jucevičienė, P., & Valinevičienė, G. (2015). A conceptual model of social networking in higher education. *Elektronika ir Elektrotechnika, 102*(6), 55-58.

Keil, M., & Johnson, R. D. (2002). Feedback channels: Using social presence theory to compare voice mail to e-mail. *Journal of Information Systems Education*, *13*(4), 295–302.

Kozan, K., & Richardson, J. C. (2014). Interrelationships between and among social, teaching, and cognitive presence. *The Internet and Higher Education*, *21*, 68–73. doi:10.1016/j.iheduc.2013.10.007

Kreijns, K., Van Acker, F., Vermeulen, M., & Van Buuren, H. (2014). Community of Inquiry: Social presence revisited. *E-Learning and Digital Media*, *11*(1), 5–18. doi:10.2304/elea.2014.11.1.5

Kumar, S., & Ritzhaupt, A. D. (2014). Adapting the Community of Inquiry survey for an online graduate program: Implications for online programs. *E-Learning and Digital Media*, *11*(1), 59–71. doi:10.2304/elea.2014.11.1.59

Lee, M. J., & McLoughlin, C. (2010). *Beyond distance and time constraints: Applying social networking tools and Web 2.0 approaches to distance learning. In Emerging technologies in distance education* (pp. 61–87). Athabasca University Press.

Lim, D. H., Morris, M. L., & Kupritz, V. W. (2014). Online vs. Blended Learning: Differences in Instructional Outcomes and Learner Satisfaction. *Journal of Asynchronous Learning Networks*, *11*(2), 27–42.

McGee, P., & Reis, A. (2012). Blended Course Design: A Synthesis of Best Practices. *Journal of Asynchronous Learning Networks*, *16*(4), 7–22.

McKerlich, R., Riis, M., Anderson, T., & Eastman, B. (2011). Student Perceptions of Teaching Presence, Social Presence and Cognitive Presence in a Virtual World. *Journal of Online Learning and Teaching*, *7*(3), 324–336.

McLaughlan, R. G. (2007). Instructional strategies to educate for sustainability in technology assessment. *International Journal of Engineering Education*, *23*(2), 201–208.

Meishar-Tal, H., Kurtz, G., & Pieterse, E. (2012). Facebook groups as LMS: A case study. *The International Review of Research in Open and Distributed Learning*, *13*(4), 33–48.

Meyer, K. A. (2004). Evaluating online discussions: Four different frames of analysis. *Journal of Asynchronous Learning Networks*, *8*(2), 101–114.

Mirabolghasemi, M., & Iahad, N. A. (2013). Social network intervention for the enhancement of students' social presence in relation to satisfaction and perceived learning. *International Journal of Web Based Communities*, *9*(2), 233–242. doi:10.1504/IJWBC.2013.053246

Osguthorpe, R. T., & Graham, C. R. (2003). Blended Learning Environments: Definitions and Directions. *Quarterly Review of Distance Education*, *4*(3), 227–233.

Özmen, B., & Atıcı, B. (2014). The Effects of Social Networking Sites in Distance Learning on Learners' Academic Achievements. *European Journal of Open. Distance and E-Learning*, *17*(2), 61–75.

Perry, B., & Edwards, M. (2005). Exemplary Online Educators: Creating a Community of Inquiry. *Turkish Online Journal of Distance Education*, *6*(2), 1–10.

Petrovic, N., Jeremic, V., Cirovic, M., Radojicic, Z., & Milenkovic, N. (2014). Facebook versus Moodle in practice. *American Journal of Distance Education*, *28*(2), 117–125.

Petrović, N., Petrović, D., Jeremić, V., Milenković, N., & Ćirović, M. (2012). Possible educational use of Facebook in higher environmental education. In *International Conference on Information Communication Technologies in Education* (pp.355-362).

Picciano, A. G. (2006). Blended learning: Implications for growth and access. *Journal of Asynchronous Learning Networks*, *10*(3), 95–102.

Pilli, O. (2014). LMS Vs. SNS: Can Social Networking Sites Act as a Learning Management Systems. *American International Journal of Contemporary Research*, *4*(5), 90-97.

Richardson, J. C., & Swan, K. (2003). Examing social presence in online courses in relation to students' perceived learning and satisfaction. *Journal of Asynchronous Learning*, *7*(1), 68–88.

Roblyer, M. D., & Ekhaml, L. (2000). How interactive are your distance courses? A rubric for assessing interaction in distance learning. *Online Journal of Distance Learning Administration*, *3*(2). Retrieved from http://www. Westga.edu/~distance/roblyer32.html

Rožac, J., Buendía, F., Ballester, J., Kos, A., & Pogačnik, M. (2012). Integration of learning management systems with social networking platforms. In *The Fourth International Conference on Mobile, Hybrid, and On-line Learning* (pp. 100-105). Athabasca University.

Rubin, B., Fernandes, R., & Avgerinou, M. D. (2013). The effects of technology on the Community of Inquiry and satisfaction with online courses. *The Internet and Higher Education*, *17*, 48–57. doi:10.1016/j.iheduc.2012.09.006

Selwyn, N. (2000). Creating aConnected'Community? Teachers' Use of an Electronic Discussion Group. *Teachers College Record*, *102*(4), 750–778. doi:10.1111/0161-4681.00076

Severson, A. (2004). *Faculty Support Required for the Implementation of a New Learning Management System*. (Master Thesis). Simon Fraser University, Canada.

Shea, P., Hayes, S., Uzuner-Smith, S., Gozza-Cohen, M., Vickers, J., & Bidjerano, T. (2014). Reconceptualizing the community of inquiry framework: An exploratory analysis. *The Internet and Higher Education*, *23*, 9–17. doi:10.1016/j.iheduc.2014.05.002

Shea, P., Hayes, S., Vickers, J., Gozza-Cohen, M., Uzuner, S., Mehta, R., & Rangan, P. (2010). A re-examination of the community of inquiry framework: Social network and content analysis. *The Internet and Higher Education*, *13*(1), 10–21. doi:10.1016/j.iheduc.2009.11.002

Shea, P., Li, C. S., & Pickett, A. (2006). A study of teaching presence and student sense of learning community in fully online and web-enhanced college courses. *The Internet and Higher Education*, *9*(3), 175–190. doi:10.1016/j.iheduc.2006.06.005

Short, J., Williams, E., & Christie, B. (1976). *The social psychology of telecommunications*. London: John Wiley & Sons.

Sinnappan, S., & Zutshi, S. (2011). Using Microblogging to facilitate Community of Inquiry: An Australian tertiary experience. In *ASCILITE-Australian Society for Computers in Learning in Tertiary Education Annual Conference* (Vol.1, pp. 1123-1135). Australasian Society for Computers in Learning in Tertiary Education.

Sohaei, S., & Iahad, N. A. (2014). Review on Social Network Sites for Teaching and Learning. *Journal of Information Systems Research And Innovation*, *7*(1), 10–16.

Stanciu, A., Mihai, F., & Aleca, O. (2012). Social networking as an alternative environment for education. *Accounting and Management Information Systems*, *11*(1), 56–75.

Stein, D. S., Wanstreet, C. E., Glazer, H. R., Engle, C. L., Harris, R. A., Johnston, S. M., & Trinko, L. A. (2007). Creating shared understanding through chats in a community of inquiry. *The Internet and Higher Education*, *10*(2), 103–115. doi:10.1016/j.iheduc.2007.02.002

Swan, K. (2003). Developing Social Presence in Online Discussions. In S. Naidu (Ed)., *Learning and Teaching with Technology: principles and practices* (pp. 147–164). London: Kogan Page.

Swan, K., Shea, P., Richardson, J., Ice, P., Garrison, D. R., Cleveland-Innes, M., & Arbaugh, J. B. (2008). Validating a measurement tool of presence in online communities of inquiry. *E-mentor*, *2*(24), 1–12.

Szeto, E. (2015). Community of Inquiry as an instructional approach: What effects of teaching, social and cognitive presences are there in blended synchronous learning and teaching? *Computers & Education*, *81*, 191–201. doi:10.1016/j.compedu.2014.10.015

Thoms, B., & Eryilmaz, E. (2014). How media choice affects learner interactions in distance learning classes. *Computers & Education*, *75*, 112–126. doi:10.1016/j.compedu.2014.02.002

Tongmak, M. (2013). Social Network System in Classroom: Antecedents of Edmodo© Adoption. *Journal of e-Learning &. Higher Education*. doi:10.5171/2013.657749

Vaughan, N., & Garrison, D. R. (2005). Creating cognitive presence in a blended faculty development community. *The Internet and Higher Education*, *8*(1), 1–12. doi:10.1016/j.iheduc.2004.11.001

Weyant, L. E. (2013). Designing online management education courses using the Community of Inquiry framework. *Journal of Instructional Pedagogies*, *12*(1), 1–14.

Wheeler, S., Yeomans, P., & Wheeler, D. (2008). The good, the bad and the wiki: Evaluating student-generated content for collaborative learning. *British Journal of Educational Technology*, *39*(6), 987–995. doi:10.1111/j.1467-8535.2007.00799.x

Chapter 2
Blended Learning:
Confronting the Challenges of Higher Education in Oman

Virendra Gawande
Sur College of Applied Sciences, Oman

ABSTRACT

Higher Education in developing countries like Oman offers several unique circumstances that suggest curriculum delivery using Blended Learning is a rational choice. Research has been conducted to identify the factors affecting blended learning adoption and to determine whether the teaching and learning process is amenable to its adoption at HEIs in Oman. Based on the findings a model, Blended Learning Acceptance Model (BLAM), has been developed. It was inferred from the findings that there is a positive relationship among the demographic factors and behavioral Intention i.e. blended-learning adoption. In addition, the teaching and learning styles also had significant influence on adoption of blended learning. BLAM was primarily intended to be used at HEIs in Oman, but may also be used in other developing countries as a reference for the adoption of blended learning.

INTRODUCTION

Application of Information and Communication Technologies (ICTs) in teaching and learning has gained much popularity among practitioners in Higher Education Institutions (HEIs) (Jones, Campbell, McNamara and Stanley, 2008). Today, ICT plays a major role in pedagogical methodology which enables flexibility and reinforces interactivity in the learning environment (Su, Bonk, Magjuka, Liu and Lee, 2005) both inside and outside the classroom. The integration of ICT such as wikis, course pages, blogs, discussion groups, emails etc. in learning and teaching in HEI significantly changed the design of the curriculum, and recently eLearning has gained a significant importance in teaching and learning experience, for both teachers and students. In addition to eLearning, the concept of blended learning (BL) has emerged, the combination of traditional classroom teaching and technology mediated instruction to be the "new traditional model" or "new normal" in the coming era of higher education (Ross and Gage, 2006; Graham, 2006; Norberg, Dziban and Moksal, 2011).

DOI: 10.4018/978-1-5225-0359-0.ch002

Although the concept of blended learning is ambiguous in terms of its conceptual definition, its advantages offer various benefits to both students and teachers (Chan, 2011). For instance, studies conducted in developed countries reported that the blended format provides flexibility, motivation, idea sharing, interaction and better communication over traditional learning formats (Fearson, Starr and McLaughlin, 2011; Cohere, 2011; Chan, 2011). These studies also concluded that online education was an important component of their long term strategy (Allen and Seaman, 2010). Despite these findings, the results were inconsistent in relation to the student's acceptance and adoption level. Adoption is the key factor that would ensure a productive use of a system, therefore understanding the user adoption of new technologies is important to their use in education. However, this is challenging, as it includes many factors that would influence behavioural intentions.

This study aimed to investigate the challenges in the emerging field of blended learning in a developing country like Oman. The study was concerned with measuring the attitude and experiences of the faculty members and students at Colleges of Applied Sciences (CAS) in Oman, as they explore the use of Learning Management Systems (LMS), as a supplement to traditional face-to-face delivery of the curriculum.

BACKGROUND

The introduction of online learning systems has increased access and flexibility in the curriculum, enhancing the communication and the learning experience; however, there are issues with fully online courses in relation to the lack of social contact, choices, and engagement (Singh, 2003; Holley and Oliver, 2010).

The concept of blended learning refers to the pedagogical paradigm of a proactive student's focus on curriculum delivery, rather than as a passive observer present in a classroom. Using the strengths of online and face-to-face learning, blended learning curriculum delivery combines the relative advantages of both the environments and research shows that blended learning is increasing in Higher Education (Chandra, 2004; Garrison and Vaughan, 2008; Graham, 2006, Picciano, 2009).

Blended learning may have different definitions (Matheos *et al.*, 2005). Allan (2007) states that there is agreement that the pedagogy is 'a mixture of face-to-face and eLearning', and according to Graham (2006), blended learning is 'the combination of instruction from two historically separate models of teaching and learning: traditional face-to-face learning systems and distributed learning systems' (Figure 1). For the purposes of this study, Graham's (2006) definition is adopted, as it is general and ignores the complexity of blended learning (Stacey and Gerbic, 2009). As a logical development from both face-to-face and online learning, Garrison and Vaughan (2008) suggested that, 'Blended learning — a design approach whereby both face-to-face and online learning are made better by the presence of the other – offers the possibility of recapturing the traditional values of higher education while meeting the demands and needs of the twenty-first century'.

Blended learning is a fundamental redesign of pedagogical practices. In blended learning design there is a paradigm change in which the emphasis is more on learning than on teaching (López-Pérez *et al.*, 2011). Pedagogy can be transformed towards more active learning with wider use of learner-centred approaches through blended learning curriculum delivery (Garrison and Vaughan, 2008; Nunan *et al.*, 2000). Vaughan (2007) states that facility usage can be increased if a greater number of students can be accommodated with less campus attendance. Garrison and Kanuka (2004) and Garrison and Vaughan (2008) reported the challenges and complexities in administration and development of blended learning

Figure 1. Development of blended learning
Source: Graham (2006)

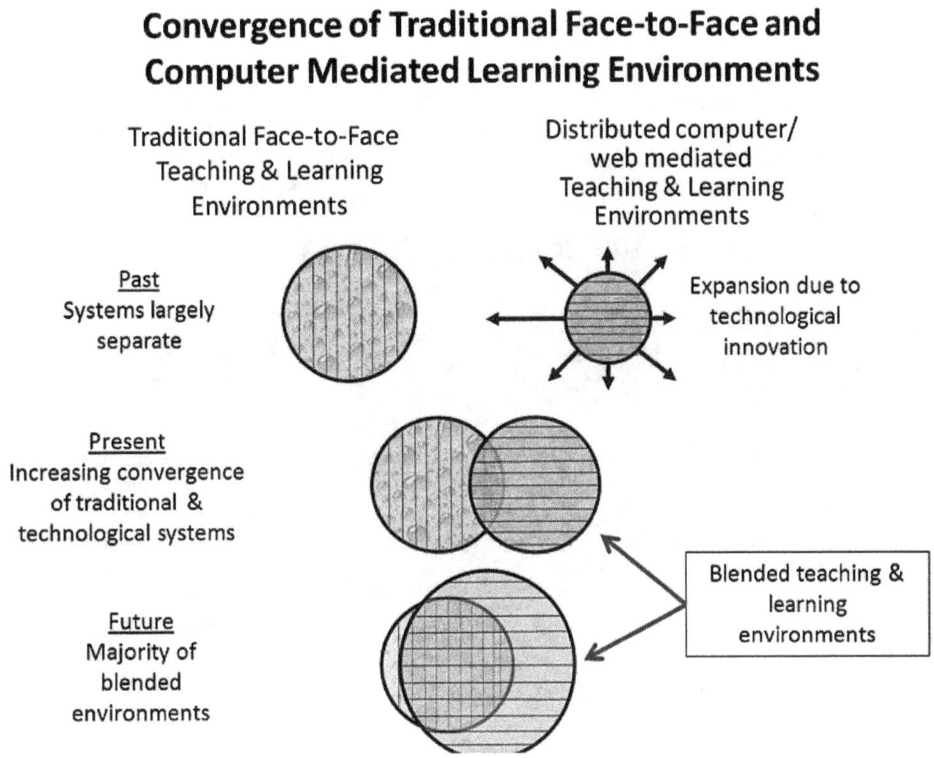

courses. These include changes to university policies, infrastructure planning, allocation of resources, scheduling training and implementation, and technical support (Garrison and Kanuka, 2004).

Blended Learning Strategies

For the purposes of this study, a blended learning system is defined as the student's access to the physical and online presence of the lecturer that enhances their communications and interactions for the purpose of achieving knowledge. The reference for this context is given by Picciano (2009) who stated that cultural and individual differences of learners, faculty and course designers should use multiple blended approaches. A major benefit of multiple modalities is that they facilitate flexible learning whilst challenging students to widen their learning experiences. Picciano's (2009) model shows that academic goals and objectives drive the course outcomes and therefore also drive the approaches and technologies employed. The author defines the issues in blended learning as the nature of the course, ICT availability and usage by faculty and students, and the social environment of the higher education institution. These elements are depicted in a model as four groups, according to the extent of technology use and physical attendance by the students (Figure 2).

In Figure 2, the first group A comprises a teaching environment where students attend lectures and tutorials with minimal use of technology. In the second group B, attendance is combined with varying degrees of technology. For group C, students meet online, however the content tools are simple, such as

Figure 2. Blended learning approaches
Source: Picciano (2009)

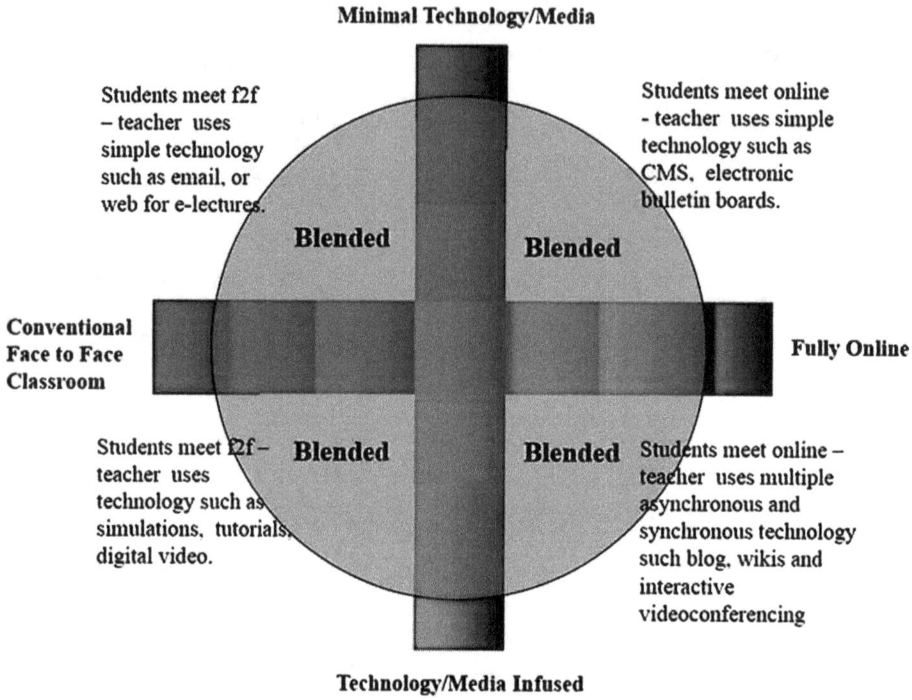

electronic bulletin boards. In the fourth group D, students meet online with an advanced use of technology, such as interactive video-conferencing (Picciano, 2009).

The design of blended learning depends on its goals and discipline. The blend could involve only one aspect or feature of online components (Chen and Looi, 2007). However, the integration of the online and face-to-face modes should support each other (Alebaikan and Troudi, 2010a, and b). According to So and Brush (2008), the poor integration of learning components in blended courses may negatively affect the approach, 'blended formats do not necessarily provide students with more interactive and flexible learning experiences. More careful analyses of learners, contexts, and technologies are needed'. A new strategy in blended learning was investigated by Chen and Looi (2007) on the effectiveness of incorporating online discussion in both on campus face to face and off-campus settings. The findings were that online discussion can offer a wider spectrum of debate, increase participation, promote cognitive thinking skills, and enhance the students' engagement in learning. Again, the lack of proximity in physical interactions and finding sufficient time for online postings were major challenges in implementing this form of debate. Nevertheless, Graham (2006) and Garrison and Vaughan (2008) described different examples of blended learning in higher education. The cases were global and the technology varied, however, the outcomes were invariably positive.

Development of Blended Learning in Oman

With the rapid development in internet technologies, distance education has evolved; in particular with web based educational tools and compressed videos. This form of flexible course delivery has given rise to a new concept called online learning, also referred to as eLearning (Ostendorf, 1997). Harasim (1989) refers to online learning as a combination of distance education and traditional face-to-face education using computer mediated instruction. Most of the education providers, including HEIs in Oman, have implemented flexible learning through eLearning technologies. Richard (2005) explained the categories of flexible learning environments as follows:

- **None or Trivial Online Presence:** It is based on traditional classroom environment with usage of electronic tools.
- **Web Supplemented:** Students use course documents provided by lecturers, communicate by emails and refer to external links for additional course information.
- **Web Dependent: Students** are required to participate in key course activities through online discussion, discussion boards, while attending the classroom.
- **Mixed Mode:** Students need to participate in online activities as a course requirement as well as attending classroom sessions. Some parts of classroom sessions can be replaced. *This method has been implemented at HEI in Oman.*
- **Fully Online:** Students need not to attend classroom lectures.

CAS in Oman adopted a mixed mode of teaching, where "students are required to participate in online activities, e.g. online discussions, assessments, online project/collaborative work, as part of the course work, which replace part of the face-to-face teaching/learning. Significant campus attendance remains" (Richard, 2005).

Blended learning is still in its formative stages in the country, and there is ambiguity regarding conceptualization of the delivery systems such as blended learning and eLearning (Alebaikan, 2010). Further, the majority of Omani HEIs, including CAS, added blended learning applications merely to supplement their traditional lecture delivery in the classroom, without a comparable reduction in attendance requirements for students or their instructors. As such, a clear policy for blended learning delivery was required, and this study has contributed to the blended learning paradigm design in Omani HEI. This study has also explored the potential for blended learning curriculum designs, gender-segregation issues and the development of the ICT sector in Oman to identify the factors that may affect the successful implementation of blended learning in the HEI.

ELearning Acceptance Model (ELAM)

eLearning is described as the process of learning promoted and assisted via information and communication technologies (Jenkins and Hanson, 2003). Hence, eLearning incorporates the utilization of ICT equipment (e.g. Internet, computer) and content developed via technology (e.g. animations, videos) to aid teaching and learning.

Acceptance of eLearning includes the adoption of technology and there are also some important issues like teaching aspects which need to be considered. Research on eLearning technology acceptance is based on the models of Technology Acceptance Model (TAM) or Unified Theory of Acceptance and Use of Technology (UTAUT), and investigates it either on instructors (Nanayakkara, 2007; Yuen and Ma, 2008) or students (Keller *et. al.*, 2008; Masrom, 2007). These researches provide clear evidence favouring centrality of characters in eLearning acceptance. It was observed that usability or effort expectancy acts as the vital element for the instructors, whereas for the students perceived usability or performance

Figure 3. ELearning acceptance model
Source: Umrani-Khan and Iyer (2009)

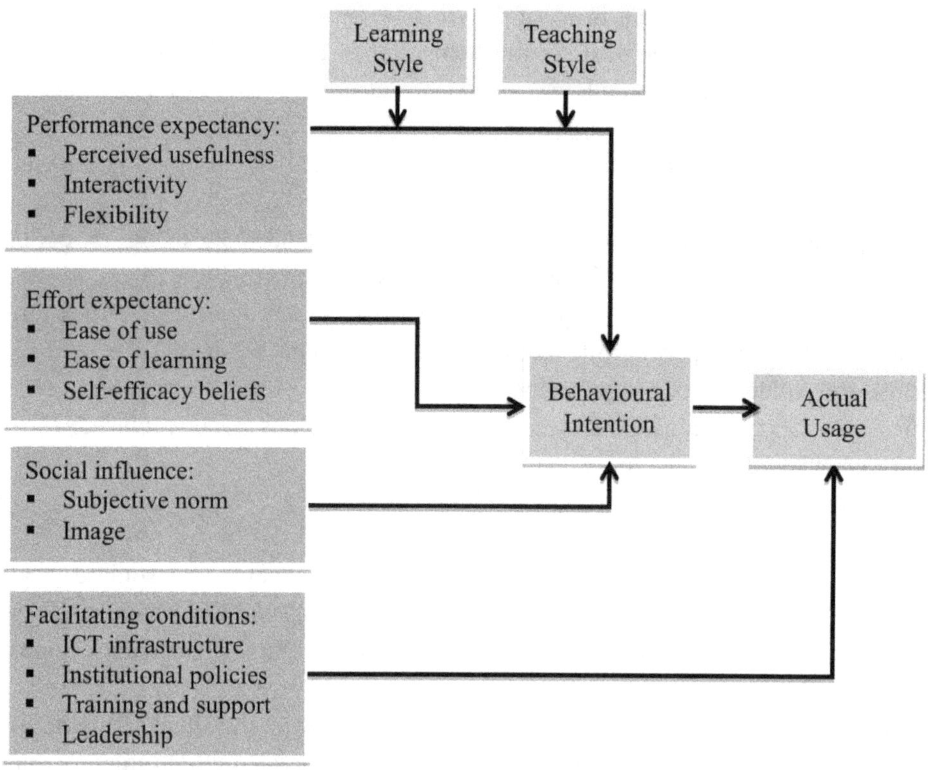

expectancy acts as the vital element (Jung *et al.*, 2008; Raaij and Schepers, 2008). Figure 3 shows the ELAM model specifically developed for the eLearning context.

Description of ELAM

Either the perspective of teachers or students is considered by the existing eLearning research. In the process of education, ELAM provides a more complete view as it considers factors that are related to each stakeholder in the education process. Moreover, the research that already exists has opted for the participants from the West or developed countries. This research provides a potential solution that addresses the restrictions in the developing countries. There is enough evidence to show that while performance expectancy is prominent for students, effort expectancy is important for teachers in their eLearning acceptance. Since assessment items are reflected in the two groups, the conclusive evidence for the difference between students and teachers will be provided by the model's empirical test. Moreover, instruments like eLearning tools or learning management systems are common to eLearning and can be easily adapted, but they have to be checked for psychometric analyses. Incorporation of learning styles into the model will be carried out by the researcher of this study and conceptualizing a mediation role will affect the performance of students' beliefs. Teaching style is regarded as same as accepting the eL-earning style by teachers. In a technology-enabled distance education program, a proposal is presented to

Table 1. Inventory of learning styles

Name	Learner's Categorisation	Assessment Instrument
Kolb Learning Style Inventory (Kolb, 1984, 1985)	Divergers (concrete, reflective) Assimilators (abstract, reflective) Convergers (abstract/active) Accommodators (concrete/active)	Learning Style Inventory (LSI), consisting of 12 items in which subjects are asked to rank 12 sentences describing how they best learn.
Dunn and Dunn-Learning Style Assessment Instrument (Dunn and Dunn, 1978, 1999)	Environmental, Emotional, Sociological, Physical factors.	i) Learning Style Inventory (LSI) designed for children grade 3-12; ii) Productivity Environmental Preference Survey (PEPS) –adult version of the LSI containing 100 items
Felder-Silverman-Index of Learning Styles (Felder and Silverman, 1988; Felder, 1996)	Sensing-intuitive, Visual-verbal, Indicative-deductive, Active-reflective, Sequential-global	Soloman and Felder questionnaire, consisting of 44 questions
Riding-Cognitive Style Analysis (Riding and Cheema, 1991; Riding, 1994)	Wholists-Analytics, Verbalisers-Imagers	CSA (Cognitive Styles Analyst) test, consisting of three sub tests based on the comparison of the response time to different items.
Honey and Mumford Learning Styles Questionnaire (Honey and Mumford, 1992)	Theorist, Activist, Reflector, Pragmatist	Honey and Mumford's Learning Styles Questionnaire (LSQ), consisting of 80 items with true/false answers
Gregoric-Mind Styles and Gregoric Style Delineator (Gregoric 1979, 1982)	Abstract Sequential, Abstract Random, Concrete Sequential, Concrete Random	Gregoric Style Delineator Containing 40 words arranged in 10 columns with 4 items each; the leaner is asked to rank the words in terms of personal preference
McCarthy-4 Mat System (McCarthy, 1980, 1997)	Innovative, Analytic, Common sense, Dynamic	
Gardner-Multiple Intelligence Inventory (Gardner, 1993)	Linguistic, Logical-mathematical, Musical, Bodily-kinaesthetic, Spatial, Interpersonal, Intrapersonal	An instrument consisting of 8 questions
Grasha-Riechmann-Student Learning Style Scale (Hruska-Riechmann and Grasha, 1982; Grasha, 1996)	Competitive-Collaborative, Avoidant-Participant, Dependent-Independent	90 items self-report inventory measuring the preferences of both high school and college students

Source: Karagiannidis and Sampson (2004)

test the ELAM model on teacher and student participants particularly to understand perceptions towards adoption of blended learning at CAS.

Table 1 lists the different constructs and instruments to assess each of the factors in the proposed eLearning acceptance model (ELAM). The following paragraph presents a brief description of each of the factors (Farida Umrani-Khan and Sridhar Iyer, 2009).

1. **Performance Expectancy (PE):** Defined as the degree to which the student and teacher believes that using the system will result in gains in the teaching and learning process. This includes three factors:
    a. **Perceived Usefulness:** Refers to the extent to which students and teachers believe that using eLearning will enhance their performance. The facets tapped are improved understanding, higher achievement, efficiency and decreased study/teaching load.

b. **Interactivity:** Refers to the extent to which eLearning facilitates interaction between students and teachers and amongst group of students. The facets measured are asking questions to students and teachers, working in collaboration and using online resources.
c. **Flexibility:** Refers to the extent to which eLearning tools and content accommodate the preference of students and teachers. The facets assessed are choosing topics in the order of interest, self-paced learning or teaching, convenience (any time-any place) and adaptability to preferred learning style of students or teaching style of teachers.

2. **Effort Expectancy (EE):** Defined as the extent to which the student and teacher believes that the eLearning tool requires effort. It includes three factors:
   a. **Ease of Use:** Refers to the degree to which the user expects the target system to be free of effort. The facets tapped are effort required, understanding of how the system works.
   b. **Ease of Learning:** Refers to the extent to which the user finds the eLearning tool easy to learn.
   c. **Self-Efficacy:** Refers to the evaluation of competence to use eLearning. Also included within this dimension is evaluation of competence of the other role, that is, students are asked to evaluate teachers' competence with eLearning and vice versa.
3. **Social Influence (SI):** Defined as the extent to which the students and teachers perceive a social pressure to use eLearning. This involves two factors:
   a. **Subjective Norm:** Taps the perception that people who are important to them (teachers, students, colleagues, head of the department/institute) think they should or should not use eLearning.
   b. **Image:** Captures the degree to which use of technology is perceived to enhance one's image or status in one's social context.
4. **Facilitating Conditions (FC):** Defined as the extent to which the students and teachers perceive institutional support to use eLearning. This includes four factors:
   a. **ICT Infrastructure:** Availability and reliability of facilities.
   b. **Institutional Policies:** Opportunities and incentives for use of eLearning.
   c. **Training and Support:** Training to become an efficient user of the technology and sustained technical assistance.
   d. **Leadership:** Role model and support from the head of the department and institute.

Behavioural intention (BI) refers to the individual's decision regarding future of technology.

Actual usage (AU) taps the variety and frequency of technology used.

It is to be noted that in the context of a developing countries the intention to use eLearning may not culminate in actual usage due to real world constraints. For example, a group of teachers and students may hold positive attitudes towards technology, but may not have access to it. Hence attitude (BI) as well as behaviour (AU) needs to be considered to tap acceptance of technology.

PE, EE and SI determine BI which in turn determines AU. However, facilitating conditions has a direct effect on AU, since it gauges the real life constraints and facilitators in transforming the intention to action. In addition to the above, the learning style of the student and teaching style of the teachers are considered as mediators affecting the relation between perceived usefulness and behavioural intention to use eLearning. These are discussed in the next section.

BLENDED LEARNING IN HIGHER EDUCATION

Statement of the Problem

Oman is a rich and fast developing country. Rapid developments have been observed in the last five years. Higher Education in Oman offers unique circumstances as follows, which indicates that the adoption of Blended Learning as a rational choice.

- Gender segregation.
- Increasing student population leading to shortage of resources including proficient teachers.
- Rapidly developing country.
- Dispersed communities.
- Rich economy.

No significant researches have been conducted with particular reference to blended learning, within HEIs in Oman. There are studies that have been done in the past but those are mainly in developed nations. This indicates a need for similar research to be done within Oman.

Blended learning is an intermediary approach to eLearning and traditional face to face classroom teaching. LMS tools like Blackboard are already in use as an initiative towards eLearning, but they have not been used efficiently or effectively across the HEIs. Another need for research is to investigate the factors affecting the adoption of Blended Learning from the perspective of students as well as faculty.

Blended learning is relevant to Oman, as the youth (under the age of 25 years) constitute the major portion of the population and there is no other way to take education to a larger scale without the intervention of technology.

Oman adopted ICT later than many developed economies, confronting challenges of providing sufficient infrastructure for its widespread population. The adoption of technologies is a challenge for the Omani government, as the younger members of society embrace social networking, and conservative sections of the community require encouragement to utilize these services.

Other challenges include increasing student numbers, gender segregation, lack of faculty members, and inadequacy of learning resources. Despite the challenges, the adoption of ICT in education is a fundamental objective for the government of Oman-for spreading and creating equality of access through computer literacy.

Most HEIs in Oman offer undergraduate degrees, and admit all Omani students. The establishment of new private universities and colleges across the country therefore enhances higher education access for the majority of Omani citizens.

However, in a traditionalist and rich society like Oman, higher education offers several unique circumstances that suggest curriculum delivery using blended learning is a rational choice. These are an increase in student numbers, coupled with an inadequate number of qualified instructors which has led to a situation where the traditional system is unable to respond, and thus, an initiative leading towards blended learning is promising.

The Government of Oman has taken cognizance of the institutional support required for resources in eLearning and formulated the national mission on education through ICT. Thus, the grounds for conducting this study are to support the government policies on blended learning, and the Ministry of Higher Education and CAS administration and faculty in delivering the government's objectives.

To the best of the researcher's knowledge, so far there are no significant studies on technology acceptance of blended learning that includes factors related to students and teachers within the same model. Thus this research study may be very significant.

Research Objectives

The purpose of this research was to identify factors affecting the adoption of blended learning and to study whether teaching and learning processes are predictive in its adoption, and develop a blended learning model to be adopted at CAS. The specific research objectives were as follows:

1. To examine-factors of *User Acceptance of Technology, Teaching styles* and *Learning styles* affecting the *Behavioural Intention (Blended Learning)* at CAS.
2. To examine the impact of *Demographic variables, Teaching styles of teachers* and *Learning style of the students* on *Behavioural Intention (Blended Learning)* at CAS.
3. To seek an association between the *Teaching style of teachers* and *Learning style of students* at CAS.

The following research questions were posed:

1. To what extent the *Teaching of styles of teachers* is considered as a mediator affecting the relationship between *User Acceptance of Technology* and *Behavioural Intention (Blended Learning)* at CAS?
2. To what extent the *Learning style of the students* is considered as a mediator affecting the relationship between *User Acceptance of Technology* and *Behavioural Intention (Blended Learning)* at CAS?
3. To what extent does the *Teaching styles of teachers* and *Learning style of the students* affect the relationship of *Demographic variables, User Acceptance of Technology* and *Behavioural Intention (Blended Learning)* at CAS?

Significance of the Research

As blended learning is in its early stage at HEI in Oman, this study seeks to provide a substantial contribution by identifying the factors that can influence the adoption of blended learning as a means of delivering curricula in Oman HEI.

1. This research was among the initial studies, in investigating the perceptions of students and instructors towards the impact of blended learning in Oman.
2. This study also contributed to the framework, principles, and guidelines for effective implementation of blended learning in the Oman.
3. The study further contributed to the knowledge for theoretical and practical aspects of blended learning design.
4. Theoretically, the study assesses the literature findings that explain the impact of blended learning design on the quality of teaching and learning.

5. Most of the existing eLearning research has considered either the perspective of teachers or students, while the present study focuses on both perspectives.
6. In addition, the earlier studies have considered participants from the west or developed countries while the present study focuses on the developing countries context.
7. This study provides insight on the motivation and engagement of students and instructors to create effective blended learning environments that enhance their interactions, teaching, and learning.

Research Methodology

The study is primarily descriptive and explanatory, employing a quantitative research method using a cross-sectional survey to elicit the attitude of the students and instructors in relation to their learning and teaching experiences in blended learning.

The study adopted *survey* and *correlational design* for this quantitative study (Creswell, 2014). The choice is justified as follows: Firstly, the research objective is to observe the user acceptance, and to identify the specific learning and teaching style of students and faculty. Therefore, a survey design procedure would be the best choice in quantitative research in which a questionnaire will be administered to target the population, in this case students and faculty. Secondly, the correlational design would be the best choice as the study does not stop with observing the trends, but will also examine the impact of user acceptance on behavioural intention and the role of learning and teaching style, and demographic factors on this relationship. Therefore, the correlational design would be the perfect choice for this research. This design ensures greater validity as it applies empirical data collection (Creswell, 2014; 2009; Creswell and Plano Clark, 2007; Doyle, Brady and Byrne, 2009). The design approach was therefore appropriate to determine convergence of perspectives and corroboration of results from different perspectives of faculty and students. Figure 4 illustrates the study framework;

In this study, the researcher adopted a multilevel relationship which involved the use of two or more sets of samples that are extracted from different levels of the study (i.e., different populations). Kemper et al., (2003) defined it as occurring "when probability and purposive sampling techniques are used on different levels of the study".

Target Area and Population

The study was conducted at CAS in Oman and both teachers and students were chosen as participants. The participants for the study, in case of students, were restricted to those who have access to the Blackboard LMS on the CAS portal, while in the case of teachers, those who are teaching in any of the three academic departments at CAS were selected, viz. Department of Information Technology (DIT), Department of Communication and Department of General Requirements using Blackboard.

Sampling Method

A convenience sampling technique was used to select both the students and teachers for the study as it helps to minimize the chances of distortion in the collected data. This choice of sampling was chosen as the research objective was to empirically test the model among students and faculty who have access to Blackboard.

Figure 4. Research framework

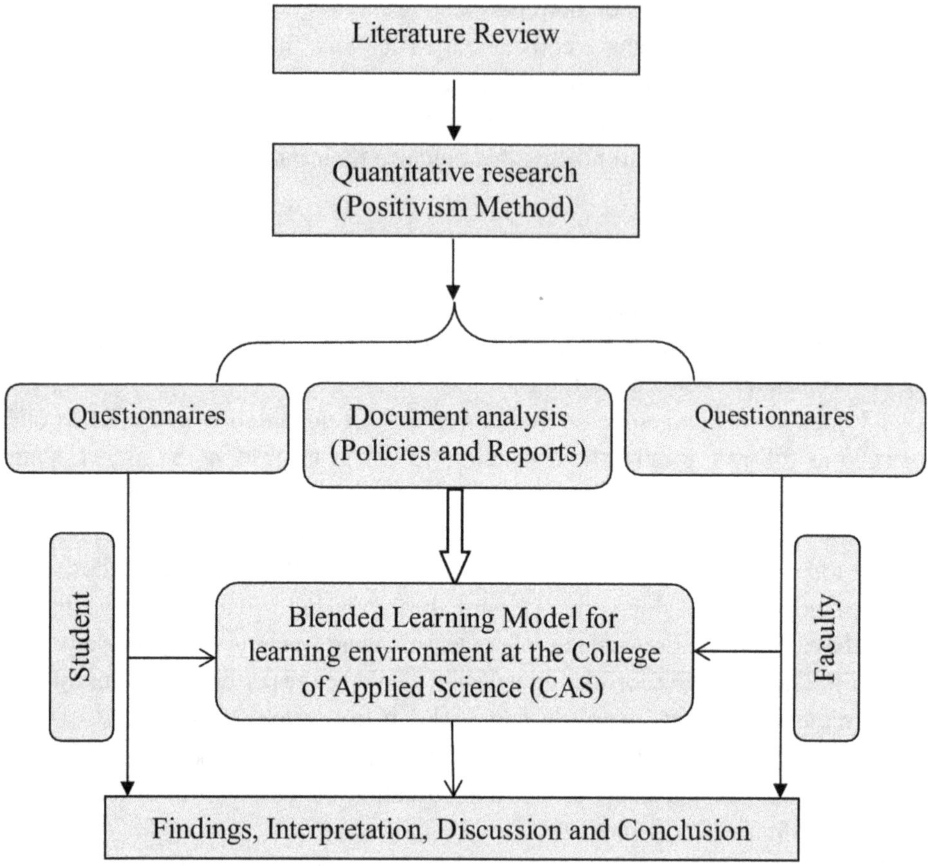

Data Analysis Procedures

Data analysis is the most essential part in a research study. The analysis of students' and faculties' responses to the sets of questionnaires was statistically analysed and computed using SPSS software (IBM Corp. version 21.0, NY: IBM Corp) and then interpreted. Descriptive statistics like frequency distributions, percentages, average mean of the scores and standard deviations for all the dependent variables are employed.

Summary of Results

Based on the research questions, 18 hypotheses were designed, some of which were also subdivided. The results are as follows:

1. The factors of Performance expectancy of *teachers,* namely a) Perceived usefulness (β= 2.494; p=0.016) b) Interactivity (β= 0.591; p<0.001) and c) Flexibility (β= 0.263; p=0.002) were significantly and positively influenced on BL. Overall, performance expectancy of teachers (β= 0.836; p<0.001) significantly influenced blended learning adoption. Hence it fulfils research question

1.1 and the formulated hypotheses H_{1a}, H_{1b} and H_{1c} were not rejected. Similarly, results were also obtained for the *students* where a) Perceived usefulness (β= 0.560; p<0.001) b) Interactivity (β= 0.706; p<0.001) and c) Flexibility (β= 0.655; p<0.001) were significantly and positively influenced on BL. Hence it fulfils research question 1.1 and the formulated hypotheses H_{5a}, H_{5b} and H_{5c} were not rejected. Overall, Performance expectancy of students (β= 0.821; p<0.001) significantly influenced blended learning adoption.

2. The factors of Effort expectancy of *teachers,* namely a) Ease of learning (β= 0.371; p=0.004) b) Ease of use (β= 0.615; p=0.007) and c) Teacher (self) efficacy (β= 0.617; p=0.003) were significantly and positively influenced on BL. Hence it fulfils research question 1.1 and the formulated hypotheses H_{2a}, H_{2b} and H_{2c} were not rejected. Overall, Effort expectancy of teachers (β= 0.933; p<0.001) significantly influenced blended learning adoption. Similarly, the factors of Effort expectancy of *students,* namely a) Ease of learning (β= 0.527; p<0.001) b) Ease of use (β=-0.110; p=0.143) and c) Students efficacy (β= 0.411; p<0.001) were partially influenced on BL. Hence it fulfils research question 1.1 and the formulated hypotheses H_{6a} and H_{6c} were not rejected, whereas H_{6b} was rejected in the study. Overall, Effort expectancy of students (β= 0.795; p<0.001) significantly influenced blended learning adoption.
3. The factors of Social influence of *teachers* namely a) Subjective norm (β=-0.090; p=0.489) b) Image of teachers (β=-0.138; p=0.422) were not significantly on BL. Hence it fulfils research question (RQ) 1.1, but the hypotheses H_{3a} and H_{3b} were rejected. However, overall, Social influence of teachers (β= 0.535; p=0.017) significantly influenced blended learning adoption. Similarly the factors of Social influence of *students* namely a) Subjective norm (β=0.522; p<0.001) b) Image of students (β=0.343; p<0.001) were significantly and positively influenced on BL. Hence it fulfils research question 1.1 and the formulated hypotheses H_{7a} and H_{7b} were not rejected. Overall, Social influence of students (β= 0.606; p<0.001) significantly influenced blended learning adoption.
4. The factors of Facilitating conditions of teachers namely a) ICT infrastructure (β= 0.058; p=0.709) b) Institutional policies (β= -0.063; p=0.682) c) Training and technical support (β= -0.210; p=0.002) and d) Leadership of teachers (β= 0.016; p=0.854) have partial influence on BL. Hence it fulfils research question 1.1 and the hypotheses H_{4c} was not rejected whereas H_{4a}, H_{4b} and H_{4d} were rejected in the study. However, overall, Facilitating conditions of teachers (β= -0.182; p=0.246) did not influence blended learning adoption. Similarly, the factors of Facilitating conditions of *students* namely a) ICT infrastructure (β= 0.060; p=0.367) b) institutional policies (β= 0.198; p<0.001) c) Training and technical support (β= 0.410; p<0.001) and d) leadership of students (β= 0.421; p<0.001) were partially influencing BL. Hence it fulfils research question 1.1 and the formulated hypotheses H_{8b}, H_{8c} and H_{8d} were not rejected. However H_{8a} was rejected in the current study. Overall, Social influence of students (β= 0.648; p<0.001) significantly influenced blended learning adoption.
5. The factors of teaching style of teachers, namely a) expert-formal-authority b) personal model-expert c) facilitator-personal model-expert d) Delegator-facilitator-expert and e) delegator were not significantly influencing BL. Hence it fulfils research question 1.2 while formulated hypotheses H_{9a}- H_{9e} were rejected. Similarly, the factors of Learning style of students namely a) participant vs. avoidant b) independent vs. dependent c) collaborative vs. competitive and d) individualistic vs. collectivistic were not significantly influencing BL. Hence it fulfils research question 1.2 and formulated hypotheses H_{10a}- H_{10d} were rejected.

6. Gender, age and years of experience of the teacher significantly influence all the five factors of teaching style. Hence it fulfils research question 2.1 and the hypotheses H_{11a} - H_{11c} were not rejected. Similarly, gender of the students significantly influences all the five factors of learning style and only the factor participant was not significant. Hence it fulfils research question 2.1 and hypothesis H_{12a} was partially rejected.
7. Age of the students significantly influences all the five factors of learning style and only the factor avoidant was not significant. Hence it fulfils the research question 2.1 and hypothesis H_{12b} was partially failed to reject.
8. Computer experience of the students do not significantly influence all the six factors of learning style. Hence it fulfils research question 2.1 and hypothesis H_{12c} was rejected.
9. Gender, age and years of experience of the teacher do not significantly influence the Behavioural intention (Blended Learning). Hence it fulfils research question 2.2 and hypotheses H_{13a}, H_{13b} and H_{13c} were rejected. However, in case of students, gender, age and years of computer experience of the students do not significantly influence Behavioural intention (Blended Learning). Hence it fulfils the research question 2.2 and hypotheses H_{14a}, H_{14b} and H_{14c} were not rejected.
10. There was a significant and positive relationship between the factors of teaching style of teachers and learning style of students. Hence it fulfils research question 3 and the hypothesis H_{15} was not rejected.
11. It was found that the relationship between the user acceptance of technology and behavioural intention was not mediated by the teaching style of teachers in the study. Hence it fulfils research question 4 and hypothesis H_{16} was rejected. However, in the case of students the relationship between some of the user acceptance of technology and behavioural intention was mediated by the learning style of students. Hence it fulfils research question 5 and hypothesis H_{17} was partially failed to be rejected.
12. It was found that the relationship between the some of the user acceptance of technology and behavioural intention was affected by the teaching style and demographic variables of teachers. Hence it fulfils research question 6 and hypothesis H_{18a} was partially failed to reject. However, in case of students, the relationship between all the factors of the user acceptance of technology and behavioural intention was not affected by the learning style and demographic variables of students. Hence it fulfils research question 6 and hypothesis H_{18b} was rejected.
13. It was found that the relationship between all the factors of the user acceptance of technology and behavioural intention was not affected by the both teaching style of teachers and learning style of students and demographic variables of both teachers and students. Hence it fulfils research question 6 and hypothesis H_{18c} was rejected.

Summary of Hypotheses Testing

Tables 2 and 3 show the statistical analysis methods used for the hypotheses testing.

Development of Blended Learning Acceptance Model

Based on the research findings and the above results of hypothesis testing, the models in Figures 5 and 6 have been developed.

Table 2. Statistical Analysis methods used

Hypotheses	Statistical Analysis method(s) used
H_1-H_{10}	Regression Analysis
H_{11}-H_{14}	Independent sample t-test and One way ANOVA
H_{15}	Correlation
H_{16}-H_{17}	Sobel mediation analysis
H_{18}	MANCOVA

Table 3. Summary of hypotheses testing

Hno.	Hypotheses	Faculty	Students
H_1/H_5	Performance Expectancy has significant influence on blended learning adoption.	Failed to reject	Failed to reject
H_{1a}/H_{5a}	Perceived Usefulness has significant influence on blended learning adoption.	Failed to reject	Failed to reject
H_{1b}/H_{5b}	Interactivity has significant influence on blended learning adoption.	Failed to reject	Failed to reject
H_{1c}/H_{5c}	Flexibility significantly increases blended-learning adoption.	Failed to reject	Failed to reject
H_2/H_6	Effort Expectancy has significant influence on blended learning adoption.	Failed to reject	Failed to reject
H_{2a}/H_{6a}	Ease of learning significantly influences blended-learning adoption.	Failed to reject	Failed to reject
H_{2b}/H_{6b}	Ease of use significantly influences blended-learning adoption.	Failed to reject	Rejected
H_{2c}/H_{6c}	Efficacy significantly influences blended-learning adoption.	Failed to reject	Failed to reject
H_3/H_7	Social Influence significantly influences blended-learning adoption.	Failed to reject	Failed to reject
H_{3a}/H_{7a}	Subjective norm significantly influences blended-learning adoption.	Rejected	Failed to reject
H_{3b}/H_{7b}	Image significantly influences blended-learning adoption.	Rejected	Failed to reject
H_4/H_8	Facilitating conditions significantly influences blended-learning adoption.	Rejected	Failed to reject
H_{4a}/H_{8a}	ICT infrastructure significantly influences blended-learning adoption.	Rejected	Rejected
H_{4b}/H_{8b}	Institutional policies significantly influences blended-learning adoption.	Rejected	Failed to reject
H_{4c}/H_{8c}	Training and technical support significantly influences blended-learning adoption.	Failed to reject	Failed to reject
H_{4d}/H_{8d}	Leadership significantly influences blended-learning adoption.	Rejected	Failed to reject
H_9	Teaching Style significantly influences the Blended Learning adoption.	Rejected	
H_{9a}	The 'expert-formal authority' teaching style significantly influences blended-learning adoption.	Rejected	
H_{9b}	The 'personal model-expert-formal authority' teaching style significantly influences blended-learning adoption.	Rejected	
H_{9c}	The 'facilitator-personal model-expert' teaching style significantly influences blended-learning adoption.	Rejected	
H_{9d}	The 'delegator-facilitator-expert' teaching style significantly influences blended-learning adoption.	Rejected	
H_{9e}	The 'delegator' teaching style significantly influences blended-learning adoption.	Rejected	
H_{10}	Learning Style significantly influences the Blended Learning adoption.		Rejected
H_{10a}	The 'participant' style of learning significantly influences behavioural intention (blended-learning adoption) than avoidance learning style.		Rejected
H_{10b}	The 'independent' style of learning significantly influences behavioural intention (blended-learning adoption) than dependent learning style.		Rejected

Table 3. Continued

Hno.	Hypotheses	Faculty	Students
H_{10c}	The 'collaborative' style of learning significantly influences behavioural intention (blended-learning adoption) than competitive learning style.		Rejected
H_{10d}	The 'individualistic' style of learning significantly influences behavioural intention (blended-learning adoption) than collectivistic learning style.		Rejected
H_{11}	Demographic factors of teachers significantly influence the Teaching Style.	Failed to reject	
H_{11a}	Gender significantly influences the teaching style of teachers.	Failed to reject	
H_{11b}	Age group significantly influences the teaching style of teachers.	Failed to reject	
H_{11c}	Experience significantly influences the teaching style of teachers.	Failed to reject	
H_{12}	Demographic factors of students significantly influence the Learning Style.		Failed to reject
H_{12a}	Gender significantly influences the learning style of students.		Failed to reject
H_{12b}	Age group significantly influences the learning style of students.		Failed to reject
H_{12c}	Computer Experience significantly influences the learning style of students.		Rejected
H_{13}	The Demographic variables of teachers significantly influence Behavioural Intention (Blended Learning).	Rejected	
H_{13a}	Gender of teachers significantly influences the behavioural intention (blended learning).	Rejected	
H_{13b}	Age group of teachers significantly influences the behavioural intention (blended learning).	Rejected	
H_{13c}	Experience of teachers significantly influences the behavioural intention (blended learning).	Rejected	
H_{14}	The Demographic variables of students significantly influence the Behavioural Intention (Blended Learning).		Failed to reject
H_{14a}	Gender of students significantly influences the behavioural intention (blended learning).		Failed to reject
H_{14b}	Age group of students significantly influences the behavioural intention (blended learning).		Failed to reject
H_{14c}	Computer experience of students significantly influences behavioural intention (blended learning).		Failed to reject
H_{15}	There is a significant relationship between the factors of teaching style of teachers and learning style of students.	Failed to reject	Failed to reject
H_{16}	The relationship between user acceptance of technology and behavioural intention is mediated by the teaching. style of teachers	Rejected	
H_{17}	The relationship between user acceptance of technology and behavioural intention is mediated by the learning style of students.		Failed to reject partially
H_{18a}	The relationship between user acceptance of technology and behavioural intention is affected by the teaching style of teachers.	Failed to reject partially	
H_{18b}	The relationship between user acceptance of technology and behavioural intention is affected by the learning style of students.		Rejected
H_{18c}	The relationship between user acceptance of technology and behavioural intention is affected by both the teaching style of teachers and learning style of students and demographic variables of both teachers and students.		Rejected

Figure 5. A model based on the results and findings
Source: Gawande (2015)

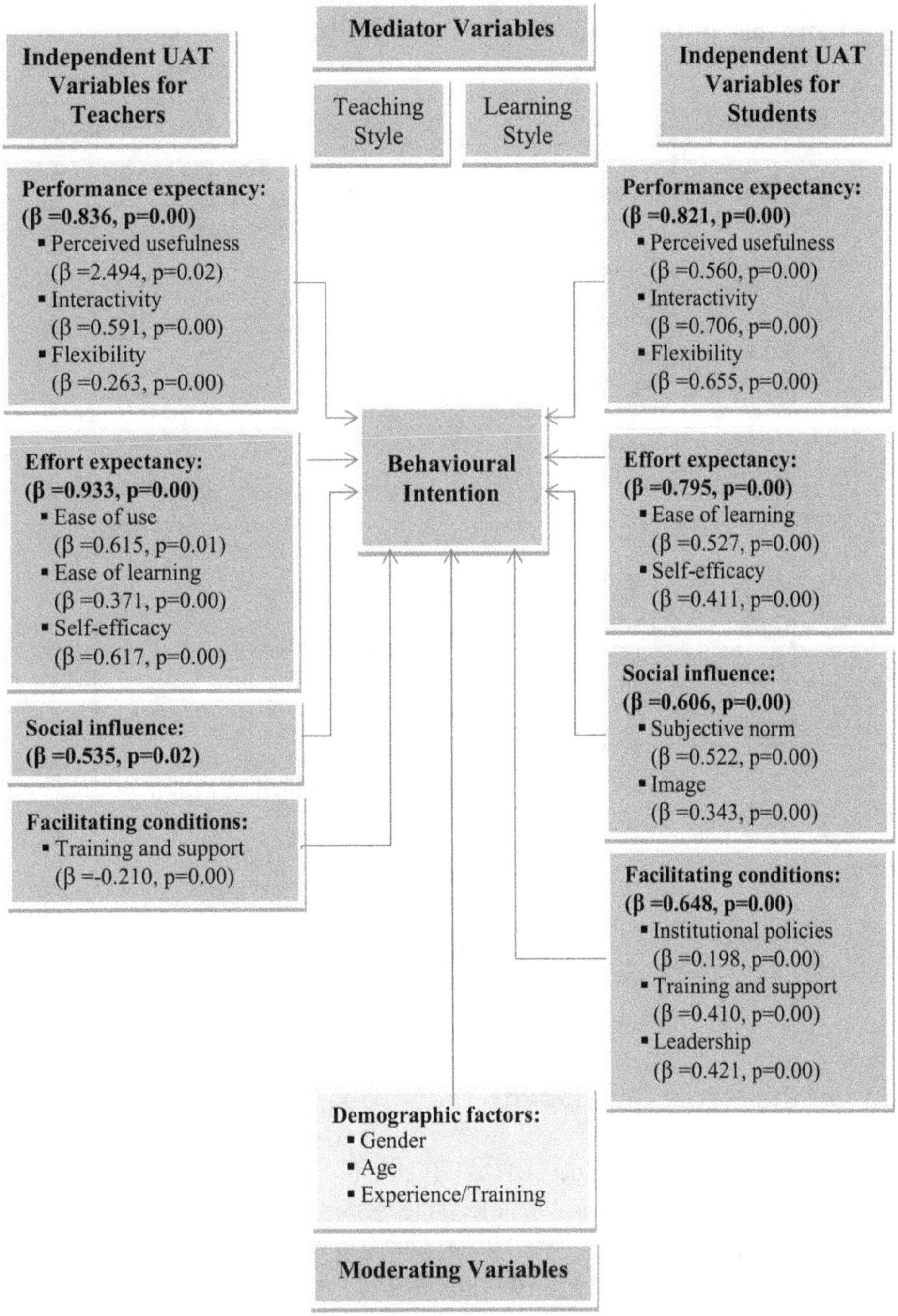

Figure 6. Blended Learning Acceptance Model (BLAM)
Source: Gawande (2015)

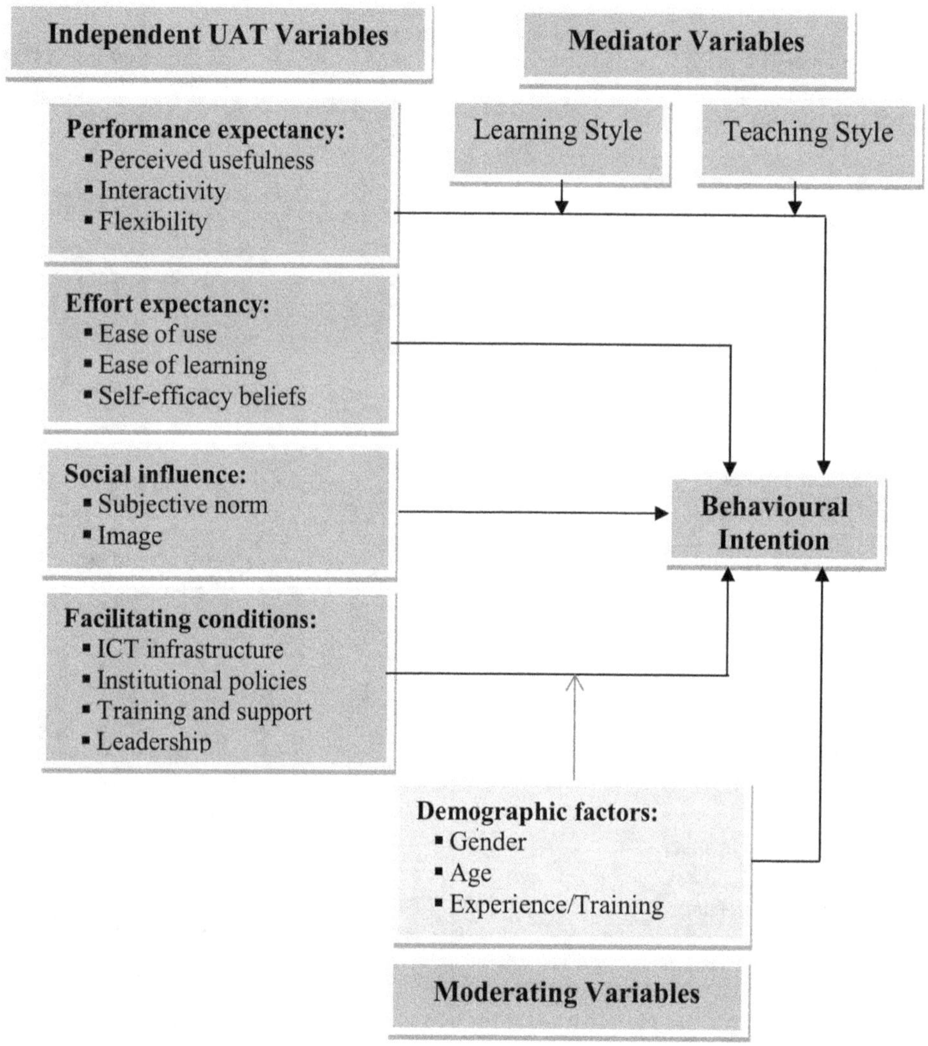

Description of the Model Developed

The Blended Learning Acceptance Model (BLAM, Figure 6) has emerged as a result of findings from this study. This model shows the relationship among the factors and the variables identified and used in this research. These independent factors were Performance expectancy, that includes the three variables Perceived usefulness, Interactivity and Flexibility; Effort Expectancy, that includes the three variables Ease of use, Ease of learning, and Self-efficacy; Social Influence, that includes the two variables Subjective norms and Image; and Facilitating Conditions, that includes the four variables ICT infrastructure, Institutional policies, Training and support and Leadership. These four factors were independent and have direct effect on Behavioural Intention (Blended Learning). The other was demographic factors which have direct influence on Behavioural Intention (Blended Learning), as well as a mediating effect on the relationship between Independent variables and Behavioural Intention, and hence it was termed as a Moderating variable. The demographic factors include the three variables Gender, Age and Experience or Training for teachers as well as students. The other two factors were Learning style and Teaching style which mediate the relationship between the Independent variables and Behavioural Intention, and hence are termed as Mediator variables.

Discussion

The present research adopted a unique application of the eLearning Acceptance Model (ELAM), (Umrani-Khan and Iyer, 2009) in Higher education Institutes (HEI) in Oman, by introducing the Gender factor and observing its influence on the behavioural Intention (Blended Learning) and hence suggesting a new model, Blended Learning Acceptance Model (BLAM) for the Colleges of Applied Sciences (CAS) in Oman. This model proves that Teaching and Learning process has a significant influence on Blended Learning adoption, which was in relation to the main research question.

From the results it was concluded that the factors related to User Acceptance of Technology were Performance expectancy, Effort expectancy, Social influence and Facilitating conditions (Umrani-Khan and Iyer, 2009), and these factors directly influence on Behavioural Intention (Blended Learning).

From the results of data analysis it was evident that demographic variables like gender, age and experience of teachers does influence the behavioural Intention of adopting blended learning. Conversely in the results the student respondents had significant positive association with gender, age and computer experience.

The findings emphasised that the teaching style such as Expert-Formal authority, Personal model-Expert-formal authority, Facilitator-Personal model-Expert, Delegator-Facilitator-Expert and Delegator does not influence blended-learning adoption among teachers. Similarly, among the students' learning style, 'Participant' 'Independent style' 'Collaborative style' of learning do not influence blended-learning adoption. Therefore it was concluded that teaching and learning style does not influence blended learning adoption.

The current study also identifies the extent of the mediating effect of teaching styles of teachers and learning styles of the students affecting the relationship between user acceptance of technology and blended learning at CAS. The teaching style of teacher was mediating the relationship between behavioural intention (blended learning) on competitive and participant. Likewise the learning style of students, significantly mediated the relationship between the behavioural intention (blended learning) on competitive and facilitators. However both teaching style of teachers and learning style of students shows a partial mediating effect between the behavioural intentions on user acceptance of technology.

The Blended Learning Acceptance Model (BLAM) is intended to be proposed at Colleges of Applied Sciences in Oman, as a model to guide the adoption of Blended Learning, however it may also be considered as a reference model for the adoption of this technology in other developing countries.

FUTURE RESEARCH DIRECTIONS

Recommendations and Implications for Further Research

In the near future research should concentrate on the development of more designs of blended learning models in such a way that it could enhance communication and interaction among the Students and Teachers.

For further research, a thorough analysis and evaluation of aspects hindering adoption is needed. This should also include an investigation among students. Although this study might have covered known knowledge at some levels of research in higher education, it was unique in the sense that it applied known knowledge in a new context. Future research can focus on the survey to explore and to create the awareness in relation to blended learning. It is important to get the views of the students and the teachers in the technological improvisation and the implementation of the techniques in terms of the blended learning.

CONCLUSION

This study aimed to develop the blended learning acceptance model specifically for Oman by relating User Acceptance of Technology (independent variables) to Behavioural Intention for both students and faculties at CAS. In addition, the study also examined the mediating role of learning and teaching style of students and teachers respectively and the demographic factors as moderators.

Blended learning was in its formative stage in Oman, and there was an ambiguity regarding conceptualization of the delivery systems such as eLearning or blended learning. Further, the majority of Omani HEI, including CAS, added blended learning applications merely to supplement their traditional lecture delivery in the classrooms, without a comparable reduction of attendance requirements for students or their lecturers. As such, a clear policy for blended learning delivery was required, and this study contributes to the debate on blended learning in Omani HEI. This study explored the potential for blended learning curriculum designs for undergraduates, and also investigated gender-segregation issues and the importance of ICT sector in Oman to identify the factors that may affect the successful implementation of blended learning.

The study was conducted at CAS in Oman, both students and teachers were chosen as the participants, including male and female. The sample size for teachers (n=30) was determined based on the target specificity, and in case of students (n=841). Convenience sampling method was adopted for students and for teachers. The developed questionnaire included 5 sections, based on the themes required for the information on the objectives of the research. The validity and reliability of the questionnaires was verified through a pilot test, after which the stability of questionnaire was validated using Cronbach Alpha.

Quantitative data collection was used in the present study. Student questionnaires were distributed to 1,010 students while the responses received ware 841. Teacher questionnaires were distributed to 40 teachers at CAS, out of which 30 responded.

The questionnaires were distributed online and collected information was statistically analysed using several data analysis methods in SPSS.

From the findings it can be concluded that the key factor Performance expectancy and Effort expectancy had significant influence over the adoption of Blended learning. Moreover, the Performance expectancy's attributes like Interactivity and Flexibility had significant influence on Blended learning adoption for both teacher and students. In addition Effort expectancy's key attributes like Ease of learning, Self-efficacy, Training and technical support had significant influence on blended-learning adoption by both teachers and students. However, ICT infrastructure attribute of Effort expectancy did not influence blended learning adoption.

Student satisfaction in blended courses was found to be influenced by factors of convenience, accessibility, and quality of interactions although ease and of relevant use, flexible delivery and richness of content were also considered to be important. The effective blending of course management and monitoring, online tutoring, student assessment and feedback were the advantages of the effective blending. Effective learning is promoted through greater flexibility that allowed the use of varied learning styles. There was enough evidence to show that while Performance expectancy was prominent for students, effort expectancy was important for teachers in the eLearning acceptance. Teaching style for teachers was regarded as the same as accepting the eLearning style by teachers. It is important to incorporate the views of students about the technologies and their implementations.

Based on the present findings, a 'Blended Learning Acceptance Model (BLAM)', was developed. This model shows the association and meditation among the variables considered in this study, and was intended to be used as a reference to implement blended learning at HEI in Oman.

Contribution of the Study to Theory

The study contributed to the literature in the following ways. Firstly, the study adds to the theoretical knowledge by validating UTAUT model (47%), where Effort expectancy, Performance expectancy and Social influence directly influence Behavioural intention. Secondly, the study developed a modified model called 'Blended Learning Acceptance Model' (BLAM) that would suit an environment where Blended Learning implementation is at an early stage and where actual usage is still not experienced by the students/teachers. The model has been validated in this study and results were presented.

Practical Implications

The study also found that Individualistic learning style of students mediates the relationship between Performance expectancy and Behavioural intention. This was one of the important finding as students from Oman in general prefer the Individualist style which means they would like to learn without interaction with teachers and other learners, and thereby strongly believe that this style of learning along with blended learning would enhance the performance, understanding and overall efficacy of the students.

The effective learning is promoted through greater flexibility that allowed the use of varied learning styles. Primary importance was the interaction and communication between students and their instructors in blended learning. Many researchers found course design and technology availability contribute to successful implementation, and student characteristics in access to ICT, competency, prior experience with online learning and attitudes toward online discussions were important issues. The present study provides a solution that addresses the restrictions in developing countries. Since assessment items were reflected in the two groups, the conclusive evidence for the difference between students and teachers will be provided by the model's empirical test. Incorporating learning style into the model and conceptualizing a mediation role will affect the performance of students' beliefs.

An interactive online approach is essential to raise student interest and to understand and appreciate the reason for online discussions. In the process of education, BLAM provides a more complete view as it considers factors that were related to each key player in the education process. Moreover, the research that already exists has opted for participants from the developed countries. The research finding has potential value for policy-makers, instructors, curriculum designers and administrators in public higher education. The study provides an evaluation of the benefits, influences, challenges, and future professional development needs of blended learning.

REFERENCES

Alebaikan, R., & Troudi, S. (2010a). Blended learning in Saudi universities: Challenges and perspectives. *Research in Learning Technology, 18*(1), 49–59. doi:10.1080/09687761003657614

Alebaikan, R., & Troudi, S. (2010b). Online discussion in blended courses at Saudi Universities. *Procedia: Social and Behavioral Sciences, 2*(2), 507–514. doi:10.1016/j.sbspro.2010.03.054

Allan, B. (2007). *Blended learning: tools for teaching and training*. London: Facet.

Allen, I. E., & Seaman, J. (2008). *Staying the course: Online education in the United States*. The Sloan Consortium.

Chandra, V. (2004). *The impact of a blended web-based learning environment on the perceptions, attitudes and performance of boys and girls in junior science and senior physics*. Perth: Curtin University of Technology.

Chen, W., & Looi, C. K. (2007). Incorporating online discussion in face to face classroom learning: A new blended learning approach. *Australasian Journal of Educational Technology, 23*(3), 307–326.

Creswell, J., & Plano Clark, V. (2007). *Designing and Conducting Mixed Methods Research*. Thousand Oaks: Sage.

Creswell, J. W. (2008). *Educational research: Planning, conducting, and evaluating quantitative and qualitative research*. Upper Saddle River, NJ: Pearson Prentice Hall.

Creswell, J. W. (2009). *Research design: Qualitative, quantitative, and mixed methods approaches*. Los Angeles, CA: Sage Publications.

Doyle, L., Brady, A., & Byrne, G. (2009). An overview of mixed methods research. *Journal of Research in Nursing, 14*(2), 175–185. doi:10.1177/1744987108093962

Garrison, D., & Kanuka, H. (2004). Blended learning: Uncovering its transformative potential in higher education. *The Internet and Higher Education, 7*(2), 95–105. doi:10.1016/j.iheduc.2004.02.001

Garrison, D., & Vaughan, N. (2008). *Blended learning in higher education: framework, principles, and guidelines*. San Francisco, CA: Jossey-Bass.

Gawande Virendra. (2015). *Development of Blended Learning model based on the perceptions of students and faculty at Higher Education Institutes in Oman. CIDE*. Thailand: Assumption University.

Graham, C. (2006). Blended learning systems: definition, current trends, and future directions. In C. Bonk & C. Graham (Eds.), *Handbook of blended learning: global perspectives, local designs*. San Francisco, CA: Pfeiffer.

Harasim, L. K. (1989). *Online education: A new domain*. New York: Pergamon.

Jenkins, M., & Hanson, J. (2003). *A guide for senior managers. LTSN Generic Centre Elearningseries, No.1*. York: Learning and Teaching Support Network.

Jones, N. (2006). e-College Wales, a case study of blended learning. In Handbook of Blended Learning: Global Per spectives, Local Designs, (pp. 182–194). San Francisco, CA: Pfeiffer Publishing.

Jung, M. L., Loria, K., Mostaghel, R., & Saha, P. (2008). E-Learning: Investigating University Student's Acceptance of Technology. *European Journal of Open, Distance and e-learning*. Available at: http://www.eurodl.org/materials/contrib/2008/Jung_Loria_Mostaghel_Saha.htm

Keller, C., Hrastinski, S., & Carlsson, S. A. (2008) Students' Acceptance of E-Learning Environments: A Comparative Study In Sweden and Lithuania. In *Proceedings of the Fifteenth European Conference on Information Systems*, (pp. 395-406).

Kemper, E. A., Stringfield, S., & Teddlie, C. (2003). Mixed methods sampling strategies in social science research. In A. Tashakkori & C. Teddlie (Eds.), *Handbook of mixed methods in social and behavioral research*. Thousand Oaks, CA: Sage.

López-Pérez, M. V., Pérez-López, M. C., & Rodríguez-Ariza, L. (2011). Blended learning in higher education: Students' perceptions and their relation to outcomes. *Computers & Education*, *56*(3), 818–826. doi:10.1016/j.compedu.2010.10.023

Masrom, M. (2007). Technology Acceptance Model and E-learning.*12th International Conference on Education*. Sultan Hassanal Bolkiah Institute of Education.

Matheos, K., Daniel, K.B., & Mccalla, I.G. (2005). Dimensions for Blended Learning Technology: Learners' Perspectives. *Journal of Learning Design*, 56-74.

Nanayakkara, C. (2007). *A model of user acceptance of learning management systems: A study within tertiary institutions in New Zealand*. Retrieved from http://citeseerx.ist.psu.edu/viewdoc/download

Nunan, T., George, R., & McCausland, H. (2000). Rethinking the ways in which teaching and learning are supported: The Flexible Learning Centre at the University of South Australia. *Journal of Higher Education Policy and Management*, *22*(1), 85–98. doi:10.1080/713678130

Ostendorf, V. A. (1997). *Teaching and learning at a distance: What it takes to effectively design, deliver, and evaluation programs*. San Francisco: Jossey-Bass.

Picciano, A. G. (2009). Blending with purpose: The multimodal model. *Journal of the ResearchCenter for Educational Technology*, *5*(1), 4–14.

Raaij, E. M. & Schepers, J. J. L. (2008). *The Acceptance and Use of a Virtual Learning Environment*. Academic Press.

Richard, G. (2005). *E-learning in tertiary education: Where do we stand?* Paris: Centre for Education Research and Innovation.

Singh, H. (2003). Building effective blended learning programs. *Educational Technology*, *43*(6), 51–54.

So, H. J., & Brush, T. A. (2008). Student perceptions of collaborative learning, social presence and satisfaction in a blended learning environment: Relationships and critical factors. *Computers & Education*, *51*(1), 318–336. doi:10.1016/j.compedu.2007.05.009

Stacey, E., & Gerbic, P. (2009). Introduction to blended learning practices. In E. Stacey & P. Gerbic (Eds.), *Effective blended learning practices: evidence-based in perspectives in ICT-facilitated education*. Hershey, PA: IGI Global. doi:10.4018/978-1-60566-296-1.ch001

Su, B., Bonk, C. J., Magjuka, R. J., Liu, X., & Lee, S.-h. (2005). The importance of interaction in web-based education: A program-level case study of online MBA courses. *Journal of Interactive Learning Research*, *4*(1), 1–19.

Umrani-Khan & Iyer. (2009). *ELAM: A Model for Acceptance and use of e-learning by Teachers and Students*. International Conference on E-Learning, Toronto, Canada.

Vaughan, N. (2007). Perspectives on blended learning in higher education. *International Journal on E-Learning*, *6*(1), 81–94.

Venkatesh, V., & Davis, F. D. (2000). A Theoretical Extension of the Technology Acceptance Model: Four Longitudinal Field Studies. *Management Science*, *45*(2), 186–204. doi:10.1287/mnsc.46.2.186.11926

Venkatesh, V., Morris, M. G., Davis, G. B., & Davis, F. D. (2003). User Acceptance of Information Technology: Toward a Unified View. *Management Information Systems Quarterly*, *27*(3), 425–478.

Yuen, H. K. A., & Ma, W. W. K. (2008). Exploring teacher acceptance of e-learning technology. *Asia-Pacific Journal of Teacher Education*, *36*(3), 229–243. doi:10.1080/13598660802232779

Chapter 3
Is Flipped Classroom a Tendency or a Fad?
The Point of View of Future Teachers in the Philippines

Michel Plaisent
University of Quebec Montreal, Canada

Filomena Dayagbil
Cebu Normal University, Philippines

Angeline M. Pogoy
Cebu Normal University, Philippines

Prosper Bernard
University Consortium Americas, Philippines

ABSTRACT

This chapter presents an overview of the challenges facing teachers and education students with regard to the growing interest in the flipped classroom, and the results of an exploratory study, conducted in the Philippines, about the intention to adopt this pedagogy by students training to be teachers. After a brief definition and a description of this pedagogical approach, a review of recent research explores the relevance and effectiveness of this method. The chapter questions if it is to be a fad or a revolution and presents the pro and cons of its implementation. A survey has been conducted with 153 fourth year education students of to determine their knowledge of this pedagogical approach, portraying their perception of the flipped classroom and associated concepts as well as their future behavior intention.

DOI: 10.4018/978-1-5225-0359-0.ch003

INTRODUCTION

The buzzword "flipped classroom" has been the center of great interest in the past few years in academic and other media (Estes, 2014; Atteberry, 2013) and is also proposed for training purposes in industry (Goodman, 2013). It is sometimes referred to as "reverse teaching", "flipped learning", "backwards classroom" or "reverse instruction" (Sunguro, 2014; Hoffman, 2014). This popularity is not only evident in the USA, but also in the Philippines (Azucena, 2014; Esperanza & Toto, 2014), which is the context of this particular study.

Yarbro, McKnight and Arfstrom (2014) recall that in a recent joint survey of the Flipped Learning Network (FLN) and Sophia Learning, 96% of 2,358 teachers recognized the term "Flipped Classroom" and 78% among them indicated they had flipped, at least partially, one or more of their lessons, which represents an increase of respectively 22% and 30% in regards to a similar survey conducted two years before. This is reported to be the case in all subjects, all grades, but more in high school; they flip 5.4% every day, 20% 3 to 4 times a week and 24% less than once a week; 89% of those who did not flip yet wish to know more about this pedagogy.

This chapter adopts the FLN (2014) definition of flipped learning which is a pedagogical approach in which direct instruction moves from the group learning space to the individual learning space, and the resulting group space is transformed into a dynamic, interactive learning environment where the educator guides students as they apply concepts and engage creatively in the subject matter. The flipped classroom strategy finds its origins in the work of pioneers, namely

- Alison King (1993) who suggested the necessity to better use the class-time to produce learning through meaning instead of being used only for information transmission.
- Eric Mazur (1997) who initiated the discussion about "peer instruction."
- Lage et al (2000), credited for having coined the term "inverting the classroom", based on their own experience at college level, as a method to create a learning environment.
- Khan (2004), creator of Khan Academy, who based his teaching on videos as a method to skip what was understood and watch again what was not clear.

After reviewing a wide range of research on the flipped classroom, Bishop and Verleger (2013) concluded that flipping was perceived as positive, despite questioning the fact that a majority of studies are based on one single group design and student perceptions. Only a few are experimental researches aimed at assessing the real outcomes of this approach. Indeed there appears to be a lack of controlled experiments to assess the real value of flipped learning with rigor, as the implementations are often anecdotal and reported in blogs instead of peer-reviewed conferences or journals (Goodwin and Miller, 2013). Despite the fact that many question its real effectiveness, many others seem to advocate this revolutionary approach (Lape, Levy & Yong, 2014), and favor towards this approach is still strong (Straumsheim, 2013).

Apart from the academic importance of the flipped classroom, there is a question regarding the readiness of the Philippines to implement this pedagogy. The fear that some could be left behind is a concern shared by Butrymowicz (2012) for the poorest students who may not be able to afford the technology costs. This opposite position is seen as inevitable according to one of the eleven teachers of a $300,000 flip project at CATCH Academy: "If students don't have access at home, they're already at a disadvantage" (Phillips, 2013). Horn (2013) also mentions this discrepancy between rich and poor districts and that the flipped classrooms are not so different from the traditional setting.

Table 1. Didactic passive learning vs. Socratic active learning (Moeller, 2015)

Lectures- DIDACTIC (Passive Learning) [Master – Disciple]	SOCRATIC Classes(Active Learning) [Shared Inquiry]
Teacher centered: based on the assumption that the teacher is the primary agent in learning	Problem centered: based on the assumption that the student is the primary agent in learning
Teacher's role: to impart the results of experience, personal study, and reflection	Teacher's role: to uncover the question that the answer hides. To be a co-learner
Primarily deductive: the usual methods are lecture, storytelling, use of analogy, and aphorism	Primarily inductive: the usual methods discussion
Test of truth: authority and experience	Test of truth: reason and evidence
Learning is the reception of ideas	Learning is a conflict of ideas: a thesis, antithesis, and a synthesis that results in new knowledge
Student's role: to be passive, open, receptive, trusting, and unquestioning	Student's role: to be active, questioning, critical, and discriminating – learning to trust one's own judgment (independent thinking)
Evaluation is factual recall of data – commonly in the form of objective tests – right and wrong answers	Evaluation is application of understanding interpretation of data – commonly in an essay, speech, journal, or a review
Ultimate goal: wisdom viewed as the internationalization of truths and beliefs	Ultimate goal: wisdom viewed as an informed ignorance (knowing what one does not know – Socratic paradox)

This study intends to present the pros and cons of the flipped classroom, before presenting the results of a survey attempting to assess the perception of Filipinos (working or studying in the education field) toward this new pedagogical approach.

BACKGROUND

In a flipped classroom, students take responsibility for their learning while teachers change classes for facilitation and moderation (Sunguro, 2014). Moeller (2015) presents the two models of learning presented in Table 1.

Mazur (2013) presents his understanding of the Flipped Classroom from these four theoretical constructs:

1. *Constructivist pedagogy* which focuses on student-centered learning, instead of the teaching; Knowledge is viewed as the consequence of the personal appropriation of the new information which the learner integrates with previous information through a reflective effort.
2. *Inquiry* is defined as a passionate quest for critical thinking and deep understanding of the world through exploration, which can take many forms, from questioning to formal investigation, in order to develop critical thinking and deep understanding.
3. *Student-centered learning* is presented as a pedagogical approach in which the teacher acts more as a planner and a facilitator for class social interactions while the student manages their learning according to their own needs. This encourages and facilitates student active learning.
4. *Active learning* happens as the consequence of the dynamic context. Student-centered learning is characterized by multiple intellectual stimulations offered from several instructional methods from reading to communicating, watching video, using computer simulations, etc.

Aspects of the Flipped Classroom

This study undertakes a wide review of the literature, expanding the inspirational work of Tolson (2014) with twenty annotations of influential researches. We revisit the main knowledge about flipped classrooms, from three points of view: the benefits, the disadvantages and the implementation success factors.

Hoffman (2014) urges us to evolve from giving a series of simple elementary tasks to new pedagogical learning environments, where the target is not only to succeed in assignments but more importantly to give an opportunity for more sophisticated knowledge applications and deeper understanding of phenomena. Indeed, according to Abernathye, Abshire, Cummings and Mason (2012), an active learning classroom model permits new learning. It creates a cooperative environment, offers technological support for collaboration and sharing and for presenting work to instructors and peers, allows instructors to coach students interactively during their activities and permits new possibilities for interaction and class structure. According to Bergman and Sams (2012), flipped instruction would permit a significant increase in student-centered activity time.

de Waard (2014) identifies the main benefits of flipping the classroom for students such as:

- Class time shifts from passive lectures to problem solving and active learning;
- Students get empowered for their own learning potential and challenges;
- Students prefer hands-on activities than lectures;
- Students can revise online material as often as needed to understand;
- Students learn how to learn (meta-cognition);
- Teachers can easily detect the challenges of students' understanding the concepts;
- Teachers can give more guidance for they have more available time for students.

Pinnock (2014) points out some other benefits in flipping classes for students, including

- It provides opportunity for differentiated learning;
- Students become independent learners;
- Active Learning is promoted that focused on developing higher order thinking skills;
- Peer interaction and collaboration on common projects is manifested;
- Individuals interact with content prior to class time;
- Learners get an opportunity for individualized attention;
- Learning becomes more prominent compared to teaching;
- The use of classroom time is improved by increasing the efficiency in coaching and feedback;
- It permits the exploration of content and the practice of necessary skills;
- Students have deeper understanding of the content, are more engaged using their practice skills, and receive feedback on their progress.

O'Neil, Kelly and Bone (2012) suggest that flipped instruction could help build a stronger community among faculty members, sharing instructional materials and comments. Roehl, Reddy, and Shannon (2013) also suggest that flipping has many advantages. It could expand class time devoted to learning, mentoring, peer-interaction and active learning.

Aceto (2013) suggests the following evaluation of flipping classes:

1. **The Pros:** Students have more control over their own learning in terms of content, time and pace. This approach favors student cooperation and focus on the learning for lessons, and content is more accessible provided there is technology access. Parents can easily see what type and style of teaching is done. Flipped classes can be more efficient, since the homework has a limited duration.
2. **The Cons:** This type of pedagogy may not fit everyone as it supposes that students have internet access at home and the required technology. This model of home task relies on good preparation at home and on the belief that the video or other materials will be used by the students. Some students may be reluctant or can have difficulties to focus on a computer screen if they are more people-oriented. It also supposes that the teachers will prepare adequate materials as a preparation for the class and that they have the technological skills to do so.

A number of other disadvantages have been identified by several authors, such as:

- Teachers need more time to prepare, though the prepared materials/activities could be oftentimes reused (Lage, Platt and Treglia 2000).
- In some schools there is lack of equipment and support to deliver quality videos (Sungoro, 2014).
- Video recording and editing requires a lot of effort and dedicated skilled staff which may not be available in some school settings (Sungoro, 2014).
- Well made videos need to focus on one single concept and be short and appealing to students, which necessitates time and effort for non specialists (Tucker, 2012).
- Too much homework (Pinnock, 2014).
- Lectures on video are monotonous (Pinnock, 2014).
- Not all students have access to technology outside of school (Pinnock, 2014).
- Filipinos prefer face-to-face interactions to computer mediated communication (Tabaque, 2014).

Evaluation through Experiments

In order to know if the flipped classroom really works, several experimental studies have been conducted, with results varying from benefits to grade improvement, appreciation by parents and students, and inconclusive outcomes.

On Its Effectiveness

1. **University Level:** In North Carolina State, one study compared students enrolled in a physics course according to 2 modalities: traditional versus scale-up (Beichner et al, 2007). The main finding was that the scale-up setting improved student skills for problem solving and their conceptual understanding as well as them showing better attitudes resulting in lower failure rates.

Talbert (2012) presented the positive impact of inverted classroom through three universities: University of California at Irvine, Miami University Ohio, and Franklin College Indiana. He reported that there was an increased success of 21% in one case, self-assessed improvement in the second case and improvement due to video in the last case.

In the University of British Columbia, Deslauriers and Wieman (2011) compared both modes of learning for Quantum mechanics concepts with 850 students and noted a difference of 18% in favor of those under the experimental learning condition with class time being reserved to applied, interactive and peer activities. At the University of Memphis, Ruddick (2012) reported a positive impact of flipped learning for a general chemistry preparation course with a difference from 48.4% to 73.7% of students getting at least a C. Moreover, Prunuske, Batzli, Howell and Miller (2012) reported that there was an implementation of flipped classroom in an undergraduate biology program at the University of Wisconsin, involving 150 students over 5 years. They found out that at least 70% of the students read the online lectures before attending classes and that these students performed better than the others, but that it did not impact on their higher-order cognitive skills.

In another study with two undergraduate groups (n1=267 and n2=271) in British Columbia University, Deslauriers, Schelew and Wieman, (2011) found that flipped learning could double the learning outcomes while increasing attendance and engagement. Wilson (2013) in Capital University (Ohio), made a comparison between blended teaching to students doing real world statistics in class and students taught in a traditional way. The experimental group performed 10% better for final grades than the students in the previous year. The difference between both groups was only 2%. Hence, students reacted positively to the new model. Similarly, Clintondale High School in Michigan turned completely to flipped classes after experimenting and finding that no student from the flipped class had less than a grade of C contrary to 13 the previous semester (Estes, 2014; Butrymowicz, 2012).

In Georgia Institute of Technology, Day and Foley (2006) experimented with flipping a traditional course with class meetings, discussions, hand-on activities for one of two groups, resulting in an increase of final grades of 4%, and students declared themselves to be satisfied. At the University of North Carolina, Chapel Hill, Mc Laughlin (2014, 2013) flipped a class with handling devices to solve applied problems and make presentations. The improvement from the previous year was 2% and the self-reported attendance increased from the previous semester. Another finding revealed in a longitudinal study of flipped classroom over 3 years showed an increase of 5.1% in the mean evaluation of the students of a course in pharmaceuticals (Meyer, 2013).

Finally, a study was conducted by Schneider, Wallace, Blikstein and Pea (2013) to determine if the sequence of learning method, first the textbooks then the open-end experimentation or vice-versa, had an impact on student performance. The study concluded that there was superiority in the exploration using flipped classes.

2. **Other Levels:** Two mathematics teachers from a high school of Niagara Falls observed an improvement at state standardized tests after flipping the class, from 71% to 83% of students passing and from 14% to 35% achieving mastery of honors trigonometry. It was noted that the performance increase may be due to the quality and intensity of support from teachers who benefit from the help of a technology integrator to prepare and post their lectures, write courses notes to complete these and make themselves available to students in real time (Yarbro et al, 2014).

Ashland Middle School as noted by Dill (2012) introduced a flipped classroom in a comparative study for a two week period and reported a homework completion increase of 17% and an increase of 10% on a grammar test. Finally less disruptive behavior was noted during the experimental period. In addition, at Bullis School in Maryland and Madeira School in Virginia, as reported by Roshan and Roshan, (2012), there was an increase of 20% in the percentage of students scoring a 4 or 5 in an Advance Placement test. It was noted that there was a lower level of anxiety in the classroom but it took a large part of the year to make videos.

3. **On Parent and Student Perspectives:** Fitzgerg (2012) reported that parents appreciate the fact that their children would be more willing to do their homework and to succeed in their courses. Most importantly, they can access the pedagogical material and better follow their children's progress. In his classes, students had more time to explore concepts and ask questions, which would help prepare for state tests. In addition, Fulton (2012) noted that a survey done with parents was indicative of a general overall positive reaction to flipped classrooms.

In Bamber Valley Elementary School, fourth graders (as reported by Stolle, 2014) noted that technology permits flipped classes which in turn motivates students. At the University of Columbia, Missouri, Stone (2012) introduced flipped classroom with elective biology students without motivation problems. Attendance was similar between the flipped group and control group. The attitudes toward this new pedagogy were very positive, but it was noted that no improvements in success were evident.

Butt (2014) reported on an experimental introduction of flipped classrooms with fifth year actuarial program students. 75% of the students were revealed to have a positive view and were mostly satisfied with the availability and low cost of material at the end of the term. In addition, Barstow High School in California had 102 students who enjoyed and had positive increases in attitudes towards the value of Mathematics, a significant increase in motivation and to a lesser degree self-confidence with inverted pedagogy (Esperanza & Toto, 2014).

Similarly, Heng (2014) reported on a study conducted with 46 students learning programming. These students assessed the flipped class as being "effective", "efficient", "helpful" and "useful", appreciating being able to review videos as often as needed. He noted a greater involvement of students who felt more empowered for their own learning. Moreover, a study conducted in an AACSB Northern University of USA with 125 millennial students in an accounting major had preferred learning by doing, while they are mostly taught by lectures, supporting the flipped classroom pedagogy (Philip & Trainor, 2014). Furthermore, Swanson School of Engineering at the University of Pittsburgh used the College and University Classroom Environment Inventory to assess the classroom environment and for the formative evaluation. Clark, Norman and Besterfield-Sacre, (2014) found that student cohesiveness, individualization, innovation, involvement and personalization differ significantly after flipping class but not their satisfaction or the task orientation.

Inconclusive

In public high school in rural southwest Louisiana, Clark (2013) compared two 9th grade algebra groups for a 7 week period and could not find a significant difference due to the flipped classroom. Similarly, an experiment was noted by Findley-Thompson (2013) with three groups of an introductory business course at Mount Saint Vincent University concluding with no evidence of difference in grades, and students had mixed perceptions of the flipped classroom.

At Texas A&M University, Missildine, Fountain, Summers and Gosselin (2013) found little differences between three modes of learning: totally flipped, partially flipped and traditional. The researchers linked the satisfaction of flipped group to increased workload.

Lastly, Lape, Levy and Young (2014) conducted a study over three years to determine the effects of flipped classrooms on student performance for the disciplines; science, technology, engineering, and mathematics in Harvey Mudd College in California. They did not find any significant difference. They concluded that this new model could be effective, but that under what conditions was still to be determined (Atteberry, 2013).

Implementation and Success Factors

For the FLN (2014), the flipped classroom relies on four aspects: First, the objectives and of the inversion must be presented and well understood by the students. Second, the approach suggests that all digital material would be shared. Third, the work in the classroom should focus on understanding through exercises, group work, projects fed by peers and teacher feedback. Fourth, the process shall be assessed by all.

In order to succeed, students must comply with the following two rules (Heng, 2014): be present in class for the learning activities and do the preparatory work prior to coming to class, a prerequisite that the teachers should verify though self-quiz by the students and tests by the teacher. However, more important than these prerequisites, they must undergo a transition (Abernathy et al., 2012):

- Moving their learning space from "inside" the classroom to "inside and outside" the classroom,
- Forget single tasks to master multitasking,
- Expand text to pictures, sound, and video,
- Maneuver from linear, logical and sequential to random access,
- Pass from the independent and individual to interactive collaborative network.

In a case study, Herreid and Schiller (2013) mention two important drawbacks: the resistance of students to working at home (even watching videos) and the amount of work needed by teachers to produce videos. But videos should not be the only tools to include in this approach and other "learning objects" should be considered (Bergmann & Sams, 2014). At this point, for the pre-class, some of the alternative tools and techniques mainly for in-class activities need to be presented and discussed for a better understanding of the implementation context and success factors:

- Reading assignments; books and newspapers (online audio or video lectures if not too long are usually preferred to book reading),
- Screen-capture software can be used to provide a static image to study more than a video due to its dynamic nature or for practice process (ex: Camtasia, Adobe Captivate, others),
- PowerPoint presentations, more or less sophisticated with animation, can keep the interest of students while giving the structure of a lesson,
- Simulation is a recognized tool to provide skills, less dangerous and costly than experimentation,
- Laptops, tablets, and smart phones are flexible tools to access information using browsers and a formidable tool for peer-learning through the use of group communication, particularly with webcams.

According to Guro (2013) careful planning is needed in order to implement the flipped classroom model in the Philippines context with the following considerations:

- Guided Practice and Mentoring Strategies.
- Classroom Management Skills (leadership online and offline).
- Accessibility of Resources (namely computers and fast internet).
- Profiling of Students (to ensure they have the prerequisites).
- Detailed Syllabus (good planning).
- Assessment and Evaluation of students' strengths.

Stakeholders include not only students and teachers but also parents and schools. For Abadines, (2012), schools are responsible for the availability of technological support and appropriate training of teachers while parents should be technology literate to support both students and schools to realize its benefit and most importantly to stimulate their children to value learning.

According to Azucena (2014), teachers are aware of technological limits, especially in poverty stricken areas. They do not have access to the Internet and do not even own computers at home. Solutions could be presented through discussions on how they could adapt ideas coupled with resourceful and innovative actions in increasing access.

There is a risk that flipped pedagogy will suffer from the same deprivation in the Philippines unless a clear political commitment is made toward resources (de Dios, 2014). According to Tabaque (2013) the Philippines is at least ten years behind with regard to m-learning and e-learning due to lack of access to tablets and computers since they are expensive, and high cost of internet. Even in richer nations, like Canada, schools regularly complain of lack of resources, namely technology resources, despite a drop in costs.

MAIN FOCUS OF THE CHAPTER

Methodology: Respondents and Data Collection

In order to get an initial picture of the situation, a questionnaire was adapted from the four pillars of FLIP (FLN, 2014). This questionnaire comprised two sections: 1) the actual knowledge of the concept of flipped classroom and three of its related expressions: reverse education, inverted pedagogy and active learning and 2) the four pillars of FLIP intended behavior such as flexible environment, learning culture, intentional content and professional educator.

The questionnaire was distributed to the pre service teachers of a university in the Philippines. This university is well known for its successful teaching and coverage of pedagogy. The data collection was made through a web survey in order to produce fast and accurate results.

Data Analysis

The website recorded 160 answers but some were rejected due to lack of responses in the questionnaire. One hundred fifty three (153) valid questionnaires were kept for further analysis with SPSS version 20 software. The gender distribution was composed of 127 (83%) females compared to 26 (17%) males, similar to the university registration record (t-test). A t-test was also used to examine each question in relation to gender and no significant difference was observed. A scale analysis (Cronbach's Alpha) was performed on the data which showed an alpha= 0.884 with possible improvement item was deleted ranging from 0.875 to 0.886 compared to .884, so it was decided that no item needed to be deleted from the questionnaire before further analysis.

Table 2 shows the main perception of students regarding the four (4) related names for flipped class.

As can be seen in Table 2, active learning is the most well-known among the concepts despite its mode of only 4 with only 12% of respondents indicating a poor or fair knowledge of this concept. The three other concepts are not so familiar to the respondents, specifically the flipped classroom which has the lowest mean.

Table 2. Knowledge of the pre-service teachers on concepts of flipped classroom

Statistics on Concepts of Flipped Classroom (1=poor, 5=excellent)	Frequency	Mean	Std. Deviation	Median	Mode
Active learning	151	3.5629	0.949	4	4
Reversed education	153	2.4118	0.950	2	3
Inverted pedagogy	151	2.3179	1.042	2	2
Flipped classroom	151	2.2053	1.015	2	1

Table 3. Knowledge of the four pillar appellations on flipped classroom

Statistics on Concepts of Flipped Classroom (1=poor, 5=excellent)	Frequency	Mean	Std. Deviation	Median	Mode
Flexible environment	152	3.5526	0.933	4	3
Learning culture	153	3.5490	0.881	4	4
Intentional content	152	2.8092	1.066	3	3
Professional educator	153	3.7059	0.986	4	4

Three of the four pillars for the flipped classroom are moderately well-known by the respondents with more than 3.5 and a median of 4 as reflected in Table 3. The notion of intentional content may be too vague to get an agreement.

The tables that follow reflect the respondents' intention to behave in compliance with the four pillars namely; flexible environment, learning culture, intentional content and professional educator.

Most respondents agree with the availability of a flexible environment in the flipped class; it is a model of classroom instruction which takes advantage of the versatility of media and asserts that there is still a place for direct instruction and scaffolding via instructional videos (Hamdan, McKnight, P., McKnight, K., and Arfstrom, K. (2013) and Mazur (2009). In the flipped class approach, this shift in the learning culture, from educator-centered to student-centered, is done by taking content introduction out of class time, and focusing only on meaningful and collaborative work (Table 4) (Lambert, 2013).

The second pillar which is relative to learning culture is shifting the focus of learning from the teacher to the students. The respondents agree strongly with this approach as shown in the result with a median and mode of 4. This indicates that the pre-service teachers have the intention to create a learning environment for their future students where blended learning takes place.

The third pillar is concerned with the intention to behave in compliance to intentional content. It focuses on the intention to develop or use material for better conceptual understanding. The result suggests that respondents are moderately in favor of this aspect as seen in the three statistics presented in Table 6. This means that the pre-service teachers are uncertain about what content needs to be taught directly, and how to best introduce that content in their video lectures or the use of technology in teaching. They are hesitant since they have limited experience in the field. An experienced and qualified educator knows the scope and content of their course and that they can decide on significant topics and how each relates to the course and the student's learning goals (Hamdan et al., 2013). There is a need to carefully select within the overarching curriculum for the course (Bergmann & Sams, 2012).

Table 4. Intention to behave in compliance with the first pillar on flexible environment

Statistics on Pillar: Flexible Environment (1=never, 5 =totally)	Frequency	Mean	Std. Deviation	Median	Mode
I intend to establish spaces and time frames that permit students to interact and reflect on their learning as needed.	152	3.3684	.889	3	3
I intend to continually observe and monitor students to make adjustments as appropriate.	153	3.5556	.873	4	4
I intend to provide students with different ways to learn content and demonstrate mastery.	151	3.7152	.867	4	4

Table 5. Intention to behave in compliance with second pillar on learning culture

Statistics on Pillar: Learning Culture (1=never, 5 =totally)	Frequency	Mean	Std. Deviation	Median	Mode
I intend to give students opportunities to engage in meaningful activities without the teacher being central.	153	3.9085	.927	4	4
I intend to scaffold these activities and make them accessible to all students through differentiation and feedback.	153	3.6732	.909	4	4

Table 6. Intention to behave in compliance with third pillar on intentional content

Statistics on Pillar: Intentional Content (1=never, 5 =totally)	Frequency	Mean	Std. Deviation	Median	Mode
I intend to prioritize concepts used in direct instruction for learners to access on their own.	152	3.4605	.868	3	3[a]
I intend to create and/or curate relevant content (typically videos) for my students.	152	3.3882	.942	3	3
I intend to differentiate to make content accessible and relevant to all students.	153	3.4183	.886	3	4

The fourth pillar of the flipped class is related to the professional role of a teacher. The pre-service teachers find this topic very important as reflected in their responses as very positive with median and mode of 4. As pre-service teachers, they are exposed in terms of content and experience in their role as future teachers. The courses offered in their baccalaureate degree in education such as the major courses and professional education courses prepared them in terms of content. They have also gained confidence in their Field Study courses and pre-service teaching experiences during their practicum. However, in flipped classrooms they should create the learning outcomes and provide expert guidance for their students. Though the flipped class is based on a student centered, collaborative and interactive model, the expertise and teaching style of the educator are vital elements (Table 7) (Hamdan et al., 2013).

These tables show that respondents agree in the majority to the principles of flipped class, as defined by FLN (2014) and intend to behave in compliance with those. As future teachers, they should find time to set student learning outcomes, prepare video lectures, select a class forum, and design flexible projects and activities (Bergmann & Sams, 2012) and be media literate and have more access to technology to achieve the goals in educating their learners.

Table 7. Intention to behave in compliance with fourth pillar on the professional educator

Statistics on Pillar: Professional Educator (1=never, 5 =totally)	Frequency	Mean	Std. Deviation	Median	Mode
I intend to make myself available to all students for individual, small group, and class feedback in real time as needed.	151	3.8212	.960	4	4
I intend to conduct ongoing formative assessments during class time through observation and by recording data to inform future instruction.	153	3.5229	.911	4	4
I intend to collaborate and reflect with other educators and take responsibility for transforming my practice.	152	3.5855	.938	4	4

In order to perform a factor analysis for the principal components, the requirements were verified. The Kaiser-Meyer-Olkin Measure of Sampling Adequacy = .860 which is much over the needed 0 .5 and is qualified "great" by Hutcheson and Sofroniou (1999). This indicates that the number of respondents is sufficient to perform the factor analysis. The Barlett's test of sphericity was performed, resulting in an approx. Chi-Square of 1444.382 with df of 171 and significance of 0.000, which is highly significant. The 3 definitions of flipped classroom, the 5 related concepts and the 11 behavioral intents were all included in the analysis. Three factors were extracted (eigenvalue over 1) with a varimax rotation that accounted for 61.5% of the variance: factor 1 is 32.06% (all and only all intended behavior), factor 2 is 14.77% and factor 3 is 14.32%. Factors 2 and 3 relate to the other appellations of flipped class and group names of behaviors.

A second analysis was also performed with only the 11 behavioral intentions. In the same manner, the Kaiser-Meyer-Olkin Measure of Sampling Adequacy measured was 0.911 which is over the needed 0.5 and is qualified of "great" by Hutcheson and Sofroniou (1999) and indicates that the number of respondents is sufficient to perform the factor analysis. The Bartlett's Test of Sphericity gives an approx. Chi-Square of 876.67 with df of 55 and significance of 0.000, which is highly significant. The factor analysis with the 11 behavioral intentions resulted into one single factor, accounting for 55.44% of the variance.

A test was made by Pearson correlation to determine if knowledge of flipped classroom would be related to behavior intention. Table 8 presents the results. It appears that except for video creation, respondents are not significantly concerned and do not intend to adopt a behavior related to flipping the classroom.

These findings may be attributed to the limited ICT access in the country and the culture of teaching in the country which is teaching face-face. Though there are already on-line classes or blended learning, there is a need for educators and researchers to verify existing theories on flipped classes. There is a need to generate new knowledge involving the role of technology within the evolving relationship between educational reform and pedagogical change in the country.

Finally, two variables were computed: Qknow which sums all knowledge answers, as a surrogate of general knowledge about flipped class and Qintend which sums up all intention items, as a surrogate of intended behavior. A correlation was found between both constructs of 0.221 (p=,009, n=138). This may be indicative that better information on this pedagogical approach could result in more intention to use this strategy.

Table 8. Correlation between the knowledge of Flipped classroom and behavior intention

Flipped Classroom (n=149 - 151)	Correlation	Sig. (2-Tailed)
I intend to establish spaces and time frames that permit students to interact and reflect on their learning as needed.	0.093	0.260
I intend to continually observe and monitor students to make adjustments as appropriate.	0.031	0.701
I intend to provide students with different ways to learn content and demonstrate mastery.	-0.043	0.604
I intend to give students opportunities to engage in meaningful activities without the teacher being central.	0.057	0.486
I intend to scaffold these activities and make them accessible to all students through differentiation and feedback.	-0.004	0.963
I intend to prioritize concepts used in direct instruction for learners to access on their own.	0.048	0.556
I intend to create and/or curate relevant content (typically videos) for my students.	0.112	0.174
I intend to differentiate to make content accessible and relevant to all students.	0.078	0.340
I intend to make myself available to all students for individual, small group, and class feedback in real time as needed.	0.022	0.794
I intend to conduct ongoing formative assessments during class time through observation and by recording data to inform future instruction.	0.066	0.418
I intend to collaborate and reflect with other educators and take responsibility for transforming my practice.	0.030	0.718

SOLUTIONS AND RECOMMENDATIONS

In this brief review of literature, a new pedagogical approach to learning, facilitated by the revolution of technology, has been revealed to have very enthusiast supporters. This is somehow surprising given the fact that the most positive studies report only a little improvement in effectiveness of learning. It requires a lot of new work for the teachers and several scientific studies have failed to detect significant differences in the effectiveness of the flipped classroom. The appreciation demonstrated by pre-service education students seems also positive but it must be considered in the light of the social desirability.

Some key elements for Filipinos, namely the face-to-face learning preference as mentioned by Tabaque (2014) should be noted along with Arimbuyuta et al (2007) that the young generation is at ease with virtual communication and also that virtual communication with webcam is relatively their new mode of communication. The remarkable growth of enrolments in e-learning in Open Universities like Dela Salle University, University of the Philippines, University of Santo Tomas and many others is a convincing witness of this new trend.

Angel de Dios (2015) mentions in her blog that procrastination and "ningas kugon", qualified by her as typically Filipino, can reduce the potential of introducing this pedagogical strategy in the Philippines. The face-to-face preferences of the Filipino learners might also be a constraint in the implementation of flipped classrooms. Hence, there is a need for educators to continually revisit a course curriculum toward better integration of technology like video content and in-class activities and asking for student feedback. An educator has the advantage of concrete data on which to suggest curriculum changes. Thus, data from a flipped course can help strengthen future iterations of the course, and if this data is shared, it can benefit the educational community as a whole (Saltman, 2011).

FUTURE RESEARCH DIRECTIONS

In reviewing the literature on the flipped classroom approach, several areas of research are worthy of further exploration. It may be important to assess how the flipped classroom could benefit hands-on courses in different courses or subjects in basic education or higher education levels, class time that could be used for experiential learning or blended learning. Moreover, several issues have not been addressed, such as the long-term requirements of video content improvement, keeping videos current, and whether the initial time investment of the flipped class affords less time to prepare. Finally, questions about video content ownership and negotiation of content rights between an institution and an educator may be a future issue of concern. These issues might not just be in the Philippines but other countries as well. As education advances, blended learning and flipped classroom approaches are moving forward in the twenty-first century, and best practices for integrating educational technology and engaging practice will continue to emerge in different schools or universities across countries.

CONCLUSION

Despite many previous results of studies, flipped classroom is not well known and/or accepted in the Philippines. The many advantages reported in the literature maybe outweighed by the limitations related to their introduction. Our empirical examination of the perception of education students shows that there is still room for improvement in the dissemination of this new pedagogical approach. However it should be noted that the fact that the study is conducted in only one university in one country constitutes a severe limit to the generalization of the results.

Nevertheless, this new pedagogy will keep necessitating new studies to get a better understanding. Theoreticians of pedagogy value the active learning which is at the center of this approach and indeed fills the gap between teaching and learning. Despite the relative success of this new method, further researches could be conducted to delve deeper into this emerging pedagogical approach in teaching.

REFERENCES

Abadines, A. (2012). *Part 2: The Future of Education: Flip it! The Filipino Teacher.* Retrieved April 4, 2012, from http://thefilipinoteacher.com/2012/04/04/part-2-the-future-of-education-flip-it/

Abernathye, K., Abshire, S., Cummings, C., & Mason, D. (2012). *A Conversation with the Lamar Academic Technology Committee: An Overview of Technology-enabled 21st Century Learning Spaces.* Retrieved April 26,2012, from http://www.slideshare.net/dianemason/academic-computingpreso4-2612kaysheryldmccfinal

Acedo, M. (2010). *Pros And Cons Of A Flipped Classroom.* Retrieved from http://www. teachthought. com/trends/10-pros-cons-flipped-classroom

Arimbuyutan, R. C., Kim, S., Song, J. G., & So, W. (2007). A Study on e-Learning for Philippines. *International Journal of Multimedia and Ubiquitous Engineering*, *4*(2), 49–53.

Atteberry, E. (2013). *Flipped classrooms may not have any impact on learning.* USA Today, Retrieved from http://www. usatoday. com/story/news/nationa/2013/10/22/flipped-classrooms-effectiveness/3148447

Azucena, J. (2014). *Studying flipped. Manila Bulletin.* Retrieved October 8, 2014, from http://www. mb.com.ph/studying-flipped/

Beichner, R. J., Saul, J. M., Abbott, D. S., Morse, J. J., Deardorff, D., Allain, R. J., ... & Risley, J. S. (2007). The student-centered activities for large enrollment undergraduate programs (SCALE-UP) project. *Research-Based Reform of University Physics, 1*(1), 2-39.

Bergmann, J., & Sams, A. (2012). *Flip your classroom: Reach every student in every class every day.* International Society for Technology in Education.

Bergmann, J., & Sams, A. (2014). Flipped learning: Maximizing face time. *T+ D, 68*(2), 28-31.

Bishop, J. L., & Verleger, M. A. (2013, June). The flipped classroom: A survey of the research. In *ASEE National Conference Proceedings.*

Butrymowicz, S. (2012). *Promise of 'flipped classroom' may elude poorer school districts. The Hechinger Report.* Retrieved June 12, 2012, from http://hechingerreport.org/promise-of-the-flipped-classroom-eludes-poorer-school-districts/

Butt, A. (2014). Student views on the use of a flipped classroom approach: Evidence from Australia. *Business Education & Accreditation, 6*(1), 33–43.

Clark, R., Norman, B., & Besterfield-Sacre, M. (2014). Preliminary Experiences with "Flipping" a Facility Layout/Material. *Proceedings of the 2014 Industrial and Systems Engineering Research Conference.*

Crouch, C. H., & Mazur, E. (2001). Peer instruction: Ten years of experience and results. *American Journal of Physics, 69*(9), 970–977. doi:10.1119/1.1374249

Day, J., & Foley, J. D. (2006). Evaluating a web lecture intervention in a human–computer interaction course. *Education. IEEE Transactions on, 49*(4), 420–431.

De Dios, A. C. (2014, June 2). *Philippine Basic Education "Pagputi Ng Uwak".* Retrieved June 2, 2014, from http://philbasiceducation.blogspot.com/search/label/Classrooms#ixzz3ONET7KRR

De Waard, I. (2014). *The Flipped Classroom. Presentation given at the Gulden Sporen College in Kortrijk, Belgium for one of their SOS sessions (pedagogical sessions). Educational researcher and practical implementer at Athabasca University and Open University of UK.* Retrieved November 20, 2014 from http://www.slideshare.net/ignatia/the-flipped-classroom-introduction-and-sources

Deslauriers, L., Schelew, E., & Wieman, C. (2011). Improved learning in a large-enrollment physics class. *Science, 332*(6031), 862-864.

Dill, E. M. (2012). *The Impact of Flip Teaching on Student Homework Completion, Behavior, Engagement and Proficiency.* University of New England.

Esperanza, P., & Toto, C. (2014, October). Does the Implementation of a Flipped Classroom Pedagogy changes the Students' Attitude towards Mathematics? In *World Conference on E-Learning in Corporate, Government, Healthcare, and Higher Education* (Vol. 2014, No. 1, pp. 552-557).

Estes, L. (2014). Flipped classrooms' get mixed reviews. *McClatchy Tribune Business News*. Retrieved April 08, 2014, from http://www.mcclatchydc.com/2012/06/12/152089/promise-of-flipped-classroom-may.html

Findlay-Thompson, S., & Mombourquette, P. (2013, July). Evaluation of a flipped classroom in an undergraduate business course. In *Global Conference on Business & Finance Proceedings* (Vol. 8, No. 2, p. 138). Institute for Business & Finance Research.

Fitzgerg. (2012). *Flipping The Classroom Eduphoria*. Retrieved from http://www.slideshare.net/fitzgerg/flipping-the-classroom-mhs-tech-camp

FLN & Sophia Learning. (2014). *Growth in Flipped Learning: Transitioning the focus from teachers to students for educational success: Transitioning the focus from teachers to students for educational success*. Retrieved May, 2014, from http://flippedlearning.org/Page/95

FLN the Flipped Learning Network. (2014). Retrieved from http://flippedlearning.org/site/default.aspx?PageID=1

Fulton, K. (2012). Upside down and inside out: Flip Your Classroom to Improve Student Learning. *Learning and Leading with Technology*, *39*(8), 12–17.

Goodman, N. (2013). Flipping the Classroom: Use online tools to provide content before the class ever meets—and still retain the value of interpersonal interaction. *Best practices- Training magazine*. Retrieved from https://trainingmag.com/content/best-practices-flipping-classroom

Goodwin, B., & Miller, K. (2013). Evidence on flipped classrooms is still coming in. *Educational Leadership*, *70*(6), 78–80.

Guro. (2013). Is the Philippines ready for the Flipped Classroom Model? *EdTech Guro Filipino Tech Educator in Ation*. Retrieved February 7, 2013, from https://edtechguro.wordpress.com/2013/02/07/is-the-philippines-ready-for-the-flipped-classroom-model/

Hamdan, N., McKnight, P., McKnight, K., & Arfstrom, K. (2013, June). *A review of flipped learning*. Retrieved from Pearson - research & innovation network: http://researchnetwork.pearson.com/wpcontent/uploads/LitReview_flippedLearning1.pdf

Herreid, C. F., & Schiller, N. A. (2013). Case studies and the flipped classroom. *Journal of College Science Teaching*, *42*(5), 62–66.

Hoffman, E. S. (2013). Beyond the flipped classroom: Redesigning a research methods course for e3 instruction. [CIER]. *Contemporary Issues in Education Research*, *7*(1), 51–62. doi:10.19030/cier.v7i1.8312

Horn, M. (2013). The transformational potential of flipped classrooms. *Education Next*, *13*(3), 78–79.

Hutcheson, G. D., & Sofroniou, N. (1999). *The Multivariate Social Scientist: Introductory Statistics Using Generalized Linear Models*. Sage Publication.

King, A. (1993). From sage on the stage to guide on the side. *College Teaching*, *41*(1), 30–35. doi:10.1080/87567555.1993.9926781

Lage, M. J., Platt, G. J., & Treglia, M. (2000). Inverting the classroom: A gateway to creating an inclusive learning environment. *The Journal of Economic Education*, *31*(1), 30–43. doi:10.1080/00220480009596759

Lape, N., Levy, R., & Yong, D. (2014). Can flipped classroom help students learn? We are trying to find out. *Slate*. Retrieved April 25, 2014, from http://www.slate.com/articles/technology/future_tense/2014/04/flipped_classrooms_can_they_help_students_learn.h

Mazur, A. D. (2013). *The Flipped Classroom: Theoretical Constructs*. Retrieved from https://ambermazur.files.wordpress.com/2013/06/flipped-classroom-theoretical-constructs-visual-with-explanation.pdf

Mazur, E. (1997). Peer Instruction: A User's Manual Series in Educational Innovation. Upper Saddle River, NJ: Prentice Hall. Retrieved from http://mazur.harvard.edu/publications.php?function=display&rowid=0

Mazur, E. (2009, November 12). *Confessions of a Converted Lecturer*. Retrieved from YouTube: http://www.youtube.com/watch?v=WwslBPj8GgI

McLaughlin, J. E., Griffin, L. M., Esserman, D. A., Davidson, C. A., Glatt, D. M., Roth, M. T., & Mumper, R. J. (2013). Pharmacy student engagement, performance, and perception in a flipped satellite classroom. *American Journal of Pharmaceutical Education*, *77*(9), 196. doi:10.5688/ajpe779196 PMID:24249858

McLaughlin, J. E., Roth, M. T., Glatt, D. M., Gharkholonarehe, N., Davidson, C. A., Griffin, L. M., & Mumper, R. J. (2014). The flipped classroom: A course redesign to foster learning and engagement in a health professions school. *Academic Medicine*, *89*(2), 236–243. doi:10.1097/ACM.0000000000000086 PMID:24270916

Meyer, R. (2013). The post-lecture classroom: How will students fare. *Atlantic (Boston, Mass.)*.

Missildine, K., Fountain, R., Summers, L., & Gosselin, K. (2013). Flipping the classroom to improve student performance and satisfaction. *The Journal of Nursing Education*, *52*(10), 597–599. doi:10.3928/01484834-20130919-03 PMID:24044386

Moeller. (2015). Didactic & Socratic models of teaching and learning. *Moeller Humanities Book*. Retrieved from http://collegeenglishbooks.com/two-models-of-teaching-learning.html

Mok, H. N. (2014). Teaching tip: The flipped classroom. *Journal of Information Systems Education*, *25*(1), 7.

O'Neil, K., Kelly, T., & Bone, S. (2012, June). We turned learning on its ear: Flipping the developmental classroom. In *World Conference on Educational Multimedia, Hypermedia and Telecommunications* (Vol. 2012, No. 1, pp. 2752-2756).

Phillips, C. R., & Trainor, J. E. (2014). Millenial students and the flipped classroom. *ASBBS Proceedings, 21*(1), 519-530.

Phillips, R. (2013). Computer giveaway designed to give freshmen running start. *The Record*. Retrieved June 18, 2013, from http://www.recordnet.com/article/20130618/A_NEWS/306180317

Pinnock, M. (2015). *Reaching More Learners: Using Flipped Classrooms*. Retrieved January 8, 2015, from http://www.slideshare.net/michelepinnock/flipped-learning-intro

Prunuske, A. J., Batzli, J., Howell, E., & Miller, S. (2012). Using online lectures to make time for active learning. *Genetics, 192*(1), 67–72. doi:10.1534/genetics.112.141754 PMID:22714412

Roehl, A., Reddy, S. L., & Shannon, G. J. (2013). The flipped classroom: An opportunity to engage millennial students through active learning strategies. *Journal of Family and Consumer Sciences, 105*(2), 44–49. doi:10.14307/JFCS105.2.12

Roshan, S., & Roshan, W. (2012). *My View: It's never too late to begin flipping your classroom. CNN Schools of Though.* Retrieved August 24, 2012, from http://schoolsofthought.blogs.cnn.com/2012/08/24/my-view-its-never-too-late-to-begin-flipping-your-classroom/

Ruddick, K. W. (2012). *Improving chemical education from high school to college using a more hands-on approach.* Academic Press.

Saltman, D. (2011, November-December). Flipping for Beginners: Inside the New classroom Craze. *Harvard Education Letter, 29*(6). Retrieved from http://hepg.org/hel/article/517

Sarah, T. (2014). *The Flipped Classroom. Research Synthesis EDET.* Retrieved June 24, 2014, from http://www.slideshare.net/sarahdtolson/the-flipped-classroom-36263899

Schneider, B., Wallace, J. M., Blikstein, P., & Pea, R. (2013). Preparing for future learning with a tangible user interface: The case of neuroscience. *Learning Technologies. IEEE Transactions on, 6*(2), 117–129.

Stolle, M. (2014). *Teacher of the Month: Bamber Valley teacher 'flips' the classroom. Postbulletin.com.* Retrieved February 1, 2014, from http://m.postbulletin.com/news/local/teacher-of-the-month-bamber-valley-teacher-flips-the-classroom/article_ebff74ad-0aa7-5a45-aed6-1a581a95edb2.html?mode=jqm

Stone, B. B. (2012). Flip your classroom to increase active learning and student engagement. In *Proceedings from 28th Annual Conference on Distance Teaching & Learning.*

Straumsheim, C. (2013). Still in Favor of the Flip. *October, 30.*

Sunguro, S. (2013). *Flipped classroom.* Retrieved October 14, 2013, from http://www.slideshare.net/stephens2042/flipped-classroom-27164701

Tabaque, M. (2014). *The Teacher of the 21st century. Is The Philippines Ready For Elearning Or Mlearning?* Retrieved from http://www.minreeva.com/blog

Tabaque, P. (2013). *The Benefits of Learning for Schools in the Philippines.* Retrieved February 3, 2013, from http://www.minreeva.com/blog/the-benefits-of-mlearning-for-schools-in-the-philippines

Talbert, R. (2012). Inverted Classroom. *Colleagues*, Article 7.

Tucker, B. (2012). The Flipped Classroom: Online Instruction at Home Frees Class Time for Learning. *Education Next, 12*(1), 82–83. http://educationnext.org/files/ednext_20121_BTucker.pdf Retrieved June 24, 2012

Wilson, S. G. (2013). The Flipped Class A Method to Address the Challenges of an Undergraduate Statistics Course. *Teaching of Psychology.*

Yarbro, J., Arfstrom, K. M., McKnight, K., & McKnight, P. (2014). *Extension of a review of flipped learning*. Academic Press.

KEY TERMS AND DEFINITION

Active Learning: A model of instruction where the focus is placed on the learner as responsible of its learning.

Education: A formal process of facilitating knowledge, skills or attitudes by the means of more or less formal pedagogical approaches.

Flipped Classroom: A pedagogical approach favoring learning thru the use of class time to integration activities and collaborative work on solving problems while the home work is devoted to instructional content.

Learning: A dynamic process of acquisition of knowledge or skills or attitudes, or their modification.

Pros and Cons: An expression that shortens the Latin expression "pro et contra" meaning the presentation of "for" and of "against" arguments.

Reverse Teaching: A synonym for flipped classroom.

Virtual Learning Environment: A computer-mediated platform that support or replace the traditional classroom educational settings.

Chapter 4
Learning about Blended Learning through Students' Experiences:
An Exploratory Study in Postgraduate Guidance and Counselling Programs in a University with Campuses in Australia and Singapore

Margaret Anne Carter
James Cook University, Australia

Cecily Knight
James Cook University, Australia

Paul Pagliano
James Cook University, Australia

Donna Goldie
James Cook University, Australia

ABSTRACT

The digital age together with the political and economic agenda to make higher education more accessible and cost effective are challenging teaching academics worldwide to rethink and redesign their pedagogical practices. The challenges include how to respond to increasingly diverse twenty first century learners who demand flexibility, and the requirements of a more service driven economy. For a program to be successful in higher education's current learnscape, teaching and learning need to optimise students' experiences and learning outcomes. The exploratory study discussed in this chapter investigates seventy-three postgraduate students' experiences with blended learning in guidance and counselling programs in one Australian university with an offshore branch campus in Singapore. The study aims to inform the future design and delivery of teaching and learning within blended learning spaces with a focus on pedagogical practices for student engagement. As such, it contributes to the body of knowledge about learning design that enhances student learning experiences.

DOI: 10.4018/978-1-5225-0359-0.ch004

INTRODUCTION

Both the Ernst and Young's "University of the Future" report (2012) and The New Media Consortium Horizon Higher Education report (2014) have recognised the digital age as a key driver in higher education institutions for twenty-first century learners. The emergent corollaries include: global mobility, the changing demographics of student candidates, the political and economic agenda making higher education more accessible and cost effective. These have coincided with a mushrooming availability of rapidly changing learning technologies. The upshot has been that technology-enhanced learning is more appealing to an increasingly diverse and more flexibility-demanding group of students within higher education (JISC, 2009; Shroff & Vogel, 2010).

The challenge is for teaching academics to implement programs that align with the demands of higher education institutions, which meet all the students' diverse learning needs and which intelligently employ the ever-changing technology. The goal is to maximise student engagement in order to achieve the highest outcomes. For programs to be successful in the current learning environment, Jordan (2015) finds that there needs to be a clearly articulated vision. Evaluating the current learnscape is an imperative precursor.

The chapter commences with a wide-ranging review of the current literature on blended learning, including a focus on counsellor education. Blended learning is where face-to-face classes are complemented, or fused, with web-based, online learning materials (El-Mowafy, Kuhn, & Snow, 2013; Garrison & Vaughan, 2011; Keppell, Suddaby, & Hard, 2011; López-Pérez, Pérez-López, & Rodríguez-Ariza, 2011). The context and the methodology for the study is then introduced. Findings and discussion follows, including the translation of research learning into practice. The limitations of the study and future research concludes the chapter.

BACKGROUND

Blended Learning

Blended learning is often defined as a combination of traditional and online teaching designs. Traditional approaches typically refer to face-to-face, synchronous teaching experiences in the form of lectures, workshops, tutorials and seminars (Oliver & Trigwell, 2005). These traditional designs are distinguished from those approaches utilising technology, such as enhanced, online and blended learning experiences (Garrison & Kanuka, 2004). At the pedagogic core of traditional methods are key elements where students learn by: reading materials (e.g., textbooks and lecture notes); guided self-teaching (developing reading lists); independent study (assignment and exam preparation); engage in personal communication (through contact with other students and lectures); use of audio-visual materials (such as films); and, learning by traditional academic teaching methods (seminars, lectures and other classes) (Peters, 2001).

Blended learning approaches contain the same pedagogic core as those in traditional approaches and provide a vehicle for thoughtful integration of the synchronous (traditional face-to-face) and asynchronous (internet assisted) learning activities (Garrison & Kanuka, 2004). Therefore it is not a simple combination of the two means of learning (Shea & Bidjerano, 2010) as blended learning requires a commitment to creating an ideal learning environment for students where pedagogy and technology is meaningfully combined (Rovai & Jordan, 2004) in order to foster engagement and high learning outcomes. There are benefits in both methods as each approach aims to deliver high quality learning outcomes and experiences for students.

Blended learning is a specific approach to learning design. It can be implemented at multilevels (curriculum, subject, activity and/or assessment) and is a relatively cost-effective method of education delivery (Keppell, Suddaby, & Hard, 2011). The core principles of delivery include flexible instruction and learning; face-to-face and online delivery modalities; integrated instructional methods across hybrid contexts; synchronous and asynchronous formats; partially instructor centred and learner controlled environments (Young, 2002; Delialioglu & Yildirim, 2007; Oh & Park, 2009; Hamilton & Tee, 2010). Teaching academics can be severely tested when redesigning curriculum content to create successful interactive hybrid learning approaches (Hamilton & Tee, 2010).

Challenges discussed in the literature include what is to be blended and the quality of the blend. Questions posed by Bath and Bourke (2011) are pertinent: "What is the value of bringing students together in a single place and time? How is it different when students are learning face-to-face vs. online/distance, or in real time vs. asynchronously?" There is no single approach to blended learning that answers these conundrums (Partridge, Ponting, & McCay, 2011), but frameworks can be utilised to provide meaningful learning experiences and outcomes to students. Two established frameworks are the 'Seven Principles of Good Practice in Undergraduate Education' (Chickering & Gamson, 1987) and the 'Community of Inquiry' (Garrison & Vaughan, 2011).

The Seven Principles (see Table 1) are a reflective practice tool for guiding professional approaches to teaching and learning. Although developed pre-online learning, nevertheless the Seven Principles are relevant to today's teaching using blended pedagogies (Chickering & Ehrmann, 1996; Partridge et al., 2011). The principles are applicable to both undergraduate and postgraduate education (Ames, 2015).

Table 1. A general overview of the seven principles of good practice in undergraduate education (Chickering & Gamson, 1987)

1. Good practice encourages contacts between students and faculty	Frequent student-faculty interaction in and out of class as a key factor in student motivation and involvement.
2. Good practice develops reciprocity and cooperation among students	Working in collaborative environments increases involvement in learning. Sharing ideas enhances understanding.
3. Good practice uses active learning techniques	Learning involves active participation in order to make education meaningful and relevant.
4. Good practice gives prompt feedback	Feedback assist student in self-reflection of their current skills and knowledge to enhance academic performance.
5. Good practice emphasises time on task	Efficiently utilising realistic amounts of time means effective teaching and learning for both students and teaching academics.
6. Good practice communicates high expectations	Withholding high expectations of students is critical for student to make the link between effort and outcomes.
7. Good practice respects diverse talents and ways of learning	Inclusive best practice that supports and enhances individual student's diversity and capitalises on their existing strengths and knowledge.

The community of inquiry (CoI) framework is rooted in a social-constructivist theoretical approach where learners make sense of their own experiences through social interaction and collaboration (Garrison & Vaughan, 2011). Learners are actively involved in the construction of knowledge, building on their existing understanding. Teaching and learning evolves from instructor-centred and content-orientated to learner-focused and process-orientated.

The CoI framework (see Figure 1) was developed in response to the identification of the crucial elements in the context of computer-mediated communication that are "pre-requisites for a successful higher education experience" (Garrison, et al., 2000, p. 87). Validated by a number of studies, it is a leading model guiding research into blended learning in higher education (Arbaugh, Bangert, & Cleveland-Innes, 2010; Partridge, Ponting, & McCay, 2011; Shea, Hayes, Vickers, & Gozza-Cohen, 2010).

The CoI is comprised of social, teaching, and cognitive presences. Social presence represents the learner's engagement in synchronous and asynchronous reciprocal interaction; cognitive presence characterizes the integration of critical thinking and purposeful discourse to promote and sustain higher levels of learning; teaching presence entails leading and managing productive hybrid learning environments, and integrating high level knowledge learning with advanced reasoning (Garrison, Anderson, & Archer, 2000). Within these spaces there exists a high level of student choice, active engagement and, self-responsibility on the part of the student. The educational experience is predicated on the overlap of the three presences.

Figure 1. Community of inquiry (Garrison & Vaughan, 2011, pp. 18-19)

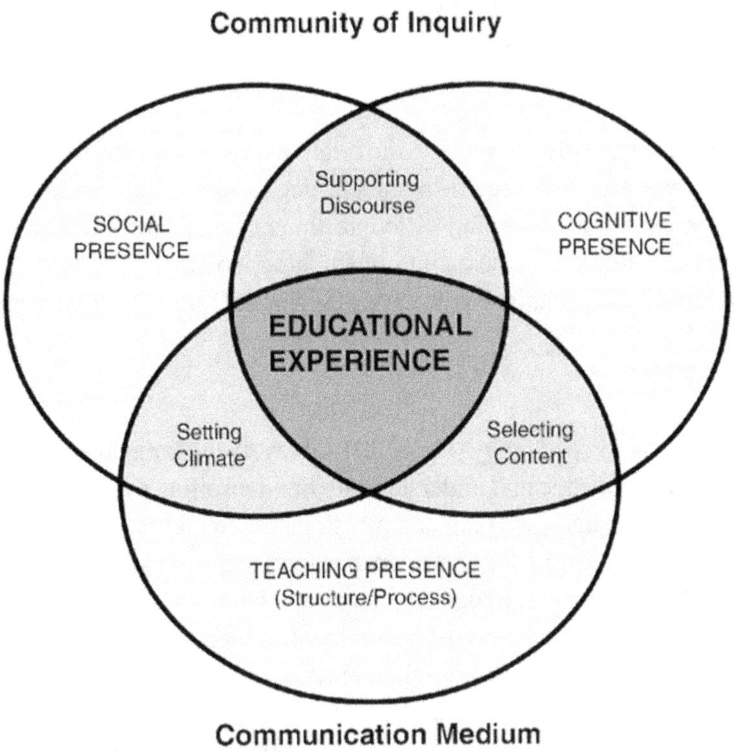

The Postgraduate Experience with Blended Learning

A growing body of research has reported that teaching academics in higher education are responding to the challenges of integrating technology-enhanced learning into teaching and learning spaces to create, support, promote, and sustain active learning (Cater, Michel, & Varela, 2012; Henrie, Bodily, & Manwaring, 2015; Kim & Bonk, 2006). No longer considered a trend, as initially presented in the literature (Cottrell & Robinson, 2003; Young, 2002), blended learning is now firmly embedded in the twenty-first century university experience (Bonk, Kim, & Zeng, 2004). As the emphasis on blended approaches has increased, so too has the focus on student engagement within these spaces (Vaughan, 2014, p. 247).

Student engagement within blended learning environments is a key goal for teaching academics in higher education (Keppell et al., 2011). Student engagement, as defined in higher education literature is "active and collaborative learning, participation in academic activities, formative communication with academic staff, involvement in enriching education experiences, and feeling legitimised and supported by university learning communities" (Coates, 2007, p. 122). Further, Henrie et al. (2015) defines engagement as the "quantity and quality of cognitive and emotional energy students exert to learn" (p. 132).

Within higher education research, levels of student engagement in the teaching and learning process is one of the "most important indicators of the quality of teaching activities" (Saritepeci & Çakir, 2015, p. 203). There is some evidence to suggest that blended learning design may enhance student engagement, through the provision of technologies to support learning, however little is known about what specific technology is leading to the observable gains (Means, Toyama, Murphy, & Baki, 2013). Conversely student isolation and limited peer interaction have been reported as potential drawbacks in the online component of blended programs (Jonas & Burns, 2010), with Hong, Lai, and Holton (2003) offering caution by highlighting the potential for high-level passivity. Some studies report higher levels of satisfaction, with learner and instructor interactions and social connectedness aligned with face-to-face teaching (Parkinson, Greene, Kim, & Marioni, 2003); but others report a stronger sense of learning and a broader range of interactions for learners and instructors with blended courses (Rovai & Jordan, 2004; Shroff & Vogel, 2010).

The literature is therefore clearly mixed as to the relative benefits of face-to face teaching, web based teaching and blended learning, with regard to student engagement. The teaching academics' competencies and capacity in effectively blending different dimensions of the curriculum are thought to be paramount, postulated to be the key to the quality of student motivation, engagement and satisfaction in blended programs (Derntl & Motschnig-Pitrik, 2005; Lawlor & Donnelly, 2010; Shroff & Vogel, 2010).

Counsellor Education

The aim of counsellor education pedagogy (Egan, 2014; Ivey, Ivey, & Zalaquett, 2014) is for students to develop and deepen their knowledge and understanding of counselling theory and practice by engaging with and reflecting on the literature, viewing and practicing micro skills in simulated and real contexts, receiving feedback, being coached in skill performance, and using self-assessment processes to reflect on competence and skill development. Integrating technology into these spaces provides multiple pathways for teaching academics and students to develop key knowledge and skills whilst at the same time, providing space for students to develop a professional identity as a global counsellor (Sells, Tan, Brogan, Dahlen, & Stupart, 2012). Many confirm that blended delivery is an appropriate and effective method for professional preparation (Wantz et al., 2003; Wilson, 2001; Kato, Shiono, Goto, &Tsuchida, 2011).

However it does not have unanimous support, with some in the counselling profession viewing it "with scepticism or resistance" (Sells et al., 2012, p. 39). Their argument is that counselling is essentially a person-to-person experience and that counsellor education online with minimal face-to-face interaction is inadequate (Benshoff & Gibbons, 2011). The rebuttal to this argument is the growing trend, internationally, of clients themselves choosing to access care (including mental health care Kummervold et al., 2002) online rather than face-to-face.

Identifying standards of competence in counsellor education using blended learning has proved to be an ethical and educational challenge (Benshoff & Gibbons, 2011; Sells et al., 2012). Regulating agreed standards in action has been a focus for accreditation bodies around the world. These professional associations set the standards for professional programs relative to their respective association standards. Higher education institutions abide by these standards in the development of curriculum material. Blended modes of education can present challenges for quality assurance, however, accreditors are responding to the academic challenges posed by internet-enabled learning (Eaton, 2001).

THE STUDY

The institution where this study was conducted hosts two campuses along the east coast of northern Australia and has an offshore branch campus in Singapore. The guidance and counselling programs reported in this study, are the Master of Guidance and Counselling (MGuidCouns) and the Graduate Certificate Career Development (GradCertCareerDev). The MGuidCouns course provides students with extended knowledge and skills in guidance and counselling. As this course is accredited by the Australian Counselling Association and the Singapore Association for Counselling, graduates can pursue careers in both countries, as well as internationally. The GradCertCareerDev, accredited by the Career Industry Council of Australia, qualifies graduates to work as career development practitioners. Students completing this program are eligible for entry to the MGuidCouns, and advanced standing is granted for two of the three subjects.

The curriculum is standardized across the tri-campuses, with 80% of the content common across the sites and 20% contextualized to local considerations. Each subject in these programs requires a 130 hour work load of study-related participation. In Singapore, the degrees are taught on a trimester basis, combining thirteen weeks of online material with twenty-five hours mandatory face-to-face teaching. Students are either residents or persons with employment or dependent passes. In Australia, the programs are taught in blended and online mode with eleven days of mandatory face-to-face classes dispersed across four of the eight core subjects.

METHODOLOGY

The aim of this research was to ascertain students' perceptions of their experiences with the blended guidance and counselling programs taught across the tri-campuses of the university. There were two online surveys, with the first being a preliminary exploration and the second survey was more detailed. The focus group drilled down on the issues elicited by the two surveys. The ethical guidelines of the university were followed with approval for the research to proceed prior to the commencement of the study.

Research Questions

The researchers were curious to know about students' experiences with blended learning in their postgraduate guidance and counselling programs. Consequently the research questions framing the study are: (1) What are the experiences of students in their blended guidance and counselling programs (2) What are the recommendations of students to improve their experiences in their blended guidance and counselling programs.

Participants

Conducted over three years six months, a total of 73 postgraduate students participated in this exploratory study. Table 2 outlines the demographic details across the data sets.

Measures

As a comprehensive literature search was unable to locate a survey measure that contained all the items of interest under investigation, the first survey was based on the conceptual frameworks of Billings

Table 2. Demographic details

Data Set	Data Set One	Data Set Two	Data Set Three
Instrument	Online survey 2011	Online survey 2014	Focus group 2015
Participant numbers	28	38	7
Gender	• Male 3 • Female 25	• Males 7 • Females 31	• Males 2 • Females 5
Age	02: 20-29 08: 30-38 14: 40-49 04: 50+	09: 20-29 07: 30-39 15: 40-49 07: 50+	03: 30-39 04: 40-49
Ethnicity	Chinese - 16 Indian 4 Australian - 3 Other - 5: French, American, Irish, Canadian	Australian – 24 Chinese – 4 Indian – 4 Other – 6	Australian - 6 Other – 1
Location	Students Singapore campus	Students cross campus: Cairns – 13 Townsville – 13 Singapore - 12	Students Australian campus Mandatory face-to-face classes
Participant Enrolment	28 students enrolled in the Master Guidance and Counselling program	35 students enrolled in the Master Guidance and Counselling program; 4 students in the Graduate Certificate Career Development program; 2 students enrolled in Master of Education, studying guidance and counselling subjects.	7 students enrolled in the Master Guidance and Counselling program
Total Enrolment	52 total students enrolled	122 total student enrolments	15 total students enrolled

(2000) and Paechter and Maier (2010). Both frameworks are structured around student perceptions of satisfaction with, and the benefits and barriers of, postgraduate studies. The survey consisted of 65 questions, comprising 18 forced-choice items, 7 binary option items, Likert-type scale items - 23 x 5 and 5 x 3 Likert-type scale items - and 12 open ended items. The open ended items were designed to generate extended responses on respondents perceptions of blended learning experiences not biased by the researcher's preconceived notions of what respondents would comment on. A sample of Likert-type scale items include: Rate your satisfaction with the block week quality of lecturing throughout your course (very satisfied, dissatisfied, neither dissatisfied or satisfied, satisfied, very satisfied); I think that my point of view was acknowledged by other participants in the class, both online and face-to-face (strongly disagree, disagree, uncertain, agree, and strongly agree). A sample of open ended items include: What are your career goals / aspirations; what are your reasons for enrolling in the JCU Singapore MGuidCouns degree.

Preliminary analysis of the data set from survey one was undertaken in order to inform the content in the second survey. This process allowed the research team to delve further with their questions pertaining to experiences with meaningful blending primarily in the online learning space. The 48 survey comprised 8 forced-choice items, 9 binary option items, 30 x 5 point Likert-type scale items, and 1 open-ended item. The open-ended item provided respondents with an opportunity to extend upon responses if needed. A sample of Likert-type scale items relating to the guidance, counselling and career development community site include: The information on the site has not extended my learning; the site provides relevant information; the site adds value to the learnjcu subject site; the site is interesting to explore (strongly disagree, disagree, uncertain, agree, strongly agree).

The analysis of the online surveys laid the foundation for the focus group interview, where participants elaborated on their experiences of the phenomenon under discussion. The 20 minute focus group interview was audiotaped and transcribed prior to analysis, thus helping to maintain the trustworthiness of the data. The researchers regarded the focus group as both a typical case sample and a sample of convenience based on the phenomenon under investigation and the availability of the participants.

Data Collection

A mixed methods approach was chosen to elicit responses from students at different stages of their studies. Students completing survey one did not complete survey two as they were a different student cohorts. Triangulation was used to support the credibility of the findings with data from one source enhancing, supporting, and complementing data from other sources. By using a pragmatic approach where the research questions remained the focus, these multiple data sources enabled the researchers to combine the strengths of both qualitative and quantitative research design while compensating for weaknesses.

Data Analysis

Quantitative data was submitted for descriptive statistical analysis using survey monkey (survey 1) and SPSS 22 (survey 2). In these surveys, the Pearson product moment correlation Coefficient (Pearson r) was performed to calculate values, either positive or negative linear relationships between two variables. The standard convention level of *Probability* $(p) < .05$ was accepted for evaluating statistical significance.

The method for identifying, analysing and reporting patterns across the qualitative data set was Braun and Clarke's (2006) systematic six phase thematic analysis coding and categorization process:

transcribing data, generating initial coding, searching for themes, reviewing themes, naming and defining themes, writing report (p. 96). This framework guaranteed a transparent audit trail to help ensure the analysis process was methodologically and theoretically sound. After strenuous re-analysis of concepts and ideas, a thematic map was generated to provide a consistent interpretation of the data set. Pertinent quotes were extracted from the qualitative data and relevant statistics from the quantitative data were used to illustrate and support the thematic analysis of participants' reality. The thematic analysis was triangulated to the literature and back to the surveys and focus group data.

FINDINGS

Eight themes were reported within the data: blend as a multidimensional concept; communication and connectedness; arranged collegiality; digital literacy; information communication technologies; Discipline knowledge; authentic support; reflecting about action.

Blend as a Multidimensional Concept

The convenience and availability of accessing subject content and resources online, and constructing their own learning without being time bound, in places and spaces of their choice, was appreciated by most of the students across the three data sets (Littlejohn & Pegler, 2007). The complimentary nature of online and face-to-face learning enabled students to be actively engaged with learning materials. Commentaries included: *I felt prepared as I was able to access all material easily, leaving me prepared for the block weeks; It gives me flexibility to plan for my learning, allowing me to schedule family, personal, work commitments and educational needs accordingly.* In the first data set there was a positive correlation between students who frequently accessed subject material online and satisfaction with the blended mode: n=.572, n=26, p=<.01. Moreover, the majority of students in this data set, 75.00% (n=21), indicated satisfaction with the face-to-face component of the blend commenting on the effectiveness of merging mandatory face-to-face classes and online learning: *Having a block week for one subject and the rest of the study period, pursuing learning and reading at one's own pace; The intensive face-to-face lectures during block week enhances understanding through engaging topics and activities.*

While students in the first two data sets valued social connectivity, working as a team and networking, the consensus of the third data set was cooperating and respecting the contribution of others was of little worth unless it involved assessment. Comments included: *I think if there was a purpose to it ... even if it's only worth, fifteen per cent, if you have that purpose for why you are talking.* This is in contrast to students in data set one and two who identified the collaborative nature of communication and sharing their learning as key to their knowledge creation and their student experiences. Students in these data sets enthusiastically remarked on studying with persons with diverse racial, ethnic and national backgrounds. Observations included: *The rich sharing by students of different backgrounds and age groups allows cross-fertilization of ideas; I like a mix of both local and Australian staff.*

Communication and Connectedness

Connectedness between cross-cultural students and academics contributed to students' satisfaction with the blended program. Explanations specific to students studying on the Singapore campus included: *Being able to interact with a very diverse group of students. This is a very strong advantage of JCU Singapore which needs to be optimized for further learning opportunities.*

Students' engagement with synchronous and asynchronous communication and collaboration shaped and developed their knowledge, understandings and skills as developing counsellors and career development practitioners. Students appreciated the blend of communication, with many remarking on the accessibility of the teaching academics, saying their "you can call me anytime, email me" mantra added value to their learning experience. While 40.75% (n=11) of students in the first data indicated they were comfortable participating in online discussions, 51.85% (n= 14) were equally comfortable interacting online with teaching academics and peers.

Online communication was dismissed by a nominal number of students (n= 1) in the third data set, advocating that face-to-face communication was the only form of communication worthy of acknowledgment: *No one got to know each other until today.* This viewpoint discounts online communication as a key contribution to knowledge creation in a blended program, with opportunities to foster professional working relationships through the social presence utilising online methods not being taken up. The majority of students however appreciated the strategic communication opportunities offered online, with 53.85% (n=14) of students in data set one indicating a high level of comfort when introducing themselves via the community blogs, surveys and discussion boards. Furthermore, students in data set two recognized the value of participating in the Blackboard Collaborate session(s) prior to attending face-to-face classes. Commentaries included: *It was still an icebreaker to come in here and oh I kind of know the name, I know someone.* While many students appreciated the self-directed nature of the blend, they noted restrictions in working online when minimal numbers of students joined the collaborate sessions or when those that did join in did not have a microphone so communication were restricted.

While students in the first data set appreciated the accessibility of teaching academics across the blended spaces, willing to discuss concepts to deepen meaning, to share aspirations, avert worries, and inform career plans, a minority of students (n=2) in the third data set were apprehensive about approaching teaching academics. Explanations included: *I've got a problem, I don't want to talk to Hilda (teaching academic) ... because, I want to look like I'm actually a mature student, who do I talk to?*

Arranged Collegiality

In response to the findings of data set one, students were allocated online partners in some subjects. This arranged collegiality was "meant to encourage greater association between [students] and to foster more sharing, learning, and improvement of skills and expertise" (Hargreaves & Fullan, 2012, p. 118). Feedback in data set two identified many students valued the arranged collegiality initiative, while others noted it as unhelpful. Commentaries included: *I had a frustrating encounter with my online partner ... we did not seem to communicate very well. Perhaps this was because we had trouble accessing work through Pebble; Partner or group work online is extremely frustrating....getting partners to engage is difficult and causes lots of stress.* Moreover, a small number of students in the third data set (n=2) perceived online partnering as imposed, superficial and inflexible, only useful if it was assessed. One recommendation proposed was to abandon partners and invite communication on the discussion board meaning students "can then interact with everyone", thereby encouraging contact between students and students with an aim of developing reciprocity and cooperation, which is supported by Chickering and Gamson (1987) as part of the Seven Principles for Good Practice.

Pebble+ became a university led teaching and learning initiative prior to the second survey being administered Whilst student engagement on the confidential platform was included as an assessment criteria in one subject, findings from the second data set indicated that while 67% (n= 24) of the students knew about Pebble+ and its use, 84% (n= 32) did not use it to store and share artifacts. Remarks across data set two and three included: *I see and appreciate the value of Pebble but using it ... was not as productive as I thought that it would be ... I ran out of time; Prefer that as adult learners we were left to negotiate our own way of collaborating; Merging Pebble+ with subject learning would increase its usage.*

According to Hargreaves and Fullan (2012), arranged collegiality is "double edged" (p. 118): on the one hand a stepping stone for students to begin working together, conversely, a forced, insincere and unproductive way of student engagement. Students need to recognize the tasks as credible, worthwhile and relevant. Seemingly good intentions directed toward promoting a collaborative culture can fail if perceived as contrived and unconnected to assessment. These findings aligned with student engagement pertaining to the energy they spend on academically purposeful activities (Kuh, 2001; Robinson & Hullinger, 2008).

Digital Literacy

The ease with which students independently accessed the Learning Management System (LMS) in the first data set was positively correlated with their satisfaction with blended learning: r=.527, n= 26, p= <.01. Explanations included: *Able to access the course materials easily and felt prepared prior to entering the block week classes.* Whilst acknowledging students were generally positive and satisfied with accessing the LMS, digital literacy impacted engagement for some students. Accounts included: *Basic computer skills are not sufficient to access this model of learning; I am not tech savvy, so sometimes I find the many online activities a bit daunting especially as I feel it is all; I find that the IT intimidating and standing in the way of my studies.* As indicated by the literature, there has been a shift in focus from developing students' technical skill and knowledge, to the learning process (Partridge et al., 2011), however this data is indicating a dualistic student need which includes developing technological knowledge for learning, and enhancing digital literacy skills required to access and use the technology.

Information Communication Technologies

Whilst students across the data sets had regular access to a computer, students in the second data set identified limited bandwidth hindering their access to the LMS and the Blackboard Collaborate sessions. Explanations included: *We had to switch the video off because it was cutting off the audio but it was just the one time; My greatest problem has been my internet connection which is via satellite and very slow; Black board collaborate and the resources on lib guides are excellent it is just my internet speed and download that is the problem; I have very limited internet access at home I cannot listen to collaborate sessions.*

Whilst identifying these barriers, the majority of students valued the Blackboard Collaborate sessions. Commentaries included: *I am relatively new to online study and I definitely enjoyed the interactive nature of the Blackboard Collaborate sessions. I had a few technical issues from my end, however the help and support available was excellent. The orientation of how to use the site in the first session was very valuable; I did feel a little less isolated with the subject I enrolled in this semester because of the collaborator sessions. They were my life saver. Discussions in the Blackboard Collaborate sessions are excellent.* It was noted in the second data set that Blackboard Collaborate sessions need to be relevant to elicit optimal engagement and *"not just outlining text chapters that we have to read anyway"*.

Whilst many students across the exploratory study recognised the collaborative nature of the Blackboard Collaborate sessions, some identified the problematic scheduling of the sessions yet appreciated the recording option meaning they could listen to the session at a more suitable time. Statements included: *Finding a suitable time is often the difficulty. It is good we can listen later but so much better when we can all interact and discuss.*

Students in the first data set identified room for improvement with the design of the LMS menu, recommendations taken on board prior to the administration of the second survey. Commentaries included: *Materials are not organized efficiently; One never knows where something will be found; Many of the lecturers did not post anything but the subject outline online; Sometimes resources or even the subject outline wasn't posted until well into the trimester.*

Discipline Knowledge

Students appreciated the "learning and inquiry process" led by teaching academics and counsellor professionals with specialized knowledge and expertise, meaningfully engaging students and contributing to a sense of connectedness. Within these collective learning spaces, real life shared standards of practice were understood, analysed, synthesised, evaluated, and integrated. There was a significant correlation in the first data set between student satisfaction and the quality of practitioner educators emphasising discipline focused inquiry based learning: $r=.537$, $n=24$, $p=< .01$. Commentaries included: *The quality of the subject and expertise of the teaching staff is above expectations. When I graduate, I will know I have been part of a very rigorous course and be confident of my competency due to the high standards expected; Learning from an experienced lecturer and applying what we learn.* These findings are indicative of developing professional capital (Hargreaves & Fullan, 2012) and social presence within the course whereby students and teaching academic develop group cohesion (Garrison & Vaughan, 2011).

Students were unanimous in their enjoyment of the subject content, tasks and assessment. Explanations included *the structure, the foundations in itself are very user friendly. The weekly tasks and that sort of thing are really helpful; I do like every activity that I do. I like the course mate; The resources on the LibGuide counselling pages are helpful and I particularly like the videos.*

Authentic Supports

Considerable emphasis is placed on academic skills and professional literacy and numeracy in the blended programs. Students are encouraged to avail themselves of student support services and obtain help through the Education librarian. A library site has been set up for the programs as well as the Guidance, Counselling and Career Development community site which provides students with a wealth of constructive information and resources. With these supports available, the researchers have pondered, have students been taught how to learn online.

The Guidance, Counselling and Career Development community site, established after the analysis of the first data set, is an online environment where students enrolled in the programs, have a shared space to engage, individually or collectively, with global counselling issues. This site compliments rather than duplicates the individual subject sites and is aimed at building counsellor knowledge. While 89%

(n=34) of students in the second data had set accessed the site, 61% (n=23) indicated it provides relevant information, and 63% (n =24) that it extended their learning.

Students in the first data set prioritised human resources and physical resources compared with web-based support, with 50.00% (n=14) naming the teaching academics as their primary source of support accessing the LMS, compared with peers (28.57%, n=8), student services (3.57%, n=1), infohelp (10.71%, n=3), and university staff (7.14%, n=2).

Remarks by students (n=3) in third data indicated the academic writing and English proficiency requirements frustrating, irrespective of the resource available. Developing one APA (American Psychological Association) format and style guide to be used across the programs was proposed. Furthermore these students were exasperated with the time learning advisors took when responding to their queries. Commentaries included: *I've spent two weeks and I'm still waiting ... I find that very difficult.* Aligned with the Seven Principles by encouraging contact between students and faculty, this commentary indicates that prompt feedback (Chickering & Gamson, 1987) is required by students in all interactions.

Reflecting about Action

Students recommended face-to-face induction days at the commencement of their studies as adding value to their student experience. While much of what they identified as content for this induction is included online on the subject site and the community site, students voiced face-to-face induction as their preference to online induction. Recommendations included: *I think they should have a session like this where you come out here for two days and you learn, how to write an academic essay, you meet the people here, you do the reference like what is expected and boom, you've got someone for life or through the whole course and you know what you are doing and what is expected; Giving a beginners course relevant to the subject and where and how to source materials.*

In conjunction with these recommendations, students suggested practices for improving the student experience in their blended programs (Table 3). The first four recommendations were enacted prior to administering survey two, and the remaining recommendations were reported in data set two and data set. Actions encompass collaborative responsibilities at the organizational, managerial, interpersonal and teaching academic level. Some of these recommendations have been enacted €, are being developed (BD), under investigation at a managerial level (UIM), and under investigation at an organization level (UIO). Under investigation at the managerial and organization levels means the decisions are beyond the jurisdiction of teaching academics.

DISCUSSION

Higher education institutions have adopted blended learning as a response to rapid changing technology and connectivity demands, student needs of ease of access and flexible learning and expectations of high quality learning experiences and outcomes, and an institutional need to deliver cost-effective courses (Benshoff & Gibbons, 2011; Henrie et., 2015; Saritepeci & Çakir, 2015). This exploratory study investigated postgraduate students' experiences with the blend in their guidance and counselling programs. Two research question were posed: (1) What are the experiences of students in their blended guidance and counselling programs; (2) What are the recommendations of students to improve their experiences in their blended guidance and counselling programs.

Table 3. One step better: Next practices

Students' Recommendations for Improving the Online Component of the Blend	Element of the Corresponding CoI	Stage of Implementation
Developing more ICT information and how to guides; Developing more 'how to learn online' information and how to guides	Teaching presence	E and BD
Standardizing LMS menus for ease of navigation	Teaching presence	E
Redesigning assessment tasks for more fluid active engagement	Teaching presence	E and BD
Posting subject content on the LMS at the commencement of the study period	Teaching presence	E
Providing more ICT support	Teaching presence	E and BD and UIO
Identifying the existing digital literacy of students and providing appropriate support	Teaching presence	BD
Organizing face-to-face induction programs	Teaching presence	UIO
Scheduling academic writing workshops at the commencement of study periods	Teaching presence	UNM
Having one authoritative APA quick guide recognized across all the subjects	Teaching presence	BD
Scheduling online orientation sessions specific to (1) the subject content and (2) navigating the LMS	Teaching presence	E
More efficient response from online learning advisors across the duration of studies	Social presence	UIO
Equity and accessibility to technology (bandwidth, downloads, data) and identifying ways to equate this	Teaching presence	UIO
Converting Blackboard Collaborate sessions to MP3 files for ease of access	Cognitive presence	UIO
Multiple formats of video files: allowing for streaming or file upload options through use of secure dropboxes	Cognitive presence	UIM
Linking Blackboard Collaborate recordings as audio files through the LearnJCU App	Cognitive presence	UIO
Compiling documents on the subject site into folders to just download one folder	Teaching presence	UIM
Looking into a way videos could be put into drop box to save time to download as they always cut out for me	Cognitive presence	UIO
Revising the online partner system making it more worthwhile and relevant	Social presence	BD
Promoting Microphone usage during the Blackboard Collaborate sessions	Teaching presence	E
Scheduling more casual 'getting to know me' activities on subject site (currently some social activities are on the community site)	Social presence	BD
Designing collaborate sessions designed for optimal engagement through utilization of analytics data of dwell time and student feedback	Teaching presence	BD
Aligning online networking with assessment	Social presence	BD

Overall, irrespective of gender, age and ethnicity, student satisfaction with the blended programs were positive. Students verified their satisfaction was influenced by the flexibility, convenience and accessibility of the curricula in the blended spaces, mandatory face-to-face classes and online learning. In alignment with the CoI framework, research has found that there are no gender differences in perceptions of any of the presences within blended programs (Swan & Ice, 2010) and instead there is a focus on the process of the construction of meaning through personal responsibility and inquiry learning (Garrison & Vaughan, 2011), regardless of gender, age or ethnicity.

According to previous studies (Osguthorpe & Graham, 2003; Twigg, 2003) students value the high level of diverse curriculum engagement, with many tasks reflecting authentic work-integrated deep learning. Negotiated tasks, aligned with subject learning outcomes, content and assessment tasks, were not reliant on the blended learning design, but were indicative of students becoming co-constructors of their learning as they engaged with realistic, relevant learning tasks. Students recommended the inclusion of additional collaborative tasks in both online and face-to-face learning to enhance and sustain social presence. Broadening collaborative learning activities to include wikis, blogs, pecha kucha's may be a mechanism for stimulating students' interest and adding value to collaborative CoI spaces across the hybrid program.

Ensuring teaching academics create a smooth interface between face-to-face and online delivery is essential. Orientation to blended learning is necessary so students learn codes of engagement as CoI members as well as hybrid learning and teaching requirements. This orientation is focused on redesigning students' experiences of the structure and approach to learning in higher education institutions. O'Quinn and Correy (2002) maintain that this mindset shift is also targeted at teaching academics as they redesign their subject for the self-paced blended learning setting … "in how educators orchestrate the act of learning" (p. 1) and in how they design effective practical blended educational environments. The initial step in this process is ascertaining and addressing teacher academics and student digital literacy needs with regard blended learning design, and responding to these needs so that academics and students develop their competencies' as they engage meaningfully in each subject's blended CoI.

Reciprocal communication is named in the literature as a factor contributing to student's positive perceptions of learning online (Wise, Chang, Duffy, & del Valle, 2004). Purposeful dialogue with peers and instructors has been linked with enhanced motivation and course satisfaction. Our research team found that interaction across the blended environments positively contributed to collaboration and reports of course satisfaction. Importantly, Ocak (2011) and Biggs (2003) declare that "good dialogue elicits those activities that shape, elaborate, and deepen understanding" (p. 13). However this potential for communication was not echoed in our study. While students reported favourably on synchronous and asynchronous interactions, they did not identify the hybrid communication medium leading to deeper levels of learning.

When online and face-to-face teachings are compared, students report higher satisfaction levels of dialogue and social connectedness with face-to-face teaching (Parkinson et al., 2003). This is contrary to Rovai and Jordan (2004) who report a strong sense of learning and a greater range of interactions for students with their instructor with blended programs.

Conrad (2005) affirmed the influential role competent teaching academics leading blended CoIs have in quality teaching and learning. Her research supports face-to-face engagement with online learning to "enhance connectedness and satisfaction" (Garrison & Vaughan, 2011, p. 25). Our findings indicated that students and teaching academics courteously engaged with each other in the blended spaces as both contexts were equally conducive to curriculum engagement.

Several researchers (Derntl & Motschnig-Pitrik, 2005; Arbaugh & Hwang, 2006; Lawlor & Donnelly, 2010) have identified one of the most influential factor in blended learning pedagogy is the competency of the instructor leading the blending. Students repeatedly acknowledged the professionalism, competency, and commitment of the teaching academic and counsellor professionals in creating and sustaining relevant and authentic, learner driven, CoIs.

Students acknowledged teaching academic based on the campus as their point of contact in any matters relating to their studies, with Singapore based students specifying this teaching academic provided a sense of belonging and identity within the program and the institution. This is reinformed in the writings of Rourke, Anderson, Garrison, and Archer (1999) who maintains identity is important for students to grow as they develop as a community of learners in hybrid learning experiences.

Students commented that the blend of the programs, enabled them wider and faster access to meaningful curriculum content and instructor support. These findings replicate those of Matheos, Daniel, and McCalla (2005), who note that instructor support in traditional classrooms, is often restricted to specific office times or classroom schedules. Whereas the amount and variety of subject material differed across subjects, many students commented on the degrees of variance in the depth of curriculum materials available on the LMS. This finding highlights the importance of the educational design being similar across subjects in the degree course, yet flexible in response to individual subject learning outcomes, curriculum content, and student needs incorporated within the CoI framework.

Whilst most students found the LMS easy to navigate, they did identify support mechanisms that would increase their familiarity with the platform. Information guides (e.g., logging into Blackboard Collaborate, navigating the Blackboard Collaborative space, engaging in Pebble+, discussion board etiquette, APA referencing) were suggested as practical ways of adding value to learner engagement within the blended programs. Similar recommendations are promoted in the writing of Cockbain, Blyth, Bovill, and Morss (2009) and will be prioritized for ongoing redesigns of the guidance and counselling and career development programs.

Commenting that blended learning is time intensive, students proposed standardizing the design structure and format of the LMS menus. One could speculate that the variation in levels of digital literacy could equate with different amounts of time needed to familiarise themselves navigating e-learning platforms. Buddying novice with experienced web users may be one way of providing collaborative support for students as they grow and develop their skills with navigating the blended environment.

While time management was considered a potential barrier to experiences with blended learning, technology was further recognised as a hurdle. Although learner flexibility and convenience is widely identified as a key factor for blended course choice by students (Graham, 2004), technological access issues can hinder access to online learning environments. Internet access is not universal due to logistical and economic factors, and these can impact a student's level of engagement with the subject materials (Univeristy of Illinois, 2010). There are specific areas currently under investigation to ensure equality of access, such as video streaming and file upload options.

TRANSLATING RESEARCH LEARNINGS INTO PRACTICE

In response to students' recommendations, the teaching academics have made changes to the "instructional attributes" (Oh & Park, 2009, p. 328) of the blended subjects. Blackboard Collaborate sessions explaining blended learning and introducing students to the digital literacy skills and ICT knowledge required are scheduled at the commencement of each study period separate to curriculum focused sessions. Blackboard Collaborate sessions are scheduled to describe how to learn online as well as to explain assessment tasks. In response to feedback on the time taken to navigate the online environment, many of the subjects now have generic LMS menus. To keep students informed with the module content, academics have developed short video vignettes describing elements of the content modules. Pebble+ is set up on a voluntary basis for students to provide reflective feedback to one another on readings, activities and assessment tasks. Students have options to choose from when introducing themselves to their subject community including "getting to know me" blogs, learning style questionnaires, and wellness quizzes.

There is opportunity to further investigate student experiences and engagement through learning analytics as, "studies have demonstrated a relationship between the frequency of student LMS usage and academic performance" (Whitmer, Fernandes, & Allen, 2012, p. 2). Learning analytics has the potential to capture direct data of learners' online activities, assuming their engagement, by providing authentic and timely evidence (Ma, Han, Yang, & Cheng, 2014, p. 27). This provides the opportunity to further investigate student interaction and behaviours within the programs.

The support of the library and the learning support staff have been enlisted to lead online information sessions on topics that are perplexing to students. To further support students develop their understanding of APA, an APA checklist is provided, directed toward students developing their skills in self-reflection (López-Pérez et al., 2011). When aligned with the Seven Principles, this provides the opportunity for prompt feedback in various ways (Chickering & Ehrmann, 1996) to support students take ownership of their learning and develop self-reflection as part of cognitive presence.

Baepler and Murdoch (2010) identify the need for early and frequent assessment as best practice in blended learning. Consequently more formal requirements pertaining to formative assessment has been adopted. By clearly communicating expectations for assessment, with explicit rubrics for making judgements, Sells et al. (2012) find this provides a respectful learning environment that is conductive for building professional working relationships. This element also addresses the teaching presence as it moderates and focuses the assessment tasks throughout each study period (Garrison & Vaughan, 2011).

CoI development within the course is foregrounded by the teaching presence. This is "essential to bring all the elements together and ensure that the community of interest is productive" (Garrison & Vaughan, 2011, p. 24). Based on our findings, there is an opportunity to explore the social presence, in particular global, professional identity development, in consideration of inter-cultural awareness and blended spaces for collaboration and question answer sessions. This element could be explored through the adoption of Web 2.0 tools, including those provided by social media. It is argued that the sociability and scalability of this media can be utilised as both tools for learning and learning spaces (Licona & Gurung, 2011).

The cognitive presence element of the framework provides a space for meaning to be constructed from experience (Garrison er al., 2000; Garrison & Vaughan, 2011). Based on our findings, there was a strong focus on the technological skills and knowledge required by the students to access the course. This was an intentional focus within the data collection tools, however, an investigation into student experiences in the realm of conceptual knowledge and skill development, in the online, face-to-face and practicum experiences, could further function to inform the blend, thus providing an indicator of student engagement practices (Henrie et al., 2015), within the cognitive presence.

LIMITATIONS AND FUTURE RESEARCH DIRECTIONS

It should be acknowledged that although the surveys and focus group data identified student experiences in the blended programs, the study did not measure the extent students' experiences influenced their sense of agency, nor if it was a barrier or an enabler of cross cultural interactions, engagement and collaborations. Despite these limitations, an important contribution of this study to the literature is that it provides a starting point for evaluating the learnscape to inform the design and implementation of meaningful blended programs.

Within this study, there is a limitation that the voices of the teaching academics are not included. In addition, there is a gap in terms of the effect of different teaching approaches within blended environments, with a focus on the perspective of teaching academics within the space. Therefore, there is a need to extend the current investigations into "approaches for blending technology with different instructional approaches" (Delialioğlu, 2012, p. 312).

CONCLUSION

To triangulate the data further, inclusion of learning analytics data can further inform the blended program. Learning analytics is defined as the "measurement, collection, analysis and reporting of data about learners and their contexts, for purposes of understanding and optimising learning and the environments in which it occurs" (Learning Analytics & Knowledge, 2011, para. one). It is a field that is used to analyse transactional data, for example, time spent engaging with online content or discussion boards within a LMS, such as Blackboard. This transactional data can either focus on specific information for one student or general patterns of interaction which has the possibility of measuring progress, inferring engagement and predicting learning outcomes (Richards, 2011). Whilst not explicitly measuring student engagement, there are identifiable patterns in student behaviours, such as interactions with course content and learning activities that are measured by learning analytics programs (Blackboard.com, n.d; Morrison, 2012), which give indicators of engagement within a blended subject. Learning analytics has the potential to capture direct data of learners' online activities, assuming their engagement, by providing authentic and timely evidence (Ma, Han, Yang, & Cheng, 2014, p. 27). In this context, extracting, analysing and synthesising checkpoint and process learning analytics from the subject sites to learn more about students' online usage patterns and interactions, knowledge processing, evolution or construction would be useful in informing the sites design and layout, in conjunction with future learning and teaching practices (Lockyer, Heathcote, & Dawson, 2013).

In alignment with investigating pedagogy, our findings further inform the blended programs in two ways. Firstly, by adding to the existing small group of studies using qualitative methods (Delialioğlu, 2012, p. 311) and investigating student experiences within blended learning environments from the perspective of teaching academics. Secondly, Henrie et al. (2015) state, there are currently few studies which examine student experiences and engagement through the lens of operational principles or core attributes provided by a teaching academic to promote greater student engagement and subsequent performance and outcomes within blended learning environments. Therefore, this study aims to inform the future designs and delivery of teaching and learning materials within meaningful blended learning spaces with a focus on pedagogical practices for student engagement. As such, it contributes to the body of knowledge about learning design that optimises the 21st century learning experience.

ACKNOWLEDGMENT

This study was funded by an internal grant from JCU Singapore, under grant number JCUS007/2011/MC. It was also funded by a teaching and learning grant from JCU Australia, *Meaningful Blending and Student Engagement Across our Master Guidance and Counselling Program*. The authors would like to thank the research assistance who worked on the project and the students who participated in this study.

REFERENCES

Ames, K. (2015). *Online Education - Reflecting on Practice*. Retrieved from https://onlineedreflections.wordpress.com/2013/02/05/reflecting-on-the-seven-principles/

Arbaugh, J., & Hwang, A. (2006). Does 'teaching presence' exist in online MBA courses? *The Internet and Higher Education*, *9*(1), 9–21. doi:10.1016/j.iheduc.2005.12.001

Arbaugh, J. B., Bangert, A., & Cleveland-Innes, M. (2010). Subject matter effects and the Community of Inquiry (CoI) framework: An exploratory study. *The Internet and Higher Education*, *13*(1), 37–44. doi:10.1016/j.iheduc.2009.10.006

Baepler, P., & Murdoch, C. J. (2010). Academic Analytics and Data Mining in Higher Education. *International Journal for the Scholarship of Teaching and Learning*, *4*(2), 1–9. doi:10.20429/ijsotl.2010.040217

Bath, D., & Bourke, J. (2011). The blending of blended learning: An experiential approach to academic staff development. In G. Williams, P. Statham, N. Brown, & B. Cleland (Eds.), Changing Demands, Changing Directions. Proceedings ascilite Hobart 2011 (pp. 133–138). Available at http://www.ascilite.org.au/conferences/hobart11/downloads/papers/Bath-concise.pdf

Benshoff, J. M., & Gibbons, M. M. (2011). Bringing Life to e-Learning: Incorporating a Synchronous Approach to Online Teachign in Counselor Education. *The Professional Counselor: Research and Practice*, *1*(1), 21–28. doi:10.15241/jmb.1.1.21

Biggs, J. (2003). *Teaching for quality learning at university: What the student does* (2nd ed.). Maidenhead, UK: Open University Press.

Billings, D. (2000). A framework for assessing outcomes and practices in web-based courses in nursing. *The Journal of Nursing Education*, *39*, 60–67. PMID:10688463

Blackboard.com. *(n.d.). Blackboard Analytics for Learn*. Washington, DC: Blackboard Inc.

Bonk, C. J., Kim, K.-J., & Zeng, T. (2004). *Future Directions of Blended Learning in Higher Education and Workplace Settings* (C. J. Bonk & C. Graham, Eds.). San Francisco, CA: Pfeiffer Publishing.

Cater, J. J. III, Michel, N., & Varela, O. E. (2012). Challenges of Online Learning in Management Education: An Empirical Study. *Journal of Applied Management and Entrepreneurship*, *17*(4), 76–96.

Chickering, A.W., & Ehrmann, S.C. (1996). Implementing the Seven Principles: Technology as Lever. *American Associates for Higher Education Bulletin, 49*(1-10), 3-6.

Chickering, A.W., & Gamson, Z.F. (1987). Seven Principles for Good Practice in Undergraduate Education. *American Associates for Higher Education Bulletin*, 3-7.

Coates, H. (2007). A model of online and general campus-based student engagement. *Assessment & Evaluation in Higher Education*, *32*(2), 121–141. doi:10.1080/02602930600801878

Cockbain, M., Blyth, C., Bovill, C., & Morss, K. (2009). Adopting a blended approach to learning: Experiences from radiography at Queen Margaret University, Edinburgh. *Science Direct*, *15*, 242–246.

Conrad, D. (2005). Building and maintaining community in cohort based online learning. *Journal of Distance Education*, *20*(1), 1–20.

Cottrell, D. M., & Robinson, R. A. (2003). Blended Learning in an Accounting Course. *Quarterly Review of Distance Education*, *4*(3), 261–269.

Delialioğlu, Ö. (2012). Student engagement in blended learning environments with lecture-based and problem-based instructional approaches. *Journal of Educational Technology & Society*, *15*(3), 310–322.

Delialioglu, O., & Yildirim, Z. (2007). Students' perceptions on effective dimensions of interactive learning in a blended learning environment. *Journal of Educational Technology & Society*, *10*(2), 133–146. Retrieved from http://search.proquest.com/docview/61934868?accountid=16285

Derntl, M., & Motschnig-Pitrik, R. (2005). The role of structure, patterns and people in blended learning. *The Internet and Higher Education*, *8*(2), 111–130. doi:10.1016/j.iheduc.2005.03.002

Eaton, J.S. (2001). Distance Learning: Academic and Political Challenges for Higher Education Accreditation. *Council for Higher Education Accreditation (CHEA), 1.*

Egan, G. (2014). *The Skilled Helper* (10th ed.). Belmont, CA: Brooks/Cole.

El-Mowafy, A., Kuhn, M., & Snow, T. (2013). Blended learning in higher education: Current and future challenges in surveying education. *Issues in Educational Research*, *23*(2), 132–150.

Ernst & Young. (2012). *University of the future: A thousand year old industry on the cusp of profound change.* Retrieved from http://www.ey.com/Publication/vwLUAssets/ University_of_the_future/$FILE/University_of_the_future_2012.pdf

Garrison, D., Anderson, T., & Archer, W. (2000). Critical inquiry in a text-based environment: Computer conferencing in higher education. *The Internet and Higher Education*, *2*(2-3), 87–105. doi:10.1016/S1096-7516(00)00016-6

Garrison, D. R., & Kanuka, H. (2004). Blended learning: Uncovering its transformative potential in higher education. *The Internet and Higher Education*, *7*(2), 95–105. doi:10.1016/j.iheduc.2004.02.001

Garrison, D. R., & Vaughan, N. D. (2011). *Blended Learning in Higher Education: Framework, Principles, and Guidelines.* Retrieved from http://jcu.eblib.com.au/patron/FullRecord.aspx?p=819029

Glahn, R., & Gen, R. (2002). Progenies in Education: The evolution of internet teaching. *Community College Journal of Research and Practice*, *26*(10), 777–785. doi:10.1080/10668920290104868

Graham, C. (2004). Blended Learning Systems: Definiton, Current Trends, and Future Directions. In C. J. Bonk & C. Graham (Eds.), *Handbook of blended learning: Global perspectives, local designs.* San Francisco, CA: Pfeiffer Publishing.

Greene, J., Caracelli, V., & Graham, W. (1989). Toward a conceptual framework for mixed-method evaluation design. *Teaching and Teacher Education*, *26*, 1530–1543.

Hamilton, J., & Tee, S. (2010). Smart utilization of tertiary instructional modes. *Computers & Education*, *54*(4), 1036–1053. doi:10.1016/j.compedu.2009.10.009

Hargreaves, A., & Fullan, M. (2012). *Professional Capital: Transforming teaching in every school*. New York: Teachers College Press.

Henrie, C. R., Bodily, R., & Manwaring, K. C. (2015). Exploring Intensive Longitudinal Measures of Student Engagement in Blended Learning. *International Review of Research in Open and Distance Learning*, *16*(3).

Hong, K., Lai, K., & Holton, D. (2003). Students' satisfaction and perceived learning with a web-based course. *Journal of Educational Technology & Society*, *6*(1), 116–124.

Ivey, A., Ivey, M., & Zalaquett, C. (2014). *Intentional interviewing and counselling. Facilitating client development in a multicultural society* (8th ed.). Belmont, CA: Brooks/Cole.

JISC. (2009). *Effective practice in a digital age: A guide to technology-enhanced learning and teaching*. Available at: http://www.jisc.ac.uk/media/documents/publications/effectivepracticedigitalage.pdf

Jonas, D., & Burns, B. (2010). The transition of e-learning. Changing the focus of educational delivery in children's pain management. *Nurse Educator*, *10*(1), 1–7. doi:10.1016/j.nepr.2009.01.015 PMID:19264550

Jordan, J.D. (2015). *What's next: Blended Learning 2.0: The Hechinger Report*. Academic Press.

Kato, C., Shiono, Y., Goto, T., & Tsuchida, K. (2011). Development of Online Counseling System and Usability Evaluation. *Journal of Emerging Technologies in Web Intelligence*, *3*(2), 146–153. doi:10.4304/jetwi.3.2.146-153

Keppell, M., Suddaby, G., & Hard, N. (2011). *Good practice report: technology-enhanced learning and teaching*. Sydney, NSW: Australian Learning and Teaching Council.

Kim, K.-J., & Bonk, C. J. (2006). The Future of Online Teaching and Learning in Higher Education: The Survey Says. *EDUCAUSE Quarterly*, *29*(4), 22–30.

Kuh, G. D. (2001). Assessing What Really Matters to Student Learning Inside The National Survey of Student Engagement. *Change: The Magazine of Higher Learning*, *33*(3), 10–17. doi:10.1080/00091380109601795

Kummervold, P. E., Gammon, D., Bergvik, S., Johnsen, J. A. K., Hasvold, T., & Rosenvinge, J. H. (2002). Social support in a wired world: Use of online mental health forums in Norway. *Nordic Journal of Psychiatry*, *56*(1), 59–65. doi:10.1080/08039480252803945 PMID:11869468

Lawlor, B., & Donnelly, R. (2010). Using podcasts to support communication skills development: A case study for content format preferences among postgraduate research students. *Computers & Education*, *54*(4), 962–971. doi:10.1016/j.compedu.2009.09.031

Learning Analytics & Knowledge. (2011). *1st International Conference on Learning Analytics and Knowledge*.

Licona, M. M., & Gurung, B. (2011). Online Multicultural Education: Asynchronous Discussions in Online Multicultural Education. *Multicultural Education, 19*(1), 2–8.

Littlejohn, A., & Pegler, C. (2007). *Preparing for blended e-learning*. New York: Routledge.

Lockyer, L., Heathcote, E., & Dawson, S. (2013, October01). (2103). Informing pedagogical action: Aligning learning analytics with learning design. *The American Behavioral Scientist, 57*(10), 1439–1459. doi:10.1177/0002764213479367

López-Pérez, M. V., Pérez-López, M. C., & Rodríguez-Ariza, L. (2011). Blended learning in higher education: Students' perceptions and their relation to outcomes. *Computers & Education, 56*(3), 818–826. doi:10.1016/j.compedu.2010.10.023

Ma, J., Han, X., Yang, J., & Cheng, J. (2014). Examining the necessary condition for engagement in an online learning environment based on learning analytics approach: The role of the instructor. *The Internet and Higher Education, 24*, 26–34. doi:10.1016/j.iheduc.2014.09.005

Matheos, K., Daniel, K., & McCalla, G. (2005). Dimensions for blended learning technology: Learners' perspectives. *Journal of Learning Design, 1*(11), 56–76.

Means, B., Toyama, Y., Murphy, R., & Baki, M. (2013). The Effectiveness of Online and Blended Learning: A Meta-Analysis of the Empirical Literature. *Teachers College Record, 115*(3).

Morrison, D. A. (2012). *How Learning Analytics Can Make Instructors More Effective in an Online Course*. Retrieved 20 July 2015, 2015, from https://onlinelearninginsights.wordpress.com/2012/10/15/how-learning-analytics-can-make-instructors-more-effective-in-the-online-course/

O'Quinn, L., & Correy, M. (2002). Factors that deter faculty from participating in distance education. *Online Journal of Distance Learning Administration, 5*(4), 1–16.

Ocak, M. (2011). Why are faculty members not teaching blended courses? Insights from faculty members. *Computers & Education, 56*(3), 689–699. doi:10.1016/j.compedu.2010.10.011

Oh, E., & Park, S. (2009). How are universities involved in blended instruction? *Journal of Educational Technology & Society, 12*(3), 327–342.

Oliver, M., & Trigwell, K. (2005). Can "Blended Learning" Be Redeemed? *E-learning, 2*(1), 17. doi:10.2304/elea.2005.2.1.2

Osguthorpe, R., & Graham, C. (2003). Blended learning environments: Definitions and directions. *The Quarterly Review of Distance Education, 4*(3), 227–233.

Paechter, M., & Maier, B. (2010). Online or face-to-face? Students' experiences and preferences in e-learning. *The Internet and Higher Education, 13*(4), 292–297. doi:10.1016/j.iheduc.2010.09.004

Parkinson, D., Greene, W., Kim, Y., & Marioni, J. (2003). Emerging themes of student satisfaction in a traditional course and a blended distance course. *TechTrends, 47*(4), 22–28. doi:10.1007/BF02763508

Partridge, H., Ponting, D., & McCay, M. (2011). *Good Practice Report: Blended Learning*. Sydney, NSW: Australian Learning and Teaching Council.

Peters, O. (2001). *Learning and Teaching in distance education*. London: Routledge Falmer.

Richards, G. (2011). *Measuring Engagement: Learning Analytics in Online Learning*. Paper presented at the Electronic Kazan 2011, Russia.

Robinson, C. C., & Hullinger, H. (2008). New Benchmarks in Higher Education: Student Engagement in Online Learning. *Journal of Education for Business*, *84*(2), 101–108. doi:10.3200/JOEB.84.2.101-109

Ross, B., & Gage, K. (2006). Global Perspectives on Blending Learning. In C. J. Bonk & C. Graham (Eds.), *The Handbook of Blended Learning: Global perspectives, local designs*.

Rourke, L., Anderson, T., Garrison, D., & Archer, W. (1999). Assessing social presence in asynchronous text-based computer conferencing. *Journal of Distance Education*, *14*(2), 50–71.

Rovai, A., & Jordan, H. (2004). Blended learning and a sense of community: A comparative analysis with traditional and fully on-line graduate courses. *International Review of Research in Open and Distance Learning*, *5*(3), 197–211.

Saritepeci, M., & Çakir, H. (2015). The Effect of Blended Learning Environments on Student Motivation and Student Engagement: A Study on Social Studies Course. *Egitim ve Bilim*, *40*(177).

Sells, J., Tan, A., Borgan, J., Dahlen, U., & Stupart, Ý. (2012). Preparing International Counselor Educators Through Online Distance Learning. *International Journal for the Advancement of Counseling*, *34*(1), 39–54. doi:10.1007/s10447-011-9126-4

Shea, P., & Bidjerano, T. (2010). Learning presence: Towards a theory of self-efficacy, self-regulation, and the development of a communities of inquiry in online and blended learning environments. *Computers & Education*, *55*(4), 1721–1731. doi:10.1016/j.compedu.2010.07.017

Shea, P., Hayes, S., Vickers, J., Gozza-Cohen, M., Uzuner, S., Mehta, R., & Rangan, P. et al. (2010). A re-examination of the community of inquiry framework: Social network and content analysis. *The Internet and Higher Education*, *13*(1), 10–21. doi:10.1016/j.iheduc.2009.11.002

Shroff, R., & Vogel, D. (2010). An Investigation on Individual Students' Perceptions of Interest Utilizing a Blended Learning Approach. *International Journal on E-Learning*, *9*(2), 279–294.

Swan, K., & Ice, P. (2010). The Community of Inquiry Framework Ten Years Later: Introducation to the Special Issue. *The Internet and Higher Education*, *13*(1-2), 1–4. doi:10.1016/j.iheduc.2009.11.003

The New Media Consortium. (2014). *The New Media Consortium Horizon Report: 2014 Higher Education Edition*. The New Media Consortium.

Twigg, C. (2003). Improving quality and reducing costs: New models for online learning. *EDUCAUSE Review*, *38*(5), 28–38.

Univeristy of Illinois. (2010). *Weaknesses of Online Learning*. Retrieved from http://www.ion.uillinois.edu/resources/tutorials/overview/weaknesses.asp

Vaughan, N. (2014). Student Engagement and Blended Learning: Making the Assessment Connection. *Education Sciences*, *4*(4), 247–264. doi:10.3390/educsci4040247

Wantz, R.A., Tromski, D.M., Mortsolf, C.J., Yoxtheimer, G., Brill, S., & Cole, A. (2003). Incorporating Distance Learning into Counselor Education Programs: A Research Study. *Cybercounseling and Cyberlearing: An Encore*, 327-344.

Whitmer, J., Fernandes, K., & Allen, W.R. (2012). Analytics in Progress: Technology Use, Student Characteristics, and Student Achievement. *Educause Review Online*, 1-7.

Wilson, S. E. (2001). *The application of technology to counselor education: Video conferencing in distance supervision.* (Doctoral dissertation), University of Texas, San Antonio.

Wise, A., Chang, J., Duffy, T., & del Valle, R. (2004). The effects of teacher social presence on student satisfaction, engagement, and learning. *Journal of Educational Computing Research, 31*(3), 247–271. doi:10.2190/V0LB-1M37-RNR8-Y2U1

Young, J. R. (2002). Hybrid Teaching Seeks to End the Divide Between Traditional and Online Instruction. *The Chronicle of Higher Education, 48*(28), 2–3.

Zhao, Y., Lei, J., Yan, B., Lai, C., & Tan, H. S. (2005). What makes the difference? A practical analysis of research on the effectiveness of distance education. *Teachers College Record, 107*(8), 1836–1884. doi:10.1111/j.1467-9620.2005.00544.x

KEY TERMS AND DEFINITIONS

Blackboard Collaborate: An online learning tool which provides synchronous learning such as secure video conferencing, interactive whiteboards, breakout rooms and document viewing and sharing. It has an asynchronous capability where sessions are recorded and can be viewed at a later date.

Blended Learning: Face-to-face (FTF) classes are meaningfully complemented, or fused, with web-based, online learning.

Counsellor Education: Involves students developing a comprehensive knowledge and understanding of diverse counselling theory and cross cultural practices.

Counsellor Professionals: Skilled counsellor practitioners employed in the field and in academia.

Digital Literacy: People's skills and knowledge in using a range of technological tools to search, use, create and evaluate information, construct new knowledge and interact with multiple platforms and formats.

Higher Education Institutions: Institutions that provide post-secondary education. They are also commonly referred to as universities or colleges.

Learnscape: A collaborative working and learning environment where teaching academics and students share ideas, develop and construct meaning, and consolidate their skills and understanding.

Offshore Campus: Higher education universities offering courses and programs to students studying at an international site.

Student Engagement: The time and energy devoted to working with and through academically purposeful activities.

Teaching Academics: Skilled teachers, educators, subject coordinators, lecturers who work in higher education institutions where teaching university students is a focus.

Section 2
Mobile Learning

Chapter 5
Improving Teaching Practice in Early Childhood Supported by Mobile Technology

Kathryn MacCallum
Eastern Institute of Technology, New Zealand

Heather R. Bell
Massey University, New Zealand

ABSTRACT

This chapter discusses the findings of an ethnographic case study investigating the implementation of mobile learning at an early childhood centre in Hawkes Bay, New Zealand. The study describes how mobile technology is being used to support children's learning and communication. The findings show that the devices are an integral part of the learning culture of the centre. The devices are being used to actively engage children in the learning environment and support teaching inquiry. As one of the early studies to investigate how mobile technology is being used in early childhood education, the current study provides pedagogically sound examples and insight on how mobile technology can be embedded into early childhood. The study is seen as a starting place for more in-depth investigations into the impact of mobile learning on young children's learning.

INTRODUCTION

Technology can be used to facilitate, share and transform the activities, roles, and relationships, with children and educators in early childhood education setting (Bolstad, 2004). Mobile devices such as iPads, have become pervasive in all areas of our daily lives and their use, in education, is becoming increasingly more common. Mobile technologies offer the potential to enhance and support children's learning and development. They also offer the potential to positively influence current pedagogy and curriculum (Bird & Edwards, 2014).

The role of mobile technology to support teaching and learning is an emerging area of interest for educational researchers. Traditionally however, most studies into mobile learning have focused on the

DOI: 10.4018/978-1-5225-0359-0.ch005

secondary or tertiary setting (Wu, Wu, Chen, Kao, Lin & Huang, 2012). These findings, however, do not necessarily translate well into early childhood sector. The role of mobile technology within early childhood education provides a unique set of challenges, due to the conventional focus of children's active play and exploration (Bird & Edwards, 2014).

Recent studies, within the early childhood sector, have started to highlight that the portability of mobile devices has provided educators with a way to enhance existing teaching practice and support young children's learning (Fagan & Coutts, 2010; Khoo, Merry, Nguyen, Bennett & MacMillan, 2013). Mobile technology has been shown to support collaborative learning, creative play, and teacher documentation, and help to build relationships with children and the wider community. However there has also been some debate on the efficacy and value of using mobile technology with young children (Verenikina & Kervin, 2011). There have been concerns that these devices may remove the focus away from active learning and may ultimately be harmful to children (see Blackwell, Lauricella & Wartella, 2014 for a discussion on some of these factors.) Continued research into mobile technology, with regard to its effective application and implications on play and pedagogy, is needed to inform the early childhood community. More research is particularly needed in providing pedagogically sound examples of effective use (Bird & Edwards, 2014, McManis & Gunnewig, 2012).

Effective use of mobile technology within education should be embedded within an appropriate pedagogy. An inquiry-based pedagogy enables teachers to asses and react to a child's learning. This approach encompasses a cyclical and recursive orientation to learning (Harwood, Bajovic, Woloshyn, Di Cesare, Lane & Scott, 2015.) Teachers need to be able to identify opportunities for active learning and engage and support children's learning through play (Hedges, 2007.) It is the role of the educator to be able to effectively embed this approach to ensure that learning is supported and not negatively impacted when incorporating and using mobile technology. There is, however, little literature on how mobile technology can be embedded into teaching and learning within the inquiry-based framework (Harwood, et. al., 2015.) In this chapter, we explore the results of an ethnographic study of an early childhood classroom which has adopted a Teaching as Inquiry approach (Timperley, Wilson, Barrar & Fung, 2007) to support good practice with mobile technology. The chapter examines how the Teaching as Inquiry model can be used to support effective teaching and learning outcomes when implementing mobile technology in the early childhood sector.

This book chapter has been split into three parts. Firstly, there is an examination of the literature on the pedagogy and practice that supports effective teaching and current use of mobile devices and how they have been embedded into early childhood. The chapter then outlines and examines how mobile technology has been adopted in the current case to improve the learning outcomes of the children at a kindergarten. Lastly, the article concludes with advice and the implications for teaching and teacher education. Specifically, this section focuses on the impact and benefits that mobile technology will have on early childhood learning and the implications on teaching and learning in the early childhood sector.

PERSPECTIVES ON CHILDREN'S LEARNING

Early childhood education services in New Zealand manage supervised education and care programmes for children aged from birth to five years (Ministry of Education, 1996.) A subset of this provision is the kindergarten system that covers the care of children from three to five and focuses on preparing them for school. The kindergarten system employees only fully qualified practitioners as part of their staffing.

Te Whāriki is the New Zealand early childhood curriculum framework and is followed by all kindergartens (Ministry of Education, 1996.) This framework is underpinned by sociocultural theory developed to support young children's learning. The framework is based on the theories of Vygotsky, where it is acknowledged that children learn through shared relationships and activities with others. This learning is reinforced through play (Brooker & Edwards, 2010.) The principles of the curriculum emphasise the importance of relationships, empowerment, holistic development and family and community. These are woven across strands of wellbeing, belonging, contribution, communication and exploration to construct the curriculum (Hedges, 2007.) The role of the educators is to make pedagogical choices to support and encourage these principles and strands and support learning through children's play.

Supporting young children's learning has resulted in a complex set of interrelated educational theories and pedagogies (see Wood & Attfield, 2005 for a discussion of most popular of these theories.) In general, most of these theories have three important components at their root, namely; the role of the educator, the children's peers, and cultural tools, activities and events. Each of these components helps to support, shape and contribute to the learning and build an appropriate learning environment.

The learning theories that are underpinned by the sociocultural perspective focus on supporting children's inquiries and interests (Hedges, 2007). In particular, the role of the adult is acknowledged. The adult is seen as playing a vital role in supporting children's curiosity and enabling them to explore their own environment. As highlighted by Hyson (2004) adults need to be able to maintain the motivation of the children, introduce them to new experiences, sustain their exploration and provide emotional security to encourage explorations. In addition, it is the adult that choses and develops the learning environment and tools that support the learning. Cultural tools (e.g., language, symbols, resources and equipment) will play a mediating role in children's learning (Vygotsky, 1978, 1986.) It is the role of the adult to ensure that the learning is authentic and relevant and help to build learning into context. This learning is considered more effective if it is done within the context of the child's interests (Bolstad, 2004.)

In addition to the vital role of the educator and the supporting tools, the children's peers also play a fundamental role in supporting and maintain learning (Hedges, 2007.) Children help and encourage each other to explore and co-construct shared meaning. This inter-relationship can help extent learning and further build the range of learning opportunities.

TEACHER KNOWLEDGE AND INQUIRY

Children are capable of learning quickly and enthusiastically. However, this learning is highly dependent on the knowledge and skills of the adult to support and encourage the learning (Hedges, 2007.) The role of the educator is to draw on appropriate teaching skills and experiences to control and manage the children's learning experience. Teachers need to draw on evidence-based approaches which are supported by teacher inquiry to help evaluate the learning situation and learner's needs (Dana & Yendol, 2003.)

Effective teaching practice is an ongoing process and is influenced by the teacher's experiences and professional learning (Campbell, 2002.) Supporting the development of effective practice requires effective reflection and willingness to adopt and change teaching approaches (Dana & Yendol, 2003.) Reflective practice helps support teacher inquiry and helps the teacher to identify and adjust their approach to best enhance the learning experiences. According to Loughran (2002) reflective practice represents a powerful tool for teacher development and remains a vital component of teaching practice.

Supporting effective reflection builds on capturing and critically examining evidence to identify learning and learning opportunities. Evidence-based approaches in education focus on collecting evidence of learning and assessing this to identify how teaching practice could be modified to better assist and extend children's learning (Timperley, 2010.) According to Timperley (2010, pg 1) this approach involves the integration and interpretation of several kinds of evidence:

- Formal and informal curriculum-relevant evidence of teachers' own students' learning that is seen as a source of information to improve teaching and learning.
- Evidence of teachers' own knowledge and skills so that the evidence about students can be interpreted in terms of the implications for teaching practice.
- Knowledge of the research and other evidence relevant to specific practices to inform further directions and changes to practice in response to the evidence.

Using an evidence-based practice approach to support teacher inquiry can help enhance teaching and help teachers to better reflect on their teaching practice. This approach requires the teacher to critically reflect and examine their own teaching practice and interpret how it could be improved to achieve better outcomes. This examination often requires in-depth conversations with others to integrate evidence from the child, teaching practice and research (Timperley, 2010.) The role of teacher inquiry often incorporates engaging teachers in knowledge communities to challenge and debate appropriate practices (Jaworski, 2006.)

In New Zealand early childhood settings the prime mechanism of documenting and assessing children's learning is through Learning Stories (Blaiklock, 2008.) Learning Stories as developed by Margaret Carr (2001), is a widely used technique to assess children's learning in early childhood centres in New Zealand. These Learning Stories are especially well suited to align with the Te Whāriki curriculum framework (Ministry of Education, 1996.) Learning Stories are designed to support teachers and young children's learning opportunities within a sociocultural context (Blaiklock, 2008.) They are typically written narratives that describe what teachers see when observing their children in particular situations. The development of learning stories involves teachers observing children and then writing a story that describes and interprets the learning of the child (Carr, May, Podmore, Cubey, Hatherly & Macartney, 2002.) These stories are shared with the child and family as a way of showing the child as a learner. Learning stories have their foundation within sociocultural theory and emphasise the importance of children's learning being embedded in their environment and social practice and activity (Blaiklock, 2008.)

THE ROLE OF MOBILE TECHNOLOGY IN EARLY CHILDHOOD EDUCATION

The use of information and communication technologies (ICT) has continued to increase in a wide number of early childhood programmes (Donohue, 2003.) In a study by Bolstad (2004, pg vii) she identified three reasons why ICT matters in early childhood education.

- ICT already has an effect on the people and environments that surround young children's learning.
- These technologies offer new opportunities to strengthen many aspects of early childhood education practice.

- There is support and interest across the whole education sector for the development and integration of ICT into education policy, curriculum, and practice.

Bolstad (2004) however, emphasises the importance of making sure that any integration of ICT in early childhood education is clearly grounded in the understanding of the purposes, practices, and social context of early childhood education.

The successful integration of technology into young children's learning depends on the teacher (Ertmer, 2005.) The teacher must select the most appropriate tool, and understand how it can be best integrated into the teaching and learning practice. Teachers need to understand how these tools can support children's learning, development, and play (Bolstad, 2004.) Teachers need to be familiar with these various tools and understand the various affordances that each can bring to the learning space. They also need to be aware of and be able integrate appropriate teaching and learning pedagogy to ensure effective practice.

Mobile technology has recently emerged as uniquely suited to support children's learning, development, and play within the early childhood environment (Fagan & Coutts, 2010; Bird, 2012; Khoo, Merry. Nguyen, Bennett & MacMillan, 2013.) These mobile devices can provide educators with a tool to both support and initiate play. Due to the portability of these devices, they afford a number of important benefits to educators to reflect, communicate and build relationships with children and the wider community.

Researchers and educators have both seen mobile technology as being uniquely situated to encourage children's interests and support teacher inquiry (Harwood, et. al., 2015; Bird, 2012; Kearney, Schuck, Burden, & Aubusson, 2012.) Mobile technology can help support and stimulate children's curiosity and support their engagement in the learning environment as an individual and as a group.

The number of studies investigating the integration of mobile technology in early childhood education has been small. However there have been a small number of studies providing examples of how early childhood centres are using these devices to support learning. These examples illustrate mobile technology use in the hands of the educators and the children themselves.

An early study by Fagan and Coutts (2010), describes a small-scale investigation into how two early childhood centres utilised iPads to support early childhood education. Their study highlighted a variety of key benefits that these devices have on children's learning. These benefits included the ability to support collaborative learning, support creativity, support teachers with their documentation and assessment of children's learning and support communication with parents and caregivers. One example of iPad use in the study was the projecting of the iPad screen onto a big screen where the children were encouraged to work on 'projects' together. The iPads were also used to document and assess children's learning through video, photos and written text. The portable nature of the devices also enabled them to be used to capture children at play. These recordings could then be shared with the children and their families. These recordings also provided teachers with an opportunity to foster and build relationships with the children's families by supporting immediate communication and sharing.

A later study by Khoo et al., (2013), showed some of the same benefits. Their study, however, focused on the affordance that mobile devices had on building Learning Stories. Learning Stories in the two early childhood centres in this study were used to expand on the ways children's learning can be documented and reported. They found that video and voice could be used to make the learning stories more engaging and illustrative.

STUDY BACKGROUND

In 2012, teachers at a kindergarten in Hawke's Bay, New Zealand, were involved in a three-month trial to investigate the potential of mobile devices within an early childhood setting. The kindergarten follows the Te Whāriki Curriculum framework (Ministry of Education, 1996) and focuses on quality early childhood education for children between the ages of three and five years of age. The kindergarten has four trained and registered early childhood teachers. They occasionally have student teachers at the centre, training to become early childhood teachers.

The kindergarten in the trial was given two smart devices (one device between two teachers.) The devices were not given to the children; rather the aim of the trial was to explore how the devices could be used to support teaching practice. The trial included regular professional development provided by a learning facilitator from Massey University (See Bell & Van Dillen, 2013, for details and outcomes of this trial.)

The teachers and facilitator met initially to scope and plan the project to ensure it aligned with their perspectives on children's learning. The trial itself consisted of four onsite visits where the facilitator observed and worked with teachers and children during the sessions, followed by after-session teacher professional development meetings. The aim was to use the mobile devices to view assessment in new ways and to see the impact on teacher practice. Two key needs for professional development emerged during the project. Initially, teachers needed skill-based learning around the effective management and use of iPads, however the teachers quickly became independent problem solvers as they gained confidence. The rewindable nature of iPad evidence (Spencer, Coutts, Fagan and King, 2013) gave different ways of looking at both student assessment and pedagogical practice in early childhood environments so the need to recognise teacher impact on learning and the implications for teaching practice emerged.

During the trial the devices were used to create two key outcomes: ebooks that captured students' drawings, photos and voices, and videos that showcased children's activity. When creating ebooks, the children would firstly draw pictures or take photos on the devices. They would then tell stories to go with their images and these oral stories were recorded and then written as captions by their teachers. These would be shared with other teachers, children, parents and extended family. Both outcomes were used to identify teaching and learning practices during the professional development meetings with the facilitator.

A Teaching as Inquiry (Timperley, Wilson, Barrar & Fung, 2007) approach was adopted to ensure the use of the devices was embedded in good teaching practice (Figure 1.) This organising framework helps teachers focus on the evidence, explore their own practice, develop next steps for themselves and their learners, and reflect on the impact of changes, growing their pedagogical understanding and knowledge as a result.

Each teacher identified a focus child and used the iPad to capture and create evidence of teaching and learning practice around this child. The learning facilitator and the teachers worked collaboratively to identify what the child could do as shown by their evidence and then determine their potential next learning goals. Strategies were then developed to effectively use the iPads to support these next learning goals. In each weekly feedback session, teachers shared what they had done as a result, their evidence and their assessment of it. This process was repeated and developed as they learnt more and strengthened their practice. Teachers all contributed to decisions about each other's next steps in their practice and each focus child's learning. This brought the collective knowledge of the team into the decision process and enriched the critical thinking that occurred.

Figure 1. Teaching as inquiry cycle
Source: Timperley, Wilson, Barrar & Fung, 2007 pg. 1.

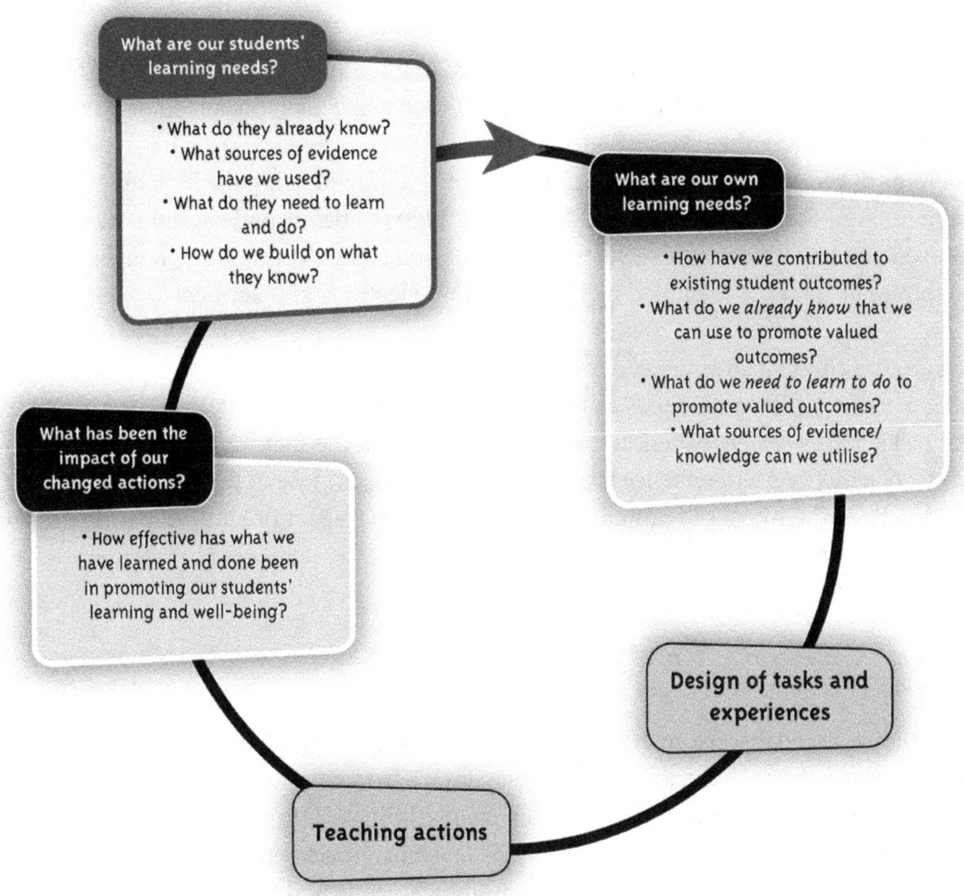

CURRENT STUDY APPROACH AND METHODOLOGY

This current study is based at the same early childhood centre as the earlier trial and addresses how the Teaching as Inquiry model has been used to develop and extend effective practice at the centre.

This study takes the form of an ethnographic case study. This approach was taken as it allowed the in-depth study of a specific culture, in this case the kindergarten, through close observation and interpretation. The aim of this study was to develop an in-depth narrative case study of the specific actions of the teachers and children. Traditionally, full-fledged ethnography typically demands long-term engagement in the field, however ethnographic case studies can be conducted over shorter spans of time to explore narrower fields of interest to help generate hypotheses (Parthasarathy, 2008.) Early childhood researchers have conducted ethnographic research in a wide range of areas (see Mac Naughton, Rolfe & Siraj-Blatchford, 2010 for a detailed discussion.)

Using a qualitative approach, the researchers visited the kindergarten to observe how teachers were using the devices and how the children were interacting with the devices. The field research involved informal interviews and direct observation of how the four teachers at the kindergarten were interacting with the children with the mobile devices. From this, data was collected through field notes and recordings (photos, audio and video.) To supplement these field observations, semi-structured interviews were conducted with the teachers, focused on refining or contextualising the observations and asking the teachers to provide examples of their use of the device to elicit their observations of how they used the devices. Additional data collected were copies of children's artefacts produced as part of the teaching and learning process using the devices.

The findings of this research were framed within and examined using the Teaching as Inquiry Cycle. The interpretation of the results within the framework helps to better understand how the full cycle has been employed and how the mobile devices have been used to support the full engagement of the teaching philosophy. This process is guided by the following questions and actions:

- What are the child's learning needs? (Learner evidence).
- What are our own learning needs as teachers? (Teacher practice evidence).
- What tasks and experiences will the learner have? (What the learner will do differently).
- What teacher actions will enhance the learner's experiences? (What the teacher will do differently).
- What has been the impact of our changed actions? (The impact or shift in learning resulting from different experiences and actions).

FINDINGS

The current study showed that the smart devices had become an embedded part of the kindergarten teaching philosophy. From the earlier trial there was one device between two teachers, now each teacher had their own iPad and they were carried by the teachers at all times. The devices had become a ubiquitous tool to immediately capture and evaluate children's learning stories. They also provided teachers with an opportunity to collaboratively reflect on the children's needs and to assess how to best support the child's needs.

From the analysis of the data collected, there was clear evidence that the Teaching as Inquiry model was being actively applied (see Figure 1.) Based on the data, two stories/exemplars were identified. The first story revolves around the child as a learner and how there was a shift in her learning as a result of the teacher's actions. The second story focuses more on the changes in teacher practice. This story illustrates how teaching practice evolved as a result of the teacher's inquiry into her impact on the learning environment.

Story 1: Lucy and the Lambs

Learner Evidence That Informed This Story

Sometimes the kindergarten had animals come and stay at the centre and, at the time of this story, the latest visitors were a couple of lambs. Generally the children were intrigued with the animals and were happy to pat and feed them, but one little girl, Lucy, showed great reluctance to go near the lambs. A few days after the lambs arrived, the teachers noticed that Lucy was getting closer to the lambs and showing a bit more interest.

Teacher Practice Evidence

Lucy's teacher decided to take this opportunity to encourage Lucy to get a bit more comfortable with the lambs. Teachers and children at this centre often use the smart devices to make books using pictures of children exploring their environment in creative, imaginative ways. These books are shared with groups to showcase how different children engage with their environment. Children have also been encouraged to create their own books, enlisting the help of their teachers to get their words onto their pages. Lucy was asked if she would like a book about her and the lambs and as she was familiar with this and comfortable with her teacher taking photos, she said yes.

Learner Experiences and Teacher Actions

As a result of her new interest and growing confidence, Lucy's teacher took pictures of her with the lambs and helped her create a story using an ebook creator application. Lucy chose the images and her teacher wrote the story with Lucy's input. Lucy also recorded a few words and sounds, such as "baa" to illustrate the images. The teacher then emailed the book to Lucy's parents who responded later that day, pleased to see what Lucy had been up to.

Impact: Shift in Learning Resulting from Different Actions

When Lucy returned to the centre the following day, she told her teacher that she had a new idea for a story. She used the smart device and started taking some photos of the lambs. She then asked her teacher to help her put the story together. Lucy recorded her voice retelling the story and the teacher placed her words into text. Lucy rewound her voice and frequently re-recorded her words until she was happy with how they sounded.

Once again the story was emailed to her parents, who were particularly pleased at how well Lucy's voice came through on the story this time. They noticed a clear progression in her use of language; from single words and sounds to clear sentences.

New Learner and Teacher Practice Evidence

Over the following few days the children had been reading "The Littlest Bird". Lucy once again came to her teacher to ask to use the device to create a story, this time she decided to make an ebook version of this story.

Learner Experiences and Teaching Actions

Instead of creating another book on her own, Lucy's teacher encouraged her to include one of the other boys at the centre, Lucas, to help her. Together, and with the aid of a stuffed chicken, they staged the various images for their story then recorded their voices to go with the pictures.

Impact: Shift in Learning Resulting from Different Actions

This time, Lucy acted more as a director and divided the speaking parts, based on the original story, between Lucas and herself. Lucy knew by now how to record her voice so together they retold the original story almost verbatim. In the recording of the ebook Lucy encouraged and supported Lucas to speak more confidently. To achieve their outcome, Lucy and Lucas needed to understand the stages of the story and to be able to recreate these in their own imaginative ways. They also had to share the tasks of director and actor, know how to use the application efficiently, and know the story well enough to be able to retell it accurately. Lucy had stepped back and supported another child to learn something she was now capable of doing on her own. This was a big step for a normally quiet, reserved child.

This time the ebook was emailed to both sets of parents and also shown at the end of the day to all the children at the centre as they waited for parents to collect them.

Story 2: Superman and Super-Kids

Learner Evidence That Informed This Story

Two boys had been observed showing a high degree of interest in superheroes for a couple of days. The boys made great use of such things as capes and props that were easily accessed and acted out situations that showed their super powers. Their conversations involved language and thinking that mimicked common perceptions of superhero behaviours. However, other children in this group were not role playing to the same degree as these two boys.

Teacher Practice Evidence

Their teacher decided to use this interest to engage her children in new learning. Having clothing and props available that could easily be adapted to suit this role had encouraged the boys to play in this way, but she knew that using the device could potentially grow their learning further. Through sharing their stories, she knew she could provide new opportunities for learning for the rest of her group.

Learner Experiences and Teacher Actions

The teacher decided to create an ebook about superheroes with her group. She started by downloading a range of superhero images. She knew that storytelling strongly supported early literacy development, so she encouraged all the children to develop a short story to go with each image, focusing mainly on the characters. She recorded each child talking about the pictures; the name or the character or something about the character, then transcribed their words into captions. The ebook was then shared with the whole group and was also printed off and available as a hard copy.

Impact: Shift in Learning Resulting from Different Actions

When the ebook was shared with all the children in the group, it sparked considerable interest and generated lively discussions as each child heard themselves talking. As a printed copy, children were able to read it on their own. Also, because it used their own words, they could retell the story using the same or similar language as was written. They were often observed finger following text, displaying initial concepts of text direction and word chunks. The teacher was also able to use it as a group story to prompt rich discussions, getting children to suggest how they could engage with each other in play in new ways.

Learner Experiences and Teacher Actions

Based on the success of the original ebook, the teacher decided to use the impetus to further engage her children but this time through active play. She set up an obstacle course comprising rope swings, monkey bars and high jumps onto mats. Each child then played a superhero of their choice - with a cape wrapped around their shoulders - and went through this course. The teacher recorded their action on her device then created movie trailers by putting together all the actions in a movie.

Impact: Shift in Learning Resulting from Different Actions

The movie was played at the end of the day when caregivers picked up their children. The movies played two roles at the centre; they allowed the children to see themselves and their peers in action, and they initiated conversation between the children and their extended family. Caregivers were able to watch the day's events as the movies provided an insight into what been happening during the day. The movies also created genuine opportunities for the teachers to have rich discussions with the caregivers, opening the door for the caregivers to talk to their children about what they had seen them doing.

Impact: Shift in Teacher Practice

These ebooks and movies were being used as rich evidence of learning and were a very important mechanism for teacher reflection. Having the opportunity to watch the stories at a later time enabled the teachers to notice, reflect and respond more directly to their own and the child's strengths. The teachers were able to identify the needs of the children and pick up learning cues that may be have been missed earlier. Since these recordings were rewindable and did not change as oral versions of stories tend to do, they provided valid and reliable evidence of children's learning that could be shared with the other teachers at the kindergarten (Spencer, Coutts, Fagan & King, 2013.)

The benefit of this was that through working collectively, the teachers could view the ebooks and movies and bring new perspectives to the understanding of the actual learning being demonstrated and the opportunities for next steps. Four different teacher lenses viewing the same material saw a significantly wider set of outcomes. The teachers were able to bounce off one another and, by bringing in their broader knowledge, the synergy created was huge. All the teachers were able to see learning in new ways and to have a stronger awareness of their colleagues' knowledge, which they could then tap into. Accessing this would help the children across the centre, as all teachers grew their own pedagogical knowledge.

DISCUSSION

The findings of this study clearly show that the smart device usage at the kindergarten had not stalled since the earlier trial. The devices had now become a ubiquitous tool for capturing and evaluating children's learning. They were providing teachers with an opportunity to effectively reflect on and support the children's needs. Teachers were sharing their videos with the children at the time they were taken and rich conversations were being held with the children. These conversations focused on what the children were able to do and where the children would themselves suggest what they could do tomorrow. Learning stories (Carr, 2001) were being created on the spot with the children. These were less time consuming and more meaningful to the teachers and the children than the paper versions teachers had been creating. Sharing these videos also raised the quieter children's profiles when their peers saw them doing different and exciting things. The devices also supported communication outside the playground. The videos and ebooks provided a great springboard for family conversations.

The two stories outlined in this study each provided a slightly different viewpoint into the influence of the teacher inquiry cycle. The first story reinforced the impact that the teacher's practice had on shifting the learning of the child. The teacher in that story supported and guided the child to expand on her own experiences and knowledge. The second story strongly emphasised the process of the reflection in teaching practice, of both the teacher as an individual, and later, collaboratively as part of a group. The reflection helped to evolve the teaching practice as the teacher evaluated and developed her actions and enhanced the learning experience for the child.

These two stories highlight the three components of effective teaching and learning pedagogy (Hedges, 2007.) The teachers in the stories each support and encourage the learning. The teacher manipulates and extends the learning activity and, with the use of the framework, they reflect and develop the learning of the child. The role of the teacher was to help the children explore their own interests. The approach was also seen in Fagan and Coutts (2010), where the teachers used technology to spark the learning. In general, in this study, learning was initiated by the teacher but the momentum and resulting learning was extended by the children. This was especially marked in the first story about Lucy.

Each story focuses strongly on the role of play and interaction with other children. Though the first story involves only Lucy at the beginning of the story, later the story evolves to also include another child who interacts with Lucy. The role of the other child not only extends Lucy's experiences it also develops and extends the other child's learning. The role of the other child was vital and important in each of these stories. In the first story, Lucy was able to build new verbal and relationship skills and develop her own skill with the tool to teach the other child how to use the device. The other child was also able to learn how to use the new tool. In this story Lucy was both the learner and the teacher. In the second story the interests of two children were exploited, however it was used to spring board the interactions of a whole group. Each story continued to also incorporate the wider community, whereby the parents and caregivers of the children were also engaged and involved in the learning.

The role of the iPad in each story was to support the learning. The technology did not take precedence and was the not the focus of the story. The same learning could have taken place in more traditional learning through paper and pen, however the multimedia aspect of the iPads have enhanced the learning. The iPads enabled the learning to be easily distributed, such as being emailed to the parents of Lucy and Lucas, or watched as a group of other children and their parents and caregivers. The iPads helped to make the learning dynamic and extend the learning into new ways not necessarily possible using traditional methods.

The iPad also helped the teacher to record evidence of the child's learning. The uses of the iPads were similar to previous studies by Fagan and Coutts (2010) and Khoo et al., (2013) where teachers used the iPads to support their documentation and assessment of children's learning. As highlighted by Timperley (2010) an evidence-based practice approach to teacher inquiry can help support teachers to better reflect on their teaching practice. The iPads provided a permanent record of the learning activities and could be viewed at a later date. By reviewing the recordings, it helped the teachers to better reflect and recall what had happened. It allowed the teachers to pick up on small indicators that may have been otherwise missed. The recordings were also used to support team communication to help draw out potential further learning opportunities, previously shown as an important factor in supporting effective reflection (Jordan, 2003.)

IMPLICATIONS FOR LEARNING OUTCOMES AND TEACHING PRACTICE

New technology is becoming more ingrained into children's lives. The integration of new technology into teaching and learning has resulted in a learning environment that is drastically different from how the children's parents and grandparents used to learn. There is a rich body of research on how young children grow, learn, play and develop (Radich, 2013.) However, there is still very little on how especially mobile technology can be effective integrated into the early childhood curriculum (Harwood et al., 2015.)

The above stories, each examined within the framework of the teaching as inquiry cycle, helped demonstrate how a considered integration of mobile technology could be embedded into early childhood curriculum. The cycle helped the teachers to examine the needs and the learning experience of the child and the impact of their teacher actions on the child's experiences and learning. The stories each demonstrated some attempt to develop the full cycle of the teaching as inquiry framework.

Radich (2013) has identified a number of principles to guide the appropriate use of technology and interactive media as tools in early childhood programs serving children from birth through age 8. The principles are clearly evident in the above stories.

- The use of the teaching as inquiry framework has supported and encouraged appropriate practices. The inquiry has a considered approach and teachers are clearly challenged to consider whether and when to integrate technology and interactive media into early childhood programs.
- The teachers are required to make a judgment to determine if and when a specific use of technology or media is age appropriate, individually appropriate, and culturally and linguistically appropriate. In the cycle they also need to consider if the approach worked and what could be done to enhance the learning next time.
- The use of the devices is generally active, hands-on, engaging, and empowering. When the device is in the child's hands the child is given control and it is used to provide adaptive scaffolds to ease the accomplishment of tasks and they are used as one of many options to support children's learning.
- Each story outlines how the devices were used to enhance children's cognitive and social abilities. The recordings of the children at play were used to support interaction with the other children and their families. The technology was used to help educators make and strengthen home–school connections.
- The interactions were playful and support creativity, exploration, pretend play, active play, and outdoor activities.
- Digital literacy was essential in this case study to guide the early childhood educators in the selection, use, integration, and evaluation of technology and interactive media. The initial trial helped set up the teachers to be comfortable with the devices they then extended their own learning through exploration and peer support.

When integrating technology into early childhood programs, educators must be aware of both the challenges and the opportunities of the particular technology (Radich, 2013.) A considered approach is needed to support the intentional leveraging of the potential of mobile technology for the benefit of every child. In this case study the adoption of the Teaching as Inquiry Cycle has helped teachers to be critical and aware of how they are engaging with the children. The technology does not initiate the learning but is used to engage and support the child's exploration. The cycle can help support the teachers to be critical in their use of the devices and help determine when it is appropriate.

The effective use of mobile technology, as framed within the cycle, does highlight the need for training with the technology and the cycle. Teachers will need support in developing their technology and media knowledge, skills, and experiences. Additional professional development opportunities will be needed to share effective practice and support the teachers' explorations with the devices.

CONCLUSION AND FUTURE RESEARCH

A growing body of literature is emerging which has started addressing how mobile technology can be effectively harnessed in the early childhood sector (Harwood, et. al., 2015; Spencer, Coutts, Fagan, & King, 2013; Khoo, et al., 2013; Verenikina, & Kervin, 2011.) However its true value and effective integration is still being determined. Future research is needed to encourage and develop this field. More research is needed to help determine the complex set of factors that will influence the successful integration of mobile learning. The factors that need to be explored include determining the content and design of the apps used, institutional and policy contexts and the quality of the digital interactions (Harwood et. al, 2015.)

Each of these stories highlighted sound pedagogical practice and how that impacted on child learning. The uses of the devices were well conceptualised and clearly identified effective practice that was grounded in appropriate pedagogy. The findings here challenge some of the fears raised earlier (Blackwell et al., 2014) about the potential for such devices to be harmful to children through taking the focus away from active learning. Properly used, the opposite appears to be the case in this situation as teachers are using them to motivate and encourage 'active learning' and exploration.

Despite being a small study, the results provide solid groundwork for showcasing effective integration of mobile technology into early childhood settings. This study provides some pedagogically sound examples of how such devices could be used to support effective teaching and learning within the early childhood sector. It is anticipated that a wider study will be initiated based on the findings of this study, involving a wider set of early childhood centers to help evolve and develop further examples of effective mobile technology practice. Further studies will also address the role of the Teacher Inquiry Cycle. Specifically the study will investigate whether the good pedagogical practice identified here was the result of adopting the model, the use of the smart devices in new ways, or a combination of both.

REFERENCES

Bell, H., & Van Dillen, F. (2013, May). iPads supporting good teaching practice. *New Zealand Education Gazette*, *92*(7), 2–3.

Bird, J. (2012). *The rabbit ate the grass! Exploring children's activities on digital technologies in an early childhood classroom.* (Unpublished doctoral dissertation). Australian Catholic University.

Bird, J., & Edwards, S. (2014). Children learning to use technologies through play: A digital play framework. *British Journal of Educational Technology*. doi:10.1111/bjet.12191

Blackwell, C., Lauricella, A., & Wartella, E. (2014). Factors influencing digital technology use in early childhood education. *Computers & Education*, *77*, 82–90. doi:10.1016/j.compedu.2014.04.013

Blaiklock, K. (2008). A critique of the use of Learning Stories to assess the learning dispositions of young children. *New Zealand Research in Early Childhood Education*, *11*, 77–87.

Bolstad, R. (2004). *Role and potential of ICT in early childhood education: A review of New Zealand and international literature*. Wellington, New Zealand: Ministry of Education. Retrieved November 12, 2014, from http://www.educationcounts.govt.nz/publications/ict/4983

Brooker, L., & Edwards, S. (Eds.). (2010). Engaging Play. Maidenhead, UK: Open University Press.

Campbell, A. (2002). Research and the professional self. In O. McNamara (Ed.), *Becoming an evidence-based practitioner: A framework for teacher-researchers* (pp. 27–36). London: Routledge/Falmer.

Carr, M., May, H., Podmore, V. N., Cubey, P., Hatherly, A., & Macartney, B. (2002). Learning and teaching stories: Action research on evaluation in early childhood in Aotearoa-New Zealand. *European Early Childhood Education Research Journal*, *10*(2), 115–125. doi:10.1080/13502930285208991

Dana, N. F., & Yendol-Silva, D. (2003). *The reflective educator's guide to classroom research: Learning to teach and teaching to learn through practitioner inquiry*. Thousand Oaks, CA: Sage.

Donohue, C. (2003). Technology in early childhood education. *Child Care Information Exchange*, *154*, 17–22.

Ertmer, P. A. (2005). Teacher pedagogical beliefs: The final frontier in our quest for technology integration? *Educational Technology Research and Development*, *53*(4), 25–39. doi:10.1007/BF02504683

Fagan, T., & Coutts, T. (2012). *To iPad or not to iPad?* Christchurch, New Zealand: CORE Education. Retrieved November 3, 2014, from http://www.core-ed.org/thought-leadership/research/ipad-or-not-ipad

Harwood, D., Bajovic, M., Woloshyn, V., Di Cesare, D. M., Lane, L., & Scott, K. (2015). Intersecting Spaces in Early Childhood Education: Inquiry-Based Pedagogy and Tablets. *The International Journal of Holistic Early Learning and Development*, *1*, 53–67.

Hedges, H. D. (2008). *Funds of knowledge in early childhood communities of inquiry*. (Unpublished doctoral dissertation). Massey University, Palmerston North, New Zealand.

Hyson, M. (2004). *The emotional development of young children: Building an emotion-centered curriculum*. Teachers College Press.

Jaworski, B. (2006). Theory and practice in mathematics teaching development: Critical inquiry as a mode of learning in teaching. *Journal of Mathematics Teacher Education*, *9*(2), 187–211. doi:10.1007/s10857-005-1223-z

Kearney, M., Schuck, S., Burden, K., & Aubusson, P. (2012). Viewing mobile learning from a pedagogical perspective. *Research in Learning Technology*, 20.

Khoo. E., Merry. R., Nguyen, N.H., Bennett. T., & MacMillan. N. (2013). Early childhood education teachers' iPad supported practices in young children's learning and exploration. C*omputers in New Zealand Schools: Learning, Teaching, Technology*, *25*(1-3), 3-20.

Loughran, J. J. (2002). *Developing reflective practice: Learning about teaching and learning through modelling*. Routledge.

MacNaughton, G., Rolfe, S., & Siraj-Blatchford, I. (2010). *Doing early childhood research*. Berkshire, UK: McGraw-Hill Education.

McManis, L. D., & Gunnewig, S. B. (2012). Finding the education in educational technology with early learners. *Young Children*, *67*, 14–24.

Ministry of Education. (1996). *Te Whãriki: Te Whãriki matauranga mo nga mokopuna o Aotearoa, Early childhood curriculum*. Wellington: Learning Media.

Parthasarathy, B. (2008). *The ethnographic case study approach*. Technology & Social Change Group, University of Washington. Retrieved from http://www. globalimpactstudy. org/2008/07/the-ethnographic-case-study-approach/

Radich, J. (2013). Technology and interactive media as tools in early childhood programs serving children from birth through age 8. *Every Child*, *19*(4), 18–19.

Spencer, K.M., Coutts, T., Fagan, T., & King, A. (2013). Connections, Diversity, Coherence: Three Vignettes Exploring Learning with iPads in Primary Schools. *Computers in New Zealand Schools: Learning, Teaching, Technology*, *25*(1–3), 38–55

Timperley, H. (2010). *Using evidence in the classroom for professional learning*. Paper presented at the Ontario Education Research Symposium, Toronto, Ontario, Canada.

Timperley, H. S., Wilson, A., Barrar, H., & Fung, I. (2007). Teacher Professional Learning and Development: Best evidence synthesis iteration. Wellington, New Zealand: Ministry of Education. Retrieved from http://www.educationcounts.govt.nz/publications/series/2515/15341

Verenikina, I., & Kervin, L. (2011). iPads, Digital Play and Pre-schoolers. *He Kupu*, *2*(5), 4–19.

Vygotsky, L. S. (1978). *Mind in society: The development of higher psychological processes* (M. Cole, V. John-Steiner, S. Scribner, & E. Souberman, Eds.). Cambridge, MA: Harvard University Press.

Vygotsky, L. S. (1986). *Thought and language* (A. Kozulin, Ed.). Cambridge, MA: MIT Press.

Wood, E., & Attfield, J. (2005). Play, learning and the early childhood curriculum. *Sage (Atlanta, Ga.)*.

Wu, W. H., Wu, Y. C. J., Chen, C. Y., Kao, H. Y., Lin, C. H., & Huang, S. H. (2012). Review of trends from mobile learning studies: A meta-analysis. *Computers & Education*, *59*(2), 817–827. doi:10.1016/j.compedu.2012.03.016

KEY TERMS AND DEFINITIONS

Early Childhood Education: Early Childhood Education provides education and care for children before they are old enough to go to primary school (birth to five years of age).

Ethnographic: A research method based on fieldwork that uses techniques such as observation, interviews and collection of artefacts to make sense of the issues being investigated.

Evidence-Based Approach: A process by which a range of information is used to help understand both what has happened and future directions for teaching and learning.

Inquiry-Based Pedagogy: Pedagogy based on constructivist theory, where learning is seen as a social process involving a mutual exploration of ideas through experiences and language.

Kindergarten: NZ Kindergartens are non-profit, community-based services managed by a regional Kindergarten Association. They cater for children between two and five years old and can have set morning and afternoon sessions for different age groups. They follow the Te Whāriki Early Childhood Curriculum.

Learning Stories: Learning Stories as developed by Margaret Carr (2001), is a widely used technique to assess children's learning in early childhood centres in New Zealand. They are written narratives that describe what teachers see when observing their children in particular situations.

Mobile Learning: Learning with and through mobile devices, such as tablets and smart phones that can be connected to the internet.

New Zealand Curriculum: The National Curriculum is composed of *The New Zealand Curriculum* and *Te Marautanga o Aotearoa* which set the direction for student learning and provide guidance for schools as they design and review their school based curriculum. They start with a vision of young people developing the competencies they need for study, work, and lifelong learning, so they may go on to realise their potential.

Sociocultural Theory: A theory that is based on the interaction between people and the culture within which they live. Vygotsky, the founder of this theory, strongly believed this interaction between child and culture - parents, caregivers and peers, was the foundation for higher order thinking.

Te Whāriki – Early Childhood Curriculum: Te Whāriki is the Ministry of Education's early childhood curriculum policy statement. It is a framework for providing children's early learning and development within a sociocultural context. It emphasises the learning partnership between teachers, parents, and families. Teachers weave a holistic curriculum in response to children's learning and development in the early childhood setting and the wider context of the child's world.

Teaching as Inquiry Cycle: Since any teaching strategy works differently in different contexts for different students, effective pedagogy requires that teachers inquire into the impact of their teaching on their students. Inquiry into the teaching–learning relationship can be visualised as a cyclical process that goes on moment by moment (as teaching takes place), day by day, and over the longer term.

Teaching Pedagogy: Teaching approaches, strategies and learning environments that are known to have a positive impact on student learning.

Chapter 6
Increasing Learner Interaction in Large-Scale Lectures by Using a Mobile Learning Application

Katja Lehmann
University of Kassel, Germany

Matthias Söllner
University of St. Gallen, Switzerland & University of Kassel, Germany

Jan Marco Leimeister
University of St. Gallen, Switzerland & University of Kassel, Germany

ABSTRACT

Universities face increasing numbers of students leading to increasingly large lectures, and decreasing interaction and feedback, which are important factors for individual learning success and satisfaction. The use of IT can help in overcoming this challenge by increasing the interaction in large-scale lectures without massively increasing the workload of lecturers. This paper presents the design, use and evaluation of a mobile-learning application aiming to increase the interaction in large-scale lectures and the success of learners. For designing the application, the authors follow a design science research approach. The authors rely on insights from interaction theory as well as requirements gathered from lecturers and students in a focus group workshop. With the implementation, large-scale lecture related limitations can be overcome and the results help lecturers to face the according challenges. The results emphasize the potential of IT for university teaching and provide transferable insights for practical use in other learning scenarios.

INTRODUCTION

Universities in many western countries face increasing numbers of students. As a result, growing numbers of learners in lectures and an unfavorable lecturer-to-students-ratio of up to 80 students per lecturer is a common situation. These large-scale lectures are characterized by high anonymity and suffer from a

DOI: 10.4018/978-1-5225-0359-0.ch006

lack of interaction - not only among learners themselves but also among learners and lecturers. Moreover collaborative learning is not feasible in traditional large-scale lectures, where learners are single learners (Lehmann and Leimeister, 2015). The results of this decreasing interaction are often deficient learning outcomes and unsatisfied learners. For example, comprehension questions regarding the lecture as well as discussions on specific topics are not feasible, as they are impractical. This development is alarming, since fundamental elements of learning success include the opportunity to ask questions and the possibility of sharing one's opinions concerning the subject matter (Picciano, 2002). Additionally, interaction and feedback are regarded as significant predictors in terms of the learning success (A. Moore et al., 1996) and positively influence the long-term satisfaction of learners (Alonso et al., 2009; Hardless et al., 2005). It is true that by being actively engaged in the learning-teaching process, learners will get a deeper comprehension of the subject matter (Evans and Gibbons, 2007). But bringing interaction in a large-scale-lecture is a widespread problem (Lehmann et al., 2015).

A promising possibility to increase the interaction without massively increasing the workload of lecturers is the use of IT. Several researchers have investigated the possibility of mobile technologies to improve the classroom situation (Ratto et al., 2003). With the aid of IT and mobile devices, interactive data can be transferred between students and lecturers in real-time, which provides potential for improving the interaction in lectures (Dyson et al., 2009; Wortmann et al., 2014), for intervening in the learning-teaching process and enriching traditional courses. The application of IT supported learning and teaching has increased recently in education (Johnson et al., 2012). The use of mobile devices is widespread. They are flexible in use, easy to use and allow synchronous and asynchronous communication (Lehmann et al., 2012). The current state of research shows that existing learning applications do consider only single types of interaction but not all three types of interaction, which are proposed by Moore (1989) and no learning application exists which consider interaction completely.

The goal of the present research is to develop a mobile-learning application to increase interaction in large-scale lectures. Particularly, by using the mobile-learning application, learners should interact more with the learning content, the lecturer, and among each other. Hence, all three interaction types proposed by Moore (1989) are addressed.

Therefore, this paper aims to examine the following research questions (RQ):

RQ 1: How should the mobile-learning application be designed to increase interaction in the learning-teaching process of a large-scale lecture?

RQ 2: Is the mobile-learning application perceived as helpful in terms of quality, usefulness and satisfaction?

RQ 3: How does the frequency of use of the mobile-learning application affect the interaction among the students in a university large-scale lecture?

In order to answer the research questions, the chapter is structured as follows. The second part of the chapter presents the research methodology. Afterwards the related work part on mobile-learning applications is described. Then, the authors rely on insights from theory of interaction in order to gather theoretical requirements and they conduct a focus group workshop to gather practical requirements. Accordingly, the sixth part describes the design and development of the mobile-learning application. Section 7 presents the results as answers to the second and third research questions, before section 8 discusses the relevant results. Finally, the chapter closes with a conclusion, the implications, as well as limitations and an outlook on future research.

Figure 1. Research approach for developing the mobile-learning application
Source: Adopted from Peffers et al. (2006)

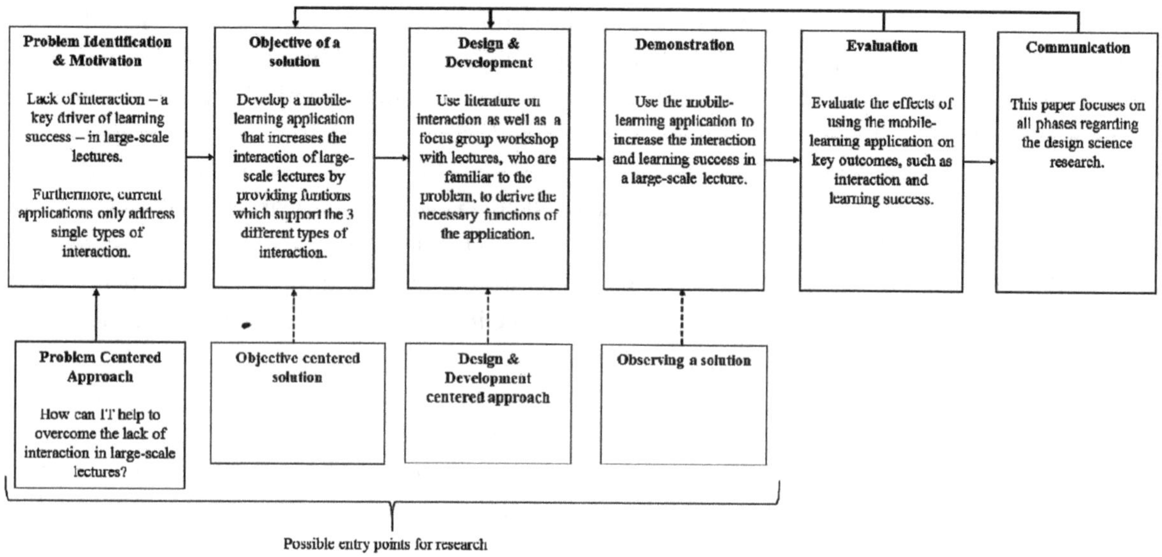

RESEARCH METHODOLOGY

The goal of the research is to develop a mobile-learning application which enables a robust and operational interaction between learners and lecturers as well as among the learners themselves in large-scale lectures. To achieve the research goal, the authors follow the design science approach (Hevner et al., 2004; Peffers et al., 2006), particularly the design science research approach of Peffers et al. (2006) (see Figure 1). Moreover, to ensure that the application addresses all important types of interaction, the authors follow Briggs (2006) theory-driven design approach, by grounding the research in theory on interaction. This paper presents details on all phases advocated by Peffers et al. (2006) for the development of the mobile-learning application. The introduction has addressed the phase problem identification and motivation. The next two sections describe the objectives of a solution phase by identifying requirements from theory of interaction and from a focus group workshop. The authors then provide details on the third phase, design and development of the mobile-learning application. Afterwards they present evaluation results which emphasize the demonstration and evaluation phase.

RELATED WORK

Mobile learning is a type of eLearning, which utilizes mobile devices, and makes learning flexible, spontaneous and portable (Kukulska-Hulme and Traxler, 2005). Hereby, the term *mobile* encompasses learning with the aid of devices outside educational institutions. The integration of such devices in a traditional course refers to blended learning (Duncan-Howell and Lee, 2007). One large advantage of mobile devices is their flexible usage: they can be used in class without any organizational effort (e.g. booting time, special computer labs) (Wessner and Dawabi, 2004). In traditional courses the use of mobile devices offers the possibilities to improve the interaction between the lecturer and the student, or that among the students. It can also help to enhance the participation of the learners, and the quality of teaching (Wessner and Dawabi, 2004).

Classtalk was one of the first applications which were used in lectures pursuing the aim to create a more interactive, learner-centered classroom. The application enhances communication in class, presented questions for small group work and collected students answers for displaying them (Dufresne et al., 1996). The project *ConcertStudeo* activates students by means of handheld-computer, an electronic blackboard and PDAs for the students with brainstorming, a quiz, voting or ranking (Dawabi et al., 2003). Using audience response systems or clickers (e.g. *Turning Point*), as they are commonly called, students are able to answer questions in the form of quizzes or self-assessments and are able to provide feedback (Kenwright, 2009). In a contribution made by Wessner und Dawabi (2004) two systems (Haake and Wessner, 2004; Roschelle et al., 2002) are examined using specific design questions (e.g. integration in the learning scenario, distribution of information, interaction). Their results are that both applications are suitable to keep the traditional course advantages but enrich those lectures with different interaction opportunities (Wessner and Dawabi, 2004). One application of the Wake Forest University (Class-In-Hand) offers students the opportunity to answer a quiz via web browser or to provide feedback to a statement on a scale. Thereby the communication is unidirectional from the students to the lecturer (Scheele et al., 2004). Another type of student-response-system was realized in form of the web application *Swatt* to provide questions in a multiple-choice, true-false, or yes-no format to the students for answering (Shotsberger and Vatter, 2001). Similar to a multiple-choice format is the method Peer Instruction created by a Harvard professor. Besides answering a short conceptual question interspersed by the teacher, the students should discuss about the question to their colleagues (Fagen et al., 2002). The application of Peer Instruction is carried out using mobile devices, i.e., in the project *Pingo* (Reinhardt et al., 2012).

Current research focuses on creating the user interface more intuitive and investigate how to use a high quality presentation of images, videos, etc. despite a small screen size without having to leave out relevant content (Kopf et al., 2011; Schon et al., 2012; Van Rijsselbergen et al., 2012). Tabata et al. (2010) focuses on an online learning application for the iPhone to support students in answering knowledge questions regardless of time and place. Clunie et al. (2012) present a platform which makes it possible to connect Android mobile devices with the Learning Management System (LMS) Moodle.

To sum up, use of digital media to enhance and to support teaching, especially large-scale lectures, is not new in IS research. But continuous improvement of lectures quality with digital media is still a trend topic. Moreover, the employment of mobile devices in lectures continues to increase significantly (Johnson et al., 2012). Thus, the number of owners of smartphones and tablets among students is growing. Gartner predicts tablet PC sales will reach 327 million worldwide in 2015. The current state of research shows that there exist a large number of applications to enhance interaction and collaboration. But none of these applications consider a set of functions integrated in a single application to enhance the three types of interaction while simultaneously guaranteeing browser and platform independence. Current standard solutions like LMS Moodle or Blackboard are not delivering liberties in design or technical solutions. The LMS Moodle is used at our university. It is not possible to implement and integrate a new function into Moodle. Furthermore anonymity is not guaranteed using Moodle. These are the reasons for the development of the mobile learning application presented here. It is an entirely new design, and is not based on an already existing application.

REQUIREMENTS FROM THEORY OF INTERACTION

The meaning of the term 'interaction' in the disciplines of sociology, education and psychology addresses the interrelation between humans and their communicative actions amongst each other (Bryant and Heath, 2000). IT applications can be used to support and enhance interaction. In this paper the prototype is a mobile-learning application which aims at increasing the interaction between learners and lecturers and among the learners themselves and the learning success in large-scale lectures. Regarding interaction, we specifically refer to the work of Moore (1989), in which the author differentiates between three types of interaction: learner-content-interaction, learner-lecturer-interaction and learner-learner-interaction. We adopt those three types of interaction for our paper and define interaction itself as learning activities, including exchange between learners, lecturers and content (M. G. Moore, 1989; Schrum and Berge, 1997).

Prior research has shown that learners who interact with their lecturers are more actively involved in the learning process (Liu et al., 2003; Wang et al., 1990). The question-answer-game is the classic form of interaction found between learners and lecturers. The lecturer can actively include the learner in teaching, assess the learning progress by means of the answers and provide direct feedback. The learners have the opportunity to contribute their ideas and thoughts, thus, also initiating new thought processes (Gagné et al., 1993; Morgan, 1991). Furthermore, interaction influences the quality of learning in a positive way. A study showing that learners with low or intermediate previous knowledge profit from a high degree of interaction and achieve higher learning results (Snell, 1999).

To increase the interaction, factors which inhibit interaction need to be eliminated. First, when attending large-scale lectures, learners often avoid interaction with lecturers because they fear embarrassment in such an impersonal setting (Ratto et al., 2003; Siau et al., 2006). Second, the seating order of learners (Roth et al., 1999), the limited time for the study unit (VanDeGrift et al., 2002) and the fact that it is hardly ever possible to involve all learners in discussions (Siau et al., 2006) inhibit interaction, especially in large-scale lectures. Some learners could fear holding up the whole auditorium with their specific question (Ratto et al., 2003). As is known from the psychology of learning, both the attention and the motivation of learners decrease after approximately 20 minutes (Smith, 2001). Thus, it is even more important to employ elements in university lectures which have activating functions. An interactive setting in the learning-teaching process can enhance learners' motivation, attention and participation in class, as well as foster greater learners' exchange (Liu et al., 2003; Sims, 2003).

The three types of interaction represent the starting point for deriving requirements. Table 1 presents the general requirements (R) we could identify from theory. These are the basis for developing our mobile-learning application. Requirements formulation from the sentence structure is based on Hull et al. (2010). The requirements formulation argues from the learners' perspective since the learners are the main focus of the mobile-learning application. Moreover, the three interaction types focus on a learner perspective as well. We base our subsequent design decisions on the constructs linked to our phenomena of interest. Summing up, we could identify seven requirements from theory to ensure that all three types of interaction are addressed by our mobile-learning application (Table 1).

REQUIREMENTS FROM A FOCUS GROUP WORKSHOP

Regarding the goal of rectifying the lack of interaction within universities' large-scale lectures, the authors initiated a lecturers' workshop with eight university lecturers with varying degrees of teaching

Table 1. Requirements from theory of interaction

Interaction Type	Description	Requirements from Theory (T)
Learner-Learner-Interaction	This interaction type takes place between one learner and others, alone or in group settings, with or without the real-time presence of an lecturer (Moore, 1989). It is described as an extremely valuable resource for the individual learning (Moore, 1989). Learners should be encouraged to cooperate with the peers during the learning-teaching process to exchange knowledge (Alavi, Marakas and Yoo, 2002; Boud and Falchikov, 2007) and for discussions to enhance motivation (Eisenkopf, 2010) and learning success (Fredericksen, Pickett, Shea, Pelz and Swan, 2000; Moore and Kearsley, 2011). This interaction type is not effectively facilitated in large-scale lectures and calls for specific IT use (Phillips et al. 1988). In collaborative assignments students learn from each other and create new knowledge mutually (Topping, 2005).	T1) Learners should be creating learning material collaboratively. T2) Learners should discuss among each other.
Learner-Lecturer-Interaction	In this type of interaction, the lecturer attempts to achieve teaching aims regarding the learning content (M. G. Moore, 1989). Moreover, the lecturer seeks to enhance the learner's interest, including self-direction and self-motivation (M. G. Moore, 1989). The lecturer still plays a significant role for learning success as well as satisfaction (Eom et al., 2006). Lectures should give advice and feedback to learners and need to retain an overview of their individual performance (Bligh, 1998). In addition, the teacher should verify which learning goals have been achieved or may not have been achieved. In interaction with lecturers, students can request clarification of unclear points and lecturers can reinforce correct interpretation (Thurmond and Wambach, 2004).	T3) Learners should receive feedback. T4) Learners should provide feedback. T5) Learners should have the possibility to ask questions regarding unclear points.
Learner-Content-Interaction	This type of interaction contains what Holmberg (Holmberg, 1986) calls the "internal didactic conversation" when learners talk to them-selves regarding the learning content (Moore, 1989). Moreover, this interaction form takes place when learners participate in class activi-ties (Thurmond and Wambach, 2004). Learning materials should be designed in an activating way to support peer learning and it should be developed with regard to instructional and multimedia design guidelines (Clark and Mayer, 2008). Assignments regarding the learning content should be integrated in class. Factors that affect the learner-content-interaction can be contact with the content (Leasure, Davis and Thievon, 2000) and participation in class discussions (Jiang and Ting, 1999).	T6) Learners should get content specific assignments. T7) Learners should get content specific assignments to discuss.

experience in large-scale lectures. The utilization of focus groups aims to generate numerous innovative ideas (Greenbaum, 1998). To guarantee a systematic collaboration, the authors chose the collaboration process design approach from Kolfschoten and De Vreede (2009). The task of the workshop was to gather ideas for activating elements for the mobile-learning application to improve the interaction in large-scale lectures at universities. The first activity performed in the workshop was brainstorming in order to collect a variety of ideas to reduce the lack of interaction. In a moderated discussion, each idea was discussed with participants, the redundant ideas were eliminated, and the remaining ideas were evaluated using a Likert scale. Afterwards, the ideas that received the highest scores in the evaluation were discussed. The discussion results in four ideas that the authors afterwards used to derive concrete requirements for the mobile-learning application. The practical requirements (P) are the following:

P1: Even in mass settings learners should have the possibility to provide feedback to the lecturer to incomprehensible learning content.

P2: Learners should be able to ask questions to their lecturer anonymously.

Figure 2. Screenshot of the functions of the learners' application

Figure 3. Screenshot of the functions of the lecturers' application

P3: Learners should create true/false-statements to deal with the learning content and exchange those with their fellow students.

P4: Lecturer should provide questions to the learners who answer and discuss them with their fellow students.

DESIGN AND DEVELOPMENT OF THE MOBILE-LEARNING APPLICATION

The mobile-learning-application aims to increase activation among a large number of learners, while allowing for a stable interactivity format ready for use even in large-scale lectures (Lehmann et al., 2012). Parallel to the development of the prototype, evaluations involving university lecturers and students took place in iterative cycles. At regular intervals the functions were designed, implemented and tested. The goal of the conducted evaluation was to monitor the graphic preparation of the individual functions. This approach is typical for the so-called *participatory design* (Pilemalm and Timpka, 2008), which is an approach that centers on the user. Consequently, the user is involved as an active participant in the planning, presentation and evaluation of the design process (Pilemalm and Timpka, 2008). Four to six people suffice for each evaluation round in order to obtain a reliable assessment of the results (Nielsen, 1994). The objective regarding the graphical preparation of the individual functions was to realize an intuitive and practical interface which is very easy to use. Lecturers, e.g., prefer to obtain all relevant information for the lecture on a single screen. This guarantees that lecturers are still able to view and control the presentation and use their notes and, at the same time, are still able to monitor the information about the current setting. Moreover it makes it unnecessary to switch between several programs. Based on Olivia (2004), a simple yet clear design should be selected in order to reduce non-data pixels and visual complexity. Every function in the mobile-learning application is represented by a simple, but very characteristic and easily recognizable symbol. This is especially important to avoid a further increase in the cognitive load of the lecturers, to not inhibit the actual transfer of knowledge. The mobile-learning application is separated in two views; one view for the learners and another view for the lecturer. The following figures 2 and 3 show an extract of the application; separated into the two views.

Specific Description of the Functions

The functions in the mobile-learning application are aimed at increasing the activation and enabling a robust and operational interaction. The design of the functions is the final result of the participatory design aiming an ease of use for the users. Students are able to use the application place synchronous or place asynchronous, since the lecture is broadcasted via livestream during the class. Access to the application for both students and lecturer is via a URL. There is no ex ante registration necessary to ensure anonymity (Bitzer et al., 2012). In Table 2 the four functions are described. The embedding of activating functions in a large-scale lecture stops passiveness during lecturing (Snell, 1999). All functions aid the realization of interaction, which Moore (1989) differentiates.

The Question Button allows for an anonymous communication between students and lecturers, thus avoiding factors which may limit interaction in the learning-teaching process. The Question Button is applicable in- and outside the in-class lecture. Parallel to the lecture, incoming questions are answered in class by the lecturer. Additionally, automatic e-mail forwarding to the lecturer is set up in the system for the Question Button. This allows students to ask questions outside the class. Outside the class, incoming questions will answer via the accompanying LMS Moodle forum.

Table 2. Description of the functions of the mobile-learning application

	Functionality	Design	Theoretical (T) and Practical (P) Requirements
Question Button	Students can send their questions to the lecturer anonymously. The questions are answered by the lecturer during class or in the class-concomitant forum. In addition, the system allows automatic email forwarding from incoming questions to the lecturer.	The number of incoming questions asked by the learners via the Question Button light up red for the lecturer and can be read from the lecturer in a separate window in chronological order.	- T5 - P2
Panic Button	Students can press the Panic Button to send a signal to the lecturer about ambiguous learning content. The button can be used during the whole lecture.	The Panic Button lights up orange on the lecturer's application if it has been clicked by a learner. If a predetermined number is exceeded, the button will light up in red as a special signal to the lecturer.	- T4 - P1
Co-Create Your Exam	To a certain time in class students generate true or false statements dealing with the content they just heard. Statements are answered and evaluated by colleagues. Moreover the lecturer gets randomly selected statements for answering and evaluating using a projector. The application allows exporting the generated statements. They are then available via the university's LMS and students can use them as exercises. To a certain extent, the statements are used for the exam, thus enhancing the incentive for learners to work with the learning content and produce high quality statements.	Using the button for Co-create Your Exam! opens new windows for in- and output in both applications. Learners and lecturers can rank statements with a five-point-star-rate.	- T1, T6 - P3
Peer Discussion	The lecturer can send single-choice questions with five predetermined answers to the students. First of all each student answers the question alone. Second, each student shall discuss the question to their colleague. Third, each student has to answer the question again. The lecturer can hereby evaluate the learning success of the students and give them direct feedback.	The results of the vote are calculated in real-time and can be shown by the lecturer using a projector. The Peer Discussion is the only one that must be activated by the lecturer and is not available for using outside class.	- T2, T3, T7 - P4

The Panic Button aims to provide lecturers a response to the transfer of knowledge. After being pressed by a student, the Panic Button lights up in orange on the lecturer application. When exceeding a predefined number the button lights up in a bright red as a special signal for the lecturer. This is in line with literature suggestions on demonstratively highlighting important information (Yigitbasioglu and Velcu, 2011). By pressing the button, the lecturer may reset this number.

Similar to the Question Button, a new window for input and output opens in both applications when using the Co-Create Your Exam. The co-created true-false statements generated by the students are uploaded in the form of accessible tasks to the LMS. Some of the statements are to be used in the exam as an incentive for the students to reflect on the learning content and to create genuine statements. In addition to these activation effects, the students create a data pool of reusable tasks. The lecturer may additionally store the student solutions in other programs in the long term by means of export features for the Question Button and the Co-Create Your Exam.

When pressing the button for Peer Discussion, new windows for input and output open on the student as well as the lecturer applications. The Peer Discussion allows the students to pick one out of five possible answers. The voting result is determined in real time and presented by the lecturer via the projector. Thus, the lecturer may assess the learning success of the students before providing according feedback. The present IT support additionally enables the lecturer to store the response assessment as a result documentation, possibly to be used for post-processing.

Co-Create Your Exam and Peer Discussion allow the students to reflect on the learning contents in class rather than merely adapting them. All four functions implement the three types of interaction differentiated by Moore (1989). The lecturer, in turn, receives important information on the current learning success of the students, allowing for a more specific recognition of deficits and a repetition of learning contents.

The lecture comprises activation phases designed by the lecturer, who unlocks parts of the activating functions (Peer Discussion only). The other activating functions (Question Button, Panic Button, and Co-Create Your Exam) are continuously available to the students. Co-Create Your Exam and the Question Button may additionally be used outside the class. The Panic Button is only useful during the lecture.

The extensive IT support requires suitable terminal equipment from the learners. In the present case, the learners could borrow a tablet or netbook for the duration of the semester to access and use the mobile-learning application during the learning-teaching process. Learners were allowed to take the mobile devices home. This guarantees a use of the mobile-learning application inside and even outside the class.

Conception of Technology

The authors designed and implemented both mobile-learning applications (learners, lecturer) as web applications. The lecturer application was developed by means of traditional web technologies/scripting languages such as PHP5, HTML5, CSS3, and JavaScript. MySQL was used as relational database management system. In order to guarantee platform independence on the part of the learners while supporting iOS and Android-based mobile end devices in addition to computers, we developed the learner application on the basis of the jQuery Mobile framework, thus allowing for the application to be used via all prevalent internet browsers as well as smartphones and tablets.

EVALUATION RESULTS AND DISCUSSION

To evaluate the mobile-learning application regarding the research questions data from two different semesters but from the same lecture were taken. The lecture is an IS introductory large-scale lecture for undergraduate students which is held every semester at a German university. In the first semester the authors conducted a pre-test of the mobile-learning application and gathered data regarding quality, usefulness, and satisfaction because the research intention was to examine whether the mobile-learning application is ready for being used. In the following semester the authors started to establish the mobile-learning application in the IS introductory lecture and gathered data to answer research question 3. The application was used over the whole semester. The goal focuses on investigating whether the frequency of use of the mobile-learning application affects the interaction among the students in the university large-scale lecture.

Results Regarding Research Question 2

To assess quality of the mobile-learning application, the authors conducted an evaluation aiming to answer the question whether the application is ready for being used in a large-scale lecture. Therefore, the authors used the application at the end of a large-scale lecture in an IS introductory lecture for undergraduate students. The students received an introduction on the functions of the mobile-learning application, could access it via their mobile devices and were asked to use it during the lecture. Participation was voluntary. The lecture was designed in a way that all functions of the application could be used. Afterwards, the students were asked to complete a questionnaire containing the necessary items for the evaluation. All items were measured using a bipolar five-point Likert response format with the endpoints labelled as “I disagree” and “I agree”. In total, 49 questionnaires of the 85 class-attendees (response rate 58%) could be used for our evaluation.

Due to authors’ interest on quality, usefulness, and satisfaction of the mobile-learning application, and the fact that there was no incentive for the students to complete the questionnaire but their goodwill, the authors decided to keep the items to a minimum and focused on the following constructs: Perceived Usefulness (PU), User Satisfaction (US), and Perceived Presentation Quality (PPQ) for the overall application. Furthermore, the authors assessed the Performance Expectancy (PE) and Intention to Use (IU) for every function to gather more detailed feedback on the quality of different functions. The results of the evaluation are presented in Table 3. To evaluate the research question, a t-test was used and the tool of analysis was SPSS 22.

Regarding the measurement of the constructs, the authors relied on scales provided by previous studies. Furthermore, the values for Cronbach’s Alpha and indicator loadings fulfilled the requested thresholds (Chin, 1998; Nunnally, 2010). Further, the mean values for all but one constructs are significantly higher than the neutral point of the scale (“neither agree nor disagree”). This indicated a high PE and IU of all but one function of the mobile-learning application. Only the Panic Button received comparably low scores for both, PE and IU. The overall application received high values for PU, US and PPQ. Consequently, the mobile-learning application in general is ready for being used in a large-scale lecture, and perceived as helpful in terms of quality, usefulness and satisfaction. But the usefulness, design and presentation of the Panic Button should be reinvestigated.

The results of the t-test regarding the Panic Button did not show satisfactory results. Possibilities could be different students’ learnings styles. Another reason could be inaccurate feedback from the

Table 3. Evaluation of measurement model and summary statistics (N=49)

Construct	No. of Items (References)	Mean	Standard Deviation	Cronbach's Alpha	t(df) = t-value
Performance Expectancy – Question Button	4 (Venkatesh et al., 2003)	3.51	1.40	.910	t(48) = 3.530***
Performance Expectancy – Panic Button		3.18	1.40	.960	t(48) = 1.220
Performance Expectancy – Co-Create Your Exam		3.92	1.13	.937	t(48) = 6.262***
Performance Expectancy – Peer Discussion		3.62	1.31	.960	t(48) = 3.919***
Intention to Use – Question Button	4 (Söllner et al., 2012)	3.64	1.32	.979	t(48) = 3.508***
Intention to Use – Panic Button		3.42	1.44	.973	t(48) = 2.192**
Intention to Use – Co-Create Your Exam		4.08	1.20	.978	t(48) = 6.517***
Intention to Use – Peer Discussion		4.08	1.10	.981	t(48) = 7.054***
Perceived Usefulness – Mobile-learning Application overall	6 (Davis, 1989)	3.99	1.01	.971	t(48) = 7.336***
User Satisfaction – Mobile-learning Application overall	7 (Arbaugh, 2000)	4.11	1.03	.967	t(48) = 7.660***
Perceived Presentation Quality - Mobile-learning Application overall	3 (Wells et al., 2011)	3.90	0.81	.950	t(48) = 8.195***

Significance with * p= < 0.05 / ** = p < 0.01 / *** = p < 0.001

teacher which didn't map on students' expression while using that button. Consequently, this button needs to ad-just. One idea from a lecturer was to adopt each panic vote to a specific slide in the lecture material. That will give a direct feedback to the learning material. That will give a direct feedback to the learning material.

After the results of the first evaluation of the mobile-learning application shows satisfactory results the authors conclude that the mobile-learning application is ready for being used in the large-scale lecture. For the following semester the authors started to establish the mobile-learning application in the IS introductory lecture and gathered data to answer the third research question.

Results Regarding Research Question 3

Research question 3 focuses on investigating whether the frequency of use of the mobile-learning application affects the interaction among the students in the university large-scale lecture. The mobile-learning application was applied in each lecture unit over the whole semester. At the semesters' end an online questionnaire was provided via the LMS to be answered by the students. Participation was voluntary. The online questionnaire was completed by 49 students, who all participated in the exam. The sample of N=49 is sufficient for obtaining representative results (Greene and Zhang, 2003). From a total of 243 exam participants the response rate is 20.2%. The constructs to measure interaction were adapted from

Table 4. Descriptive results: Comparison of mean value and standard deviation

Construct	No. of Items (References)	Mean	Standard Deviation
Frequency of Use: Question Button	-	1.96	1.19
Frequency of Use: Panic Button		1.61	0.99
Frequency of Use: Co-Create Your Exam		3.53	1.34
Frequency of Use: Peer Discussion		3.45	1.40
Interaction	13 (Siau et al., 2006)	3.20	0.73

Siau et al.(2006). All items were measured using a bipolar five-point Likert response format ranging from 1 = "I disagree" and 5 = "I agree". To measure frequency of use the learners were asked to assess their use of each activating function from 1 = "not very often" to 5 = "very often".

Table 4 presents the descriptive results with mean value as well as standard deviation with respect to the frequency of use of each activating function as well as mean value and standard deviation for the interaction.

The frequency of use for Question and Panic Button is low. By contrast, Co-Create Your Exam and Peer Discussion suggest a good frequency of use. Both latter functions were applied directly during each lecture, which accounts for the better values. The mean value of interaction appears in a good center span.

In order to assess to what extent the use of the activating functions contributes to the prediction of interaction, the authors conducted a multiple linear regression according to the inclusion method (predictors: the use of the four activating functions, criterion: interaction). The basic requirements for the implementation of a multiple linear regression were considered (Ziegler and Bühner, 2009) and adequately met. All tolerance values are significantly higher than .4 and thus well above the critical value of .1 (Brosius, 2008).

Table 5 illustrates the according regression results (interaction). With an overall R^2 of 0.58 (corrected $R^2 = .545$; $F(4.44) = 15.362$; $p < .001$; N=49), the verification was successful. The model is significant.

The construct of interaction features a Cronbach alpha of $\alpha=0.869$ (13 items). All item scales were thus confirmed to be reliable. The according value could be assessed as good (Brosius, 2008).

The R^2 value of 0.58 confirms a good prediction of the interaction by means of using the four activating functions of the mobile-learning application. The overall model is statistically significant (the significance level is less than 0.05), so it can be assumed that the predictors contribute to the explanation of the dependent variable (interaction). All regression coefficients (B) are positive, meaning that all independent variables positively affect each dependent variable.

The predictors Question Button, Co-Create Your Exam, and Peer Discussion contribute to the predictability of the criterion interaction. These three variables are significant in contributing to the prediction of the dependent variable. The independent variable Question Button is only marginally significant in this context. The beta value presented in the penultimate column of Table 5 confirms that Peer Discussion most significantly affects the interaction. Co-Create Your Exam provides the second largest contribution to the variance verification. Only the Panic Button does not independently contribute to the prediction of interaction.

Table 5. Multiple linear regression for the prediction of interaction

Construct	Non Standardized Coefficients		Standardized Coefficients	
	B	Standard Error	Beta	T
(Constant)	.939	.375		2.505
Question Button	**.139**	**.071**	**.226**	**1.945***
Panic Button	.062	.078	.085	.795
Co-Create Your Exam	**.202**	**.077**	**.370**	**2.628****
Peer Discussion	**.209**	**.074**	**.400**	**2.824****

Comment: Multiple linear regression (method: inclusion). Predication of interaction: $R^2 = 0.58$, corrected $R^2 = .545$; $F(4,44) = 15{,}362$; $p < .001$; N=49.

The functions Co-Create Your Exam and Peer Discussion are directly implemented into the learning-teaching-environment by the lecturer. In addition, these activating functions encourage the students to thoroughly reflect on the learning contents of the lecture. Additionally, both functions promote an IT-mediated communication among the students as well as between the students and the lecturer, which constitutes a possible explanation for the fact that these two variables contribute more effectively to the prediction of interaction. In return, the Question and Panic Button are permanently available to the students in- and outside of class, but they are not actively implemented into the teaching by the lecturer. The tentative use of the Panic Button indicated by the mean value in Table 4 already suggests that this feature hardly contributes to predicting interaction. The use of the Question Button only slightly contributes to the prediction of the interaction, as suggested by the low mean value in Table 4. This low significance may be explained by the fact that the lecturer does not actively implement this function into the teaching, since it merely allows for a unidirectional communication between students and lecturer.

To sum up, three of the four activating functions of the mobile-learning application contribute to the prediction of interaction in the large-scale lecture.

LIMITATIONS, FUTURE RESEARCH, AND CONCLUSION

The evaluation results demonstrate that the use of the mobile-learning application positively affects the interaction in a university large-scale lecture. Since the data are gathered in a real setting, it cannot precisely prove that an increase in interaction will solely result from learners' use of the mobile-learning application during the learning-teaching process. A field study is subject to several confounders, so changes in interaction could also arise from other external effects (Coolican, 2014). Usually, the external validity is higher in field studies against a smaller internal validity. The interaction is a complex variable, difficult to measure, and can be affected by various effects – even outside the use of the mobile-learning application. By means of a text, the students were made aware of the fact that all questions and information in the online questionnaire were concerned with the mobile-learning application. Still, some students might have given statements on the lecture in general without directly referring to the mobile-learning application, constituting another limitation of the present study results. Another limitation of the research methodology is the self-selection bias (Gautier and Klaauw, 2012). Another issue that limits the results is regarding the response rate of 58%. It is possible that only students who liked the mobile-learning application participated in the evaluation and the remaining 42% didn't like the mobile-learning application. This effect would bias and limits the results. However, after the evaluation had ended specific analyses of that cannot make.

The mobile-learning application is going to be established in the didactic concept of the IS introductory lecture. Additionally, the application is offered to other departments to conduct additional evaluations. The goal will be to extensively test the application and collect feedback for further development.

For the theoretical implication, the study contributes to the learners' activation and engagement in the learning-teaching process. Thus, the results contribute to the didactical theories of IS research by providing insights on a learner-centered approach integrating interaction into university large-scale lectures. The results can also be applied to further learning-teaching environments. The use of the mobile-learning application is valuable and enhances the didactical approach not only for university large-scale blended learning lectures, but also for e-learning lectures (e.g., massive open online courses (MOOCs)) (Janson et al., 2014) or traditional learning-teaching environments or other new contexts of interest.

The results of this paper are of practical relevance for researchers, lecturers, and practitioners alike, since they illustrate a way to use the mobile-learning application to time- and resource-efficiently support and enhance communication, exchange, and interaction in a large-scale lecture. According to service research (Leimeister, 2012), the concepts of IT support (Leimeister, 2014) as well as customer integration a scalability of university teaching is facilitated. The study reveals a way to engage and activate learners in large-scale lectures in spite of limited resources, specifically to address the interaction regarding the learners and the learning content, between the learners and the lecturer as well as the interaction among the learners themselves. The results indicate practical as well time- and resource-friendly guidelines for the design of large-scale lectures and at the same time antagonize the challenges of large-scale lectures.

REFERENCES

Alonso, F., Manrique, D., & Viñes, J. M. (2009). A moderate constructivist e-learning instructional model evaluated on computer specialists. *Computers & Education*, *53*(1), 57–65. doi:10.1016/j.compedu.2009.01.002

Arbaugh, J. B. (2000). Virtual classroom characteristics and student satisfaction with internet-based MBA courses. *Journal of Management Education*, *24*(1), 32–54. doi:10.1177/105256290002400104

Bitzer, P., Lehmann, K., Hirdes, E. M., & Thillainathan, N. (2012). Managing the Masses - Developing an Educational Dashboard for Lecturers in Large-Scale Lectures. In *European Conference on Information Systems (ECIS)*.

Bligh, D. A. (1998). *What's the Use of Lectures?* Eastbourne: Intellect books.

Briggs, R. O. (2006). On theory-driven design and deployment of collaboration systems. *International Journal of Human-Computer Studies*, *64*(7), 573–582. doi:10.1016/j.ijhcs.2006.02.003

Brosius, F. (2008). *SPSS 16*: Hüthig Jehle Rehm.

Bryant, J., & Heath, R. (2000). *Human communication theory and research: concepts, contexts, and challenges*. Mahwah, NJ: Lawrence Erlbaum Associates.

Chin, W. W. (1998). The partial least squares approach for structural equation modeling. In G. A. Marcoulides (Ed.), *Modern methods for business research* (p. 437). Mahwah, NJ: Lawrence Erlbaum Associates Publishers.

Coolican, H. (2014). *Research methods and statistics in psychology*. New York, NY: Psychology Press.

Davis, F. D. (1989). Perceived usefulness, perceived ease of use, and user acceptance of information technology. *Management Information Systems Quarterly, 13*(3), 319–340. doi:10.2307/249008

Dawabi, P., Dietz, L., Fernandez, A., & Wessner, M. (2003). Using PDAs to support face-to-face learning. In Computer Support for Collaborative Learning. Academic Press.

de Clunie, G. T., Serrão, T., Monteiro Braz, J.-L., Serrão, T., Rangel, N., & Castillo, A. (2012). Developing an Android based learning application for mobile devices. In *Telematics and Information Systems (EATIS), 6th Euro American Conference on*. doi:10.1145/2261605.2261641

Dufresne, R. J., Gerace, W. J., Leonard, W. J., Mestre, J. P., & Wenk, L. (1996). Classtalk: A classroom communication system for active learning. *Journal of Computing in Higher Education, 7*(2), 3–47. doi:10.1007/BF02948592

Duncan-Howell, J. A., & Lee, K.-T. (2007). *M-Learning–Innovations and Initiatives: Finding a place for mobile technologies within tertiary educational settings*. Auckland.

Dyson, L. E., Litchfield, A., Raban, R., & Tyler, J. (2009). *Interactive classroom mLearning and the experiential transactions between students and lecturer*. Ascilite.

Eom, S. B., Wen, H. J., & Ashill, N. (2006). The Determinants of Students' Perceived Learning Outcomes and Satisfaction in University Online Education: An Empirical Investigation*. *Decision Sciences Journal of Innovative Education, 4*(2), 215–235. doi:10.1111/j.1540-4609.2006.00114.x

Evans, C., & Gibbons, N. J. (2007). The interactivity effect in multimedia learning. *Computers &amp. Education, 49*(4), 1147–1160.

Fagen, A. P., Crouch, C. H., & Mazur, E. (2002). Peer instruction: Results from a range of classrooms. *The Physics Teacher, 40*, 206.

Gagné, E. D., Yekovich, C. W., & Yekovich, F. R. (1993). *The cognitive psychology of school learning*. New York: HarperCollins College.

Gautier, P. A., & Klaauw, B. (2012). Selection in a field experiment with voluntary participation. *Journal of Applied Econometrics, 27*(1), 63–84. doi:10.1002/jae.1184

Greenbaum, T. L. (1998). *The handbook for focus group research*. Thousand Oaks, CA: Sage.

Greene, W. H., & Zhang, C. (2003). Econometric analysis. Prentice Hall.

Haake, J. M., & Wessner, M. (2004). Kooperative Lernräume. In *CSCL-Kompendium. Lehr- und Handbuch zum computerunterstützten kooperativen Lernen* (pp. 109–117). München: Oldenbourg Wissenschaftsverlag GmbH. doi:10.1524/9783486593976

Hardless, C., Nilsson, M., & Nuldén, U. (2005). 'Copernicus' experiencing a failing project for reflection and learning. *Management Learning*, *36*(2), 181–217. doi:10.1177/1350507605052557

Hevner, A. R., March, S. T., Park, J., & Ram, S. (2004). Design science in information systems research. *Management Information Systems Quarterly*, *28*(1), 75–105.

Hull, E., Jackson, K., & Dick, J. (2010). *Requirements engineering*. Springer Science & Business Media.

Janson, A., Ernst, S.-J., Lehmann, K., & Leimeister, J. M. (2014). Creating Awareness and Reflection in a Large-Scale IS Lecture–The Application of a Peer Assessment in a Flipped Classroom Scenario. In *4th Workshop on Awareness and Reflection in Technology-Enhanced Learning (ARTEL 2014) to be held in the context of EC-TEL 2014*.

Johnson, L., Adams, S., & Cummins, M. (2012). *The NMC horizon report: 2012 higher education edition*. Austin, TX: The New Media Consortium.

Kenwright, K. (2009). Clickers in the Classroom. *TechTrends*, *53*(1), 74–77. doi:10.1007/s11528-009-0240-7

Kolfschoten, G. L., & De Vreede, G. J. (2009). A design approach for collaboration processes: A multi-method design science study in collaboration engineering. *Journal of Management Information Systems*, *26*(1), 225–256. doi:10.2753/MIS0742-1222260109

Kopf, S., Haenselmann, T., Kiess, J., Guthier, B., & Effelsberg, W. (2011). Algorithms for video retargeting. *Multimedia Tools and Applications*, *51*(2), 819–861. doi:10.1007/s11042-010-0717-6

Kukulska-Hulme, A., & Traxler, J. (2005). *Mobile learning a handbook for educators and trainers*. London: Routledge.

Lehmann, K., & Leimeister, J.-M. (2015). Theory-driven Design of an IT-based Peer Assessment to Assess High Cognitive Levels of Educational Objectives in Large-scale Learning Services. In *23rd European Conference on Information Systems (ECIS 2015)*.

Lehmann, K., Oeste, S., Janson, A., Söllner, M., & Leimeister, J. M. (2015). Flipping the Classroom – IT-unterstützte Lerneraktivierung zur Verbesserung des Lernerfolges einer universitären Massenlehrveranstaltung. *HMD Praxis der Wirtschaftsinformatik*, 1-15.

Lehmann, K., Thillainathan, N., Bitzer, P., & Leimeister, J. M. (2012). Performance Dashboard für Dozenten in der universitären Lehre. In *Multikonferenz der Wirtschaftsinformatik*. Braunschweig, Deutschland: MKWI.

Leimeister, J. M. (2012). *Dienstleistungsengineering und-management*. Berlin, Heidelberg: Springer DE. doi:10.1007/978-3-642-27983-6

Leimeister, J. M. (2014). Collaboration Engineering. IT-gestützte Zusammenarbeitsprozesse systematisch entwickeln und durchführen. Heidelberg Springer DE.

Liu, T.-C., Liang, J.-K., Wang, H.-Y., Chan, T.-W., & Wei, L.-H. (2003). Embedding educlick in classroom to enhance interaction. In *International conference on computers in education (ICCE)*.

Moore, A., Masterson, J. T., Christophel, D. M., & Shea, K. A. (1996). College Teacher Immediacy and Student Ratings of Instruction. *Communication Education*, *45*(1), 29–39. doi:10.1080/03634529609379030

Moore, M. G. (1989). Editorial: Three types of interaction. *American Journal of Distance Education*, *3*(2), 1–7. doi:10.1080/08923648909526659

Morgan, N., & Saxton, J. (1991). *Teaching, questioning, and learning*. New York: Routledge.

Nielsen, J. (1994). Estimating the number of subjects needed for a thinking aloud test. *International Journal of Human-Computer Studies*, *41*(3), 385–397. doi:10.1006/ijhc.1994.1065

Nunnally, J. C. (2010). *Psychometric Theory 3E*. New York: Tata McGraw-Hill Education.

Oliva, A., Mack, M. L., Shrestha, M., & Peeper, A. (2004). Identifying the perceptual dimensions of visual complexity of scenes. In *The 26th Annual Meeting of the Cogn. Sci. Soc.*

Peffers, K., Tuunanen, T., Gengler, C. E., Rossi, M., Hui, W., & Virtanen, V. (2006). The design science research process: a model for producing and presenting information systems research. In *Proceedings of the First International Conference on Design Science Research in Information Systems and Technology (DESRIST 2006)*.

Picciano, A. G. (2002). Beyond student perceptions: Issues of interaction, presence, and performance in an online course. *Journal of Asynchronous Learning Networks*, *6*(1), 21–40.

Pilemalm, S., & Timpka, T. (2008). Third generation participatory design in health informatics—Making user participation applicable to large-scale information system projects. *Journal of Biomedical Informatics*, *41*(2), 327–339. doi:10.1016/j.jbi.2007.09.004

Ratto, M., Shapiro, R. B., Truong, T. M., & Griswold, W. G. (2003). The ActiveClass project: Experiments in encouraging classroom participation. In Computer support for collaborative learning. doi:10.1007/978-94-017-0195-2_57

Reinhardt, W., Sievers, M., Magenheim, J., Kundisch, D., Herrmann, P., & Beutner, M. (2012). "PINGO: peer instruction for very large groups. In *Seventh European Conference on Technology Enhanced Learning*. doi:10.1007/978-3-642-33263-0_51

Roschelle, J., Patton, C., & Pea, R. (2002). To unlock the learning value of wireless mobile devices, understand coupling. In Wireless and Mobile Technologies in Education. doi:10.1109/WMTE.2002.1039214

Roth, W. M., McGinn, M. K., Woszczyna, C., & Boutonne, S. (1999). Differential participation during science conversations: The interaction of focal artifacts, social configurations, and physical arrangements. *Journal of the Learning Sciences*, *8*(3-4), 293–347. doi:10.1080/10508406.1999.9672073

Scheele, N., Wessels, A., & Effelsberg, W. (2004). Die interaktive Vorlesung in der Praxis. In DeLFI 2004: Die 2. e-Learning Fachtagung Informatik.

Schon, D., Klinger, M., Kopf, S., & Effelsberg, W. (2012). MobileQuiz–A Lecture Survey Tool using Smartphones and QR Tags. *International Journal of Digital Information and Wireless Communications*, *2*(3), 231–244.

Schrum, L., & Berge, Z. L. (1997). Creating Student Interaction within the Educational Experience: A Challenge for Online Teachers. *Canadian Journal of Educational Communication*, *26*(3), 133–144.

Shotsberger, P. G., & Vatter, R. (2001). Teaching and learning in the wireless classroom. *Computer*, *34*(3), 110–111. doi:10.1109/2.910902

Siau, K., Sheng, H., & Fui-Hoon Nah, F. (2006). Use of a classroom response system to enhance classroom interactivity. *IEEE Transactions on Education*, *49*(3), 398–403. doi:10.1109/TE.2006.879802

Sims, R. (2003). Promises of interactivity: Aligning learner perceptions and expectations with strategies for flexible and online learning. *Distance Education*, *24*(1), 87–103. doi:10.1080/01587910303050

Smith, B. (2001). Just give us the right answer. In H. Edwards, B. Smith, & G. Webb (Eds.), Lecturing: Case studies, experience and practice (pp. 123-129). London: Kogan.

Snell, Y. S., & Linda, S. (1999). Interactive lecturing: Strategies for increasing participation in large group presentations. *Medical Teacher*, *21*(1), 37–42. doi:10.1080/01421599980011

Söllner, M., Hoffmann, A., Hoffmann, H., Wacker, A., & Leimeister, J. M. (2012). Understanding the Formation of Trust in IT Artifacts. In *The International Conference on Information Systems (ICIS)*.

Tabata, Y., Yin, C., Ogata, H., & Yano, Y. (2010). An iPhone quiz system for learning foreign languages. In *Informatics in Control, Automation and Robotics (CAR), 2010 2nd International Asia Conference on*. doi:10.1109/CAR.2010.5456691

Thurmond, V., & Wambach, K. (2004). Understanding interactions in distance education: A review of the literature. *International Journal of Instructional Technology and Distance Learning*, *1*(1), 9–26.

Van Rijsselbergen, D., Poppe, C., Verwaest, M., Mannens, E., & Van de Walle, R. (2012). Semantic Mastering: Content adaptation in the creative drama production workflow. *Multimedia Tools and Applications*, *59*(1), 307–340. doi:10.1007/s11042-010-0710-0

VanDeGrift, T., Wolfman, S. A., Yasuhara, K., & Anderson, R. J. (2002). Promoting interaction in large classes with a computer-mediated feedback system. In Computer Science Department and Microsoft Research.

Venkatesh, V., Morris, M. G., Gordon, B. D., & Davis, F. D. (2003). User Acceptance of Information Technology: Toward a Unified View. *Management Information Systems Quarterly*, *27*(3), 425–478.

Wang, M. C., Haertel, G. D., & Walberg, H. J. (1990). What influences learning? A content analysis of review literature. *The Journal of Educational Research*, *84*(1), 30–43. doi:10.1080/00220671.1990.10885988

Wells, J. D., Valacich, J. S., & Hess, T. J. (2011). What signal are you sending? how website quality influences perceptions of product quality and purchase intentions. *Management Information Systems Quarterly*, *35*(2), 373–396.

Wessner, M., & Dawabi, P. (2004). Interaktionsunterstützung für die Präsenzlehre-Szenarien und Design. In DeLFI 2004: Die 2. e-Learning Fachtagung Informatik.

Wortmann, F., Hirdes, E. M., Lehmann, K., & Zülich, J. (2014). IT-Unterstützung von Blended Learning Konzepten. In K. Möller & W. Schultze (Eds.), *Produktivität von Dienstleistungen*. Wiesbaden: Springer Gabler.

Yigitbasioglu, O. M., & Velcu, O. (2011). A review of dashboards in performance management: Implications for design and research. *International Journal of Accounting Information Systems*.

Ziegler, M., & Bühner, M. (2009). *Statistik für Psychologen und Sozialwissenschaftler*. Pearson Deutschland GmbH.

Chapter 7
A Remote Mobile Experiment in Brazilian Public Basic Education

Priscila Cadorin Nicolete
Federal University of Santa Catarina, Brazil

Juarez Bento da Silva
Federal University of Santa Catarina, Brazil

Marta Adriana da Silva Cristiano
Federal University of Santa Catarina, Brazil

Simone Meister Sommer Bilessimo
Federal University of Santa Catarina, Brazil

Giovanni Ferreira de Farias
Federal University of Santa Catarina, Brazil

Silvio Serafim da Luz Filho
Federal University of Santa Catarina, Brazil

ABSTRACT

The STEM subjects (Science, Technology, Engineering, and Mathematics) are very important for education, but the lack of experimental laboratories for these subjects in a school might decrease the interest of its students in STEM fields. This chapter explores these issues in the Brazilian context where, in order to address this issue, remote experiments are used to share real experiments manipulated through the Internet. Teachers and students can use remote laboratories, equipped with real experiments, to put in practice theoretical concepts learned in class. This chapter presents a report on a pilot project that aims to explore the use of Mobile Remote Experimentation (MRE) by teachers and students of public high schools in Brazil. It involves the use of mobile devices to access remote experiments in STEM subjects through the Internet. The report demonstrates the effectiveness of using such educational resources to improve pedagogical results by applying the TPACK (Technological Pedagogical Content Knowledge) model to measure the impact of MRE by STEM teachers.

DOI: 10.4018/978-1-5225-0359-0.ch007

INTRODUCTION

The current educational processes are subject to continuous changes, resulted from the complexity and diversity of contemporary societies, in which technological evolution has allowed new ways of building knowledge. The integration of Information and Communication Technologies (ICT) in activities of daily life has implications for education, leading to new forms of teaching and learning (J. Silva & Salinas, 2014). Thus, the scope of pedagogy should offer effective educational responses and introduce varied and flexible methodologies, in order to overcome the new training challenges of the twenty-first-century students.

The profile of students who join Basic Education pushes teachers and educational administrators to create new teaching methods. They use learning environments that are compatible with the way in which the students learn new concepts in their daily routine. This is a new reality that must be transposed into the educational environment since the students are heterogeneous regarding their capabilities, motivations, interests, cultural background, and social and individual subjectivity. Considering that the students present a diversity of learning requirements, they also need to individualize their education process and access the knowledge through various and specific teaching methods, according to their particular way of learning.

Therefore, it is essential to extend this technology perspective not only to the classroom but also to the school, and provide the opportunity for thinking as a process of action (doing) and the creation of knowledge (knowing). It is also important to pay attention to feedback, from both students and teachers, allowing customization to create more effective connections among classrooms and a global and dynamic environment. In this context, the production of educational content for digital platforms is an important aspect to be considered to hold students' attention and achieve meaningful teaching and learning goals. Thus, when talking about education, there must be reflections on the use of digital and interactive platforms, such as forums, chat rooms, use of explanatory videos, and video conferencing. Further, there must be the use of mobile learning, virtual and remote laboratories, and other resources to support teachers in their teaching methods.

Among the many forms of innovation in the educational process, the Mobile Learning stands out in the current scenario. Corroborating with this statement, UNESCO (United Nations Educational Strategic and Cultural Organization) stresses that the use of mobile technologies leverages new educational opportunities. Its use is also justified by the increasing use of such devices by students and teachers, making the process of teaching and learning simple and ubiquitous (Unesco, 2014). It is worth mentioning that mobile devices, which are already in common use by teachers and students, allow the study, the experimentation, and the interaction exceeds the classrooms, libraries, and laboratories.

Mobile Learning and other trends, such as cloud computing and the virtual and remote laboratories, are pointed out in the editions of 2013 and 2014 of the Horizon Report: K-12 Edition (Johnson, Adams Becker, Estrada, & Freeman, 2014). All these technologies turn to mobile learning since they rely on devices that are already in the routine of students and teachers, such as mobile phones or tablets, encouraging them to bring their own devices to the educational environment.

When bringing these concepts to the Brazilian reality, the scholars can make the following question: "To what extent the Brazilian teachers are prepared to use these mobile technologies in their teaching routine effectively?" On the one hand, there are students with good technology skills, stimulated by the use of the devices to study more diligently. However, the teachers are considered unprepared to take advantage of this technology, featuring a major challenge for effective use of ICTs.

The main scholar's challenge is to differentiate how much the teacher knows about the technology from how he or she can integrate this technology effectively to the teaching context. It is necessary to observe the context of interaction, the used technology, and content available. The ubiquitous learning is the result of the integration of these factors, highlighting the content rather than the technology.

MAIN FOCUS OF THE CHAPTER

In this sense, this study presents a report on a pilot project that seeks to explore a new educational opportunity in Brazil, which is the Mobile Remote Experimentation (MRE). This is the result of the intersection of the associated features of Mobile Learning and the Remote Labs. MRE aims to use software applications to grant access to remote laboratory labs, with real instrumentation, through mobile devices. This allows the increase of the flexibility in the scope of collaborative teaching and learning activities, providing an approximation of students and the real world through the use of mobile devices. This approach also aims to arouse the interest of vocational basic education students for careers in STEM areas (Science, Technology, Engineering, and Mathematics) and scientific/technological research.

This project is structured around two axes: training and integration. The training enables the teachers for the use of the involved technologies. A diagnosis based on the concepts of TPACK (Technological, Pedagogical, Content Knowledge) supports the training process, in order to optimize the use of the technology by the teachers. The integration involves the adaptation of online OER (Open Educational Resources) for the specific teaching and learning activities that must be performed by teachers and students with the use of mobile devices. The integration counts on the collaborative support for the development of teaching plans and online educational activities with the use of OER available for the project. The students use tablet computers provided by the research project, and the OERs are developed by Remote Experimentation Laboratory (RExLab).

The RExLab emerged from the concepts of the remote lab in 1997 at the Federal University of Santa Catarina (UFSC). Currently, it counts on a network of 12 universities spread out across five countries. One of its objectives is to meet the need for the social appropriation of science and technology, popularizing scientific and technological knowledge. One expected outcome of its initiatives is encouraging young people to integrate to the scientific and technological careers. Moreover, it aims to integrate science education to the educational process by promoting the improvement and modernization of education at all levels, emphasizing actions and activities that enhance and promote creativity, experimentation, and interdisciplinarity.

An MRE pilot project has been implemented in four Brazilian public schools, which involves 287 teachers and 3,201 students. Physics, Biology, and Mathematics are the subjects that have been the main focus of the project during 2014. The research project prioritized 17 teachers, and 1,339 students enrolled in subject matters related to STEM areas. Therefore, it presents the technology integration of the subject of Physics in high-school classes in the Brazilian basic education, involving one of the partner schools.

This study also describes the methodology used to perform educational activities by the teachers and students with mobile devices and the architecture of the remote laboratory RExLab available for the participating schools. Moreover, it depicts the training methodology according to the model TPACK, which aims to understand and describe the types of knowledge necessary for a teacher to update his or her teaching practice with the mentioned technology.

It is important to cite that, in 2011, the Organization of Ibero-American States for Education, Science, and Culture (OEI) selected this research project as one of the four projects most innovative in Brazilian education. The project, which is sponsored by Fundación Telefónica (Telefónica Foundation), was cited in the NMC Horizon Report: Technological Outlook for Brazilian Universities in 2014. The report mentions the RExLab as an example of the use of remote laboratories, already in activity (Johnson, Becker, Cummis, & Estrada, 2014).

MOBILE LEARNING

The inclusion of ICT in the educational environment offers new educational perspectives. It enables varied teaching strategies and provides several ways of learning, while allowing the development of new skills to be developed in the social context, considering the new scenarios of the knowledge society.

One of the ways to use ICT in the educational environment is through the use of mobile devices, also known as Mobile Learning. The term mobile learning (or m-learning) is a concept associated with the use of mobile technology in education. It can be considered as the intersection of "mobile computing" and "e-Learning" to produce an educational experience anywhere and anytime (Hofmann, 2006) in a broad context of resource use (Welsh, Wanberg, Brown, & Simmering, 2003).

The mobile learning has the potential to become an integral part of teaching and learning processes in primary and secondary education, as long as the increasing use of mobile devices by students. With touching and interactive screens, stimulated by the Internet, tablets and smartphones become important tools for teaching, learning, collaboration and continuous productivity. Moreover, the sales of mobile phones, as well as the hardware power, grow rapidly due to cost reduction. As shown by the National Telecommunications Agency of Brazil (ANATEL), Brazil ended the month of December 2014 with 280.7 million mobile phones, i.e., around 138 cell / 100 inhabitants. Furthermore, the demand for smartphones (cell phones with touching screen) increased 82.2% in sales during 2014, whereas the sales of conventional cell phones decreased 35% over the same period (Teleco, 2014). According to UNESCO (2014), mobile devices represent the interactive ICT most present around the globe. For the first time in history, the number of mobile devices with Internet will surpass the world's population (Unesco, 2014).

Mobile devices in the hands of students and teachers can represent a saving in investment in technological equipment used in educational institutions (J. B. d. Silva, Rochadel, Schardosim, & Fidalgo, 2014). Thus, mobile technology can enhance and enrich educational opportunities for students in different environments (Unesco, 2014). In addition, mobile devices arouse great interest among young people, primarily because of their mobility, in being constantly connected, ease of communication, sharing content and location information almost instantly. The mobile learning is still capable of promoting the knowledge sharing that goes beyond the school grounds. It can bust collaboration and cooperation among pupils-students, students-teachers, friends, and family, anywhere, and at any time. Moreover, this sharing practice becomes the knowledge acquired in the field of science into something as close to reality as possible. As Sharples and Taylor, and Vavoula (2005) note, the key point is not the technology but the concept of mobility, such as physical mobility, technological mobility, and social mobility.

The educational institutions need to adapt to the changes taking technology as a tool to help them transcend traditional classes. This can happen by offering educational alternatives, in which students have an open window to their training inside and outside the school. As the result, it allows the students to strengthen the acquired skills and to reach other ones that give them an independent and effective learning, involving technological, methodological, corporate, and social skills.

Mobile Remote Experimentation

The literature demonstrates three different types of laboratories commonly used in the teaching of science and technology areas: the classroom laboratory (hands-on), the virtual lab, and the remote lab.

The laboratory classroom is conventionally employed in classroom courses, in which the student directly manipulates the lab materials in the same space and time as their colleagues, all supervised by the teacher. The virtual lab is based on simulations in which the student does not interact with instruments and devices, but with computational representations of reality.

The remote labs are distant from the students. However, even at a distance, it allows the student to control the instruments and devices remotely, through some interface that performs the mediation between the student and the devices and equipment (2003). According to Corter et al. (2007), this feature makes the remote experimentation laboratories at least as efficient as the classroom labs in the teaching of specific concepts. After all, the elements and labs are real in the remote lab, and only its access is virtual.

Thus, the remote experimentation (RE) may be defined as real labs accessed over the Internet, which allow students to have free access to them at any time (Paladini, Silva, Alves, Fischer, & Alves, 2008). The RE is a common feature in higher education institutions, and it is often used to complement lectures in the areas of science, technology, and engineering.

RE's enable students to observe dynamic phenomena that are often difficult to explain through written material problems (Ma & Nickerson, 2006). In addition, it increases the motivation of students to develop a realistic approach to solving problems. It happens because, unlike the virtual laboratories, where all processes are simulated, the remote labs allow interaction and analysis with real processes (2013) (J. B. d. Silva, 2006).

Figure 1 shows the lab called "Switchboard," accessed by a mobile device. It deals with an electrical circuit with links organized in series, in parallel, and mixed. This type of lab aims to allow access to a real physical system over the Internet (RExLab). Students from various educational institutions can make the laboratory activity remotely and thus acquire a more complete picture of the physical phenomena any time. The RE allows greater time flexibility and increases practical activities in the subject matters (RExLab).

Using an ordinary laboratory is attractive and fascinating to science teaching. However, due to the insufficient number of regular laboratories, the usual alternative in the school environment is the use of virtual laboratories or simulators, which only return values and observations recorded or calculated previously. This result simulation fails to demonstrate important aspects of reality and the actual application of the lab with other natural factors.

The first records of remote labs were in 1990 and since then they have been widely used all over the world, mainly for educational purposes. By 2011, there were nearly 60 remote labs all over the world (Kofman & Concari, 2011). The Athabasca University uses remote labs for the learning of introductory physics. The theoretical part of algebra-bases physics courses in Athabasca University is based in printed materials and web pages, similarly to distance learning courses. Then, remote labs are used to supplement the ability to supply materials for the students. They developed a prototype introductory mechanics lab that allowed many of the instructional features of "on-campus" labs to be presented to students over the Internet (Connors, Bredeson, & Al-Shamoli, 2011). The Obafemi Awolowo University (OAU) from the

Figure 1. Screenshot of the mobile application RExMobile
Source: http://www.rexlab.ufsc.br

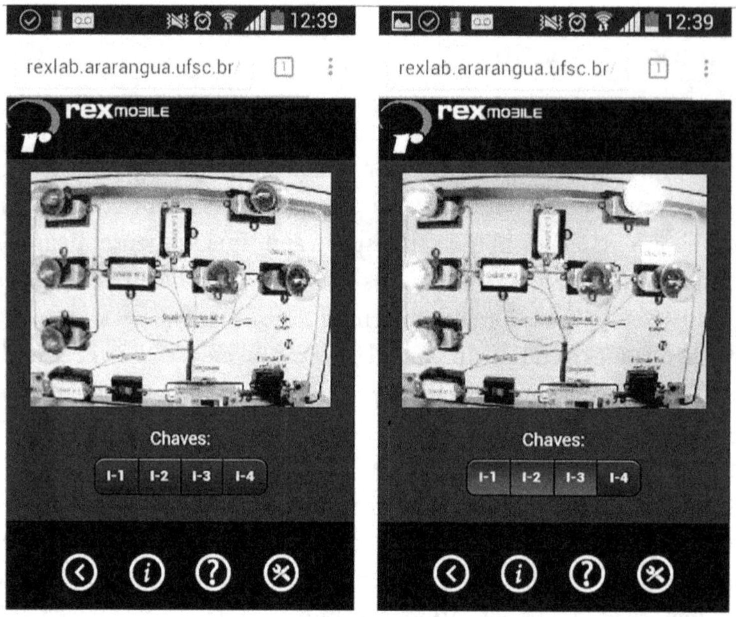

Western part of Nigeria in West Africa uses, since 2005, the iLab developed at Massachusetts Institute of Technology (MIT) (Kehinde, Ayodele, Akinwale, & Osasona, 2011). Currently, the students can enjoy a set of experiments developed at MIT and developed by OAU's remote experimentation team.

MIT iLab is an instance of remote laboratories initially developed at MIT and introduced to Africa in 2005, in partnership with three Universities: Makerere University (MARK), Uganda, University of Dar es Salaam (UDSM), Tanzania, and Obafemi Awolowo University (OAU) of Nigeria (Tickodri-Togboa, Mwikirize, Tumusiime, & Musasizi, 2011). According to Kehinde, Ayodele, Akinwale, and Osadona, "the iLab system design makes it ideally suited to Africa's bandwidth constrained environments" (2011).

Bringing the concepts of Mobile Learning to Remote Lab creates a new perspective on the educational activities: the Mobile Remote Experimentation (MRE). The MRE is described by Costa (2005) as the use of mobile devices to control labs remotely, clearly demonstrated at the intersection of these concepts, as shown in Figure 2. In addition, with a reduced number of current operating systems, it is possible to develop various portable applications. The system described by Al-Zoubi and Tahat (Al-Zoubi, Tahat, & Hasan, 2005) is an example of these applications, used to send and receive data from labs using short text messages (SMS). Another example is the application described by Zubia Garcia-Lopez-of-Ipiña (Zubía & Alves, 2011), which uses AJAX to control remotely what is called WebLabs.

Fidalgo, Rochadel, and Silva (2013) feature a mobile remote experimentation solution for electronics courses with the remote lab VISIR. This is an open and remote lab dedicated to experiments with electrical and electronic circuits. The user interface replicates a physical breadboard with components and instruments front panels, being possible to wire the desired circuit and analyze its behavior with several instruments. This system is presently available using HTML5 and CSS3 web interface in order to support educational activities compatible with different mobile devices and operating systems, such as Android, iOS, and Windows Phone.

Figure 2. Conceptual representation of Mobile Remote Experimentation (Zubía & Alves, 2011)

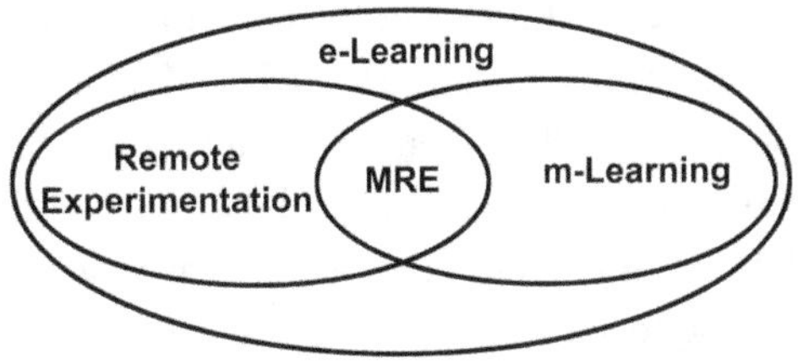

The MRE combines features and concepts associated with Mobile Learning and Remote Lab. The applications allow the user to perform remotely controlled labs with the differential interaction with real instruments via a mobile device. It provides an excellent opportunity for teachers and students from developing countries to demonstrate theories in the classroom. It is especially interesting because the low availability of laboratories and the lack of resources available in the public schools.

According to the Brazilian School Census 2013, less than half (44%) of public schools in Brazil offer computer lab, in which there are, in average, only eight computers dedicated for educational use. Only 8% of public schools in the country have this kind of science laboratories (INEP/MEC, 2013b). This situation has negative implications for higher education, such as low enrollment and increasing dropout, especially in the STEM areas.

The former students from high schools with limited lab resources demonstrate deficiency in STEM subjects, such as calculus, algebra, and physics. Consequently, they contribute to the formation deficit in engineering courses. This educational deficiency is one factor that motivates high-school dropout. According to the Higher Education Census of 2013, for each 10,000 Brazilian students, 22% are registered in courses in the fields of Science, Mathematics, and Computer Science. Further, only 8.8% of pupils effectively start the course programs while only 2.7% conclude them (INEP/MEC, 2013a). Because of the lack of awareness about the real usefulness of scientific subjects, some students are not interested in continuing their studies in scientific areas. Other ones find difficulty in their education because the lack of interaction with the practice aspects of the theory learned in schools.

Therefore, the importance of developing a low-cost and affordable alternative for schools implement science labs. With the use of MRE, it is possible to ease the barriers faced by students and teachers by the low availability of lab resources. On the one hand, Remote Lab can overcome these difficulties and deficiencies related experimentation. Furthermore, the use of mobile devices, which are widespread among students, can mitigate the shortcomings in terms of computer equipment in schools.

Methodology

The research was structured around two axes: a formative axis and an integration axis. The first axis aims at training teachers on Information and Communication Technology (ICT). The second axis regards the integration of the used technology with the teaching/learning activities.

In order to integrate ICT and teaching/learning activities, the research team makes a diagnosis of the technological infrastructure and related services available in a school.

Since the preparation of teachers for the appropriate use of the educational technology is the key factor for the project success, the project's second strategy is training the teachers on the used technology. The main teacher concern regards to their knowledge self-assessment to manage them to use the technology effectively, which is not used isolated from the pedagogical context. Therefore, it is essential to have knowledge involving pedagogy, content, and technology, forming the learning model called TPACK (Technological Pedagogical Content Knowledge). The TPACK is a model developed and updated by Mishra and Koehler (Mishra & Koehler, 2006) from Shulman's 1987 proposal (Shulman, 1986), which is based on the integration of pedagogical knowledge and content to lead pedagogical processes.

To conduct the pedagogical strategies related to teachers, primarily the research team performs the diagnosis in order to understand how teachers perceive the use of ICT in their classes. Then, the course content is specified and developed to meet the results of the diagnosis, and the teachers are trained to use them in their respective classes.

The collection of quantitative data is done by applying two questionnaires. One of them is the "Teacher Profile", which consists of 20 questions, and the other is the "TPACK Diagnosis", prepared and adapted from the Mishra and Koehler TPACK model with 36 questions. The TPACK diagnosis questions are arranged on a Likert scale of five points. It assesses the extent to which participants agree or disagree with statements about their perceptions of the relationship between technology and education.

After, the questionnaires are performed categorizing items in subscales according to the TPACK domain: Pedagogical Knowledge (PK), Content Knowledge (CK), Technological Knowledge (TK), Pedagogical Content Knowledge (PCK), Technological Content Knowledge (TCK), Technological Pedagogical Knowledge (TPK), and TPACK. The data obtained in this phase delineates the training activities for teachers. This step includes courses, workshops, and lectures provided to the teachers. The workshops and courses aim to empower and motivate teachers to use and integrate the technology into their respective teaching and learning processes. Moreover, they are used to support the development of educational plans envisioning the use of such methods and tools (especially mobile learning).

The course for training teachers uses the blended model, conciliating the autonomy of the teacher with the tutor supervision. Also, the teachers can use a Learning Manage System (LMS) provided by the RExLab not only to follow up the training activities but also to construct their respective courses, which will be delivered to their students later.

With teacher training, the research team draws up the activities related to the axis of integration of technology in teaching activities, counting on the teacher of each subject as a partner. The interaction between the teachers and the RExLab researchers also allows that project fellows develop new content or lab experiments from the specifications and requirements based on the teacher needs.

When the training and content design phases are completed, the RExLab grants the access to the remote labs for the teachers. Further, the teachers receive twenty-one tablet computers to distribute them among their students in the classroom. The teachers still receive technical advice from the RExLab fellows, and support from high-school fellows, who receive scholarships provided by the project.

At this point of the project, the students have access to the LMS, to the mobile devices, and to the remote labs provided by the RExLab. Then, the traditional classes suffer a transformation, and the teacher has a new way of teaching. This puts into practice one of the fundaments of this project: providing a constructive learning, in which the protagonist of the educational process is the student, through experience sharing among the project participants.

The pilot project has been implemented in four public basic schools[1] located in Araranguá city, Santa Catarina, Brazil. It involves 287 teachers from various areas, and 3,201 students. However, the classes of Physics, Chemistry, Biology, and Mathematics were more beneficiated. After all, the focus of the project is in the STEM areas, where the availability of laboratory resources and the practical activities

Figure 3. Macro view of the MRE structure

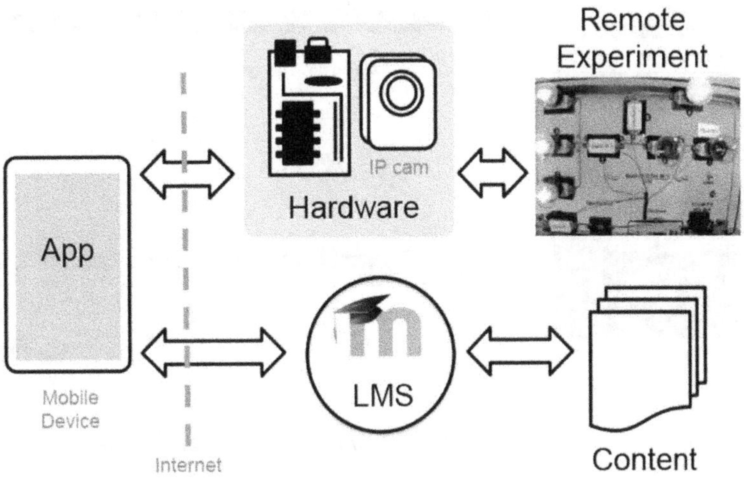

have a strong impact on training. This document will present the results in the discipline of physics at Prof. Maria Garcia Pessi Basic Education School (EMGP).

Overview of the Architecture Implemented in the Pilot Project

Figure 3 shows the macro view of the structure developed by the RExLab in order to implement a model for the use of MRE. The structured architecture has five main elements that support the MRE: Mobile-Device Application (App), Remote Experimentation (RE), Learning Management System (LMS), Content and Hardware, and its onboard software.

The block called WebApp consists of an application for mobile devices named RExMobile, which is used by students or teachers to manipulate the remote lab. Alternatively, they can use a web application that runs on any computer browser with Internet access. Figure 4 shows the interface for (a) RExMobile users and (b) web browser users.

The remote lab is the real physical system, observed and manipulated via the Internet with the use of computing devices, such as personal computers, laptops, tablet computers, and smartphones. There are different remote labs running in the RExLab currently:

- **Photoelectric Effect:** Allows the study of energy transformations, checking conversion of solar energy into electrical and mechanical energy, photovoltaic effect, semiconductor, and verifying the selectivity of the operation as the spectral region of the incident radiation.
- **Electric Table:** For practices related to associations in series, parallel and mixed in alternative-circuit networks.
- **Heat Spreading Means:** Is used to study the propagation of heat by means of conduction, convection, and radiation. Further, it is used to compare the degree of thermal isolation between different materials.
- **Thermal Conduction in Steel Bars:** Is used to study the heat propagation means through thermal conduction in different metals.

Figure 4. User interface of a lab on optics for (a) RExMobile and for (b) web browsers

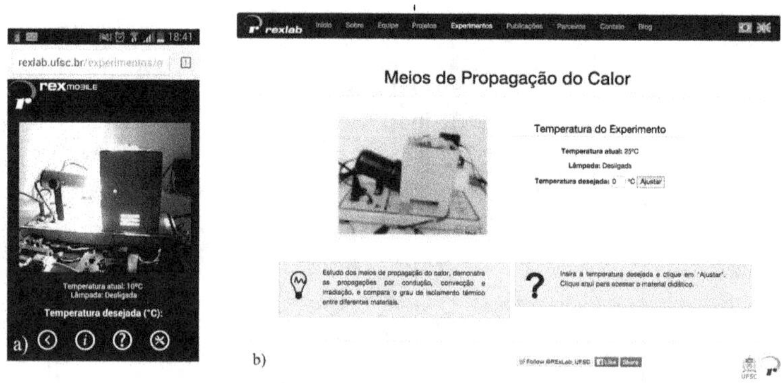

Figure 5. Remote labs on Physics available for observation and manipulation through the RExMobile app or web browser

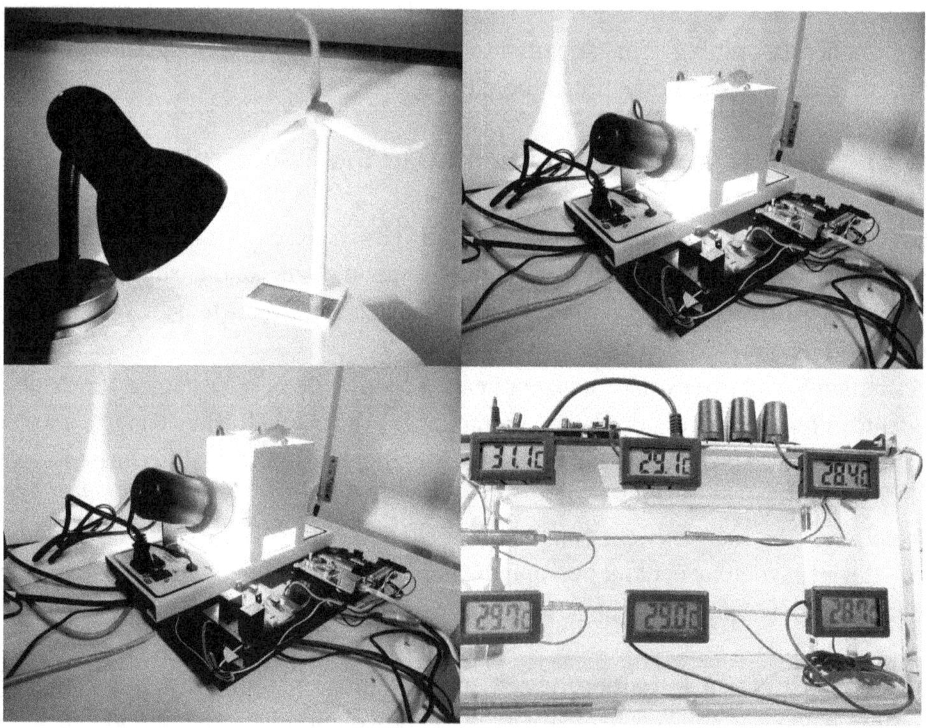

Figure 5 shows the set of physics labs with their respective web cameras and hardware dedicated to control the physical devices and transmit the live video of them.

The LMS used for the management of subjects/courses and remote labs is Moodle (Modular Object-Oriented Dynamic Learning Environment). Moodle was chosen as the LMS of the project due to its modularity and flexibility, as well as being a tool totally web-based. Moreover, Moodle is a system licensed

Figure 6. Content and activities delivered through the LMS
Source: http://rexmobile.ufsc.br/moodle/

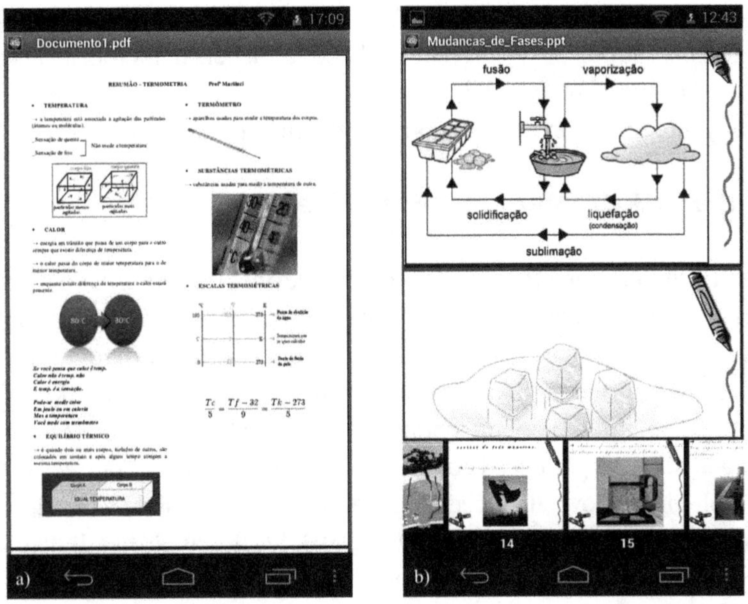

as free software that provides assistance to teachers to create online learning communities managed by them. It is an important factor since one of the project fundamentals is to provide a constructive learning. Thus, the learning protagonist is the student, through experience sharing among participants, as opposed to the traditional model, which implies in knowledge transfer from the teacher to the student. Furthermore, Moodle can still give its users a more easy design and content delivery, full management of the learning process, student assessment, access control, and technological support different pedagogical models.

The last element of the architecture of the project is the educational content, which is developed by the RExLab together with teachers from partner schools. They are open digital content that intends to provide technical, methodological, and educational support for the courses. Moreover, they aim to facilitate the integration of lab resources and syllabuses. Figure 6 shows examples of school courses content in (a) PDF (Postscript Document Format) and (b) PowerPoint format.

The hardware that controls the remote labs is based on either a low-cost platform called Raspberry Pi or an open hardware platform called Web Micro Server (WMS). The WMS platform was developed by the RExLab in order to prioritize cost-effective and open source solutions. It is a standalone device that can connect electrical devices to a computer network, equipped with sensors and actuators. A video streaming generated by an IP camera broadcasts the lab through the network. The onboard software consists of a version of Linux (operational system), PHP code and database, generating a user interface supported by HTML5, CSS3, and XML scripting languages, in a multi-layer architecture (M. S. Silva, 2012). Like the rest of the architecture elements, the hardware and onboard software prioritize the use of low-cost hardware and open software.

PRESENTATION AND DISCUSSION ON THE RESULTS

From the methodological framework developed for the project, the results presented below are divided into two sections: the first section presents results related to the teacher's profile and their perception of using technology in education from the TPACK analysis. The second section presents specifically the process of technology integration in the classroom in the discipline of physics to high school classes.

Application of the TPACK Method for the Teachers

As already stated in this chapter, the TPACK methodology defines that the effective integration of technology to teach a specific subject or content requires the understanding of the following three components: Technology, Education, and Content. A teacher that can relate these different knowledge components is distinguished by having different characteristics, which is wider than the knowledge of a specific subject teacher (a scientist, a musician, or sociologist), of a technology expert (computer scientist), or of an education specialist (an experienced educator). The TPACK model defines the complex relationships between technological knowledge, pedagogy, and content (Mishra & Koehler, 2006; Koehler & Mishra, 2009; Landry, 2010; Cruze, 2011; Madeira, 2010; Stoilescu, 2011). It displays the intersection between them as the most appropriate structure for effective educational practice, as shown in Figure 7.

Overlooking this effective way to teach, the research treated the questionnaire in two stages. The first stage was the development and implementation of a questionnaire consisting of twenty questions. This aimed to delineate the teacher profile as well as their perception of the use technology in their teaching. The second stage consists of a questionnaire with 36 items, adapted from the study "Survey of Teachers Knowledge of Teaching and Technology", which was developed by Schmidt et al. (2009). These 36 items were displayed according to Likert 5-point scale to determine how much the teachers agree with statements about their knowledge of the integration of technology in education. It was done in order to identify the factors that influence the degree of integration of technology in the classes of each teacher that participate in the project. The questionnaires were made available in Moodle for teachers who took part in the training as a prerequisite to enable access to the LMS.

The actions involving teachers were defined in three stages: (1) data collection according to the principles of TPACK; (2) categorization of questions according to the TPACK area; and (3) categorization of the results by type of knowledge (PK, CK, TK, PCK, TCK, TPK and TPACK). In this chapter, the data presented are specifically related to the Basic Education School Prof. Maria Garcia Pessi (EMGP), located in the southern state of Santa Catarina, Brazil. Table 1 presents the characteristics of the infrastructure of that school and Figure 8 shows its TPACK results.

The average score for TPACK was 3.02, with a Standard Deviation (SD) equal to 0.11, and the Coefficient of Variation (CV) of SD was 3.5%, in a range from 0 to 5. The highest mean score of the subscale was PK (Medium (M) = 3.88; SD = 0.06 and VD = 4.8%). The lowest average score was the subscale TCK (M = 2.67, SD = 0.81, and CV = 30.5%), as shown in Table 2.

Some results were expected, such as the negative trend of the technological aspects, with the result 2.78 for Technological Knowledge (TK). However other results stood out, such as the evaluation of the intersections between two knowledge components only. Teachers demonstrate a tendency to neutrality compared the Pedagogical Content Knowledge (PCK) that makes the pedagogical relationship as a mean for teaching specific content, averaging 3.26. The Educational Technological Knowledge (TPK), averaging 2.98, also presented neutrality demonstrating a blurring of the teachers' knowledge about the

Figure 7. Different knowledge components of TPACK model
Source: http://www.tpack.org

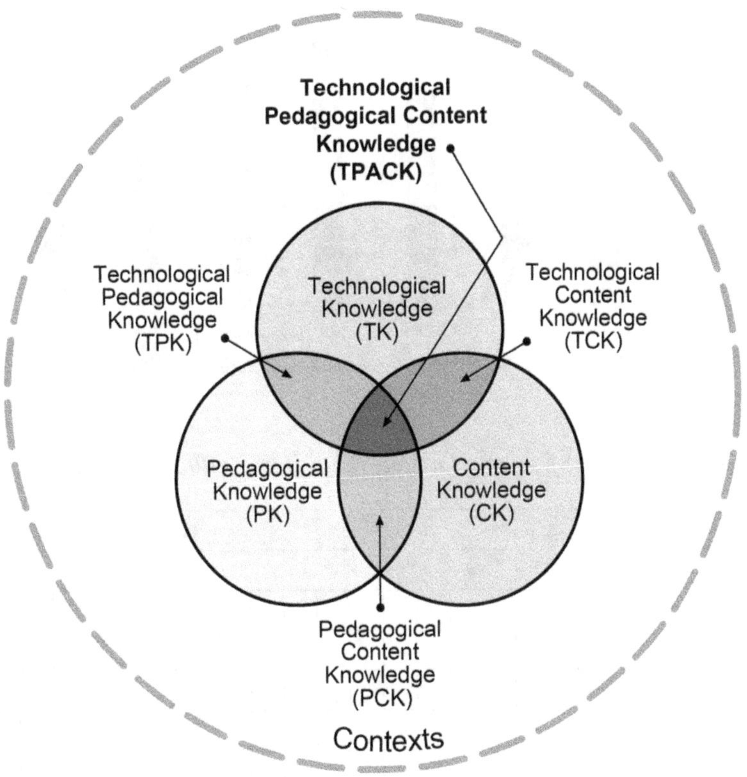

technologies and their potential in teaching, learning, and how education can be changed by the application of the involved technology.

Another relevant aspect is the average 2.67 for Technological Content Knowledge (TCK). This demonstrates a negative trend in relation to the knowledge of technology and the disciplinary content, and how to choose which technologies can enhance or limit the process of teaching and learning. In other

Table 1. Infrastructure of the EMPG

Number of staff	114
Number of students	1.752
Does it have library?	Yes
Does it have computing lab?	Yes
Does it have Science lab?	No
Does it have Internet connection?	Yes
What is the Internet connection bandwidth?	2 Mbps
How many computers are available for students?	11
How many computers are available for administration?	6

Figure 8. TPACK results for the Prof. Maria Garcia Pessi Basic Education School

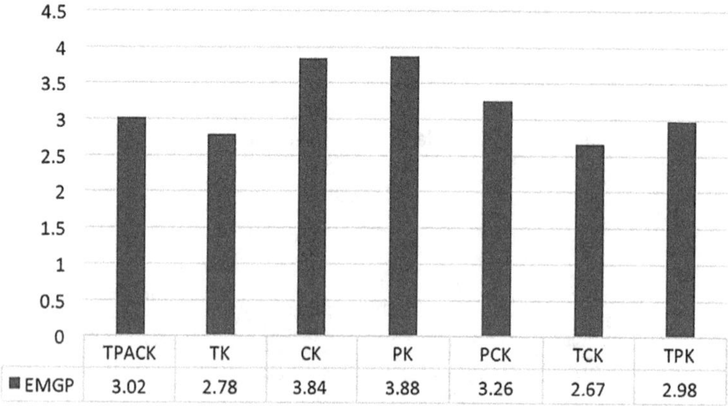

Table 2. TPACK results for the Prof. Maria Garcia Pessi Basic Education School

Research	Subscale	Average	Standard Deviation	Perception																					
				Low																					High
TPACK	Full Scale	3.02	0.11											X											
	TK	2.78	0.44									O													
	CK	3.84	0.06															O							
	PK	3.88	0.19															O							
	PCK	3.26	0.88												O										
	TCK	2.67	0.81								O														
	TPK	2.98	0.21											O											

words, when integrating the Content and Teaching Knowledge components, the tendency to neutrality is reduced.

With these results, it can be seen clearly that the concomitant use of different knowledge components still is not a common habit among teachers. This combination of different knowledge components could increase significantly the quality of education mediated by technological means.

The use of TPACK questionnaire intends to provide to the teachers a greater perception of their knowledge about the integration of technologies. Hence, they can develop proposals for pedagogical and didactic actions in order to enrich their classes integrating these tools. These initiatives aim to make them understand the necessity of integrating knowledge components and, with the help of technological resources, provide classes closer to the students' reality, promoting a teaching method that is motivating, modern, and efficient.

Regarding the use of technology, it was noticed that the teachers that integrate technology into their lessons develop an excessive ICT-centered planning. They tend to focus more on the technological tool than on the learning context supported by the technology. Harris & Hofer (2009) use the TPACK model to describe the knowledge that teachers must acquire to develop educational planning properly.

Integration of Technology in the Subject of Physics

In the Prof. Maria Garcia Pessi Basic Education School (EMGP), the research team has been working with four teachers of the courses in Physics, Chemistry, and Biology (STEM areas). They comprise 942 students divided into 29 classes. From this point on, the text reports the activities in Physics in grades 10, 11, and 12, which have respectively 170, 196, and 30 enrolled students, divided into 14 classes, all of them taught by the teacher Carine Heck.

The involved students had access to online content used in class and performed related activities on the LMS Moodle. Further, they developed practical labs remotely through computers, smartphones, or tablets. Moodle made available questionnaires, forums, and lessons, all related to the observation of the remote labs and the contents studied during the course. The teacher, her fellows, and the project development team developed these activities. The teacher has used a wide variety of resources in the construction of teaching materials that include: the features of the LMS Moodle, online videos, simulators, and remote lab, as Figure 9 exemplifies.

In the grade-11 courses, the teacher worked with digital learning materials related to heat propagation processes, heat exchange, and optics. For practical activities related the content "heat propagation," the teacher used two remote labs available in the RExLab. Figure 10 shows the access via smartphone, the two remote labs, called "Heat Propagation" and "Light-Electric Energy Conversion."

The first lab, presented in Figure 10 (a), contributes to the study of the means of heat propagation. It shows the heat propagation modes: conduction, convection, and radiation, besides comparison of the degree of thermal insulation between different materials. The second lab, shown in Figure 10 (b),

Figure 9. Examples of online content developed by the teacher on Moodle for students of the grades 10, 11, and 12
Source: http://rexmobile.ufsc.br/moodle

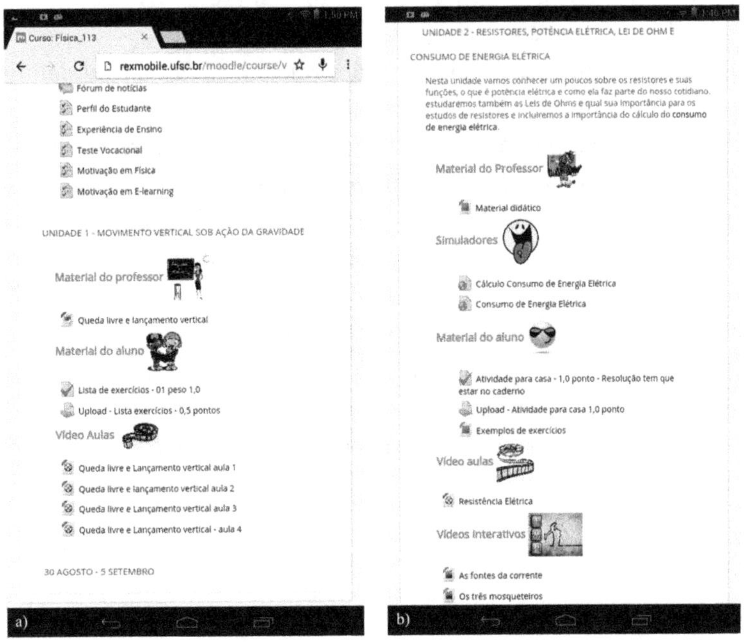

Figure 10. Remote labs "Heat Propagation" and "Light-Electric Energy Conversion"
Source: http://www.rexlab.ufsc.br

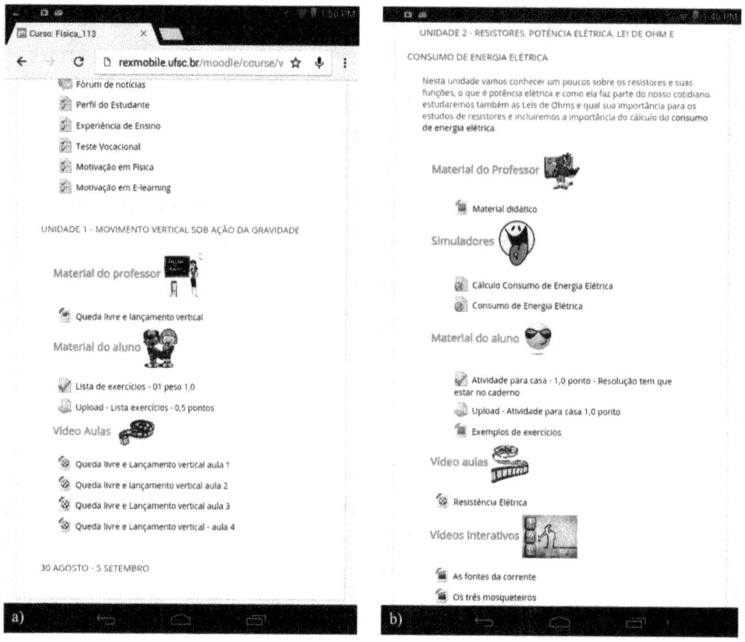

collaborates with the studies of the energy transformations, verification of solar energy conversion into electrical and mechanical energy, the photovoltaic effect, among other related concepts (RExLab).

Some students reported being surprised to see that the phone really controlled those labs and described the experience as "incredible." Another point highlighted by the students is interactivity of mobile devices. They point out the importance of immediate feedback they received from the activities carried out in the virtual environment. Thus, it allows students to find problems of comprehension quickly and to review explanations of important concepts (Unesco, 2014).

Another technological resource used by the teacher was the creation of video tutorials of real labs, developed by the students, in order to develop practical classes about optics. The videos teach how to build and operate a Newton's disk and the hole dark camera labs, and were produced using the students' smartphones' cameras. The videos were edited and made available on YouTube under the supervision of the teacher. The results were published on Moodle, as shown in Figure 11.

Figure 12 shows grade 11 students accessing educational content of the course of Physics in the classroom. To do so, more than 80% of students used their own smartphone, as indicated in Figure 13. For those who did not have a mobile device, 21 tablet computers were available for use in the classroom.

The project works in a continuous manner, in order to establish a schedule of weekly meetings with teachers of STEM areas. In these meetings, the project execution team, researchers, and fellows discuss with teachers about the development of educational content and resources to be used in the courses.

Figure 11. Video produced by students
Source: www.rexmobile.ufsc.br/moodle

Since the project is at an early stage and it is still in progress, it was not possible to measure in detail the impact of the project related to student learning. However, regarding the teaching perception, the involved teacher reported that the integration of ICT in Physics classes had a very positive result, describing the experience as being motivating for everybody. The teacher also said that the contents presented in Physics for high school are considered discouraging and complex by teachers and students, mainly due to lack of science labs in public schools, which precludes the realization of practical classes. The good performance that the students presented in class during and after the experience was extremely satisfying for the teacher. From the teacher's perspective, the students had a higher yield in these contents when compared to classes in the previous school year.

Figure 12. Students access online content and remote labs through mobile devices

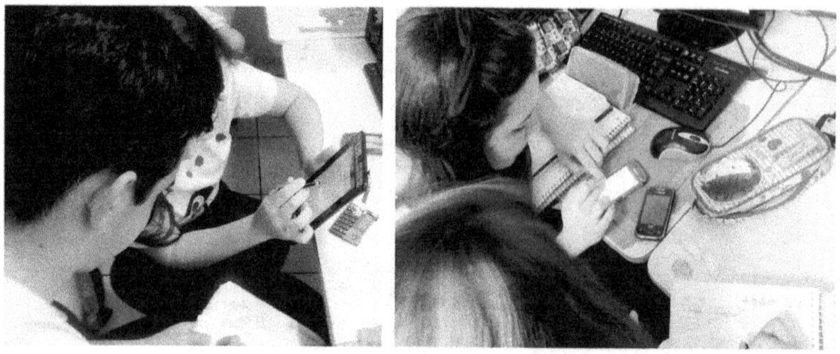

Figure 13. Students who used their own smartphones in Physics classes

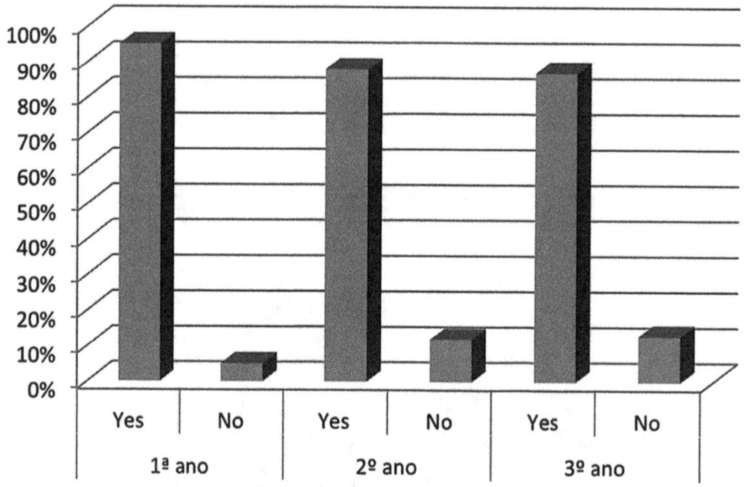

CONCLUSION

The integration of technology with the educational environment should not be a subject still to be developed in schools today but just updated. The Brazilian public schools should have structured labs with equipment chosen from a wide range of options. However, the public basic education needs more attention in the Brazilian reality. Unfortunately, the schools do not have the basic resources for science experimentation; their students cannot count on equipment or laboratories, not only in the computing area but also in science and equivalent areas.

The students' reality also does not help them to overcome this institutional barrier. A study conducted with the students who participated in the project indicated that 11% of them do not have Internet access in their homes. However, another highlighting aspect is that about 80% of these students had smartphones. This fact was very helpful so that all students could access the resources offered by the project, either by computers, by smartphones, or by the tablet computers provided by the RExLab.

Despite the technological shortage faced by Brazilian students in public schools, it is possible to make them overcome such difficulties with the RExLab project. The use the mobile devices, either owned by students (smartphones) or provided by the school (tablet computers), allows the students to enrich their comprehension of the theoretical concepts learned in the classroom by accessing the MRE. Furthermore, the MRE promotes student access to computer applications, labs, online content, and online interaction by mixing the distance education concepts with remote labs. The fact of all the project resources can be accessed from anywhere, at any time, enhances the teaching and learning processes, although within a reality yet so lacking in modern solutions in education.

Another important aspect to be observed in this process is the teacher. As seen from the TPACK results applied in this project, most teachers still do not know how to harness ICT to support teaching and learning processes. A considerable number of teachers still believe that the integration of technologies in their teaching practice requires technocentric planning without understanding exactly how to prepare their classes using the computational tools as a helper of the act of teaching. It can become a problem as long as the teacher is a fundamental piece in the pursuit of making a modern education and, at the same time, being close to the reality of the students. In this context, RExLab projects advocate the constant provision of training and advising the teachers. Ultimately, a successful MRE project deployment must involve teachers ready to use tools and devices. Further, they must be able to prepare lesson plans that can envision the integration of technology in their respective teaching contexts.

In order to continue the activities proposed in the current document, mainly aimed at technological training and support for teachers using ICT for education, the RExLab currently approved by the Brazilian National Council for Scientific and Technological Development (CNPq) the project entitled "Proposal for a methodological strategy for integrating technology in teaching STEM subjects in Basic Education of the public network". This project has financial resources to support the training and the integration of ICT in schools for a period of three years.

Regarding the remote labs, the prospect is to develop and implement four new labs, and rebuild the existing four, implementing a more efficient architecture to increase the reliability and stability of resources. After deployed this new expansion of the RExLab, it is expected to provide eight remote labs, mobile application, and educational content to support validation. This project is supported by the Brazilian Research and Education Network (RNP), and is entitled "GTMRE - Workgroup on Mobile Remote Experimentation".

REFERENCES

Al-Zoubi, A., Tahat, A., & Hasan, O. (2005). *Mobile virtual experimentation utilizing SMS*. Paper presented at the 4th IASTED International Conference on Communications, Internet, and Information Technology.

Connors, M., Bredeson, C., & Al-Shamoli, F. (2011). Distance Education Introductory Physics Labs: Online or In-Home? In J. G. Zubía & G. R. Alves (Eds.), *Using Remote Labs in Education* (pp. 309–321). Bilbao: University of Deusto.

Corter, J. E., Nickerson, J. V., Esche, S. K., Hassapis, C., Im, S., & Ma, J. (2007). Constructing reality: A study of remote, hands-on, and simulated laboratories. *ACM Transactions on Computer-Human Interaction*, *14*(2), 7, es. doi:10.1145/1275511.1275513

Costa, R. (2005). Tele-Experimentação Móvel (Mobile Remote Experimentation)-Considerações sobre uma área emergente no ensino à distância. *Journal of Scientific Activity at ISEP, 15*.

Cruze, C. I. (2011). *TPACK: a framework for educators to implement technology in their classrooms*. Academic Press.

Fidalgo, A. V. S., Rochandel, W., & Silva, J. B. d. (2013). *Remote experimentation using mobile devices*. Paper presented at the Engineering Education (CISPEE), 2013 1st International Conference of the Portuguese Society for.

Harris, J., & Hofer, M. (2009). *Instructional planning activity types as vehicles for curriculum-based TPACK development*. Paper presented at the Society for Information Technology & Teacher Education International Conference.

Hofmann, J. (2006). Why blended learning hasn't (yet) fulfilled its promises: Answers to those questions that keep you up at night. In The handbook of blended learning: Global perspectives, local designs, (pp. 27-40). Academic Press.

INEP/MEC. (2013a). *Censo da Educação Superior no Brasil 2013*. Author.

INEP/MEC. (2013b). *Censo Escolar Brasileiro 2013*. Retrieved from http://portal.inep.gov.br/basica-censo

Johnson, L., Adams Becker, S., Estrada, V., & Freeman, A. (2014). NMC Horizon Report: Edição K12 (K-12/2014 ed.). Austin, TX: Estados Unidos.

Johnson, L., Becker, S. A., Cummis, M., & Estrada, V. (2014). 2014 NMC Technology Outlook for Brazilian Universities: A Horizon Project Regional Report. Academic Press.

Kehinde, L. O., Ayodele, K. P., Akinwale, O. B., & Osasona, O. (2011). Remote Labs in Education. The Obafemi Awolowo University Experience. In J. G. Zubía & G. R. Alves (Eds.), *Using Remote Labs in Education: Two Little Ducks in Remote Experimentation* (pp. 81–111). Bilbao: University of Deusto.

Koehler, M., & Mishra, P. (2009). What is technological pedagogical content knowledge (TPACK)? *Contemporary Issues in Technology & Teacher Education*, *9*, 70.

Kofman, H. A., & Concari, S. B. (2011). Using Remote Lab for Physics Teaching. In J. G. Zubía & G. R. Alves (Eds.), *Using Remote Labs in Education: Two Little Ducks in Remote Experimentation* (pp. 293–309). Bilbao: University of Deusto.

Landry, G. A. (2010). *Creating and validating an instrument to measure middle school mathematics teachers' technological pedagogical content knowledge*. TPACK.

Ma, J., & Nickerson, J. V. (2006). Hands-on, simulated, and remote laboratories: A comparative literature review. *ACM Computing Surveys*, *38*(3), 7, es. doi:10.1145/1132960.1132961

Madeira, M. C.-A. (2010). *The development of pedagogical content knowledge in science teachers: New opportunities through technology-mediated reflection and peer-exchange*. University of Toronto.

Mishra, P., & Koehler, M. (2006). Technological pedagogical content knowledge: A framework for teacher knowledge. *Teachers College Record*, *108*(6), 1054. doi:10.1111/j.1467-9620.2006.00684.x

Nedic, Z., Machotka, J., & Nafalski, A. (2003). *Remote laboratories versus virtual and real laboratories*. Paper presented at the Frontiers in Education, 2003. FIE 2003 33rd Annual. doi:10.1109/FIE.2003.1263343

Paladini, S., Silva, J. B. d., Alves, G. R., Fischer, B. R., & Alves, J. B. M. (2008). *Using remote lab networks to provide support to public secondary school education level*. Paper presented at the Computational Science and Engineering Workshops, 2008. CSEWORKSHOPS'08. 11th IEEE International Conference on. doi:10.1109/CSEW.2008.76

RExLab. (n.d.). *Experimentação Remota*. Retrieved from www.rexlab.ufsc.br

Schmidt, D. A., Baran, E., Thompson, A. D., Mishra, P., Koehler, M. J., & Shin, T. S. (2009). Technological pedagogical content knowledge (TPACK) the development and validation of an assessment instrument for preservice teachers. *Journal of Research on Technology in Education*, *42*(2), 123–149. doi:10.1080/15391523.2009.10782544

Sharples, M., Taylor, J., & Vavoula, G. (2005). Towards a theory of mobile learning. *Proceedings of mLearn 2005, 1*(1), 1-9.

Shulman, L. S. (1986). Those who understand: Knowledge growth in teaching. *Educational Researcher*, *15*(2), 4–14. doi:10.3102/0013189X015002004

Silva, J., & Salinas, J. (2014). *Innovando con TIC en la formación inicial docente: aspectos teóricos y casos concretos*. Santiago.

Silva, J. B. d. (2006). *A Utilização da experimentação remota como suporte para ambientes colaborativos de aprendizagem*. Academic Press.

Silva, J. B. (2013). Aplicação de dispositivos móveis e experimentação remota para o ensino de Física na Educação Básica. In R. Famelli & A. P. P. Nogueira (Eds.), *Educação no Século XXI - Mobilidade* (Vol. 5). São Paulo: Fundação Telefônica.

Silva, J. B. d., Rochadel, W., Schardosim, J. P., & Fidalgo, A. V. S. F. (2014). Adaptation Model of Mobile Remote Experimentation for Elementary Schools. Tecnologias del Aprendizaje. *IEEE Revista Iberoamericana de*, *9*(1), 28–32.

Silva, M. S. (2012). jQuery Móbile–Desenvolva aplicações web para dispositivos móveis com HTML 5, CSS#, AJAX, jQuery e jQuery UI. São Paulo: Novatec.

Stoilescu, D. (2011). *Technological Pedagogical Content Knowledge: Secondary School Mathematics Teachers' Use of Technology*. University of Toronto.

Teleco. (2014). *Estatísticas de Celulares no Brasil*. Retrieved from http://www.teleco.com.br/ncel.asp

Tickodri-Togboa, S. S., Mwikirize, C., Tumusiime, A. A., & Musasizi, P. I. (2011). iLabs: Revolutionizing Teaching and Learning at Makerere University. In J. G. Zubía & G. R. Alves (Eds.), Using Remote Labs in Education: Two Little Ducks in Remote Experimentation (pp. 112-153). Bilbão: University of Deusto.

UNESCO. (2014). *Diretrizes de políticas para a aprendizagem móvel*. UNESCO.

Welsh, E. T., Wanberg, C. R., Brown, K. G., & Simmering, M. J. (2003). E-learning: Emerging uses, empirical results and future directions. *International Journal of Training and Development*, *7*(4), 245–258. doi:10.1046/j.1360-3736.2003.00184.x

Zubía, J. G., & Alves, G. R. (2011). *Using Remote Labs in Education: Two Little Ducks in Remote Experimentation* (Vol. 8). Bilbao: Universidad de Deusto.

ENDNOTE

1 The schools involved in the project are the following: Prof. Maria Garcia Pessi Basic Education School (EMGP); Otávio Manuel Anastácio City School (EOMA); Jardim das Avenidas City School (EMJA); Apolônio Ireno Cardoso Basic Education School (EBAIC).

Chapter 8
Investigating the Daily Use of Mobile Phones as Tools to Enhance mLearning for Local Cultural Subjects in the Context of Malaysian Universities

Shamsul Arrieya Ariffin
Sultan Idris Education University (UPSI), Malaysia

ABSTRACT

With the advent of mobile technology, students and academics are using mobile devices, such as mobile phones, for personal daily activities. Due to the heavy overseas 'Mat Salleh' content from English speaking countries, such as the U.S.A. and the United Kingdom this contributes to the lack of content. In fact, there is a lack of local content for local cultural subjects. Furthermore, there are limited qualitative studies that take into consideration the potential of students in the Malaysian universities to make use of mobile phones as tools to enhance mobile learning or mLearning, particularly for studying local cultural subjects. Interviews and focus group discussions were conducted for this largely qualitative study. Thematic analysis was then conducted on the data collected. The main themes that emerged from this study comprise multimedia usage and software for daily use. Thus, this study contributes to understanding by revealing future potential, including suggestions for the use of the multimedia functions of the mobile phones for better learning outcomes.

BACKGROUND

Nowadays mobile learning (or mLearning) with the use of mobile devices, such as mobile phones, as an educational tool, has become a worldwide phenomenon. For instance in Malaysia the use of mobile phones has been widely adopted by university students (Hussin et al., 2012). Before proceeding further with this study, it is appropriate to understand the background of this study context with the focus on Malaysia.

DOI: 10.4018/978-1-5225-0359-0.ch008

Lack of Local Content

Despite the urgency for the development of local content for mobile phones by the previous Malaysian minister (Lim, 2005), the study by Ariffin, Dyson and Hoskins-McKenzie (2012) identified that there is a serious lack of local cultural content. Unfortunately, the findings highlighted that local cultural content was not popular among developers. English language or 'Mat Salleh' content is more widely available via mobile phones in overseas countries, such as parts of Europe and the USA ('Mat Salleh' is a colloquial Malay word referring to English speaking content.) This leads to difficulty in accessing content pertaining to local cultural subjects for learning about the local culture in the Malaysian context. In Malaysia, it is important to follow the Malaysian Communications and Multimedia Commission (SKMM) guidelines, which is a government department that controls content development rules, such as avoiding words that are vulgar or treating others with disrespect (SKMM Guidelines, 2012). This is also in accordance with the national cultural policy that highlights the use of the main culture of the indigenous people and traditions, which is the Malay culture and behaviour (Mastor, Jin, & Cooper, 2000). The Malaysian government supports mobile learning through the use of Information Communication Technology (ICT), such as mobile learning. The recent Malaysian Development Plan for Education 2015-2025 indicates government interest in utilizing smart phone devices in education. Thus, strategizing these policies with mLearning content development should improve the outcome, including addressing the lack of content in Malaysia.

Research Aim

One of the aims of this study is to discover the potential to reduce the lack of local content for local cultural subjects or Local Cultural Studies (LCS) through the assistance of mobile devices as a tool for mLearning. For convenience, the term LCS has been used to represent subjects that relate to Malay culture as the local culture, such as History, Local Culture and National Heritage, and Cooking. Furthermore, there are limited studies in the area of local culture in the Malaysian context, particularly in understanding the experience of users with mobile devices. Malaysian university students could benefit from the use of mobile learning or mLearning in learning LCS. This area of study in the Malaysian context is under-researched. Therefore, this study explores the motivation of the users for the use of mobile devices, such as mobile phones, in their daily life, irrespective of whether their use is for educational purposes, as well as their potential use for the study of local culture.

Mobile Learning Definition for this Study

Dyson and Litchfield (2011) defined mLearning as learning facilitated by mobile technologies. This can be extended to different settings, for instance, to Malaysian universities for the study of local culture. However, this study defines mobile learning, as the use of mobile technology and devices, particularly students' own mobile phones, in facilitating the learning of LCS. This is because, according to Hussin et al. (2012), although other suitable devices may support mobile learning activities in the process of generating local content, mobile phones are universally owned by Malaysian university students. In this study, mobile phones constitute a mobile device that can be used as a tool for mLearning for potentially improving the outcome for educational purposes.

The Importance of mLearning

MLearning is seen as a potential catalyst to transform learning from a traditional to a more active approach in supporting learning in any country in the world including Malaysia. Kukulska-Hume (2010) suggested that academics and students should be aware of the potential for mLearning. It is vital to understand the impact of information technology (Myers & Avison, 2002) on the students and academics with the advent of mLearning. Their perspectives should be acknowledged with regards to the sustainability of mLearning (Ng & Nicholas, 2012); hence, the purpose of the investigation is to understand the established practices of the academics and students in using mobile devices, such as mobile phones, for personal daily use.

Mobile Phones as a Tool for Learning

Mobile phones are an appropriate tool for learning local culture for the Malaysian context. Hussin et al. (2012) state that the ownership of mobile phones by university students is 100%. Their study indicates a higher use of mobile phones for communication, multimedia usage, software and hardware utilities for the Malaysian context in the local universities. Pettit and Kukulska-Hulme (2007) explained the possibilities in bridging formal and informal learning, which can be applied from the learners' experience of personal daily usage of mobile devices, such as a hobby, to that of education. Ally (2009) revealed that the skills derived from the use of mobile devices for personal use could be applied in many different contexts of learning. This exploratory approach to self-engaging in the use of mobile devices could be embedded in students' learning. Another example is Kukulska-Hulme, Traxler and Pettit (2007) who highlighted the possible benefits of student-generated activities for meaningful education.

Activity of mLearning with Mobile Devices

Mobile devices, such as mobile phones or PDAs, enhance the mLearning activity irrespective of the situated context. Ahmed and Parsons (2013) demonstrated the effectiveness of using mobile devices for mLearning to study science subjects. On the other hand students participated actively in using mobile devices in learning Chinese culture in Taiwan, as demonstrated by Hwang and Chang (2011). Uzunboylu, Cavus and Ercag (2009) demonstrated that students perform better in using mobile phones in capturing photos while learning about the local environment. In addition, Looi et al. (2010) discussed the potential of mLearning in a seamless environment that benefits both the learner's informal and formal learning. Vavoula et al. (2009) demonstrated the success of using mobile phones as mobile devices in learning about local art and museums in a situated context. Similarly, students were able to actively use mLearning in a situated context with the assistance of mobile devices (Dyson et al., 2008; Herrington & Herrington, 2007).

In the Malaysian context, using mobile devices, such as mobile phones, can be extended to learning local cultural subjects that have similar activities, particularly in using the multimedia functions of the mobile devices. Hence, this indicates the potential activity of mLearning with mobile devices, such as mobile phones for local cultural subjects.

Why mLearning for Local Cultural Subjects?

Malaysia is using a blended approach, for example, using mobile phones with other devices, such as laptops and personal computers or tablets. Siraj and Nair (2007) in their study in Malaysia argued that mLearning should include personal computers to interface with mobile devices in blended learning situations. In the meantime, Wu et al. (2012) considered that mobile phones and PDAs are the two most used mobile devices for mLearning. More recently, Abas and Hussain (2013) conducted a survey on the successful use of tablets among academicians at Malaysian universities. However, learning using mobile phones for other areas, such as LCS in the Malaysian context, has been overlooked and unexplored. This can be discovered by investigating the potential daily use of mobile phones in the context of Malaysian universities.

mLearning activities lend themselves to the study of local cultural subjects in a variety of ways. Students can benefit through activities in learning local culture from cultural mobile applications (Brady & Dyson, 2010. In fact, student-generated content could happen anywhere, whether in the classroom through practical activities using mobile phones, such as recording videos and taking photos, or outside of the classroom (Shih et al., 2012). On the other hand a research by Ariffin and Dyson (2015) proposed four main culturally appropriate design principles that can be used when developing the local content of mobile devices, such as mobile phones: the use of the local language Bahasa Malaysia or bilingual, local philosophy, local aesthetic values, and content that is suitable for the local context. Therefore, in the Malaysian context this research helps in determining the potential for generating local cultural content as one of the approaches to reducing the lack of local content for local cultural subjects.

Local Malaysian Universities

In Malaysia tertiary or higher education is the responsibility of the Ministry of Education (MoE). At the tertiary level of education, universities offer courses leading to the award of certificates, diplomas, first degrees and higher degrees in academic and professional fields. Normally, the course duration for a bachelor degree programme is three years and courses are provided by both the public and private education sectors. There are about 18 public universities that are funded by the government. In public universities in Malaysia, there are faculties that teach Malay culture, such as in the field of art and design, performance and music, humanities and history subjects. Students take these subjects as major subjects or as an elective subject. The majority of students in local universities are Malay students, and they represent the majority of the culture in Malaysia. Most students involved in this research come from rural areas in the 'Kampung' or villages where there is a lack of telecommunication facilities, such as network coverage.

This article is organized in different sections. It begins with the background of this research, which focuses on the Malaysian context in the universities and why it is important to conduct research on the study of local culture with the advent of mobile technologies. The article covers the research methodology, which is largely qualitative; collecting data from interviews with academics and students via focus group discussions. The data for the results are organized using NVivo software and analysed using Thematic Analysis. The discussion section explains the value added for the daily use of mobile phones with the use of multimedia functions. Finally, this article concludes with the recommendation of using mobile phones for educational purposes in LCS.

RESEARCH METHODOLOGY

The study is largely qualitative in finding the in-depth meanings, and understanding the perspectives from the participants from the interviews and focus group discussion (Mertens, 2010). There are two main Malaysian universities involved in this research. There are two reasons why these universities are chosen and why they are considered ideally suited for this research. Firstly, these universities play an important part in the teaching and learning of local Malay culture. The second reason is a pragmatic one, in that the researcher already has established contacts with academics at these universities. Most of the students recruited are Malay students. The high adoption of mobile phones by university students in Malaysia made this research focus on mobile phones. The study explores the users' daily experience in using mobile phones. Academics and students were selected using purposive random sampling. Additionally, focus group discussions were selected due to their convenience in facilitating discussions with the Malaysian students. Students were expected to communicate better using this approach. This qualitative study undertook 15 focus groups with 127 students and 15 academics.

Due to pragmatic reasons, the inductive approach is adopted for this study. According to Blaikie (2009), the inductive research approach consists of four main stages: all data are recorded without deductions to their relative significance; these data are analysed, compared and classified, without using hypotheses; from this analysis, generalisations are inductively drawn as to the relations between the facts; and these generalisations may require further assessment.

In addition, the researcher's methodology for examining the data from the interviews and focus groups discussions was that of thematic analysis (Braun & Clarke, 2006). The themes were derived after subjecting the data to the thematic analysis process. They pointed out that the study of thematic analysis involves familiarising oneself with the transcripts, initialising the code, and making sense of the themes. Additionally, in this research, the participants' answers were audio recorded. The audio was then transcribed into the Malay language (Bahasa Malaysia). Occasionally, the researcher had to listen to the recordings several times in order to obtain an accurate transcription. Subsequently, the Malay language transcription was translated into English. Generating themes requires processes, such as coding, combining the coding into groups of themes, and themes being reorganised if required (Saldaña, 2009). Similarly, Auerbach and Silverstein (2003) pointed out that the coding stages are non-linear processes and can loop backwards, with alterations to the coding and themes as the development of themes carries on. The themes were organised again to guarantee the understanding of the data and that the analysis was coherent with the thematic context.

The semi-structured questions asked in this study were based on the users' daily use of mobile devices, such as their own mobile phones. Some of these questions were inspired from Petit and Kukulska-Hulme (2007) to identify valuable experiences of the students from the daily use of mobile phones. Furthermore, this approach has been elongated in the local Malaysian university context. Rather than taking the adoption of mobile phones due to their appeal (Rogers, 2010), this study takes into consideration all the meaningful and valuable experiences of using mobile phones for potential learning activities as alluded to by Pettit and Kukulska-Hulme (2007). The semi-structured open-ended questions are as below:

- What are the users' experiences in using mobile devices such as mobile phones in their daily life?
- Which mobile devices such as mobile phones do they use and why do they use them?
- What are the users' experiences in using the applications and services of mobile devices such as mobile phones in their daily life?

- What are the users' experiences in using the applications and services of mobile devices, such as mobile phones, for educational purposes?
- Why and how do they use such mobile applications and services for educational purposes?

Focus Group Discussion Procedures

The initial focus group discussions were conducted to collect data concerning their pre-existing perspectives and experiences of mLearning, which are their daily practices. Most of the students were predominantly Malay university students. In total, there were 15 focus groups with 127 student participants. The ages of the students ranged from early twenties to twenty-five years old. Focus group discussions were selected as an appropriate method for data collection from the students (Barbour, 2008), as Malaysian students are more responsive when they discuss in groups than when they are asked questions individually. Focus group discussions were conducted with the students from a wide range of humanities subjects – Local Culture and National Heritage, History, Cooking, Creative Movement, Drama and Theatre, Education Action Research, Batik Textiles, Malay Wood Craft, Malay Drums, and Management – all of which related to the study of local culture. The focus groups were organized according to subject (table 1).

Interview Procedures

Most of the academic participants taught subjects related to LCS. These interviews were conducted with fifteen academic participants. Most of the academics were from a wide range of predominantly humanities subjects; Local Culture and National Heritage, History, Cooking, Drama and Theatre, Education Action Research, Batik Textile, Malay Wood Craft, Malay Drums, Education Technology, Ceramics, Visual Art, Linguistics and Distance Learning. A Science academic was included to consider the likelihood of using mLearning to incorporate LCS in Science subjects. Meanwhile three did not include Malay culture in the way they were being taught but the academics had used some sort of mLearning and were included for this reason; these were Education Technology, Linguistics and Distance Learning. They were included to investigate the potential of mLearning embedded with LCS in their subjects. The ages of the academics ranged from late twenties to sixty years old. All of the participants were Malay academics. They had at least a few years teaching experience and at least a master's degree in order to qualify for a teaching

Table 1. Focus Group

Subject	Code	Subject	Code
Management	FG1	Education Action Research 2	FG11
Local Culture and National Heritage 3	FG2	Cooking 1	FG8
Local Culture and National Heritage 1	FG3	Cooking 2	FG9
Local Culture and National Heritage 2	FG4	Batik Textiles	FG10
Drama and Theatre 1	FG5	Malay Drums	FG13
Drama and Theatre 2	FG12	Creative Movement	FG14
History	FG6	Malay Wood Craft	FG15
Education Action Research 1	FG7		

Note: 'FG' = Focus Group, for example FG1 = Focus Group 1

Table 2. Academic

Subject	Code	Subject	Code
Cooking	Ac1	Malay Drum	Ac9
Local Culture and National Heritage	Ac2	Long Distance Technology	Ac10
Drama and Theatre	Ac3	Batik Textiles	Ac11
Education Action Research	Ac4	Linguistics	Ac12
History	Ac5	Ceramics	Ac13
Visual Art	Ac6	Learning Technology	Ac14
Creative Movement	Ac7	Malay Wood Craft	Ac15
Science	Ac8		

Note: 'Ac' = Academic, for example Ac1 = Academic 1

position. The academics had a variety of exposure to different teaching and working environments. A summary of the details of the backgrounds of the academics is illustrated in Table 2.

FINDINGS

Most of the participants owned mobile devices, such as mobile phones, and could use the mobile device functions for personal daily use. In this study, the participants conveyed the reasons why they owned mobile phones and why they used mobile phones for personal daily use. They used mobile phones for activities, such as for leisure and hobbies. The participants reported using mobile devices with brands, such as Nokia, Sony Ericsson, Samsung and BlackBerry, which they could afford to purchase. Very few mentioned using iPads or tablets when this study was conducted. In addition, participants used mobile phones for various reasons due to the affordability of the multimedia functions of mobile devices, such as for photography, audio sound, video, software applications, its free services and Internet connection. Additionally participants' relation with family members and friends could be much closer even from a long distance by using mobile phones. Examples:

I can interact with my daughter in Indonesia so she will not feel lonely. We exchange photos via MMS service. [Academic Comment]

I use my mobile phone to make video calls to see my husband and my son. [Student Comment]

Meanwhile some participants remained unaware of their potential for mLearning. As mLearning is new, they guessed about the use for mLearning due to limited exposure: "I really don't know. It is not really for learning for me but to play games" [Student Comment]. Consequently, their experience was limited to mostly simple mLearning, such as communicating using SMS and phone calls. One academic argued: "For mobile learning we need to go beyond SMS, beyond email. Are they ready for that? That is the issue [Academic Comment]".

There is also suggestion for students to become content producer. An example:

They will be the content provider. They share the ideas together with other friends, where, in the end, I learn from them too. [Academic Comment]

Therefore the important themes that emerged from this research are multimedia usage and software for daily use. These are the themes that emerged from the affordances of mobile devices for personal use and educational purposes.

Multimedia Usage

The participants used mobile devices, such as phones for the personal convenience of the applications that matched their daily task objectives. For example, academics used mobile phones for their software multimedia functionalities: photography, video recording and audio recording. Students were found to be more active than the academics in using the multimedia of the mobile phones: "My students are more advanced than their own lecturer with their own latest up to date mobile phones" [Academic Comment]. Most students use the photography, and audio recording and video recording functions from their own mobile phones.

Photography

Photos are captured by participants to photograph personal events that could later be reused. Most students informed that they benefit from the use of the photography function of the mobile phone camera to capture special moments for their own personal interest, as stated by the participants in focus groups FG2, FG3, FG5, FG6, FG7, FG8, FG9, FG10, FG11, FG12, FG13, FG14 and FG15: "I can use Bluetooth after capturing pictures of my family and friends. I then use Bluetooth to transfer the photos to my friends. I don't need to pay for Bluetooth [Student Comment]." Likewise, academics Ac1, Ac5, Ac14 and Ac15 stated that they use the photography function of the mobile phone camera to assist them to capture personal digital photos for leisure. A few educational examples:

We have done a subject called Malaysian studies. One of our fieldwork research subjects is on vandalism. For the assignment, we went to the largest bus station in Kuala Lumpur 'Pudu Raya'. We took lots of pictures of vandalism at the bus station using our Sony Ericsson mobile phones... [Student Comment]

In our Sewing course we had to take photos of the sewing process from the beginning until the end. We include the photos taken by the mobile phones in our assignment report in order to support our final report. [Student Comment]

The photos that I had taken using mobile phones are still pictures during stage presentation or at the interesting concert sets to inspire good ideas for the students. [Academic Comment]

Video Recording

Video recording is one of the most common functions for personal usage on mobile devices, such as mobile phones. Most students from focus groups FG3, FG5, FG6, FG10, FG11, FG13, FG14 and FG15 informed that they use the video recording utility of their mobile phones for personal use to capture a special memory of an event that could be watched later.

Similarly, academics Ac7 and Ac8 indicated that they use the video recording utilities of mobile phones for leisure and also because of the good quality of the video files. They benefited from using mobile phone video recording for personal use. A few educational examples:

..I taught the foreign students from Korea the meaning of 'Gurindam'. I taught them how to pronounce, recite and develop this traditional poetry while the video recorded my performance. Later they could practice the 'Gurindam' by accessing the videos. [Student Comment]

... For example, in the garden, we planted the Roselle tree. From the recording we could better understand the detailed process of planting the Roselle tree, the square feet needed for the land, how much fertiliser was needed. [Student Comment]

We can video record the performance and upload to YouTube. From there we can get connected to access the video to watch and practice as a model and explore our dance movements using the video in a self-exploratory learning approach. [Student Comment]

We had a stage show called the Court of Justice or 'Mahkamah Keadilan'; a piece of our local theatre performance. In the theatre production, my students recorded their performances using their own mobile phones for revision. [Academic Comment]

Audio Recording

The audio recording facility is utilized by the participants for personal use. The students from focus groups FG3, FG7, FG8, FG11 and FG15 highlighted that they use their mobile phones for audio recording in terms of leisure for recording songs. In addition, academic Ac14 highlighted that she uses her mobile phone to assist in audio recording as a means of leisure for her own personal use. Thus, participants can reuse the audio recording files for future usage. A few educational examples:

We went to the local market of Siti Khatijah in Kelantan and interviewed the small business local traders. We interviewed them concerning how they did business using our own mobile phones for our assignment. [Student Comment]

I recorded all the songs on the mobile phone. Next, I used Bluetooth to transfer the files to my students. I told them that these were the songs I wanted them to practice. They accepted the file transfer and composed their activities according to their creativity with the songs that I had given them. [Academic Comment]

When we learn music it is for lifelong learning and not only inside the classroom. Students need to practice; therefore, students will record their demo and reuse it as reference. [Academic Comment]

Software for Daily Use

Participants benefit from the use of the mobile software applications that are already available on their own mobile phones for daily use to manage their own daily tasks. Two broad categories of software application for daily use are those for personal organization and those for personal development.

Personal Organization

Participants utilize mobile applications for their daily activities, such as using the GPS, alarm and for finding the direction to pray. The students in focus groups FG2 and FG11 use the alarm function to keep alert and to wake up from sleeping. Likewise, the students from focus groups FG2, FG3, FG4, FG10, FG11 and FG15 reported that they use the GPS application to assist them to find locations and directions while travelling on the road. The participants in FG11 reported using the mobile software to find the direction of the 'Kiblah' in Mecca for praying. Additionally academics Ac1 and Ac14 reported using the GPS for personal usage for finding directions. Academic Ac5 used the mobile phone alarm function and diary software as a reminder. Academics Ac12 and Ac14 stated that they use the calendar in their mobile phone. Therefore, the participants gain from using mobile phones for personal organization, particularly for finding directions, as a diary, calendar and for the alarm service.

Personal Development

The participants use mobile devices to support information searches and personal development in their private lives and with respect to their hobbies. For instance, they gather information for their own personal development during their free time: learning English, Jawi and the Holy Quran, Google Searches, and websites. For example, English language learning, such as dictionary and translation, is very popular amongst students. Likewise, focus groups FG2, FG3, FG4, FG5, FG7, FG10, FG12 and FG15 use the bilingual dictionary on their mobile phones to assist them in translating words from English to Bahasa Malaysia. Focus group FG5 use mobile phones to assist in learning the pronunciation of English.

Islam, as the official religion in Malaysia, is practiced widely by the Malay Moslems including reading the Holy Quran and learning Jawi writing. Focus group FG9 informed that they use mobile phones to assist them in their personal education in learning Jawi. Additionally, the students in focus groups FG6, FG7 and FG15 utilise mobile phones to assist themselves in learning Al-Quran, the Holy Book of Moslems.

Students also search for personal information using mobile devices. Focus groups FG2, FG3, FG4, FG5, FG6, FG11, FG12, FG13 and FG14 reported using mobile phones for Google searches for personal usage to find information that they want. Participants use websites, such as blogs and wikis, for their own personal development and leisure. In addition, focus groups FG2, FG5, FG9 and FG15 indicated that they access and surf websites via mobile phones for their own leisure to find information. Focus groups FG10 and FG12 access blogs via mobile phones for their own interests. Focus groups FG2, FG6 and FG14 also use wiki via mobile phones to assist their own personal development, such as finding out about their personal interests. Students also use mobile applications for specific personal purposes, such as hobbies. FG15 reported using a fishing application via mobile phones as guidance to correctly learn fishing methods. Focus groups FG2, FG3, FG4 and FG12 use online news services to assist them to catch up with the latest news from online newspapers.

Academic Ac3 used a mobile phone to assist him to search for the meaning of words using the online English Dictionary software, while academic Ac7 uses mobile software to assist him to read the Holy Al Quran. Academics Ac4 and Ac12 reported that they use their own blog for personal usage and interest. Therefore, the participants profited from the use of mobile phones for personal development and personal interest. This suggests the further potential of mobile phones for educational purposes. Therefore, all these findings concerning daily usage reveal the potential to use mobile phones for educational purposes.

DISCUSSIONS

Malaysian university academics and students could potentially use the multimedia functions of mobile phones, such as video content production, to reduce the challenge of the lack of local content to be applied for LCS. However mLearning has been taken for granted without much focus on a more sophisticated approach for mLearning activities. There was limited awareness of mLearning; only a basic use of mobile phone practices. For example, they are not aware of the other potential of multimedia for generating content through the use of video.

This research demonstrates the possibility of adopting the use of mobile phones with multimedia functions to undertake daily activities that could be applied to educational uses, such as for the local cultural subjects. The researcher discovered that all the participants own mobile phones and use mobile devices, such as mobile phones, for personal interest and leisure. As the participants carry mobile phones with them most of the time, they also use the utilities on the mobile devices, such as multimedia functions. Hence this study stresses that the personal use of mobile phones has cultural influences, and adds meaningful value for daily tasks and activities, as well as for the multimedia functions by taking advantage of the educational purposes for a better learning outcome.

Cultural Influences

Academics and students use mobile phones to connect with their family and friends. This shows that family and sense of community plays an important role in their lives. Likewise, family ties play an important role in the Malaysian culture. The use of multimedia functions of mobile phones, such as SMS, MMS and video calls, affirmed such cultural values for the Malaysian context of usage. This aligns with the studies of Hussin et al. (2012), which indicates higher usage of the multimedia functions of these mobile phones.

Academics and students in Malaysia use mobile phones to learn more about Islam. For example, they can search the time they are going to pray from the mobile phones. In addition, they also learn to recite Al Quran correctly through the use of mobile phones. This happens through the use of special Islamic mobile applications provided by the content developer. This aligns with the national cultural policy, which stresses the importance of Islam as the mainstream culture (Mastor, Jin, & Cooper, 2000). On some occasions, students use mobile phones to unwind from the stresses of studies, such as by listening to songs, watching videos and playing games, while others communicate via SMS, MMS and social media.

Academics and students in Malaysia speak Malay as their mother tongue. They use English as their second language in the university whether for communication or learning purposes, and will definitely want to know more about English since most of the written material online is provided in English.

Therefore, a bilingual translator or dictionary application is crucial to assist them in their learning of both languages.

Adding Meaningful Value for Daily Tasks and Activities

Mobile phones have great potential in adding meaningful value, irrespective of whether they are used for personal leisure or undertaking tasks related to the daily activities of the participants. In addition, the researcher identified that participants use mobile devices for personal daily use, such as for communicating, which has the potential for mLearning for their subjects, such as the study of local culture. For example, students communicate with family, friends and their loved ones. They use them for personal development, such as finding meaning in the English online dictionaries and also for learning Islamic religious matters. However, in this study, the use of mobile devices is not a matter of appeal with the devices, as argued by Rogers (2005), instead, this study is in accordance with the argument by Pettit and Kukulska-Hulme (2007) who claimed that the personal use of mobile phones could add meaningful value rather than simply be a mere object of obsession.

Multimedia Functions and its Learning Potential

One of the affordances of mobile phones is the use of its multimedia functions that are available on most of the participants' mobile phones. In fact, there is huge potential for mLearning for a better educational outcome, as all students in the Malaysian universities own mobile phones, as pointed out by Hussin et al. (2012). For example, in this study, students were found to use the multimedia functions – video recording, audio recording and photography – on their mobile phones for personal daily use and learning informally. They could use mobile phones for student generated content activities that could improve productivity and learning outcome for the subjects (Dyson et al., 2008), such as LCS. Consequently, this local content development using mobile devices must be in accord with local policies, such as the Malaysian content development guidelines (SKMM Guidelines, 2012) and National Cultural Policy (Mastor, Jin, & Cooper, 2000). As a result, the usage of these multimedia functions could be harnessed for mLearning for the study of local culture, such as by recommending that the students become content producers in producing videos. This practice potentially promotes sustainability, and could reduce the challenges of the lack of local content that was highlighted earlier by Ariffin, Dyson and Hoskins-McKenzie (2012). The future content development could be associated with Culturally Appropriate Design Guidelines (Ariffin & Dyson, 2015), such as the use of Bahasa Malaysia in the content, and local content that is suitable for local subjects in LCS.

CONCLUSION

This study contributes to the potential use of mobile devices – mobile phones – as tools for educational or mLearning purposes. This adds new understanding for mLearning in the study of local culture in the context of Malaysian universities. For instance, this research investigates the perspectives of Malaysian participants in local universities concerning the daily use of the multimedia function on mobile phones. From this research, the participants have known to have some sets of skills in using mobile devices such as mobile phones used in their daily life. These skills are transferrable for educational purposes. The

Table 3. MLearning Strategies for LCS

	LCS Subjects	Topic (Scenario/ Context)	Learning Objectives	Strategies	Multimedia Functions of Mobile Phone Affordances for Content Production (Skills)		
					Video	Audio	Photography
1	History	Malaysian Archaeological Sites (Peninsular Malaysia)	To Identify the Archaeological Sites and their Cultural Artefacts	Interview the Cultural Experts at the Archaeological Sites	✓	✓	
				Take Photographs of Artefacts at Archaeological Sites			✓
			To Describe the Archaeological Sites and their Cultural Artefacts	Voice Narration describing the Archaeological Artefacts/ Sites		✓	
			To Demonstrate Understanding and Knowledge about Malaysian Archaeological Sites by Presenting the Assignment Report	Produce Malaysian Archaeological Historical Sites Report	✓		
2	Cooking	Malay Traditional Cooking (Perak State in Malaysia)	To Identify the Cooking Ingredients and the Cooking Process	Take Photographs for the Traditional Malay Food Preparation and Cooking Process			✓
			To Describe the Traditional Cooking Recipes	Voice Narration describing the Malay Traditional Cooking Process		✓	
			To Demonstrate Understanding and Knowledge of Local Traditional Cooking by Presenting the Assignment Report	Produce Traditional Malay Cooking Video and Report	✓		
3	Local Culture and National Heritage	Malay Traditional Palace (Malaysia)	To Identify Historical Information about the Malay Traditional Palace and its Cultural Artefacts	Interview Guardian of the Malay Traditional Palace	✓	✓	
				Take Photographs of the Malay Traditional Palace			✓
			To Describe the Architectural Components of the Malay Traditional Palace	Voice Narration about the Palace Architecture		✓	
			To Demonstrate Understanding and Knowledge about the Malay Traditional Palace by Presenting the Assignment Report	Produce Report describing the Malay Traditional Palace Architecture	✓		
4	Drama and Theatre	Malay Drama (Malaysia)	To Demonstrate Acting Skills Knowledge by Performing for the Malay Drama	Rehearse Acting Part from Self-Taken Recorded Video for the Malay Drama	✓		

guidelines for the proposed mLearning strategies to improve learning outcome using the multimedia functions and affordances of students' owned mobile phones are illustrated in Table 3.

The daily use leads to greater awareness and a wider adoption of mLearning to improve learning outcomes for LCS subjects in the Malaysian university context, such as through the usage of the multimedia functions of mobile phones. This is the first time mLearning research for the study of local culture for Malaysian universities has been conducted. Since some of the students already use multimedia functions with their mobile phones in their daily lives, this indicates the potential to create video content for local cultural subjects. This is one of the prospects found in this study in order to reduce the lack of local content by having the students themselves become content producers, which is to be conducted as future work.

REFERENCES

Abas, Z. W., & Hussain, R. M. R. (2013). *Acceptance of tablet devices for learning and teaching in institutions of higher learning: The Malaysian perspective*. New York: Routledge.

Ahmed, S., & Parsons, D. (2013). Abductive science inquiry using mobile devices in the classroom. *Computers & Education, 63*, 62–72. doi:10.1016/j.compedu.2012.11.017

Ally, M. (Ed.). (2009). *Mobile learning: transforming the delivery of education and training*. UBC Press.

Ariffin, S. A., & Dyson, L. E. (2015). Culturally Appropriate Design of Mobile Learning Applications in the Malaysian Context. In *Cross-Cultural Design Applications in Mobile Interaction, Education, Health, Transport and Cultural Heritage* (pp. 3–14). Springer International Publishing. doi:10.1007/978-3-319-20934-0_1

Ariffin, S. A., Dyson, L. E., & Hoskins-McKenzie, D. (2012). Content is king: Malaysian industry experts' point of view on local content for mobile phones'. *Journal of Mobile Technologies. Knowledge in Society, 2012*, 1–9.

Auerbach, C. F., & Silverstein, L. B. (2003). *Qualitative data: An introduction to coding and analysis*. New York: New York University Press.

Barbour, R. (2008). *Doing focus groups*. Sage Publications Ltd.

Blaikie, N. (2009). *Designing social research*. Polity Press.

Braun, V., & Clarke, V. (2006). Using thematic analysis in psychology. *Qualitative Research in Psychology, 3*(2), 77–101. doi:10.1191/1478088706qp063oa

Dyson, L. E., Lawrence, E., Litchfield, A. J., & Zmijewska, A. (2008). M-fieldwork for information systems students.*Proceedings of the Forty-First Annual Hawaii International Conference on System Sciences (HICSS)* (pp. 1-10).

Dyson, L. E., & Litchfield, A. (2011). Advancing Collaboration between M-Learning Researchers and Practitioners through an Online Portal and Web 2.0 Technologies. *International Journal of Mobile and Blended Learning, 3*(1), 64–72. doi:10.4018/jmbl.2011010104

Herrington, A., & Herrington, J. (2007). Authentic mobile learning in higher education. Australian Association for Research in Education (AARE) 2007 Conference (pp. 1-9).

Hussin, S., Manap, M. R., Amir, Z., & Krish, P. (2012). Mobile learning readiness among Malaysian students at higher learning institutes. *Asian Social Science*, *8*(12), 276–283. doi:10.5539/ass.v8n12p276

Hwang, G. J., & Chang, H. F. (2011). A formative assessment-based mobile learning approach to improving the learning attitudes and achievements of students. *Computers & Education*, *56*(4), 1023–1031. doi:10.1016/j.compedu.2010.12.002

Ismail, I., Idrus, R. M., & Gunasegaran, T. (2010). Motivation, psychology and language effect on mobile learning in Universiti Sains Malaysia. *International Journal of Interactive Mobile Technologies*, *4*(4), 31–36.

Kukulska-Hulme, A. (2010). Learning cultures on the move: Where are we heading? *Journal of Educational Technology & Society*, *13*(4), 4–14.

Kukulska-Hulme, A., Traxler, J., & Pettit, J. (2007). Designed and user-generated activity in the mobile age. *Journal of Learning Design*, *2*(1), 52–65. doi:10.5204/jld.v2i1.28

Lim, K. Y. (2005). *Minister speech*. Retrieved June 7, 2011, from http://www.tmforum.org/WhyAttend/2719/home.html?linkid=31446

Looi, C. K., Seow, P., Zhang, B., So, H. J., Chen, W., & Wong, L. H. (2010). Leveraging mobile technology for sustainable seamless learning: A research agenda. *British Journal of Educational Technology*, *41*(2), 154–169. doi:10.1111/j.1467-8535.2008.00912.x

Mastor, K. A., Jin, P., & Cooper, M. (2000). Malay culture and personality. *The American Behavioral Scientist*, *44*(1), 95–111. doi:10.1177/00027640021956116

Mertens, D. M. (2010). *Research and evaluation in education and psychology*. Sage Publications Inc.

Myers, M. D., & Avison, D. (Eds.). (2002). *Qualitative research in information systems: a reader*. London: Sage Publications.

Ng, W., & Nicholas, H. (2013). A framework for sustainable mobile learning in schools. *British Journal of Educational Technology*, *44*(5), 695–715. doi:10.1111/j.1467-8535.2012.01359.x

Pettit, J., & Kukulska-Hulme, A. (2007). Going with the grain: Mobile devices in practice. *Australasian Journal of Educational Technology*, *23*(1), 17–33.

Rogers, E. M. (2010). *Diffusion of innovations*. Simon and Schuster, Inc.

Saldaña, J. (2009). *The coding manual for qualitative researchers*. London: Sage.

Shih, W. C., Tseng, S. S., Yang, C. C., Lin, C. Y., & Liang, T. (2012). A folksonomy-based guidance mechanism for context-aware ubiquitous learning: A case study of Chinese scenic poetry appreciation activities. *Journal of Educational Technology & Society*, *15*(1), 90–101.

SKMM Guidelines. (2012). *Content code*. SKMM, Malaysia. Retrieved January 2, 2013 from http://www.skmm.gov.my/Resources/Guidelines/Content-Code.aspx

Uzunboylu, H., Cavus, N., & Ercag, E. (2009). Using mobile learning to increase environmental awareness. *Computers & Education*, *52*(2), 381–389. doi:10.1016/j.compedu.2008.09.008

Vavoula, G., Sharples, M., Rudman, P., Meek, J., & Lonsdale, P. (2009). Myartspace: Design and evaluation of support for learning with multimedia phones between classrooms and museums. *Computers & Education*, *53*(2), 286–299. doi:10.1016/j.compedu.2009.02.007

Wu, W. H., Jim Wu, Y. C., Chen, C. Y., Kao, H.-Y., Lin, C. H., & Huang, S. H. (2012). Review of trends from mobile learning studies: A meta-analysis. *Computers & Education*, *59*(2), 817–827. doi:10.1016/j.compedu.2012.03.016

Chapter 9
Create, Transform, and Share:
Empowering Creativity and Self-Expression through Mobile Learning

Maria Ranieri
University of Florence, Italy

Isabella Bruni
Italian Association for Media Education (MED), Italy

ABSTRACT

This chapter explores the potential of mobile learning for creativity in formal and informal contexts of learning with a focus on media production and self-expression. In doing so it attempts to move beyond binary views around the nature of creativity and the role of technologies for creative learning. It presents the literature on how creativity and its relationship with technologies have been conceptualized, especially in education, and provides the theoretical underpinnings that supported the study. In particular, it refers to the Vygotskyan perspective of creativity as a transformative (social) process of culture and the self, and looks at digital technologies' affordances to reflect on their potential for learning. It describes three projects addressing young people and entailing the creation of digital artefacts through mobile devices. It highlights learners' and teachers' perspectives and shows how mobile devices can serve as cultural resources that young people may use for meaning making and transforming themselves.

INTRODUCTION

Over the past decade, reflecting on the role of technologies in fostering creativity has become central to much of the debate around creativity in education (Ferrari et al., 2009; Kampylis and Berki, 2014; Lucas et al., 2013). The rapid development of digital technologies has led to the proliferation of tools that young people are increasingly appropriating for their own use in everyday life. In particular, the recent rise in popularity of mobile and social media practices, from texting and sharing multimedia to social networking and online gaming, has aroused the discussion around the affordances of technologies for creative learning. On the one hand, there are scholars who claim that the use of ICTs by children

DOI: 10.4018/978-1-5225-0359-0.ch009

will lead to a reduction in critical and divergent thinking. For example, in their critique of social media, Friesen and Lowe (2012) assert that the commercial nature of services such as Facebook would limit the opportunities to express disagreement, thus favouring conformism rather than dissent or creative learning. On the other hand, the potential of new digital media for participation, creativity and media production is emphasized: as Jenkins and colleagues (2009) point out: "We are moving away from a world in which some produce and many consume media toward one in which everyone has a more active stake in the culture that is produced" (p. 12).

These positions seem somewhat to reflect previous debates around the impact of computers on children's creativity, whereby some scholars (e.g., Cordes and Miller, 2000, p. 4) advocated "a heavy diet of ready-made computer images" to limit the risk of reducing children's imaginative thinking, while other scholars (e.g. Clements and Sarama, 2001) provided evidence of an increase in creativity and an improvement in peer relations due to positive interactive experiences with certain software. More generally, the views around the potential of technology for creativity in education can be situated between the extremes of 'apocalyptic', i.e. belief in a superior elitist culture at a distance from the media, and 'integrated', i.e. trust in the intrinsic and fascinating power of technologies to transform society (see Eco, 1964/1994); the rhetoric and practices of policy makers, individual schools, teachers or educators may be positioned at different points along the continuum.

In this chapter we want to move beyond these binary visions and support the view that mobile and social technologies can be seen as cultural and learning resources available to people in everyday life (Pachler et al., 2010; Buchem et al., 2013); from the Vygotskyan perspective, media production is based on the use of those resources and, like any creative activity, it entails the transformation of the 'creator' to some extent. With this in mind we examine in the first part of the chapter the literature relating to creativity, learning and technologies in education. In the second part, we present and analyze three projects about the use of mobile devices in and out of school for media production, and then discuss the implications of these experiences for creativity, transformation and learning from the point of view of young people and also teachers/educators.

Literature Review

Creativity and Learning

Though creativity is a word widely used in many fields of human activity and society, it remains very often undefined, while discourses around it frequently draw on implicit conceptualizations and naïve visions (Runco, 2003). Common usage is full of sentences confirming this trend. For example, creativity is commonly associated with unique individuals such as Leonardo da Vinci or Einstein, implicitly assuming that creativity is a personal ability pertaining only to extraordinary people. Or, it is linked to a specific domain such as the arts, for example when people conceive of creativity as typical of great painters or film makers. Or, it is also considered as a natural and innate feature of talented people with the consequence that only those people can be creative, whilst the rest of humankind is condemned to be uncreative with no chance for learning to be creators. These common views around creativity have been sometimes qualified as myths or misconceptions (see e.g. Ferrari et al., 2009). As we shall see below, at times they reflect some conceptualizations that can also be found in the literature, while at other times they contrast with the results of empirical studies. So, what is creativity and how do creativity and learning interact in education?

Analyzing existing research undertaken in this field, a rather complex and sometimes contradictory picture emerges (Jahnke, 2011). Indeed, though all agree that creativity entails originality, novelty and value (Sternberg, 1999), scholars provided different conceptualizations of this notion and, at times, they come to opposite definitions. Nevertheless, following previous analysis (Cachia et al., 2010; Ferrari et al., 2009; Jahnke, 2011; Sternberg and Lubart, 1999; Villalba, 2008), we can identify some main trends. Briefly, earlier studies on creativity have been carried out within the psychometric approach, which considers creativity as a quality that can be measured and is possessed by everyone (Guilford, 1950). Though the idea of measuring creativity is highly arguable, this view has historically had the merit of broadening the scope of creativity by applying it not only to eminent people, but to all individuals. In the psychoanalytic approach, the emphasis shifts to the origins of creativity which is seen as a manifestation of the unconscious for artistic purposes (Eigen, 1983), while in the mystical approach, creativity is viewed as the result of divine inspiration. This last view is a recurring approach in the artistic domain and it has been influential in the field of education, since many parents and teachers conceive of children's creativity in terms of artistic talent, thus neglecting the relevance of creativity in other areas (Sharp, 2004).

A rather dominant approach is the focus on the end-product, where creativity is interpreted as a process resulting in a product and is viewed in opposition to reproductive experiences (Taylor, 1988). Here, 'creativity' and 'creations' very often overlap (Ferrari et al., 2009). However, in recent decades a very prominent approach has been the cognitive one which sees creativity as a cognitive and thinking skill or process and as a mental representation (Sternberg and Lubart, 1999). Recently, the collaborative approach has moved the emphasis from individuals to groups, teams and organizations (Fischer et al., 2004), recognizing not only the role of individuals in the origins of creative ideas, but also that of groups: since creativity results from interaction and collaboration with other individuals (Csikszentmihalyi and Sawyer, 1995), this can be referred to as 'collaborative creativity' (Hermann, 2009).

Given that there has been a plethora of (not always consistent) definitions, several questions about creativity are still unanswered. In their extensive literature review on the (social, political, philosophical and scientific) discourses relating to creativity, Banaji et al. (2010) identified four main themes which are currently under discussion, highlighting their implications for teaching and learning. First, there is a fundamental disagreement among scholars as to whether creativity is an internal cognitive function, or an external social and cultural phenomenon. Secondly, the pervasive nature of creativity is also being debated: is creativity a ubiquitous characteristic of human activity, or an exclusive ability of special individuals or groups in certain fields? Another issue concerns the impact of creativity on society: does creativity inevitably produce social improvement, or can it generate disruption, conflict, and even antisocial effects? Finally, there are still debates about the relationship between creativity on the one hand, and teaching/learning on the other hand. Since our chapter focuses on creativity in education, this last issue deserves a deeper explanation. As in other fields, the term creativity is often used in education but is rarely defined. For example, as Beghetto (2005) observes, teachers might ask their students to be creative in their productions or evaluate their activities as creative, without providing any clarifications about the meaning of creativity, and this lack of explicit definitions may lead teachers and students to identify creativity only with artistic talent and personal peculiarities. In addition, there are some tensions between creativity and formal education that must be considered. One of them relies on what we can call the 'institutional grammar' of the school. For instance, Beghetto (2007) found that teachers prefer standardized answers rather than unique ones, partly because they associate creativity with nonconformity and disruptive behavior which can threaten teachers' authority and their control of the classroom (see also Beghetto and Kaufman, 2010). Another factor that may generate contrasting understandings of the relationship between creativity and teaching/learning, refers to the way in which creative learning is related to play. As Banaji and colleagues (2010, p. 74) explain, notions of creativity as skill-based to prepare students for work can be opposed to 'game-based learning' which criticizes the idea of learning as mechanistic and joyless work, and sees dynamic forms of learning in play. At the same time, many scholars underline the fundamental role of teachers in facilitating learners' creativity: they are seen as those experts who have the responsibility to assess the creativity of students' learning outputs (Wyse and Spendlove, 2007), as the key elements for the building of an appropriate climate for creative learning (Sharp, 2004) or to find an adequate balance between structure and freedom of expression (Beghetto, 2005).

Although the debate is still ongoing, in this chapter we make two main assumptions about the nature of creativity and its relationship with culture and learning:

A1: Creativity is Both Individual and Social

Though we are used to thinking of creativity as a peculiarity of 'geniuses', creativity cannot be associated only with single individuals. As argued by Fischer et al. (2005, p. 485), "Much human creativity arises from activities that take place in a social context in which interactions with other people and the artefacts that embody group knowledge are important contributors to the process". In other words, creativity is not limited to the mind of an individual, but it is a relational process that occurs in the interaction between individuals and society.

A2: Creativity is a Transformative Process of Both Culture and the Self

The idea of creativity as a social process has been theorized within the tradition of cultural psychology which conceives of social activity and cultural resources as key components of the creative process (see e.g. Bruner, 1990; Engeström, 1996). This approach has its roots in Vygotsky (1931/1978), according to whom play is inherently creative since, through the manipulation of culturally significant artefacts, children develop the ability to understand and use symbols. The internalization of this ability, and the combination of imagination with thinking in concepts, generates true creativity, defined as a lifelong ability to transform cultural resources and one's own identity (Banaji and colleagues, 2010).

We believe that these two assumptions can help us to understand the interplay between technologies and creativity in/for education. More specifically, if technologies are seen as cultural resources of everyday life (Pachler et al., 2010a) and if creativity is strongly tied to culture and learning, technologies can play the role of amplifiers of creative processes. In the next sections, we shall attempt to highlight these connections by deepening the analysis of the relationship between creativity and digital technology, and by analyzing some examples of use of mobile technologies to support creativity and learning. In doing this, we will try to unlock the potential of mobile media for creative learning, and to explore the role of teachers and teaching in supporting creative processes and empowering creativity.

Creativity and Digital Technology

When approaching the reflection around the relationship between creativity and technologies, a fundamental question that generally comes to the fore is whether the use of technology is inherently creative. Loveless (2002) identifies some features of technology which can facilitate creativity, and "enable users to do things that could not be done as effectively, or at all, using other tools" (p. 3). Among them he counts provisionality, interactivity, capacity, range, speed and automatic functions. However, to avoid simplistic views which overemphasize the positive influence of technology on creativity, it is worth considering these features as affordances of digital technologies (Conole and Dyke, 2004; Howells, 2005), which means seeing them as potential opportunities and constraints that emerge in relation to human agency and purpose. When we look at the effects of technologies on creative practices, we cannot expect a uniform outcome in all situations, but we have to take into account a variety of factors like the type of digital tool, the way and purpose it is used for, the context of use and so forth. Therefore, the capacity to use technological devices should be meant not only in terms of abilities to use particular technologies, but also as the capacity to evaluate their appropriateness and being open to new developments.

Also in the educational field, the adoption of digital tools does not necessarily mean that teaching and learning become more creative. What counts more is understanding their potential in specific social, cultural and psychological circumstances (Banaji et al., 2010; Tolmie, 2001), rather than limiting oneself to their use for their own sake. Through information and communication technology (ICT), learners and teachers can engage in learning experiences which can support creativity by providing opportunities for exploration and play with materials and information, for taking risks and making mistakes in a non-threatening atmosphere, enabling reflection, access to resourceful environment, and flexibility in time and space (Claxton, 2000; Craft, 2000; Davies, 1999).

Loveless (2002, 2007) suggests categorizing the activities which exploit digital technologies' affordances in relation to the elements featuring creative and learning processes. ICT can be used to develop ideas, both supporting the early stage of thinking with tools for brainstorming and mapping, and in the trying out of ideas with the use of programming languages, digital manipulatives and simulations or in problem-solving tasks (Clements and Sarama, 2003; Resnick et al., 2000). Technologies can play an important role for learners in making connections with other projects and resources as well as in collaborating with others during the creative work in progress: blogs, wikis, social network sites and other services of Web 2.0 enable communication with individuals and groups external to the classrooms in ways that were limited before the digital age (Davies and Merchant, 2009). Technology also plays a distinctive role for creation and editing processes, allowing the development of multiliteracies: digital tools can sustain meaningful expression through multimodality, providing the opportunity to trace development and revise the production process. Lastly, technology can enable the publishing and presentation of works to different publics: the awareness of audience is crucial for the creative process since it makes learners reflect more deeply on the level of originality and value of their work (Lachs, 2000; Atherton, 2002).

Another aspect to be considered is that digital technologies can support creativity both in traditional settings - like classrooms and schools - and in online learning environments such as virtual worlds or knowledge forums. An area of significant development is also represented by mobile technologies, which challenge our understanding of creative practices by bringing together both physical and virtual experiences of space and interactions (Farman, 2012; Rudd et al., 2006). Indeed in physical and/or virtual environments, creative activities carried out with new communication devices question traditional learning organizations in terms of pedagogy, curriculum, space, time, and staff. However, it is important to underline that while the use of technology for creative work is often perceived as a liberation from constraints, the emphasis should be on pursuing meaning-making in projects that enable pupils to develop their ideas over time, with opportunities to both complete carefully structured tasks and engage in open-ended experimentation (Reid et al., 1993).

Mobile Learning and Creativity: An Overview of Three Projects

The projects reported in this section refer to three diverse experiences addressing young people and entailing the creation of digital artefacts through the use of mobile devices. They are different in terms of context, media used and outputs. As far as the context is concerned, the experiences took place in a range of formal and informal settings: project 1 was implemented in school as a curricular activity, project 2 was realized as an additional activity during the annual school trip, while project 3 was carried out in an informal context within an after school program.

All the projects were characterized by a combination of Web 2.0 tools and mobile devices: the experience carried out at school (project 1) was based on the production of podcasts, the experience implemented in and out of school (project 2) mainly revolved around the use of smartphones, while the project developed in the after school program (project 3) integrated the use of social network sites with that of mobile phones.

The three projects led to the production of different types of digital artifacts: project 1 resulted in the creation of some radio sketches relating to curricular contents; in project 2, students realized a series of geotagged audioguides devoted to places visited during the field trip; in project 3 participants wrote some collective and personal narratives shared through Facebook.

We decided to focus on these experiences for more than one reason. Firstly, since they differ in contexts, media used and outcomes, they can be viewed as representatives of three different situations (formal, mixed, informal) of digital media projects that teachers/educators may face. The analysis of these cases provides insights into creative practice in different settings and places. Secondly, and most importantly, though all the projects are based on media production through mobile devices, our focus is not limited to the analysis of final products, but also to the realization process itself. With this in mind, we look at these experiences to find clues and signs of how the idea of creativity as a collaborative transformative process supported by digital technologies can take place, from self-expression (project 1) to innovative media practices and sharing (project 2) and self-representation/identity (project 3). As we will see, learners modified their representation and self-expression attitudes or their relation with the environment and the technological devices, while being always involved in a teamwork based activity of collaborative learning.

Project 1: Podcasting in the Classroom

The first experience was carried out within the European project Ensemble, a transnational cooperation between the University of Florence and public and private bodies from Italy, France and the UK.[1] Overall the project aimed at testing, both in Italy and France, the potential for mobile learning to promote European citizenship through common mobile devices such as 'old generation' mobile phones and MP3 players. In Italy the testing phase was carried out in Prato, where the second largest Chinese immigrant community in Italy lives with more than 45,000 people.

The testing phase involved 66 students aged 13-15, 10 teachers and 60 parents, coming mainly but not exclusively from China. It was carried out between February and May 2010 with the specific aim of encouraging the socio-cultural integration of second generation students and improving their ability to understand Italian contents and their capacity for self-expression (especially, oral and digital communication). To achieve these purposes four learning modules on citizenship and intercultural education were designed and implemented through podcasting.

Podcasting refers to any audio/video content delivery approach based on web syndication protocols. According to several scholars, it provides increased flexibility, user control, and portability (Abdous et al., 2009; Thorne and Payne, 2005), and contributes to students' learning by increasing motivation (Oliver, 2005). Three main uses of podcasting in education can be distinguished (McGarr, 2009): *substitutional podcasting*, which consists of audio recordings of lectures made available to students to review teaching material; *supplementary podcasting*, which involves the provision of additional material to help students deepen their understanding of specific topics; and *creative podcasting*, which sees students creating their own podcasts to develop a better understanding of a topic (e. g., Bartle et al., 2011; Lee et al., 2008). Though it is the least common form, creative podcasting seems to be more effective than substitutional or supplementary podcasting (Carvalho et al., 2009; Shamburg, 2009). In particular, it has been underlined that by turning students into "knowledge creators" (McGarr, 2009), creative podcasting helps them develop collaborative knowledge-building skills (Lee et al., 2008) as well as critical thinking skills (Frydenberg, 2008), and under certain circumstances it can foster reflective (Lazzari, 2009) and deep learning (Pegrum et al., 2015).

In our project, podcasts were conceived as supplementary tools used to provide additional materials or classroom summary but also as creative artefacts made by learners, involving them in both listening and production. For each unit, students were asked to listen to four podcasts out of school and then to discuss in pairs the key issues of the unit in the classroom. The discussion was prompted by the teacher as a preliminary step towards content creation. Students were then engaged in the production of their own podcast, with the aim of developing their self-expression skills within a collaborative process that was conceived as an opportunity to bridge cultural and linguistic differences.

Results

At the end of the process, several results were gathered from observations, students' and teachers' answers to questionnaires and students' productions. For reasons of space we limit our attention here to some key considerations. Firstly, though almost all students did own mobile technologies (93% mobile phone, 79% mp3 player), most of them (47 pupils out of 66) declared they had never heard of podcasts before. Nevertheless, they found it easy to download podcasts and they enjoyed the listening activity; as they commented, "studying with podcasts was more interesting because of the use of ubiquitous learning modalities". They particularly appreciated the opportunity for creating their own products with classmates, since this was "an original method to work with peers" and "understand what they really think". Indeed, through open discussion and exchange of ideas they got to know each other better and felt free to speak into the microphone and record their voices. In addition, almost all students stated that the learning activity was truly enriching in so far as it enabled them to successfully experiment with the use of technologies not only for fun but also to express themselves. For example, one student observed: "I learnt things in a different way and I had the opportunity to show others what I've learnt". At the same time, some students found podcast creation too challenging.

As far as the teachers were concerned, they expressed positive reactions to the use of podcasts since they support understanding of the content and reduce barriers to learning, thus augmenting students' motivation. Only a few teachers were sceptical about the project's effectiveness because of the difficulties that some students, with low levels of proficiency in the Italian language or with very low technical skills, encountered during the process.

In terms of products, the students created a total of 13 podcasts lasting between 2 and 4 minutes. Audio transmissions followed a radio sketch format, where content is conveyed through the dramatization of a real life situation with dialogues. Contexts of interaction were declared at the beginning or were suggested through background sounds, consistent with the different soundscapes. Radio dramas created by the students included both episodes about their daily lives – oral exam, afternoon break while doing homework, etc. – and imaginary situations, like the original idea of interviewing a personified nation. Each product appears as a result of a compromise between formal educational needs (e.g., the reference to curricular content) and students' inspiration, which had a positive impact on diction and dialogue interpretation making the activity more attractive. One of the most relevant results was, indeed, that even second-generation students actively participated, despite some difficulties, in the production of radio drama showing increased levels of self-esteem and self-expression.

Project 2: Geotagging Audio-Contents In and Out of School

The second experience refers to a media literacy education project on digital citizenship involving a network of schools in Tuscany and funded by the Italian Ministry of Education.[2] The aim of the project was to promote students' awareness of practices of media use. In this section, we focus on the activities carried out between March and April 2012 with a first-year class of high school students (i.e. aged 14-15 years), based on the use of location based services to create geolocated content during the annual school trip.

In the past few years, the advancement of wireless and mobile technologies has stimulated reflections on learning environments and the possibility of combining resources from both the real and the digital world (Farman, 2012; Wu et al., 2013). In particular, the location awareness functionality that is built in to many current mobile devices makes it possible to track learners' actual positions and access relevant information related to their location, opening up new opportunities for teaching and learning (Barak and Ziv, 2013; Brown et al., 2011; Chu et al., 2010). Location-based services can enhance both formal and informal learning (De Lucia et al., 2012; FitzGerald, 2012); in the school context they have been used during museum visits and field trips and proved to be effective in motivating students through browsing or creating content (Barak and Ziv, 2013; Brown, 2010) and in connecting classroom and outdoor activities (Vavoula et al., 2009). Several location-based learning approaches include games and gamification with positive impacts on students' achievement and engagement (Su and Cheng, 2015; Chen et al. 2015). Melero et al. (2015) also underlined the importance of involving teachers in the design of location-based activities and proposed learning analytics visualizations as a tool to support teachers' design practice and students' self-assessment in mobile learning.

In our project, a two-step activity was planned to experiment with the creation of geo-localized narratives during a 3-day visit to the Circeo National Park (a natural protected area with archaeological sites located in South Rome). The first activity consisted of keeping a diary of the trip, posting on a common Facebook profile: each student could freely use his/her own mobile device to publish photos or comments in real time, to document the outdoor learning experience. The second activity was more structured and consisted of creating geotagged audioguides relating to the places that mostly attracted students' attention during the visit. The class was divided into five groups, each of them recorded a short audioguide, summarizing the information received during the visit and adding personal comments. To realize these products, a location-aware application named "Broadcastr" was installed on the smartphones of some students (Figure 1). This app, available both for iOS and Android devices, automatically detects spatial

Figure 1. Two snapshots of the phone screen running Broadcastr: on the left the map where users can locate contents (four audioguides are visible), on the right an example of products player

coordinates and allows the user to record audio content associated with that place, accompanied by a picture and a brief written description. Once the recording is uploaded, any user can locate the content on the map, listen to it and share it via social network sites. Various audio tracks can also be grouped into a thematic collection; at the time of the implementation of the project some authoritative collections were available, focused on tourism topics such as natural parks and artistic cities. A pre-visit lesson was dedicated to training in the use of the new app and listening to previous collections in order to share a common audioguide format.

Results

Teachers reported some concerns expressed by the students during the planning stage: the school trip was mainly considered a playful experience, for this reason some resistance to carrying out structured activities had emerged. However, the opportunity for using mobile phones and testing location-based services was intriguing and the idea of creating a public collection of audioguides was an adequate motivating factor to overcome such concerns.

In the post-visit lesson, students declared they had particularly appreciated the diary activity on the common Facebook profile, because "it was nice during the trip to post pictures about us and the tour, to share opinions about the places that we visited". Overall, they preferred uploading images rather than writing comments: about twenty pictures were published, depicting landscapes or images of funny backstage moments. The Facebook page became "a way of interacting with people who didn't participate in the field trip", and this communication bridge seemed to stimulate content publication, in order to involve peers and teachers. Unexpectedly, the profile has been most widely used once they were back at school, whereby many students used it to share their feelings, add images and comments.

As planned, audioguides were produced during the second day of the trip. Five products where realized; they are short, lasting about 30 seconds, and quite formal in both content and language. Students preferred to synthesize the information provided by the official park guide, rather than record personal remarks. However, expressions conveying a sense of being in the place are recurrent (e.g., "we are close to the archaeological ruins of Villa Giulia... coming inside, you can see the remains of the baths..."; or also "we are in the banqueting hall of the Domitian Villa, on the shores of Lake Sabaudia..."). The geolocated work was considered by the students as one of the most challenging activities, since they found it "hard to be the guide, speaking into a cell phone". Nevertheless, the educational purpose of the activity was positively perceived: "this activity helped us understanding some functions of a device that we always have in our hands – the mobile phone – and discovering new applications". Also some technical difficulties with the app were reported: unfortunately, phone coverage in the park was scarce while Broadcastr requires internet connectivity to register content, and did not allow asynchronous geo-tagging. To overcome this difficulty, the teacher made available her own mobile phone which had network coverage. Moreover, this episode prompted an educational activity about social rating: students shared their own reviews in the application market and then tested and assessed other apps for location-based production.

Project 3: Mobile Storytelling in an After-School Program

The last experience is Piagge Mobili, a project carried out in 2011 in Le Piagge, a low-income district of Florence (IT) with a number of young people at risk of social exclusion. It was part of a programme promoted at an informal level by the local community of Le Piagge. It involved 15 children aged 11-15 with the aim of testing the potential of mobile phones as personal resources to develop narrative skills, self-esteem and self-representation through the technique of mobile storytelling.

Digital media have become powerful tools available for young people to represent themselves and explore their identities through *appropriation* and production (Bachmair and Pachler, 2013): mobile phones are today the most widely diffused user-generated content device and social networking sites are the privileged environments for life and content sharing (Livingstone et al., 2014; Papacharissi, 2011). Referring to digital storytelling (Lambert, 2012), a technique for multimedia narrative production widely used today in education, we call 'mobile storytelling' a subset of 'digital storytelling' based on the use of mobile phones (Ranieri and Bruni, 2013a). To create their digital narratives, learners collaboratively plan a story, either fictional or not, create and edit the story using mobile phones, and finally share it through social networking sites (Ranieri and Bruni, 2013b; see also McGreen and Sanchez, 2005). Though studies on the use of mobile devices to support digital storytelling are still limited, the interest towards the development of interactive and locative storytelling is growing (see Nack and Waern, 2012). For example, Lombardo and Damiano (2012) adopted adaptive methods to face interaction and movement, so that a story arch can be planned according to what users are doing and how they are moving, while Hansen, Kortbek and Grønbæk (2012) developed a system to help authors produce location-based story structures for outdoor environments. Other authors designed and tested a storytelling app based on mixed reality environments in which to tell imaginative stories (see e.g. Nuri et al., 2014).

In our case, we adopted a simpler solution based on the combination of cell phones and social networking sites. More specifically, we followed an action-research approach centred on three cycles of action, involving educators and researchers in the process of reflection and (re)planning. Initially, participants were asked to use their cell phones to take pictures and videos of themselves, their families and their neighbourhood, as a starting point for storytelling. All the content was shared via Bluetooth and a projector. A collective Facebook profile was created as a space for products and the publication of the stories. Stories were written collectively using a shared mobile phone. Each participant wrote a text message with a piece of a story, sent it for publication on the Facebook profile and passed the device to the next peer.

The Facebook page was constantly refreshed, in order to follow the story in real time. Relay writing was chosen to break down resistance to participation and to promote attention, respect, and narrative coherence. At the end, the group selected one story to be dramatized and video recorded with mobile devices, simulating the work of a real movie company.

Results

At the beginning of the Piagge Mobili project, participants showed great difficulties in representing themselves directly and some refused to be photographed or to publish contents on Facebook because – as stated by an educator – "it is the mirror of their identity, their space for peer interaction". However, the use of multimedia as an informal resource seems to have helped them explore themselves and investigate their bodies. The initial refusal disappeared progressively, until they got to the video shooting when educators stated that "children were very uninhibited in front of the camera", and immediately asked to publish the recordings online.

Another result reflects the positive role that media played in stimulating participation of teens. As revealed from observations, when asked to talk orally about themselves, "nobody wanted to record his/her voice, but with mobile phones they immediately got involved". In particular, they had never connected phones and social networking sites, but this possibility stimulated their curiosity so they started writing sentences and discovered the cell phone's multimedia functionalities. Educators believe that writing stories through mobile phones opened the way to creativity and collaboration, turning silence into play and discovery.

A total of six stories were created, including three stories written through SMS and three hand-written. In both modes there were plenty of abbreviations, vernacular words and grammar mistakes, as well as typical expressions of short text messaging and some stylistic features of the fable genre. The stories were characterized by brevity, accompanied by an extreme simplicity of plot and merely sketched characters. In most cases, the protagonists were children or adolescents, but had neither names nor specific qualities, with the exception of aesthetic and gender attributes (Figure 2). The topics were related to the condition of adolescence: some stories were linked to pupils' direct experience, whilst others referred to imaginary constructions.[3] Through this mixing of reality and imagination, pupils expressed their feeling of living in a suburban area and developed their capacity for representing themselves through 'acts of creativity'.

Figure 2. A screenshot of the Facebook profile of the group, while the story writing via text message is ongoing. This is a translation of the posts following the sequence of their publication: "Let's try to make a story.../Once upon a time there were some guys/Three pretty girls with beautiful long hair and two crazy boys, really nice /One girl had blonde hair and two had brown hair. They all were in a relationship. They were going to the sea by car"

DISCUSSION

To what extent can we qualify the practices described above as creative? To what extent have teachers/ educators been important for creative learning? What influence did the mobile media have on participants' practices of media production?

It is widely recognized that evaluating creativity is a rather complex task, particularly in the context of education, where there are no common models and tools to assess creative performances (Lucas et al., 2013). Nevertheless, we will try to highlight some of the emerging aspects of creativity, learning and technology that characterised the mobile learning experiences, focusing on learners' and teachers' perspectives.

From the learners' point of view, two elements appear to be particularly significant. Firstly, a cross-cutting component, that applies across all projects, can be identified, namely the challenging nature of the task requiring participants to rethink (and transform) their familiar uses of mobile devices. Learners clearly perceived the request to use their devices for intentional learning tasks as an unfamiliar way of approaching them. Indeed, since mobile technologies are mainly designed for communication, entertainment and consumption, their use for learning is not obvious (Seipold et al., 2013) and it is not surprising that, at the beginning, some students in all projects have shown unwilling attitudes towards the use of mobile phones to achieve learning activities. However, after some initial resistance, learners engaged with the creation of new connections by using their previous expertise on mobile technologies for new purposes and linking it with new concepts. In other words, they were involved in a sort of re-appropriation process of their mobile phones resulting in new (cultural) practices of media use with implications for identity transformation. Indeed, in project 1, the creation of podcasts provided learners with the opportunity to look at technologies with different eyes and to express themselves in a creative way, favouring reflection and learning improvement; in project 2, learners acknowledged that the use of mobile phones to create audioguides proved to be challenging, but also enriching because it allowed them to explore some new features and transform their common – and sometimes repetitive – practices; in project 3, learners clearly used the media to mutually explore their bodies and represent themselves being involved in a collective process of meaning creation. Moreover, in all projects, there has been – though to different extents – a combination of learners' pre-existing expertise of media use (i.e., their familiarity with mobile devices and social networking sites) and new media practices; this can be compared to the transfer of knowledge from one area to a new domain: as we know, this is one of the most challenging tasks for learners, especially for young people, but it is crucial to transform students into "knowledge creators" (McGarr, 2009; Pegrum et al. 2015), going beyond mere memorisation. From this point of view, and similar to other studies (Bachmair and Pachler, 2013; McGarr, 2009; Lazzari, 2009; Pegrum et al., 2015), we found that focusing on media creation can stimulate students' reflection, re-appropriation and self-expression, especially when participants' habitus of learning (see Kress and Pachler, 2007) are connected to specific patterns of learning and teaching, where learners' expertise provides the ground for transformation under the guidance of a facilitator.

Another aspect refers to the social dimension of the media practices the learners were engaged with. All the learning activities proposed in the three projects were group based. Therefore, there is obviously a social trait that applies to all the experiences. However, what was particularly relevant was that learners used the media to link each other, to be connected, to keep doing things together, reflecting a very common practice amongst young people, which is that of using mobile phones for social networking (Livingstone et al., 2014). The encounter between this common practice of socialization through mobile media and the unusual task of using them to create and share media products resulted in a form of 'collaborative creativity' (Hermann, 2009; Fischer et al., 2005) based on the social and collective resignification of very familiar media practices. Looking at the literature, this idea echoes the already mentioned notion of mobile devices as cultural resources (see Pachler et al., 2010a), where popular media of everyday life are seen as learning resources. From this perspective, the use of Facebook in project 2 is particularly significant: the social network was used to keep connected but, at the same time, it prompted students to create and share content in a public space. More generally, like other studies (e.g., Davies and Merchant, 2009; Loveless, 2007; Vavoula et al., 2009), digital tools enabled the communication in and out of school to support learners in making connections, and provided students with the opportunity to address different audiences, thinking about the value of their productions.

Moving to teachers/educators, our experiences allowed us to reflect on creative teaching and its distinctive features, which we mainly identified with what Beghetto (2013) defined as teachers' capacity to respond to 'unexpected curricular micromoments' or what Hobbs (2013) indicated as teachers' capacity of 'improvization and strategic risk-taking'. According to Hobbs, this capacity must be intended as "a set of socioemotional and experiential competencies" that teachers need "to create a learning experience from an unpredictable situation", especially when using digital media (Hobbs, 2013, p. 6). Very often the use of digital media in education (in and out of school) is seen by teachers as particularly challenging, since it generates a variety of classroom management issues such as "freedom, creativity and collaborative learning", that may not guarantee sufficient levels of "structure, scaffolding and support" (Hobbs, 2013). By pandering to such concerns, both creativity and technologies would be banned from education, unless we recognize in teachers' capacity of risk taking a key component to transform unpredictable situations into opportunities for creative learning. In our cases, there have been two unpredictable situations that teachers were able to transform into a collaborative creative activity: in project 2, the lack of internet coverage on students' phones led the teacher to lend her mobile phone and made students concentrate on a social activity of ranking and testing/discovering apps for location-based production; in project 3, the low availability of suitable mobile phones led educators to use one unique device at the beginning as a social tool for developing the capacity for building stories in a collaborative way, by adding ideas and texts to previous ideas and texts.

Finally, consideration needs to be given to teachers'/educators' appreciation of the role of the media in promoting identity transformation. For example, the activities developed in project 3 seem to have influenced the relationship between the learners and their community. Indeed, learners began unexpectedly to speak in public situations prompted by the community and for the first time they exchanged their views with adults. Subsequently, they also asked the community to fund the project for another year. Briefly, children seem to have been stimulated in a process of self-transformation and awareness, which has led them to change their relationship with the context in terms of increased levels of visibility, communication and self-representation.

CONCLUSION

The relationship between creativity and technologies has received considerable attention from researchers and policy makers in recent years. It is widely accepted that creative skills are crucial for work and life in the 21st century and that ICT can be a driver for creative learning (Cachia et al., 2010). However, much of the research in this area is characterized by contrasting definitions leading to binary oppositions that see creativity as individual or social, originating from nothing or transformative, innate or taught and learnt, intuitive or skills-based, etc. (Banaji and Burn, 2007). To overcome these oppositions, we adopted a socio-cultural approach to the understanding of creative learning processes. Such an approach allows us to grasp how cultural resources such as mobile technologies (Pachler et al., 2010b) may be transformed into creative acts by learners.

Based on these assumptions, this chapter analyzed three mobile learning experiences that engaged groups of young people in the production of digital artefacts through mobile devices in and/or out of school. In one case, students were involved in a school-based activity of creating audio podcasts about intercultural and citizenship education; in another case, they were asked to produce geotagged audioguides during a field trip, bridging formal and informal contexts of learning; and in the last case, a group of adolescents within an after-school programme participated in a collective experience of digital storytelling through mobile devices. Though the experiences were carried out in different contexts and with different media, some common results have emerged. In particular, on the part of the learners, a sort of 'collaborative creativity' (Hermann, 2009) took place in their media landscape, prompted by a series of learning inputs aimed at supporting the transformation of common and familiar media practices – which might risk being stereotyped and repetitive – into creative self-expression, innovative media use and creative self-representation. On the part of the teachers/educators, their ability to take risks and support unplanned events proved to be important to foster creativity.

The study had, of course, some limitations as it involved small groups of learners for a relatively short time and no follow up interviews were conducted to evaluate the effects of the activities in the long run. In addition, we did not report in detail all the processes that are important to record the impact on learning and change. However, the analysis of these practices provided us with sufficient ground to identify some main topics for future research, particularly considering the school context. On the one hand, a deeper understanding of the influence of different situations on the strategy of risk taking would probably lead to a better integration of creativity into the curriculum. As is well known, formal contexts are so rigid that creativity very often risks being excluded, while learning in informal contexts entails to some degree a certain level of flexibility which can evolve into creative practices. At the same time, an area which deserves more attention is teacher education which, up to now, has often not focused on the perspective of developing the creativity of learners. Trapped in stereotypes and binary visions of creativity, teachers really risk missing the opportunity to support the transformation of learners' cultural experiences of media use into new processes of media production and creative work.

ACKNOWLEDGMENT

This chapter is mainly based on the paper Ranieri, M., & Bruni, I. (2013). Empowering Creativity in Young People Through Mobile Learning: An Investigation of Creative Practices of Mobile Media Uses In and Out of School. International Journal of Mobile and Blended Learning, 5(3), 17-33. We thank the publisher for giving permission to use the paper for this chapter.

REFERENCES

Abdous, M., Camarena, M., & Facer, B. R. (2009). MALL technology: Use of academic podcasting in the foreign language classroom. *ReCALL*, *20*(1), 1–20.

Atherton, T. (2002). Developing ideas with Multimedia in the Primary Classroom. In A. Loveless & B. Dore (Eds.), *ICT in the Primary Classroom*. Buckingham, UK: Open University Press.

Bachmair, B., & Pachler, N. (2013). Appropriation and composition in a culture characterised by provisionality. In M. Böck & N. Pachler (Eds.), *Multimodality and Social Semiosis: Communication, Meaning-Making and Learning in the Work of Gunther Kress* (pp. 211–220). New York: Routledge.

Banaji, S., & Burn, A. (2007). Creativity through a rhetorical lens: Implications for schooling, literacy and media education. *Literacy*, *41*(2), 62–70. doi:10.1111/j.1467-9345.2007.00459.x

Banaji, S., Burn, A., & Buckingham, D. (2010). *The Rhetorics of Creativity: A literature review*. London: Creativity, Culture and Education.

Barak, M., & Ziv, S. (2013). Wandering: A Web-based platform for the creation of location-based interactive learning objects. *Computers & Education*, *62*, 159–170. doi:10.1016/j.compedu.2012.10.015

Bartle, E. K., Longnecker, N., & Pegrum, M. (2011). Collaboration, contextualisation and communication using new media: Introducing podcasting into an undergraduate chemistry class. *International Journal of Innovation in Science and Mathematics Education*, *19*(1), 16–28.

Beghetto, R. A. (2005). Does Assessment Kill Student Creativity? *The Educational Forum*, *69*(3), 254–263. doi:10.1080/00131720508984694

Beghetto, R. A. (2007). Does creativity have a place in classroom discussion? Prospective teachers' response preferences. *Thinking Skills and Creativity*, *2*(1), 1–9. doi:10.1016/j.tsc.2006.09.002

Beghetto, R. A. (2013). Expect the unexpected: Teaching for creativity in the micromoments. In M. Gregerson, J. C. Kaufman, & H. Snyder (Eds.), *Teaching creatively and teaching creativity* (pp. 133–148). New York: Springer Science. doi:10.1007/978-1-4614-5185-3_10

Beghetto, R. A., & Kaufman, J. C. (2010). Broadening conceptions of creativity in the classroom. In R. A. Beghetto & J. C. Kaufman (Eds.), *Nurturing creativity in the classroom*. Cambridge, UK: Cambridge University Press. doi:10.1017/CBO9780511781629.010

Brown, D. J., McHugh, D., Standen, P., Evett, L., Shopland, N., & Battersby, S. (2011). Designing location-based learning experiences for people with intellectual disabilities and additional sensory impairments. *Computers & Education*, *56*(1), 11–20. doi:10.1016/j.compedu.2010.04.014

Brown, E. (Ed.). (2010). *Education in the wild: contextual and location-based mobile learning in action. A report from the STELLAR Alpine Rendez-Vous workshop series*. Learning Sciences Research Institute, University of Nottingham.

Bruner, J. (1990). *Acts of Meaning*. Cambridge, MA: Harvard University.

Buchem, I., Jahnke, I., & Pachler, N. (Eds.). (2013). Guest editorial preface. In Special Issue on Mobile Learning and Creativity: Current Concepts and Studies. *International Journal of Mobile and Blended Learning*, *5*(3), 1–3.

Cachia, R., Ferrari, A., Ala-Mutka, K., & Punie, Y. (2010). *Creative Learning and Innovative Teaching. Final Report on the Study on Creativity and Innovation in Education in the EU Member States*. Seville: EC JRC-IPTS.

Carvalho, A. A., Aguiar, C., & Maciel, R. (2009). A taxonomy of podcasts and its application to higher education. In H. Damis & L. Creanor (Eds.), *"In dreams begins responsibility"—choice, evidence and change. The 16th Association for Learning Technology Conference (ALT-C 2009)*, (pp. 132–140). Oxford, UK: Association for Learning Technology.

Chen, C.-H., Liu, G.-Z., & Hwang, G.-J. (2015). Interaction between gaming and multistage guiding strategies on students' field trip mobile learning performance and motivation. *British Journal of Educational Technology*, n/a. doi:10.1111/bjet.12270

Chu, H., Hwang, G., Tsai, C., & Tseng, J. C. R. (2010). A two-tier test approach to developing location-aware mobile learning systems for natural science courses. *Computers & Education*, *55*(4), 1618–1627. doi:10.1016/j.compedu.2010.07.004

Claxton, G. (2000). A map of learning to learn. In B. Lucas & T. Graeny (Eds.), *Learning to Learn: Setting the Agenda for Schools in the 21st Century*. London: Campaign for Learning.

Clements, D. H., & Sarama, J. (2003). Strip Mining for Gold: Research and Policy in Educational Technology - A Response to "Fool's Gold". *Educational Technology Review*, *11*(1), 7–69.

Conole, G., & Dyke, M. (2004). What are the affordances of information and communication technologies? *ALT-J. Research in Learning Technology*, *12*(2), 113–124. doi:10.1080/0968776042000216183

Cordes, C., & Miller, E. (2000). *Fool's gold: A critical look at computers in childhood*. College Park, MD: Alliance for Childhood.

Craft, A. (2000). *Creativity across the primary curriculum: framing and developing practice*. London: Routledge.

Csikszentmihalyi, M., & Sawyer, K. (1995). Creative insight: the social dimension of a solitary moment. In R. J. Sternberg & J. E. Davidson (Eds.), *The Nature of Insight* (pp. 329–363). Cambridge, MA: Bradford/MIT.

Davies, J., & Merchant, G. (2009). *Web 2.0 for Schools: Learning and Social Participation*. New York, NY: Peter Lang.

Davies, T. (1999). Taking Risks as a Feature of Creativity in the Teaching and Learning of Design and Technology. *The Journal of Design and Technology Education*, *4*(2), 101–108.

De Lucia, A., Francese, R., Passero, I., & Tortora, G. (2012). A Collaborative Augmented Campus Based on Location-Aware Mobile Technology. *International Journal of Distance Education Technologies*, *10*(1), 55–73. doi:10.4018/jdet.2012010104

Eco, U. (1964/1994). *Apocalypse Postponed* [Apocalittici e integrati]. R. Lumley (Ed.). Bloomington, In: Indiana UP.

Eigen, M. (1983). A Note on the Structure of Freud's Theory of Creativity. *Psychoanalytic Review*, *70*(1), 41–45. PMID:6410431

Engeström, Y. (1996). Development as breaking away and opening up: A challenge to Vygotsky and Piaget. *Swiss Journal of Psychology*, *55*(2/3), 126–132.

Engeström, Y. (1999). *Lernen durch Expansion*. Marburg: BdWi- Verlag.

Farman, J. (2012). *Mobile Interface Theory*. New York: Routledge.

Ferrari, A., Cachia, R., & Punie, Y. (2009). *Innovation and Creativity in Education and Training in the EU Member States: Fostering Creative Learning and Supporting Innovative Teaching. Literature review on Innovation and Creativity in E&T in the EU Member States (ICEAC)*. Seville: EC JRC-IPTS.

Fischer, G., Giaccardi, E., Eden, H., Sugimoto, M., & Ye, Y. (2005). Beyond binary choices: Integrating individual and social creativity. *International Journal of Human-Computer Studies*, *63*(4-5), 482–512. doi:10.1016/j.ijhcs.2005.04.014

Fischer, G., Scharff, E., & Ye, Y. (2004). Fostering Social Creativity by Increasing Social Capital. In M. Huysman & V. Wulf (Eds.), *Social Capital and Information Technology* (pp. 355–399). Cambridge, MA: MIT Press.

FitzGerald, E. (2012). Creating user-generated content for location-based learning: An authoring framework. *Journal of Computer Assisted Learning*, *28*(3), 195–207. doi:10.1111/j.1365-2729.2012.00481.x

Friesen, N., & Lowe, S. (2012). The questionable promise of social media for education connective learning and the commercial imperative. *Journal of Computer Assisted Learning*, *28*(3), 183–194. doi:10.1111/j.1365-2729.2011.00426.x

Frydenberg, M. (2008). Principles and pedagogy: The two PS of podcasting in the information technology classroom. *Information Systems Education Journal*, *6*(6), 3–11.

Gauntlett, D. (2011). *Making is connecting. The social meaning of creativity from DYI and knitting to YouTube and Web 2.0*. Cambridge, UK: Polity Press.

Guilford, J. P. (1950). Creativity. *The American Psychologist*, *5*(9), 444–454. doi:10.1037/h0063487 PMID:14771441

Hansen, F. A., Kortbek, K. J., & Grønbæk, K. (2012). Mobile Urban Drama: Interactive storytelling in real world environments. *New Review of Hypermedia and Multimedia*, *18*(1-2), 63–89. doi:10.1080/13614568.2012.617842

Herrmann, T. (2009). Design heuristics for computer supported collaborative creativity. In *Proceedings of the 42nd Hawaii International Conference on System Sciences* (pp. 1-10). IEEE Computer Society.

Hobbs, R. (2013). Improvization and strategic risk-taking in informal learning with digital media literacy. *Learning, Media and Technology*, *38*(2), 182–197. doi:10.1080/17439884.2013.756517

Howells, J. (2005). *The Management of Innovation and Technology: The Shaping of Technology and Institutions of the Market Economy*. London: Sage.

Jahnke, I. (2011). How to Foster Creativity in Technology Enhanced Learning. In B. White, I. King, & Ph. Tsang (Eds.), *Social Media Tools and Platforms in Learning Environments: Present and Future* (pp. 95–116). New York: Springer. doi:10.1007/978-3-642-20392-3_6

Jenkins, H., Clinton, K., Purushotma, R., Robinson, A., & Weigel, M. (2009). *Confronting the Challenges of Participatory Culture: Media Education for the 21st Century*. Chicago, IL: The John D. and Catherine T. MacArthur Foundation.

Kampylis, P., & Berki, E. (2014). *Nurturing creative thinking*. International Bureau of Education (UNESCO) in collaboration with the International Academy of Education (IAE). Retrieved on 25th July 2015 from: http://www.ensembleproject.org/modules/wfdownloads/viewcat.php?cid=78

Kress, G., & Pachler, N. (2007). Thinking about the 'm-' in mobile learning. In T. Hug (Ed.), *Didactics of Microlearning. Concepts, Discourse, and Examples* (pp. 139–154). Münster: Waxmann.

Lachs, V. (2000). *Making Multimedia in the Classroom: A Practical Guide*. London: Routledge.

Lambert, J. (2012). *Digital storytelling: capturing lives, creating community*. New York, NY: Routledge.

Lazzari, M. (2009). Creative use of podcasting in higher education and its effect on competitive agency. *Computers & Education*, *52*(1), 27–34. doi:10.1016/j.compedu.2008.06.002

Lee, M. J. W., McLoughlin, C., & Chan, A. (2008). Talk the talk: Learner-generated podcasts as catalysts for knowledge creation. *British Journal of Educational Technology*, *39*(3), 501–521. doi:10.1111/j.1467-8535.2007.00746.x

Livingstone, S., Mascheroni, G., Ólafsson, K., & Haddon, L. (2014). *Children's online risks and opportunities: comparative findings from EU Kids Online and Net Children Go Mobile. EU Kids Online*. London, UK: LSE.

Lombardo, V., & Damiano, R. (2012). Storytelling on mobile devices for cultural heritage. *New Review of Hypermedia and Multimedia*, *18*(1-2), 11–35. doi:10.1080/13614568.2012.617846

Loveless, A. (2002). *A Literature Review in Creativity, New Technologies and Learning: A Report for Futurelab*. Bristol: Futurelab.

Loveless, A. (2007). *Creativity, technology and learning – a review of recent literature. Report 4 – Update*. Bristol: Futurelab.

Lucas, B., Claxton, G., & Spencer, E. (2013). *Progression in Student Creativity in School: First Steps Towards New Forms of Formative Assessments*. OECD Education Working Papers, No. 86. OECD Publishing.

Madden, M., Lenhart, A., Duggan, M., Cortesi, S., & Gasser, U. (2013). *Teens and Technology 2013*. Washington, DC: Pew Research Center's Internet & American Life Project.

McGarr, O. (2009). A review of podcasting in higher education: Its influence on the traditional lecture. *Australasian Journal of Educational Technology*, *25*(3), 309–321.

McGreen, N., & Arnedillo Sanchez, I. (2005). Mobile Phones: Creative learning tools. In P. Isaias, C. Borg, P. Kommers, & P. Bonanno (Eds.), *Mobile learning 2005* (pp. 241–245). Malta: International Association for Development of the Information Society Press.

Melero, J., Hernández-Leo, D., Sun, J., Santos, P., & Blat, J. (2015). How was the activity? A visualization support for a case of location-based learning design. *British Journal of Educational Technology*, *46*(2), 317–329. doi:10.1111/bjet.12238

Nack, F., & Waern, A. (2012). Mobile digital interactive storytelling—a winding path. *New Review of Hypermedia and Multimedia*, *18*(1-2), 3–9. doi:10.1080/13614568.2011.641418

Nuri, K., Cansu Cigdem, A., & Kursat, C. (2014). User Study of a New Smart Toy for Children's Storytelling. *Interactive Learning Environments*, *22*(5), 551–563. doi:10.1080/10494820.2012.682587

Oliver, B. (2005). Mobile blogging, 'Skyping' and podcasting: Targeting undergraduates' communication skills in transnational learning contexts. *Microlearning*, *107*(4), 587–600.

Pachler, N., Bachmair, B., & Cook, J. (2010a). *Mobile learning: structures, agency, practices*. New York: Springer. doi:10.1007/978-1-4419-0585-7

Pachler, N., Cook, J., & Bachmair, B. (2010b). Appropriation of mobile cultural resources for learning. *International Journal of Mobile and Blended Learning*, *2*(1), 1–21. doi:10.4018/jmbl.2010010101

Papacharissi, Z. (2011). *A networked self: identity, community and culture on social network sites*. New York, London: Routledge.

Pegrum, M., Bartle, E., & Longnecker, N. (2015). Can creative podcasting promote deep learning? The use of podcasting for learning content in an undergraduate science unit. *British Journal of Educational Technology*, *46*(1), 142–152. doi:10.1111/bjet.12133

Ranieri, M., & Bruni, I. (2013a). Mobile storytelling and informal education in a suburban area: A qualitative study on the potential of digital narratives for young second generation immigrants. *Learning, Media and Technology*, *38*(2), 217–235. doi:10.1080/17439884.2013.724073

Ranieri, M., & Bruni, I. (2013b). Racconto e mi racconto con il cellulare. In A. Parola, A. Rosa, & R. Giannatelli (Eds.), *Media, Linguaggi, Creatività* (pp. 143–165). Trento: Erickson.

Reid, M., Burn, A., & Parker, D. (2002). *Evaluation Report of the Becta Digital Video Pilot Project.* Report prepared for Becta. Retrieved from www.becta.org.uk/research/reports/digitalvideo/index.html

Resnick, M., Berg, R., & Eisenberg, M. (2000). Beyond Black Boxes: Bringing Transparency and Aesthetics Back to Scientific Investigation. *Journal of the Learning Sciences*, *9*(1), 7–30. doi:10.1207/s15327809jls0901_3

Rudd, T., Gifford, C., Morrison, J., & Facer, K. (2006). *What if...? Re-imagining Learning Spaces.* Bristol: Futurelab.

Runco, M. A. (2003). Education for Creative Potential. *Scandinavian Journal of Educational Research*, *47*(3), 317–324. doi:10.1080/00313830308598

Runco, M. A. (2007). *Creativity: theories and themes: research, development, and practice*. Amsterdam, London: Elsevier Academic Press.

Seipold, J., Pachler, N., Bachmair, B., & Döbeli Honegger, B. (2013). Mobile Learning – Strategies for planning and implementing learning with mobile devices in secondary school contexts. In N. Pachler & M. Leask (Eds.), *M-Learning. Learning to teach using ICT in the secondary school: a companion to school experience* (pp. 185–204). New York: Routledge.

Shamburg, C. (2009). *Student-powered podcasting: teaching for 21st-century literacy*. Eugene, OR: ISTE.

Sharp, C. (2004). Developing Young Children's Creativity: What can we learn from research? *Topic*, *32*, 5–12.

Sternberg, R. J. (1999). *Handbook of creativity*. Cambridge, UK: Cambridge University Press.

Sternberg, R. J., & Lubart, T. I. (1999). The concept of creativity: Prospects and paradigms. In R. J. Sternberg (Ed.), *Handbook of creativity* (pp. 3–15). Cambridge, UK: Cambridge University Press.

Su, C.-H., & Cheng, C.-H. (2015). A mobile gamification learning system for improving the learning motivation and achievements. *Journal of Computer Assisted Learning*, *31*(3), 268–286. doi:10.1111/jcal.12088

Taylor, C. W. (1988). Various Approaches to and Definitions of Creativity. In R. Sternberg (Ed.), *The Nature of Creativity: Contemporary Psychological Perspectives* (pp. 99–121). New York: Cambridge University Press.

Thorne, S. L., & Payne, J. S. (2005). Evolutionary trajectories, internet-mediated expression, and language education. *CALICO Journal*, *22*(3), 371–397.

Tolmie, A. (2001). Examining learning in relation to the contexts of use of ICT. *Journal of Computer Assisted Learning*, *17*(3), 235–241. doi:10.1046/j.0266-4909.2001.00178.x

Vavoula, G., Sharples, M., Rudman, P., Meek, J., & Lonsdale, P. (2009). Myartspace: Design and evaluation of support for learning with multimedia phones between classrooms and museums. *Computers & Education*, *53*(2), 286–299. doi:10.1016/j.compedu.2009.02.007

Villalba, E. (2008). *On Creativity: Towards an Understanding of Creativity and Innovation.* Luxemburg: European Commission / JRC. Retrieved on 25th July, 2015 from: http://publications.jrc.ec.europa.eu/repository/bitstream/JRC48604/eur_on%20creativity_new_.pdf

Vygotsky, L. S. (1978). *Mind in Society.* Cambridge, MA: Harvard University Press.

Wu, H., Lee, S. W., Chang, H., & Liang, J. (2013). Current status, opportunities and challenges of augmented reality in education. *Computers & Education, 62*, 41–49. doi:10.1016/j.compedu.2012.10.024

Wyse, D., & Spendlove, D. (2007). Partners in Creativity: Action Research and Creative Partnerships. *Education 3-13, 35*(2), 181-191.

ENDNOTES

1 More information on the project aims and results is available at: http://www.ensemble-project.eu.

2 For a detailed description of the project, see: http://www.oltrelospecchio.net.

3 For a more detailed discussion of the results, see Ranieri and Bruni (2013a).

Section 3
Mobile and Blended Learning

Chapter 10
Journalism and Law 2.0:
Collaborative Curriculum Redesign

Thomas Cochrane
Auckland University of Technology, New Zealand

Helen Sissons
Auckland University of Technology, New Zealand

Danielle L. Mulrennan
Auckland University of Technology, New Zealand

Vernon Rive
Auckland University of Technology, New Zealand

ABSTRACT

This chapter extends the authors' 2013 IJMBL article that covered the establishment of a lecturer Community Of Practice (COP) leading to two journalism 2.0 project iterations in 2011 and 2012. Since 2012 the project has had an increasingly wide impact across the journalism degree curriculum, including the establishment of a specific third year mobile journalism course in 2013, and the integration of mobile social media into over six journalism courses within the degree. In 2014 the journalism COP was also broadened to include two law lecturers from the business faculty of the university to support the integration of new pedagogies within the law degree curriculum. This led to the exploration of new pedagogies within two law degree courses in 2014, including international environment law, and law and media studies. This illustrates the potential of a COP model for supporting the brokering of pedagogical transformation across the boundaries of different (but associated) learning contexts.

INTRODUCTION

The initial impetus for this project was a growing realization that journalism was in crisis (Hall, 2005; Hirst, 2011; McChesney & Nichols, 2010) that was highlighted by the importance of mobile social media within the 'Arab Spring'. It became a question of; how should traditional journalism respond to a world where consumer preference is for music that is now distributed via the Internet rather than purchased

DOI: 10.4018/978-1-5225-0359-0.ch010

on CDs, video that is streamed either live or on demand rather than played on DVD, and news that is distributed via a host of social media channels such as Twitter, YouTube, Facebook and viewed on mobile devices such as the iPad rather than by traditional print media? One of the responses has been to lay off staff. Internationally, thousands of full-time journalists in traditional roles have lost their jobs over the past five years (Guskin, 2013; Ponsford, 2014). Since 2008, Fairfax media has reduced its workforce by more than 3,000 people in Australia and New Zealand as a direct result of the impact of digital and mobile social media technologies social media uptake (http://tvnz.co.nz/business-news/fairfax-axe-1900-jobs-video-4935480) (Hope & Myllylahti, 2013). In February 2012 the Guardian (Rusbridger, 2012) presented their view of how journalism is changing in response to the impact of mobile and social media within a fictitious revamp of the classic three little pigs fairy tale shared on YouTube http://www.youtube.com/watch?v=vDGrfhJH1P4. In our original paper, we argued that not only is traditional journalism in crisis but journalism education also cannot ignore the implications of the impact of mobile social media.

In response to these emerging issues, the authors decided to explore the possibilities and implications of journalism 2.0 within the context of journalism education. Journalism 2.0 as defined by the authors of this paper involves the exploration of two parallel aspects of social media in journalism education: exploring the embedding and modelling of social media in the delivery and pedagogy of a journalism course, and the exploration of the use and impact of social media on the practice of journalism in authentic contexts. This is based in the collaboration between an educational technology expert and experienced journalism lecturers within the framework of a community of practice (COP) investigating the potential of social media in the context of journalism education. Beginning in 2011, this was the first foray into integrating the use of social and mobile web 2.0 within the journalism curriculum, and the technology steward's previous mobile web 2.0 experiences in a variety of educational contexts were used to broker examples of mobile web 2.0 pedagogy to the journalism lecturers.

In 2011 we established the enhancing journalism education (EJE) mobile social media community of practice (COP) that continued throughout 2012 and 2013 (Cochrane, Mulrennan, Sissons, Pamatatau, & Barnes, 2013). The impact of the journalism community of practice's reified activities were used to broker pedagogical change throughout the journalism department and has resulted in significant collaborative curriculum redesign, reconceptualizing learning from a predominantly teacher-directed pedagogy towards student-determined heutagogy. Based upon our experiences we proposed a critical framework for supporting and implementing mobile social media for pedagogical change within journalism education (Cochrane, Antonczak, Guinibert, & Mulrennan, 2014; Cochrane, Sissons, Mulrennan, & Pamatatau, 2013). In this paper we explore the application of this framework to the collaborative redesign of several journalism papers in 2014 and we also extended our framework to a new context – that of law education and its crossover with journalism education. Journalism and law education overlap in contexts such as courtroom reporting, and environmental law and ethics. Thus in 2014 we established a combined journalism and law lecturer community of practice that built upon the work of establishing a mobile journalism paper in 2013. We called the 2014 combined community of practice MoJomLaw (mobile journalism/mobile law) which informed the development of mobile and social media within a variety of journalism and law curriculum areas, including: Journalism and media relations, journalism law and ethics, journalism theory and practice, television journalism, and environmental law. Our paper outlines the pragmatic application of our mobile social media framework to these two contexts, driven by heutagogy (Hase & Kenyon, 2001, 2007) – a pedagogical paradigm that facilitates a move from teacher-directed to student-determined learning, representing a reconception of both teaching and learning. We conclude that nurturing an interdisciplinary community of practice involves significant commitment, but yields creative collaboration in curriculum design.

While there is a significant body of literature exploring the pedagogical affordances of web 2.0 (McLoughlin & Lee, 2010) and mobile learning (Traxler, 2011), these two are seldom explicitly linked, and there was limited available literature on the pedagogical impact of these tools within the specific context of journalism or law education, see for example Ashton (2009). The term Web 2.0 was first coined by O'Reilly (2005) to denote the interactive, user-centred evolution of the World Wide Web from the static information and content delivery focus of early websites. Social media has become the preferred term to encapsulate the user-centric and social constructivist nature of a growing number of web 2.0 platforms and start-up services that focus upon sharing user-generated content and collaborative online experiences. Examples of the use of social media can be found in a wide variety of educational contexts including: Product Design (Withell, Cochrane, Reay, Gaziulusoy, & Inder, 2012), Business studies (Oldfield & Herrington, 2012) Language learning (Hockly, 2012), and health (Giodano & Giodano, 2011). Drawing on these wider contexts the lecturer community of practice began by exploring learning theories and pedagogical frameworks that can inform, support, and critique these innovations in communication studies and journalism. These included: Social Constructivism, Communities Of Practice, Authentic Learning, Pedagogy 2.0, Mlearning and Learner Generated Contexts.

BACKGROUND

This section introduces the foundational learning theory and pedagogical frameworks that inform the choice and use of social media within the project, including: Social Constructivism, Communities Of Practice, Authentic Learning, Pedagogy 2.0, Mlearning, and Learner Generated Contexts. These were chosen to inform the design of the project as they provide a critical framework for the context of journalism education where communication, teamwork and initiative are essential capabilities for twenty first century graduates.

Social Constructivism

According to theorists such as Vygotsky (1978) learning is a social process involving student exploration guided by more experienced peers. This interaction with experts provides learners with the potential to learn more than they could individually, increasing their zone of proximal development. Social constructivism forms a theoretical foundation for social learning theories and frameworks such as communities of practice and pedagogy 2.0 that are enabled by the use of social media (Wenger, White, & Smith, 2009; Wenger, White, Smith, & Rowe, 2005). Kirschner et al (Kirschner, Sweller, & Clark, 2006) critique constructivism as a fashionable but inefficient way to teach in comparison to guided instruction, however in reply a more balanced view was presented by Tobias and Duffy (2009) who argue that education is much more than memorizing teacher-delivered content. The relevance of social constructivism as an epistemological foundation for journalism education has been argued by educators such as Poerksen (2011), and Irby (1992).

Communities of Practice

Communities of practice is a social learning theory (Lave & Wenger, 1991; Wenger, 1998; Wenger, et al., 2009) that emphasises the process of developing membership in a community from initial peripheral participation to full participation as core members of the learning community. Based upon an apprenticeship model of teaching and learning, community of practice theory provides a foundation for critiquing and analysing Journalism education in a profession that has traditionally involved learning on the job from the more experienced Journalists, and earning your stripes in a competitive industry. Key concepts in community of practice theory include technology stewardship (Wenger, et al., 2009), the production of boundary objects as reified activities of the community of practice (Star, 2010), and transferring practice across community of practice boundaries via brokering (Wenger, 1998).

Authentic Learning

According to Herrington and Herrington (2006), higher education is largely characterised by delivery of decontextualized theory. Herrington, Mantei, Herrington, Olney and Ferry (2008) argue that new technologies such as mlearning require new pedagogies that leverage the affordances of new technologies for new forms of interaction rather than replicating current teaching practice on new devices. Authentic learning (A. Herrington & Herrington, 2007; J. Herrington & Oliver, 2000; J. Herrington, Reeves, & Oliver, 2009) is based upon situated learning (Brown, Collins, & Duguid, 1989) where explicit links are made between theory and practice using real-world (authentic) scenarios to represent and critically reflect upon on-the-job experiences. Mobile social media provide powerful tools for enabling authentic learning environments.

Pedagogy 2.0

Pedagogy 2.0 (McLoughlin & Lee, 2007, 2010) is a term coined to link social learning approaches with social media use. Pedagogy 2.0 forms a critical framework for integrating social media into education, focusing upon learner-generated content and learner-generated contexts rather than the delivery of teacher-generated content. Pedagogy 2.0 utilizes social media for "participation in networked communities of learning, personalization of the learning experience, and learner productivity in the form of active knowledge creation and innovation" (McLoughlin & Lee, 2010, p. 46).

Learner Generated Contexts

User-generated content can be enhanced via mobile devices allowing capturing of real life experiences within the actual context or location of an activity or event. This means we can bridge both formal and informal learning spaces and experiences beyond the classroom. Learner-generated contexts is a framework that bridges the gap between teacher-directed pedagogy and student-directed heutagogy (Luckin et al., 2010; Pachler, Bachmair, & Cook, 2010). Thus the focus moves from teacher delivered content to enabling student creativity and negotiation of learning outcomes and activities. This involves designing learning activities that are based around student teams working creatively in context-based scenarios. In the context of journalism education an example scenario would be a change from lecturer delivery of case studies of the impact of social media on journalism to student curation and critique of the social media source material produced during the development of a particular news story at the actual scene.

Mlearning

Mobile learning (mlearning) encompasses a wide range of pedagogies utilizing mobile devices (Pachler, et al., 2010) and a wide range of research methodologies (Wingkvist & Ericsson, 2011). However, the researcher's belief is that mlearning is a powerful tool for enabling learner-generated content and learner-generated contexts (Cochrane, 2010, 2011; Cochrane & Bateman, 2010). The ubiquitous nature of cell phones and the increasing student ownership of smartphones create a foundation for bridging both the digital divide and bridging formal and informal learning contexts (Vavoula, 2007), and as a catalyst for pedagogical change (Kukulska-Hulme, 2010). In particular, the rapid rise of Twitter uptake (McGiboney, 2009) is firmly associated with its essential link with cell phone use, and has become the defacto first source of news in many situations. Thus we argue that mlearning provides a platform for authentic journalism education, and an opportunity to reinvent interaction in and beyond the classroom.

METHODOLOGY

A participatory action research approach was adopted (Swantz, 2008), focusing upon pedagogical change while allowing the project to develop within a series of reflective interventions across the courses involved. The research was supported by a strategic framework developed by the researcher from the implementation of over 40 mlearning projects (Cochrane, 2012a, 2012b; Cochrane & Bateman, 2013; Cochrane, Black, Lee, Narayan, & Verswijvelen, 2012). This framework has four stages: the establishment of a supporting community of practice of lecturers, exploring new pedagogies afforded by mobile social media, the implementation of student-generated content and student-generated contexts via mobile social media, and the development of reflective teaching practice enabled by peer-reviewed publication of the project outcomes.

Research Questions

In preparation for the development of the introduction of social media to the journalism courses, the course lecturer travelled to the UK and the US and spoke in depth to journalism industry professionals at the BBC (British Broadcasting Corporation) in the UK and the Los Angeles Times in the US. She also spoke to leading journalism schools in Cardiff and Arizona. In addition, she has interviewed an online editor at a major organisation in New Zealand and a news editor at a local paper. This information was used to inform the project research questions:

1. How can the journalism curriculum be redesigned to respond to the impact of mobile social media on the news industry?
2. How can we then broker these pedagogical changes across departmental boundaries?

Data Gathering and Analysis

All participants in the project (lecturers and students) signed ethics consent forms allowing anonymous use of their social media sites (such as Wordpress blogs, Twitter, Storify, YouTube channel, archived Google Plus Hangouts, and Learning Management System statistics) for analysis and identification of emergent themes throughout the project. The perceived potential benefits and pitfalls of the project were discussed during the first class between the lecturers and students with the researcher present to answer any further questions. Participants were encouraged to establish a mobile eportfolio consisting of a selection of mobile social media that were explored throughout the project. These were collated in the first iterations of the project (2011-2012) using rich site summaries (RSS) via Google Reader, and then subsequently via the use of hashtag curation tools (such as TAGSExplorer and TAGBoard). A pre-project survey was completed by the participants to establish their previous experience of mobile social media. A focus group discussion provided feedback at the end of each semester of the project, and several students were selected to present reflections upon their project experiences to a wider group of department lecturers at the end of each semester.

Participants

The combined mobile journalism and mobile law lecturer community of practice (MoJoMLaw) was established in 2014 to build upon the momentum achieved within the journalism department throughout 2011-2013, and broker this pedagogical change into the context of the Law curriculum. The goal of the community of practice was to explore the potential of a cross-faculty collaboration to support pedagogical innovation, particularly in the shared overlap of the two curriculums. This followed an initial attempt to establish a law lecturer community of practice in 2013 that was useful as a pilot project but left significant room for further development. The MoJoMLaw community of practice was established in order to provide support for further pedagogical innovation in the law curriculum, and to provide cross-faculty examples of the potential of such innovative strategies. The core of the MoJoMLaw community of practice consisted of two journalism lecturers, two law lecturers, and an academic advisor as a technology steward. All participants were supplied with an iPhone 5S and an iPad mini. University ethics consent had been received as part of the 2013 projects. The MoJoMLaw community of practice met semi regularly face to face at a café that provided free WiFi and good coffee for nurturing the social fuel to sustain the community of practice. A Google Plus Community was established as a virtual community hub for the community of practice (http://bit.ly/mojomlaw), and a common hashtag was used to collate and curate the social media stream (for example using Tagboard to curate Twitter, Vine, Google Plus, Instagram at https://tagboard.com/mojomlaw/) around the project (#mojomlaw).

The MoJoMLaw community of practice also explored the relevance of blogging for creating student eportfolios, and invigorating lecture participation via facilitating backchannel conversations using Todaysmeet or Twitter. Another area of exploration included new modes of classroom presentation and collaboration via wireless screen mirroring from mobile devices. This was facilitated by the design and deployment of Mobile Airplay Screens (MOAs) (Cochrane, Munn, & Antonczak, 2013; Withell, et al., 2012).

ANALYSIS

The impact on the journalism and law curriculums were measured on the basis of the new pedagogical strategies that were enabled by the integration of mobile social media into the curriculum guided by the researcher's mobile social media framework, as outlined in the following sections.

Pedagogical Strategies

The state of journalism and law education at our university in 2014 can be summarised as two ends of a pedagogical continuum that moves from a focus upon a canon of teacher-directed content towards a focus upon empowering creativity and student-directed contexts. Luckin et al., (2010) refer to this continuum as the Pedagogy – Andragogy – Heutagogy (PAH) continuum, building upon Hase and Kenyon's (2001) articulation of the concept of heutagogy as student-determined learning.

Table 1. maps the Pedagogy-Andragogy-Heutagogy continuum onto the affordances of new technologies that have emerged over the last twenty years and are being utilized to varying degrees in higher education: from the impact of the world wide web and the introduction of online Learning Management Systems (LMS), to the emergence of social media, and the growing ubiquity of mobile devices. Thus a technological continuum that ranges from sedentary desktop computing to student-owned mobile devices maps onto a potential pedagogical continuum from teacher-directed to student-determined heutagogy. While there are examples of Journals dedicated to law education (such as "The Law Teacher"), there are few examples in the literature of the impact of social media or mobile devices on law education. In an article titled "The twenty-first century law student", Bone (2009) reports on a survey of students from nine UK undergraduate law degrees that law students "appreciate innovations by law teachers involving technology such as podcasts, wikis or blogs even though this [is] not widespread" (p. 244). Law education resides predominantly in a didactic pedagogical paradigm that has remained largely unchanged since the beginnings of universities. In comparison there has been a recent virtual tide of publications reflecting upon the impact of mobile social media upon the democratization of news reporting (Hirst, 2011; McChesney & Nichols, 2010) and a resultant impact upon journalism education (Ashton, 2009; Poerksen, 2011). We argue that law education can learn from the changes in journalism education, particularly in the areas of curriculum crossover.

Table 1. The post web 2.0 continuum

1995	2005	2013
• Web 1.0 • Teacher • LMS • Content delivery • PowerPoint • Pedagogy	• Web 2.0 • Student • ePortfolio • Student-generated Content • Slideshare • Andragogy • Social learning • Building learning communities	• Mobile • Collaboration • Connectivism • Student-generated Contexts • Mobile Social Media • Heutagogy • Creativity • Active participation in professional communities

A Mobile Social Media Framework for Creative Pedagogies

Since introducing our mobile social media framework in our earlier paper (Cochrane, Sissons, et al., 2013) we have refined the framework and identified three key elements of the framework: the need to establish and nurture supporting communities of practice, a focus upon curriculum redesign integrating new pedagogical strategies, and the provision of an appropriate supporting technical infrastructure. The theoretical framework is essentially a blend of associated new pedagogies aligned with the Pedagogy-Andragogy-Heutagogy continuum. These include: Puentedura's (2006) Substitution, Augmentation, Modification, Redefinition (SAMR) framework of technology adoption, Sternberg, Kaufman and Pretz's (2002) conception of three levels of creativity (reproduction, incrementation, and reinitiation), and supporting ontological pedagogies (2003) that move beyond knowledge construction towards reconceptualizing learners as active members of their chosen field. Kukulska-Hulme (2010) argues that mlearning provides a powerful catalyst for new forms of learning, and Laurillard (Laurillard, 2007) emphasizes the critical role of lecturers in designing new learning experiences enabled by mobile devices. Hockly (2012) argues that mlearning provides a vehicle for redefining learning activities in a heutagogical paradigm, and we illustrate examples of what this means in practice in Table 2, outlining our mobile social media framework for creative pedagogies.

Table 2. A mobile social media framework for creative pedagogies (modified from Luckin et al., 2010)

	Pedagogy	Andragogy	Heutagogy
Locus of Control	Teacher	Learner	Learner
Activity Types	Content delivery Digital assessment Teacher delivered content Teacher defined projects	Teacher as guide Digital identity Student-generated content Student negotiated teams	Teacher co-learner Digital presence Student-generated contexts Student negotiated projects
Cognition Level	Cognitive	Meta-cognitive	Epistemic
SAMR (Puentedura, 2006)	Substitution & Augmentation Portfolio to eportfolio PowerPoint on iPad Focus on productivity Mobile device as personal digital assistant and consumption tool	Modification: Reflection as VODCast Prezi on iPad New forms of collaboration Mobile device as content creation and curation tool	Redefinition: In situ reflections Presentations as dialogue with source material Community building Mobile device as collaborative tool
Creativity (Sternberg et al., 2002)	Reproduction	Incrementation	Reinitiation
Knowledge production context	Subject understanding: lecturers introduce and model the use of a range of mobile social media tools appropriate to the learning context	Process negotiation: students negotiate a choice of mobile social media tools to establish an eportfolio based upon user-generated content	Context shaping: students create project teams that investigate and critique user-generated content within the context of their discipline. These are then shared, curated, and peer-reviewed in an authentic COP
Mobile social media affordances	Enabling induction into a supportive learning community	Enabling user-generated content and active participation within an authentic COP of leaners	Enabling collaboration across user-generated contexts, and active participation within a global COP
Ontological shift (Danvers, 2003)	Learning about: Reconceptualising mobile social media: from a social to an educational domain	Reconceptualising the role of the teacher	Learning to become: Reconceptualising the role of the learner

Hase and Kenyon (2007) argue that pedagogical change requires a catalyst, and we argue that mobile social media provides such a catalyst. However changing pedagogical cultures and up skilling lectures to engage with the affordances of mobile social media in their own contexts requires a supporting professional development framework. We have found that establishing communities of practice of lecturers interested in exploring pedagogical innovation in their curriculum contexts is a powerful way of supporting and sustaining pedagogical change (Cochrane, 2014; Cochrane & Narayan, 2013). Baron and Corbin (2012) make a strong case for communities of practice in law education as a way of instilling in law students a sense of professionalism, ethics, and a sense of becoming part of a wider professional community. In a similar fashion Ashton and Irby (2009; 1992) both argue for the relevance of conceptualizing journalism education as a process of induction into the wider professional journalism community of practice.

CASE STUDIES FROM MOJOMLAW 2014

In this section we discuss initial applications of collaborative curriculum redesign as an outcome of the MoJoMLaw community of practice implementing our mobile social media framework for creative pedagogies. The impact of this is discussed in relation to three journalism courses and three law courses.

Journalism and Media Relations

This is a continuation of the course that pioneered social media within the journalism and Public Relations curricula. The use of Storify and the building of digital profiles as taught in this course has proved very successful, and been taken on by other course tutors. We also introduced Todaysmeet into the course as a formative live back channel conversational forum. Table 3 outlines the changes to assessment practices as a result of implementing our framework for mobile social media within the course context.

Journalism Law and Ethics

The course began using Storify in 2013 to replace the traditional essay as well as bringing in a course blog on Wordpress. Both were popular with the students. In 2014 we also began informal links with the law school that provided two lecturers to give two lectures in the course, as well as two joint classes where a well-known lawyer and a leading judge gave guest talks. This has built into the beginnings a collaborative research project about the effect of social media on reporting the courts and the jury process. Our findings will feed back into our teaching of social media practice and ethics in this course and

Table 3. Comparison of original and redesigned assessment activities

Previous Assessment Criteria	Redesigned Assessment Criteria
Assessment 1: Essay of 1,500 words typed up and handed in as hard copy.	Assessment 1: Essay of 1,500 words collated in and published on Storify.com and including material from at least three different social media platforms.
Assessment 2: In-class writing test	Assessment 2: In-class writing test

Table 4. Comparison of original and redesigned assessment activities

Previous Assessment Criteria	Redesigned Assessment Criteria
Assessment 1: Students each took part in a series of ethical scenarios through online proprietary software. They could compare their responses to other students but could not discuss these.	Assessment 1: Students in groups of three select a case that has been considered by one of the media regulators. They then present that case in full to the class using Prezi or any online presentation tool of their choice and lead a class discussion. Following the class the group members each write up the case including their response to it in a post on the Wordpress blog set up for the class. All other class members comment on one of the posts.
Assessment 2: Essay of 1,500 words typed up and handed in as hard copy.	Assessment 2: Essay of 1,500 words collated in and published on Storify.com and including material from at least three different social media platforms.
Assessment 3: An in-class law test.	Assessment 3: An in-class law test.

others in communications in the future. Table 4 outlines the changes to assessment practices as a result of implementing our framework for mobile social media within the course context.

Journalism Theory and Practice

The course looks at the effect of social media and multiskilling in journalism on the news product. Its readings are shared via Google Docs and assessment involves a blog (Wordpress), a presentation (Prezi) and a research essay (Storify). The researchers are continuing to explore ways in which the course can be developed to explore social media tools and take learning outside the classroom in ways that empower the students. Table 5 outlines the changes to assessment practices as a result of implementing our framework for mobile social media within the course context.

Judicial Review

The paper provides an introduction and overview of the key principles of judicial review of administrative action and legal controls on delegated legislative power, including substantive and procedural restraints on the exercise of public power and the regulation of relations between the citizen and the state. Relevant to this project, the paper includes consideration of various legal and non-legal means for resolution of public law disputes, including use of digital/social media (including websites, blogs and other social media platforms), as well as the use of such digital platforms in legal research in the public law area. Throughout semester one 2014 we explored new classroom presentation and interaction practices, and

Table 5. Comparison of original and redesigned assessment activities

Previous Assessment Criteria	Redesigned Assessment Criteria
Assessment 1: Participation in a blog hosted by Blackboard	Assessment 1: Participation in a blog hosted by Wordpress
Assessment 2: A group presentation	Assessment 2: A group presentation using Prezi or other online presentation tool
Assessment 3: Essay of 2,500 words typed up and handed in as hard copy.	Assessment 3: An essay of 2,500 words collated in and published on Storify.

Table 6. Comparison of original and redesigned course activities

Previous Course Criteria	Redesigned Course Activities
Course Content: Set readings and cases studies linked to the LMS and printed	Course Content: Collated and published as iBook and PDF for mobile access Production of a series of supporting PODCasts
Classroom Presentation: PowerPoint from lectern PC Q&A	Classroom Presentation: Wireless screen mirroring from iPad Mini Live interaction with source materials Guest lecturers Active Q&A
Assessment 1: A written legal opinion on a hypothetical scenario (30%)	Assessment 1: A written legal opinion on a hypothetical scenario, submitted online and marked using digital PDF annotations utilizing voice-dictation technology (40%)
Assessment 2: Final exam (70%)	Assessment 2: Final exam (60%)

a 'flipped' classroom mode of content delivery to student owned mobile devices. This included the introduction to the course of a suite of mobile accessible multimedia content uploaded to the LMS in a new course content category "iMaterials". Production of the iMaterials suite included, on a pilot basis, iBooks containing the relevant week's course readings and supporting case studies, a PDF version of the iBook for non iOS devices, and a supporting short video PODCast from the course lecturer. This new mode of class preparation enabled deeper interaction during the face-to-face classroom lecture, and the introduction of wireless screen mirroring from the lecturer's iPad facilitated a highly interactive lecture presentation style and a departure from the sedentary habits of PowerPoint presentations. The production of a virtual pre-recorded lecture was also explored. Student feedback on these new classroom practices and mobile content delivery were very positive. Table 6 outlines the changes to lecture practices as a result of implementing our framework for mobile social media within the course context. This will be built upon in future iterations by a redesign of course assessment activities as well.

Media Law

Established in 2014 this new elective paper looks at the common law and statutory framework and self-regulatory regimes which prescribe what can and cannot be published, broadcast or said, including the remedies available to those who have been harmed by any publication, broadcast or statement and the defences available to the publishers, broadcasters or authors. Relevant to this project, part of the in-class and assessment programme will include students providing timely advice to a newspaper / journalist / blogger about potential law and ethical breaches, for example defamation or privacy or lack of balance with respect to a published article or statement using digital mobile devices including wireless screen-mirroring to MOAs where appropriate. The assessment programme for this paper, which ran for the first time in Semester 2 of 2014, implemented the framework for mobile social media discussed above. Drawing upon the use of Todaysmeet in the journalism curriculum, weekly Media Law classes included the use of a Todaysmeet backchannel conversation to engage students in classroom debate.

Table 7. Comparison of original and redesigned assessments

Prior Course Activities and Assessment Criteria	Redesigned Course Activities
All course materials saved as PDFs and uploaded to the Blackboard LMS Assessment 1: Research paper that explores a current international environmental law issue (60%)	All course materials hosted on an interactive Wordpress.com site: http://autlawiel2014.wordpress.com Assessment 1: Research paper (45%) Presentation of research via either PowerPoint or creation of an online eportfolio using Wordpress.com that explores a current international environmental law issue (15%)
Assessment 2: Open book exam (40%)	Assessment 2: Open book exam (40%)
	Formative Assessment: Creation and presentation of team-based multimedia scenarios in class.

International Environmental Law

This elective course provides an introduction to the law governing the use, development and protection of natural resources in New Zealand, with a particular focus on the relevance, impacts and opportunities for businesses of the country's international environmental law regime. Relevant to this project, the paper includes legal considerations for media and reporting aspects of major international environmental issues and reporting of environmental litigation and use of digital/social media (including websites, blogs and other social media platforms) as a tool for resolution of environmental and public law disputes. Table 7 outlines the prior and redesigned assessment practices that were introduced in 2014 as a result of implementing our framework for mobile social media within the course context.

DISCUSSION

Key issues identified so far in this cross-curriculum collaboration include: a focus upon curriculum re-design integrating new pedagogical strategies, the need to establish and nurture supporting communities of practice, and the provision of an appropriate supporting technical infrastructure.

Integrating New Pedagogical Strategies into the Curriculum

The project has had a significant impact on assessment strategies: by using a social constructivist foundation formative feedback and peer assessment strategies become more important than in the previous teacher-directed pedagogies implemented in the courses, and these were explored and extended further in 2012. The implications of the 2011 projects informed the social media integration plan for 2012 courses, resulting in a social media integration framework for 2012 and onwards.

Specific examples of the reified activity of the mobile social media COP from 2011 to 2014 are collated and compared in Table 8.

For example, during 2012 we extended the use of Twitter as a way of brainstorming in journalism by asking the students to tweet their story ideas. The 140-character limit was used to test the clarity of the news angle and therefore the robustness of the idea, before the story was written as a blog post or online story and then pitched to followers via Twitter. One concern was that story ideas might need to be kept private until the story was written, otherwise the student could get "scooped" by another jour-

Table 8. Comparison of the Journalism 2.0 and MoJoMLaw project iterations

Year	2011	2012	2013	2014
Project Title	Journalism 2.0	EJETeam2012 Storifying journalism	MoJo Mobilizing Journalism	MoJoMLaw
Project Hub	AUTonline, the Blackboard Learning Management System (LMS)	Wordpress http://ejeteam.wordpress.com	Wordpress	Google Plus Community
Participants	N=20 students N=1 lecturers	N=70 students N=3 lecturers	N=70 students N=3 lecturers	N=100 students N=4 lecturers
Mobile Devices	Student-owned devices	Student-owned devices	Student-owned devices	Student-owned devices
Pedagogy	Andragogy	Heutagogy	Heutagogy	Moving from Pedagogy towards Heutagogy
Project Focus	Integrating mobile social media in the classroom	Exploring students as Social media reporters	Integrating mobile social media across the Journalism curriculum	Cross discipline collaboration and curriculum redesign
Web 2.0 Tools used for collaboration	Twitter QR Codes Polleverywhere	Wordpress Storify Twitter Skype Facebook Google Plus Polleverywhere	Wordpress Storify Twitter Google Hangout Facebook Google Plus Todaysmeet	Wordpress Storify Twitter Google Hangout Facebook Google Plus Todaysmeet
Selected outputs	(Cochrane, Sissons, & Mulrennan, 2012)	(Cochrane, Sissons, et al., 2013)	(Cochrane, Mulrennan, et al., 2013)	(Cochrane, et al., 2014)

nalist. This can be mitigated using a private chat platform such as Todaysmeet that was introduced into the curriculum in 2013. Three specific examples of the initial pedagogical impact of the exploration of social and mobile web 2.0 included:

- The personal engagement with Twitter as a virtual community building tool and collaboration tool by the lecturers and students, rather than the previous theoretical case studies only. Thus modelling an authentic use of these tools and appropriate etiquette.
- The use of Google Docs to collaboratively write, edit, and share this research paper itself, including previewing and critiquing its development on iPads during the weekly lecturer community of practice while drinking coffee in local cafes – nurturing a social atmosphere to the community of practice.
- The revamping of the traditional academic essay by the use of Storify.com and original rich media source material.

Establishing and Nurturing Supporting Communities of Practice

The initial 2011 community of practice comprised of one course lecturer and the technology steward proved to be a fruitful strategy for designing and supporting pedagogical change within the journalism course. The journalism lecturer community of practice continued and expanded in 2012, bringing into the community of practice two additional lecturers and funding teaching release time of one of the lecturers to take on the role of Learning and Teaching Fellow within the department to support and broker the integration of social and mobile social media within the curriculum. Thus the integration of social media and mobile social media within the journalism course was explored in depth in 2012, with a focus upon the lecturers modelling the use of mobile social media. This then led to the development and implementation of a new mobile journalism course introduced into the curriculum in 2013. Following on from this we sought to broker a new collaborative curriculum design approach into the law curriculum in 2014 via the establishment of the MoJoMLaw community of practice.

Wenger (1998) identified the need to nurture communities of practice and the role of brokering the culture and activities of a community of practice to other communities of practice and the wider community. More recently Wenger et al., (2009) have highlighted the critical importance of the role of technology stewards within communities of practice to guide the appropriate use of technology to support the communication and collaboration that sustains a community of practice. Creating a sense of community between two distinct departments in two different faculties was one of the initial hurdles of the project. Finding a common weekly time for the community of practice to meet was not easy, but once established these meetings became rich times of sharing ideas and brainstorming collaborative activities. Examples of collaborative events between the journalism and law students included guest lectures given by a chief high court judge and top media law lawyer and author of a leading textbook for journalists and media law students. These speakers covered judicial interaction with the media, legal and ethical aspects of good and bad journalism practice and related issues. Both journalism and law students were invited to these events, and a Todaysmeet backchannel was used to facilitate a cross-discipline conversation around the event and the issues highlighted (http://www.youtube.com/v/G6g2RCiR2e0?).

Provision of an Appropriate Supporting Technical Infrastructure

The cross-faculty sharing of classroom practice in the MoJoMLaw community of practice encouraged active exploration of new classroom practices and forms of interaction – including the use of wireless screen mirroring, the linking to expert guest input via live in-class video conferencing, and the use of a class backchannel conversation via Todaysmeet. This involved partnering with the institution's information technology department to open wireless access to screen mirroring and media streaming protocols across the teaching spaces used in both journalism and law. Wireless screen mirroring was enabled by three different approaches: installing Apple TVs in several of the lecture audio-visual presentation systems for large-scale interactive presentations, installing the Airserver App on lecturer's laptops to be used in lecture theatres without installed Apple TVs – the Airserver App allows multiple wireless screen connections facilitating a more collaborative and student participatory presentation, and finally the production of wireless screen-mirroring MOAs (Cochrane et al., 2013; Cochrane & Withell, 2013) for use as second screen back channel displays and portable student team workstations.

Indicative Student Feedback

The impact on student engagement, learning and participation within a more authentic learning community was evidenced in the level of critical reflection upon their learning experiences in 2014 particularly within their personal blogs that they established for the Mobile Journalism and International Environmental Law courses respectively. Of the 21 law students in 2014 only 10% of students had blogged prior to the course or engaged with social media in the context of their education. In 2014 all of the law students created and shared their own Wordpress blog as an eportfolio for the course. As the journalism COP had already established significant impact on the journalism curriculum prior to 2014 the impact on student learning in 2014 was evidenced in significant reconceptions of student identity from learners to becoming active participants within a professional journalism community. For example a journalism student summed up their experiences in a blog post as follows:

I've learnt how to use #Hashtags, as well as new mediums like Vine, Instagram, Storify etc. Without this paper, I wouldn't have even known some of these exist. But like I said, it's important for me as a professional going forward, to have a good online digital identity. It's important to plan out the content I want to post, think about the timing of it, and raise the ethical questions associated with it, to ensure I don't do anything that could harm my reputation.

Our professional identity will be with us long term, so to build a good, solid foundation now is important... Social media is NOT a phase, so it won't go away. We need to continue working with our online digital identities, through various communities, to ensure we stay in touch, we stay masters of our subject, and remain a point of contact/interest for readers/fans/viewers/etc.

I've certainly extended my communities of practice through doing this paper. I've extended it professionally ... I'm sure that with further development, I'll continue to grow in an online sense. And sometimes, the online self is just as important as the 'real' self. (Student blog post, 2014)

FUTURE RESEARCH DIRECTIONS

The sharing of ideas was mutually beneficial to the context of journalism education with interest sparked around the production of supporting iBooks and PODCasts for lectures. The next steps will involve further rethinking and redesigning appropriate assessment activities within the law curriculum as a result of exploring the mobile social media framework in lecturer practice. While the combined MoJoMLaw community of practice facilitated cross-pollination of ideas and innovative teaching practice between the two contexts it was difficult to sustain a cross-faculty community of practice. Finding an available weekly timeslot for the community of practice participants was almost impossible, and there was a significant disparity in the teaching and research workloads expected from the two faculties. In the end the MoJoMLaw community of practice was sustained predominantly through a shared online community via Google Plus split, while the physical community of practice split into two weekly groups that each met with the researcher as the technology steward who then brokered the sharing of ideas between the two groups.

CONCLUSION

The exploration of the integration of social media, and in particular mobile social media tools in the journalism curriculum has had significant impact upon the classroom experiences of the lecturers and the students. Scaffolded by a supportive community of practice, the lecturers have been empowered to try new approaches to move beyond teaching about social media in journalism to modelling and integrating the use of social media within authentic contexts in and beyond the classroom fostering student-generated content and student-generated contexts. Instead of attempting to put up barriers to the crisis created for journalism by user-generated news tools, the project demonstrates that journalism education can instead embrace and harness the potential of these tools in authentic scenarios. Building upon this experience we have found that it is possible to broker these pedagogical changes into new curriculum contexts via the establishment of a cross-discipline community of practice. Through the experience of the MoJoMLaw community of practice we have found that nurturing an interdisciplinary community of practice involves significant commitment, but yields creative collaboration in curriculum design, and empowers the exploration of new classroom practices.

REFERENCES

Ashton, D. (2009). *The professional in the age of the amateur: Higher education and journalism on-the-job*. Retrieved from http://theendofjournalism.wdfiles.com/local--files/danielashton/Daniel Ashton.doc

Baron, P., & Corbin, L. (2012). Thinking like a lawyer/acting like a professional: Communities of practice as a means of challenging orthodox legal education. *The Law Teacher*, *46*(2), 100–119. doi:10.1080/03069400.2012.681176

Bone, A. (2009). The twenty-first century law student. *The Law Teacher*, *43*(3), 222–245. doi:10.1080/03069400903332496

Brown, J. S., Collins, A., & Duguid, P. (1989). Situated cognition and the culture of learning. *Educational Researcher*, *18*(1), 32–41. doi:10.3102/0013189X018001032

Cochrane, T. (2010). Beyond the Yellow Brick Road: Mobile web 2.0 informing a new institutional elearning strategy. *ALT-J. Research in Learning Technology*, *18*(3), 221–231.

Cochrane, T. (2011). Exploring the Pedagogical Affordances of Mobile Web 2.0. In L. Chao (Ed.), Open Source Mobile Learning: Mobile Linux Applications (pp. 196-215). Hershey, PA: Information Science Reference (an imprint of IGI Global).

Cochrane, T. (2012a). Critical success factors for transforming pedagogy with mobile Web 2.0. *British Journal of Educational Technology*. 10.1111/j.1467-8535.2012.01384.x

Cochrane, T. (2012b). Secrets of mlearning failures: Confronting reality. In D. Hawkridge, S. Verjans, & G. Wilson (Eds.), *Proceedings of ALT-C 2012 - a confrontation with reality: The 19th international conference of the Association for Learning Technology* (pp. 123-134). ALT Association for Learning Technology.

Cochrane, T. (2014). Critical success factors for transforming pedagogy with mobile Web 2.0. *British Journal of Educational Technology*, *45*(1), 65–82. doi:10.1111/j.1467-8535.2012.01384.x

Cochrane, T., Antonczak, L., Guinibert, M., & Mulrennan, D. (2014). *Developing a mobile social media framework for creative pedagogies*. Paper presented at the 10th International Conference on Mobile Learning, Madrid, Spain.

Cochrane, T., Antonczak, L., Guinibert, M., Withell, A., Mulrennan, D., & Mountfort, P. (2013). *Collaboration unplugged: Herding a flock of MOAs.* Paper presented at the Electric Dreams: 30th ascilite Conference, Sydney, Australia.

Cochrane, T., & Bateman, R. (2010). Smartphones give you wings: Pedagogical affordances of mobile web 2.0. *Australasian Journal of Educational Technology*, *26*(1), 1–14.

Cochrane, T., & Bateman, R. (2013). A mobile web 2.0 framework: Reconceptualizing teaching and learning. In M. Repetto & G. Trentin (Eds.), Using network and mobile technology to bridge formal and informal learning (10th ed.; pp. 57-92). Oxford, UK: Chandos Publishing.

Cochrane, T., Black, B., Lee, M., Narayan, V., & Verswijvelen, M. (2012). Rethinking elearning support strategies. *International Journal for Academic Development,* 1-18.

Cochrane, T., Mulrennan, D., Sissons, H., Pamatatau, R., & Barnes, L. (2013). *Mobilizing Journalism Education.* Paper presented at the International Conference on Information Communication Technologies in Education (ICICTE 2013), Crete, Greece.

Cochrane, T., Munn, J., & Antonczak, L. (2013). *Design thinking for mlearning: Herding a flock of MOAs.* Paper presented at the 3rd Mobile Creativity and Innovation Symposium, Auckland, New Zealand.

Cochrane, T., & Narayan, V. (2013). Redesigning professional development: Reconceptualising teaching using social learning technologies. *Research in Learning Technology*, *21*(0), 1–19. doi:10.3402/rlt.v21i0.19226

Cochrane, T., Sissons, H., & Mulrennan, D. (2012). Journalism 2.0: Exploring the impact of Mobile and Social Media on Journalism Education. In I. A. Sánchez & P. Isaias (Eds.), *Proceedings of the IADIS International Conference on Mobile Learning 2012* (pp. 165-172). Berlin, Germany: IADIS International Association for Development of the Information Society.

Cochrane, T., Sissons, H., Mulrennan, D., & Pamatatau, R. (2013). Journalism 2.0: Exploring the impact of Mobile and Social Media on Journalism Education. *International Journal of Mobile and Blended Learning*, *5*(2), 22–38. doi:10.4018/jmbl.2013040102

Cochrane, T., & Withell, A. (2013). *Do 21st Century students dream of electric sheep? A mobile social media framework for creative pedagogies.* Paper presented at the Electric Dreams: Proceedings 30th ascilite Conference, Sydney, Australia.

Danvers, J. (2003). Towards a radical pedagogy: Provisional notes on learning and teaching in art & design. *International Journal of Art & Design Education*, *22*(1), 47–57. doi:10.1111/1468-5949.00338

Guskin, E. (2013). *Newspaper newsrooms suffer large staffing decreases.* Retrieved from http://www.pewresearch.org/fact-tank/2013/06/25/newspaper-newsrooms-suffer-large-staffing-decreases/

Hall, S. (2005). Introduction: Hidden in plain sight - journalism's critical issues. In A. Hall (Ed.), *Journalism: critical issues* (pp. 1–19). Berkshire, UK: Open University Press.

Hase, S., & Kenyon, C. (2001). From Andragogy to Heutagogy. *ultiBASE Articles,* 1-10. Retrieved from http://www.psy.gla.ac.uk/~steve/pr/Heutagogy.html

Hase, S., & Kenyon, C. (2007). Heutagogy: A child of complexity theory. *Complicity: an International Journal of Complexity and Education*, *4*(1), 111–118.

Herrington, A., & Herrington, J. (2006). What is an Authentic Learning Environment? In A. Herrington & J. Herrington (Eds.), Authentic learning environments in higher education (pp. 1-13). Hershey, PA: Information Science Publishing. doi:10.4018/978-1-59140-594-8.ch001

Herrington, A., & Herrington, J. (2007). *Authentic mobile learning in higher education.* Paper presented at the AARE 2007 International Educational Research Conference. Retrieved from http://researchrepository.murdoch.edu.au/5413/

Herrington, J., Mantei, J., Herrington, A., Olney, I., & Ferry, B. (2008). New technologies, new pedagogies: Mobile technologies and new ways of teaching and learning. In A. Farley & D. Holt (Eds.), *Proceedings of ASCILITE 2008* (pp. 419 - 427). ASCILITE.

Herrington, J., & Oliver, R. (2000). An instructional design framework for authentic learning environments. *Educational Technology Research and Development*, *48*(3), 23–48. doi:10.1007/BF02319856

Herrington, J., Reeves, T., & Oliver, R. (2009). *A guide to authentic e-learning.* London: Routledge.

Hirst, M. (2011). *News 2.0: Can journalism survive the Internet?* Crows Nest, NSW, Australia: Allen & Unwin.

Hockly, N. (2012). Mobile learning. *ELT Journal*, *67*(1), 80–84. doi:10.1093/elt/ccs064

Hope, W., & Myllylahti. (2013). Fairfax in trouble. *Foreign Control Watchdog 132.* Retrieved from http://www.converge.org.nz/watchdog/32/index32.html

Irby, J. (1992). *Creating the culture of journalism.* Paper presented at the the Annual Meeting of the Association for Education in Journalism and Mass Communication, Montreal, Quebec, Canada.

Kirschner, P., Sweller, J., & Clark, R. (2006). Why Minimal Guidance During Instruction Does Not Work: An Analysis of the Failure of Constructivist, Discovery, Problem-Based, Experiential, and Inquiry-Based Teaching. *Educational Psychologist*, *4*(2), 75–86. doi:10.1207/s15326985ep4102_1

Kukulska-Hulme, A. (2010). Mobile learning as a catalyst for change. *Open Learning: The Journal of Open and Distance Learning*, *25*(3), 181–185. doi:10.1080/02680513.2010.511945

Laurillard, D. (2007). Pedagogcal forms of mobile learning: framing research questions. In N. Pachler (Ed.), *Mobile learning: towards a research agenda* (pp. 33–54). London: WLE Centre, Institute of Education.

Lave, J., & Wenger, E. (1991). *Situated Learning: Legitimate peripheral participation.* Cambridge, UK: Cambridge University Press. doi:10.1017/CBO9780511815355

Luckin, R., Clark, W., Garnett, F., Whitworth, A., Akass, J., & Cook, J. (2010). Learner-Generated Contexts: A Framework to Support the Effective Use of Technology for Learning. In M. Lee & C. McLoughlin (Eds.), *Web 2.0-Based E-Learning: Applying Social Informatics for Tertiary Teaching* (pp. 70–84). Hershey, PA: IGI Global. doi:10.4018/978-1-60566-294-7.ch004

McChesney, R., & Nichols, J. (2010). *The death and life of American journalism: The media revolution that will begin the world again.* New York: Nation Books.

McGiboney, M. (2009). Twitter's Tweet Smell Of Success. *Nielsen Online*. Retrieved from http://blog.nielsen.com/nielsenwire/online_mobile/twitters-tweet-smell-of-success/

McLoughlin, C., & Lee, M. (2007). Social software and participatory learning: Pedagogical choices with technology affordances in the Web 2.0 era. In R. J. Atkinson, C. McBeath, S. K. A. Soong & C. Cheers (Eds.), Proceedings of Ascilite 2007, ICT: Providing Choices for Learners and Learning (pp. 664-675). ASCILITE.

McLoughlin, C., & Lee, M. (2010). Pedagogy 2.0: Critical challenges and responses to web 2.0 and social software in tertiary teaching. In M. Lee & C. McLoughlin (Eds.), *Web 2.0-Based E-Learning: Applying Social Informatics for Tertiary Teaching* (pp. 46–69). Hershey, PA: IGI Global. doi:10.4018/978-1-60566-294-7.ch003

Pachler, N., Bachmair, B., & Cook, J. (2010). *Mobile learning: Structures, agency, practices*. London: Springer. doi:10.1007/978-1-4419-0585-7

Poerksen, B. (2011). *The creation of reality: A constructivist epistemology of journalism and journalism education* (A. Kock & W. Kock, Trans.). Exeter, UK: Imprint Academic.

Ponsford, D. (2014). Labour force survey suggests surge of 12,000 in number of UK journalists since 2009. *Press Gazette*. Retrieved from http://www.pressgazette.co.uk/print/content/labour-force-survey-suggests-surge-12000-number-uk-journalists-2009

Puentedura, R. (2006). *Transformation, Technology, and Education*. Retrieved 18 February, 2013, from http://hippasus.com/resources/tte/puentedura_tte.pdf

Rusbridger, A. (2012, 29 February). *The Guardian: a world of news at your fingertips* [The Guardian's Three Little Pigs open journalism ad campaign]. Weblog posted to http://www.guardian.co.uk/help/insideguardian/2012/feb/29/open-journalism-at-the-guardian

Star, S. L. (2010). This Not a Boundary Object: Reflections on the Origin of a Concept. *Science, Technology & Human Values*, *35*(5), 601–617. doi:10.1177/0162243910377624

Sternberg, R. J., Kaufman, J. C., & Pretz, J. E. (2002). *The creativity conundrum: A propulsion model of kinds of creative contributions*. Philadelphia: Psychology Press.

Swantz, M. L. (2008). Participatory Action Research as Practice. In P. Reason & H. Bradbury (Eds.), *The SAGE Handbook of Action Research: Participative Inquiry and Practice* (2nd ed.; pp. 31–48). London: SAGE Publications. doi:10.4135/9781848607934.n8

Tobias, S., & Duffy, T. (Eds.). (2009). *Constructivist instruction: Success or failure?* London: Routledge.

Traxler, J. (2011). Introduction. In J. Traxler & J. Wishart (Eds.), Making mobile learning work: case studies of practice (pp. 4-12). Bristol, UK: ESCalate, University of Bristol, Graduate School of Education.

Vavoula, G. (2007). Learning Bridges: A role for mobile technologies in education. *Educational Technology Magazine*, *47*(3), 33–37.

Vygotsky, L. (1978). *Mind in Society*. Cambridge, MA: Harvard University Press.

Wenger, E. (1998). *Communities of Practice: Learning, Meaning, and Identity*. Cambridge, UK: Cambridge University Press. doi:10.1017/CBO9780511803932

Wenger, E., White, N., & Smith, J. (2009). *Digital Habitats: stewarding technology for communities*. Portland, OR: CPsquare.

Wenger, E., White, N., Smith, J., & Rowe, K. (2005). Technology for Communities. In L. Langelier (Ed.), *Working, Learning and Collaborating in a Network: Guide to the implementation and leadership of intentional communities of practice* (pp. 71–94). Quebec City: CEFIRO.

Wingkvist, A., & Ericsson, M. (2011). A survey of research method and purposes in mobile learning. *International Journal of Mobile and Blended Learning*, *3*(1), 1–17. doi:10.4018/jmbl.2011010101

Withell, A., Cochrane, T., Reay, S., Gaziulusoy, I., & Inder, S. (2012). Augmenting the Design Thinking Studio. In M. Brown (Ed.), ASCILITE 2012: Future challenges, sustainable futures (pp. 1071-1081). Wellington, New Zealand: ASCILITE.

KEY TERMS AND DEFINITIONS

Community of Practice: Defined by Wenger (1998) as a group of learners interested in a common issue that meet together regularly.

Heutagogy: Self-determined learning.

Journalism 2.0: Journalism practices that leverage the use of social media and user-content sharing, for example Twitter.

MOA: Mobile Airplay Screens – designed to enable wireless screen-mirroring of mobile devices.

Mobile Social Media: Social media sites and platforms that are formatted for use with mobile devices.

MoJoMLaw: The combined mobile journalism and mobile law community of practice.

The Pedagogy-Andragogy-Heutagogy Continuum: A range of pedagogical approaches from teacher-directed to student centred, and student-determined learning.

Chapter 11
Upgrading a Numerical Methods Course into New Mobile Technologies for Mathematical Education:
An Approach Based on Flexibility and Skill Development

Francisco Javier Delgado-Cepeda
Tecnologico de Monterrey, Mexico

ABSTRACT

Quality and diffusion of science and engineering education is a world challenge for the coming decades. It will require better and differentiated schemes of instruction able to reach the desired standards to combine science, technology and applied knowledge to solve future problems. Current vision is still limited in spite of the incipient use of educative technology in the universities, and despite its continuous growth based on mobile technologies in recent years. This chapter presents an updated vision of the educative evolution in a Numerical Methods course based on the integration of Math and Physics curricula, together with a Blended Learning approach. It describes the current movement towards a flexible scheme powered by mobile applications, the experience gained in the last ten years, and a strong collaborative faculty.

INTRODUCTION

Two decades ago, university faculty were commonly reluctant about the use of technology in education. Nevertheless, as time has gone by and technology has been improved, education has inevitably been a focus in the development of information technology. Today, there is a growing market around technology applied to education. In parallel, there is currently an assumption about education for the digital

DOI: 10.4018/978-1-5225-0359-0.ch011

natives' generation (people born after 1980 who supposedly have a strong affinity with technology) which is pushing the educational technology arena. Prensky (2001) states in this sense that they are native speakers of the digital language of computers, video games and the Internet. As an almost immediate conclusion, people born before those years (digital immigrants) have adopted these technologies later in their life, exhibiting in many cases a limited affinity and an uncomfortable feeling with it, particularly in education. Prensky clarifies that some immigrants are going to adapt to their digital environment better than others, but he suggests that they will never compare with digital natives (cited by Sadowski, 2014).

To summarize, several authors (Jardines, 2009; Gray, Thomas and Lewis, 2010; Hobbes, 2013) have stated some relations between the general history of distance learning and the chronology of past generations, in particular in terms of the sensorial stimulus and the learning styles developed in each case in their learning experiences (Internet World Stats, 2014). These chronological changes are imperceptible, but today they are configuring education deeply. Very recently in this history, mobile technologies have become accessible and spread across all educative levels. This accessibility, not always present in their predecessor technologies, has generated a realm of flexibility around learning processes (Johnson et al, 2013).

Johnson, Adams, and Haywood (2011) stated a very fast adoption of mobile technologies in education. Actually, with lots of educative Apps and other applications to develop educational products, the prospect is open to teacher innovation by carrying out several educative trends supported by it. Still, from the author point of view, Higher Education is still an undecided terrain in terms of clarity and overwhelming usage of learning technologies. Nevertheless, it is having an incipient tendency to be explored and increased in recent years. In some senses, Higher Education appears sometimes as the last bastion of traditional education. In contrast, as Laurillard (2005, 2008) has noted, mobile technology is solving different needs to enhance the professional learning of disciplines: abilities for employment, high quality courses and flexibility to study. Mathematical higher education is, at least, one of the most open disciplines for mobile learning technologies, due possibly to its timely skill development at the operational level (more than in terms of extensive competencies development).

Several years ago, the author began introducing complementary technologies in a Numerical Methods course, in addition to those oriented exclusively to develop the planned course skills (Excel, Mathematica, Python, etc.). As a result, in 2011, several technologies were used to start the mobile management as a part of a mobile learning institutional effort (Delgado, 2013). Those complementary resources were first oriented to cover detected weaknesses in the learning process. Over time, while educative mobile tools were improved and spread, they have brought the opportunity to use them as a parallel support of learning management. Recent years have been rich in terms of these developments, and an exponential growth of these tools has been seen. The continuous search and experimentation to apply them in the course and in other educative projects have brought changes in the old educative approach.

The aim of this work is to provide an updated version of a previous report (Delgado, 2013), relating the advances in the use of mobile technologies in the course over the last four years. The first attempt to introduce mobile technologies was reported then. Today, these technologies have shaped the course and the viability to offer it as a pure online version is being considered (based on the growing blended learning experience gained). This report is centred not only on improved technologies, but in the course design and in the comparison between blended and face to face evaluation activities. Also, on an insight, to introduce a more flexible approach for this course in terms of a hybrid learning version. In the second section, a review and a discussion about changes in the last four years in the learning technologies are presented. It states the report research objectives. In the third and fourth sections, the new design is presented, based on the outcomes for the student impact of blended strategies. In the fifth and sixth sections, a reflection about possible trends for a flexible learning approach is conducted, including the proposed future research directions.

BACKGROUND

Requirements from industry and society on engineering education are commonly based on the advancements in technology. Engineering education has been subject to continuous adaptation by introducing new information technologies. Subsequently, teachers in those areas have been continuously reinventing and proposing new models of education. The use of professional software to simulate, plan or calculate engineering processes is now common in the university, often promoting a computer lab class together with the theoretical class. Online education is a broad term to include online learning activities (scheduled instructions, tool based activities and delivering), both in the context of face to face learning and blended learning. They are part of a fully planned sequence of online activities (pure online learning). In both extreme contexts, these activities include partners and faculty interactions (Conole et al, 2008). In 2011, coincidentally with the original report (Delgado, 2013), the author became involved in an institutional effort to develop and promote the use of mobile technologies in the higher education faculties (Delgado, 2014). This project brought a boost for these kinds of technologies and associated educative trends in the university. They were capitalized as an effort supported with a proper permanent but dynamic teacher training program and a technology centre to share and build up projects on these trends.

Numerical Methods courses are mandatory in the engineering programs. This course is scheduled after the theoretical Math and Science courses. For the Math courses, it is an additional opportunity to review and apply concepts with a numerical perspective. For the Physics courses, the Numerical Methods course brings the possibility to analyse more complex problems out of the analytical treatment, connecting it with Engineering. Numerical analysis has evolved with technology. Its traditional teaching has evolved into the spread of computer systems. The fast expansion of technology has adapted the curricula and the teaching methodology. The spread of programming languages together with professional software and the necessity for the inclusion of real applications had demanded a transition from a Numerical Analysis course into a Computer Simulation one. It has reduced the gap between the academic and the professional practice. Nevertheless, in the practice, these changes introduce more complexity in the learning process, despite being closer to experience.

The use of associated software, multimedia resources and online activities (Laurillard, 2002) has been increased in higher education. The inclusion of professional technologies in class for the programming of numerical procedures (using Excel, Mathematica or Python) is a common practice. Today, it continues with other colleagues as a standard (with Matlab and Math Studio as additional alternatives). In the first approach, these resources were directly included in Blackboard (2015) in a closed access course design. Later, the use of Winksite, a basic mobile site developer platform was reported as a parallel resource integrating media (Delgado, 2013). Finally, a simple platform was adopted to support periodic quizzes in class or online: Socrative (2015).

Today, people absorb media on-the-go. Whether for simple skills or professional knowledge dissemination (conferences, courses and seminars), online video repositories let us easily share educative videos as podcasting or screencasting, together with computer tools for webcam or screen recording. An important component in the first design was a YouTube channel for this course. It was originally included as a complementary resource for the face to face sessions. Surprisingly, this resource let students follow at their own pace the construction of codes for the numerical methods being taught. Moreover, this effect was not limited to enrolled students, instead, most views were due to informal students in other groups, universities and countries. Decades ago, educative broadcasting was a very common practice in the universities to connect with experts in other locations. Although those efforts were created to reproduce face to face instruction, they offer an advantage over the face to face exposition: repetition. As Stocks (2014) remarks: "this way of learning incorporates those things to be taught, but in a way that is easy to absorb and understand". Media channels are an important ubiquitous based resource, and in the future they will surely be part of the course development into an alternative purely online version.

The evolution of the Numerical Methods course into a Computer Simulation course through computer technologies (Delgado & Martínez, 2011) was centred on visualization and simulation of more complex problems in engineering. Later, the use of mobile technology scaffolded the face to face learning under a blended scheme (Delgado, 2013). Today, the course is evolving into a flexible learning approach planned as alternative paths to teach in terms of the students' interests. Still, this aspect is questioned because theoretical math and programming skills require a rigid individual process.

Nowadays, mobile technologies combined with minimum human intervention have seen the closest approach to automated education in the concept of Massive Open Online Courses (MOOCs). This trend is under evaluation by educators, education reformers and entrepreneurs as the next step in higher education. These platforms still have some troubles when the connectivity of thousands of students collapses (Straumsheim, 2014). Despite the technical issues, the model is being evaluated in terms of its universality. For the Numerical Methods course, the restrictions mentioned in the previous paragraph could dismiss this model. Nevertheless, there are alternative models, such as Massive Open Online Research (MOOR), basically a reduced MOOC with emphasis on research, or the Small Private Online Course (SPOC), an online course based only on a segment of the university to fit local requirements and normally including face to face interactions (Goral, 2014). In these, goals and recipients are different; MOORs are oriented to students working together to solve a practical problem while a SPOC provides more directed instruction (both are valuable in a Numerical Methods course). In these approaches, students can be located in different places across the world. University of San Diego's Jacob's School of Engineering launched a Bioinformatics Algorithms major online course in Coursera, featured mainly as a MOOR. It gives to the students a chance to work on targeted research projects under the guidance of a notable researcher (Hosler, 2014).

Educative online platforms or Virtual Learning Environments (VLE's) have existed for almost 20 years or so. Platforms such as Blackboard, Moodle, Academy and others are used to deliver contents, activities, applications and the course schedule. In the previous report (Delgado, 2013), Blackboard was the main container for the course resources, for managing several processes, but limited with respect to the inclusion of external open mobile resources.

As an update of (Delgado, 2013), the current inspiration is the inclusion of mobile tools to support education. They continue enriching the positive learning impact of technology in the course and contributing with more flexible learning schemes. It is related with how technology is substituting certain spaces which were exclusive from face to face education, evaluation particularly. A related aspect is the new dimension of open resources for informal users. Thus, in the terms exposed in the aim of this work, the natural development for the course is the objective to evolve it into a hybrid course (a flexible and open combination of instruction modalities based on modern educational trends). The objectives of this update are:

1. To report and to analyse the evolved design for the course in terms of new mobile tools involved.
2. To depict the newest interactions behind it by means of its blended learning activities.
3. To analyse the hidden effectiveness of blended activities as representative evaluations and promotion predictors (for the last two years).

The background motivation of this paper is to analyse the possible evolution of the course as a flexible version, combining alternative paths for each student, proposing alternative models of instruction. In it, each student ideally should select their resource learnings, their pace and their evaluation route.

MAIN DEVELOPMENTS IN RECENT YEARS: CONTEXT UPDATED

Dissemination of Mobile Technologies

The first report (Delgado, 2013) for the mobile innovations in the course happened to meet with the mobile learning project in the institution. In that report in was the inclusion of Winksite as a lean complementary platform to gather a mobile repository. Since then, there was an exponential dissemination of mobile applications, including those for education. Most of those tools actually tend to embed their products (evaluation, gamification, delivery, etc.) to gather them in a mobile site. Nevertheless, at the time of the first report, mobile applications rarely did that. Actually, tools for mobile site development have had a fast evolution, bringing ease, aesthetics and functionality, motivating teachers to create mobile sites for specific purposes rather than solely concentrating on their course activities. This functionality lets them include several resources to reach almost any mobile device, promoting ubiquity.

Blackboard materials were developed several years ago, but version changes in this VLE unfortunately affecting the designs, sometimes irretrievably. Meanwhile, new site developers in combination with mobile applications exhibit more stability, enabling the introduction of custom-made resources. During 2014, the use of Weebly, Jotform, Google Drive and Classmarker let us replace almost all the functionalities provided by Blackboard for the Numerical Methods course in an effort to create an approach based on open resources. This fact led the author to develop other educative projects as support for other courses: Calculus, Math and Physics Widgets and a Mobile Physics Lab (Delgado & Santiago, 2014; Delgado et al, 2015) by combining appropriate mobile educative resources on integrated mobile sites, an evolution of the previous design. Figure 1 shows the evolution in the inclusion of these technologies for the Numerical Methods course.

Nevertheless Blackboard is a robust platform to manage academic content and evaluation, with the ability to include an extended number of external applications, not always possible in closed platforms, the goal pursued in the institutional mobile learning project. For these reasons, Weebly site developer was selected as the basis for the current design of the course (Weebly, 2015) substituting together both Blackboard and Winksite. Thus, the inclusion of resources (video, PDF documents, Mathematica files and specific HTML code, all of them embedded and ready to use in the same site) was possible, bringing new horizons in the design. Widgets based on platforms like Wolfram Alpha and Mathtab were introduced to allow online practice and visualization (Delgado et al, 2015). In these, several numerical methods were carried out to let students understand their parameters and operability, bringing an easier alternative for App Inventor (Delgado, 2013). Similarly, videos broadcast on the YouTube channel were included there. Both communities were integrated: the students enrolled in the courses and the informal

Figure 1. Timeline evolution in terms of class placement, resources carried out and e-learning management for the VLE's used

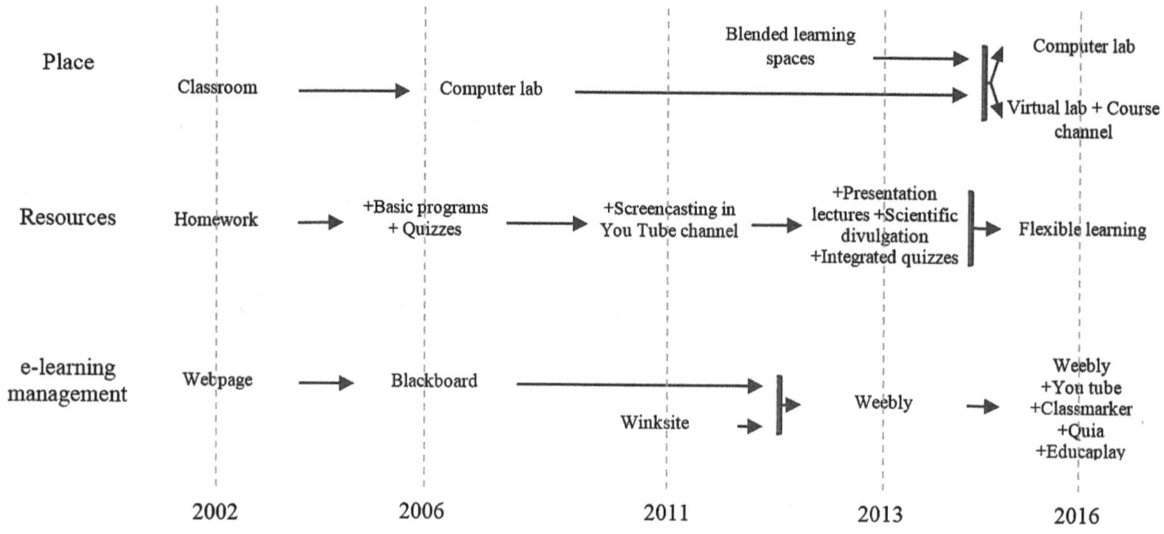

users from the internet community, almost sharing the same resources. In parallel, a Facebook group was created. Evaluation (formative and summative) was included with Socrative (2015) and Classmarker (2015). Sometimes, summative evaluations have been applied completely online on the Classmarker platform instead of the classical in-class exams. As before, a multiplatform mobile electronic book prevails still (Delgado, 2013) including extended repositories (simulations, programming codes, projects and applied complex problems). This e-book is now available on Amazon and the iTunes store, including versions for iPad and other tablets.

At end of the day, premises established by Conole et al (2008) about didactic planning were fulfilled in the first report (Delgado, 2013): a layered design combining several technologies to match learning objectives and operability. This vision is updated in Figure 2. As was stated previously (Delgado, 2013), layers in the technology scaffolding attempt to take care of diverse learning needs of students: use of technology, programming practice, use of materials in repositories and face to face sessions inclusively.

This course, which normally involves several skills and competencies to be developed, could now be addressed in a hybrid scheme based on a near construction (understood as the possibility to select the adaptive learning path to become promoted in the course). This is an important aspect to recall in this work because this blended learning approach has improved the learning outcomes.

Attending to the Current Characteristics of Users

As was stated previously, Numerical Methods courses are normally mandatory in the most engineering programs at the university. In the modern version of this discipline, it has evolved together with technology in a valuable practice to solve and simulate engineering and scientific problems. This perspective should be transmitted to the students. In the same trend, its traditional teaching has evolved dramatically with the spread of computer systems and their affordability, changing its learning approaches over the last ten years. Today, new generations have, at least partially, a distorted vision of applied technology.

Figure 2. Blended approach: a) Course objectives supported by the main technological resources; secondary resources take care of weaknesses introduced by the new goals and the technology added; b) Main technology tools for learning activities, complementary contents and supplementary support. Update adapted from Delgado (2013)

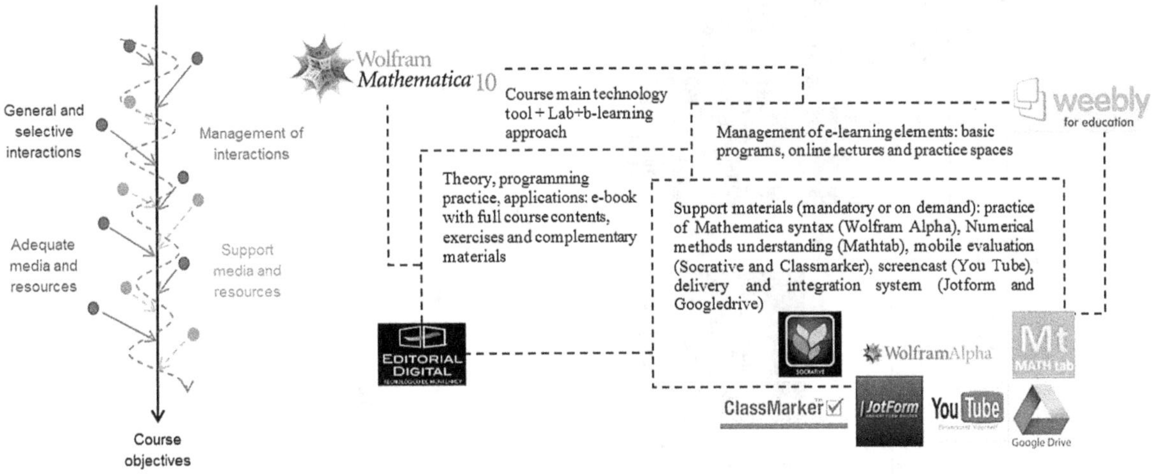

a) b)

As final users of improved technological solutions, it is sometimes difficult to understand how technology is based on abstract, complex and extended concepts developed through history. Compared with the students from four years ago, the current ones are more accustomed to appeal to electronic resources and to search for information through mobile means.

Departing from the most traditional form, the Numerical Methods being depicted has evolved into a modern version including professional software for programming and a contextual orientation, but still mainly developed under a face to face learning strategy, recently evolved into a blended learning approach. This course is a landmark in the third part of engineering programs because it lets us, in principle, integrate the math and physics curricula with that for engineering sciences. The examples and the projects included are oriented to several disciplines, and the diversity of engineering programs enrolled is wide: industrial, mechanical, electrical, chemical and biotechnological. In the current design, several sections attempt to include activities and resources based on professional focusing. Thus, the sections of the mobile site involve not only activities and resources, but take care of students' interests too. The following subsections depict briefly the contents and the relationships of each section in the site.

Resources

This section was constructed containing a subsection for each theme in the course (Figure 3). Each one uses a three stage methodology of independent learning: a) online notes and lecture (brief class in slides), b) a screencast summarizing the programming contents and the plan for each theme (videos in the YouTube channel), and c) one or two downloadable Mathematica materials (Mathematica files as advance and/or support to the class, and related with the screencast presented). Three elements were constructed as a combined strategy to replace what might be missed from the face to face instruction.

Figure 3. Resources for each theme in the course: online slide based class, screencasts and downloadable Mathematica files

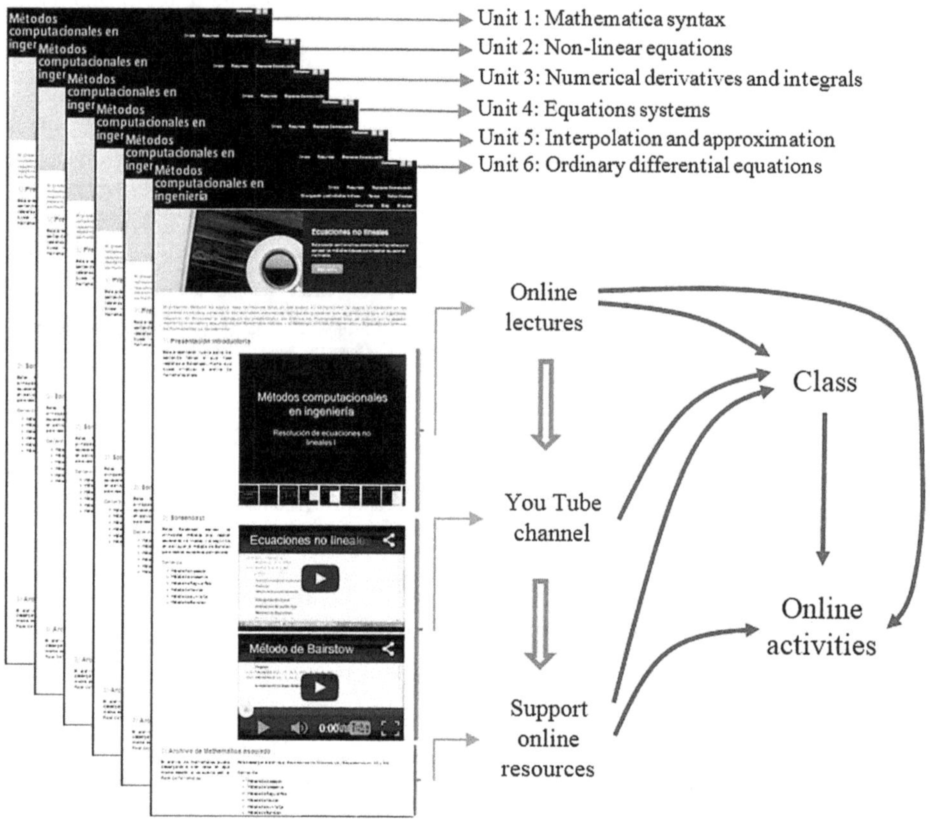

There is a twofold relationship between the class and some elements in this section. Sometimes, slides could be partially used in class to reproduce extended calculations for a specific numerical method; similarly, Mathematica files could be used as basic codes to modify and to refine in class. Slides support the online teamwork. All these elements together are expected to replace the class, at least partially, with a minimum of teacher tutoring, as occasionally happens with some student missing the class.

Daily Class Notes

Daily class notes are downloadable files published as complementary notes for the students, including final programs and examples revised there. They are allocated by the teacher in this section (Figure 4) grouped by class and theme. These notes are valuable together with the online slide based class in the Resources section. In general, this resource is a support for the evaluations and the homework by recovering important elements in the computer practice.

Figure 4. Daily class notes are based on the developments constructed in class by the teacher in the lab practice

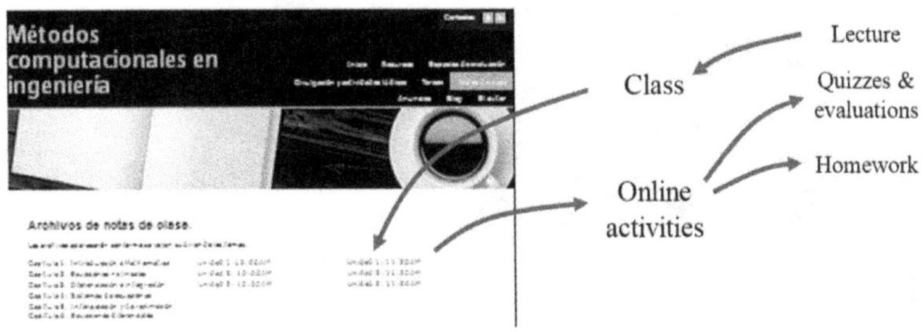

Homework and Online Work

Individual homework through the course is published here based completely on the e-book. In addition, teamwork is normally based on online activities and they are published in this section (and in the announcements section), requiring students to send a Mathematica file or a PDF report. Teamwork is planned in terms of the course advancement in the semester period. As in the evaluation space, forms developed in Jotform are included and integrated with Google Drive to drop each homework (individually or by team). The final integration automatically generates an e-mail submission notification to the teacher and an acknowledgement receipt for each sender. Afterwards, the teacher can evaluate each activity from their e-mail by reviewing a link to the Google Drive repository and then give feedback to the student or group of students involved.

These elements direct and address the student work through the semester, helping them to not miss assignments. Figure 5 shows the general aspect of these elements in this section, showing not only the delivery actions, but the classical chronology of them. Lectures and lab practice in the class normally generate elements for the individual homework based on the minimum level of skill development. Meanwhile, some online activities, typically built as teamwork, develop integrated skills in the sense of professional competencies. Finally, in this work, the question raised here is about whether complementary materials in the Resources section could substitute for face to face instruction, at least for some advanced students.

Evaluation Spaces

This section concentrates on the access and delivery of formative and summative evaluations. Normally, formative evaluations are developed in the Socrative application and they are used to evaluate the partial advancement of the individual homework, or as a previous practice for the summative evaluations. Summative evaluations are commonly done in Classmarker. Correspondingly forms developed in Jotform are included there and integrated to Google Drive, as in the homework section, to send the corresponding Mathematica file generated in each activity. It automatically sends to the teacher an e-mail notification, an acknowledgement receipt for the student and finally it closes the feedback cycle when the teacher grades the activity (Figure 6). In Classmarker each evaluation can be differentiated by student (different questions with similar complexity level), so the question about the possibility to replace the in-class evaluation with the online one arises here. This is one of the main questions given in the growing spectrum of blended activities in the course. If these evaluations could be combined effectively to reproduce and to

Figure 5. Homework section combining the information of each individual and the team assignment together with a delivery form integrated with Google Drive

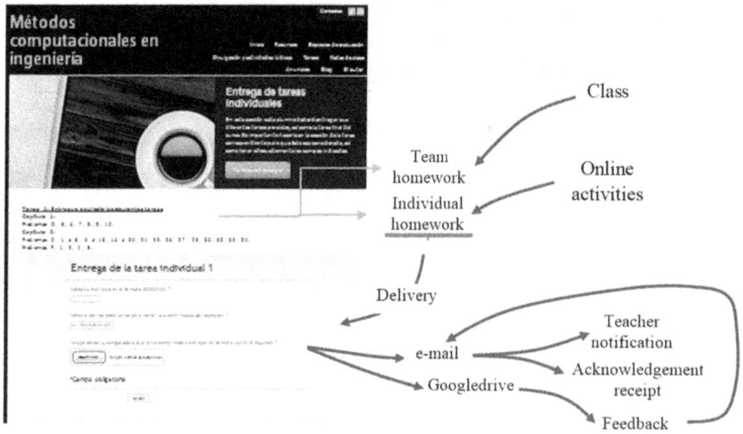

Figure 6. Evaluation section showing the access to Socrative and Classmarker applications with an associated delivery form integrated into Google Drive

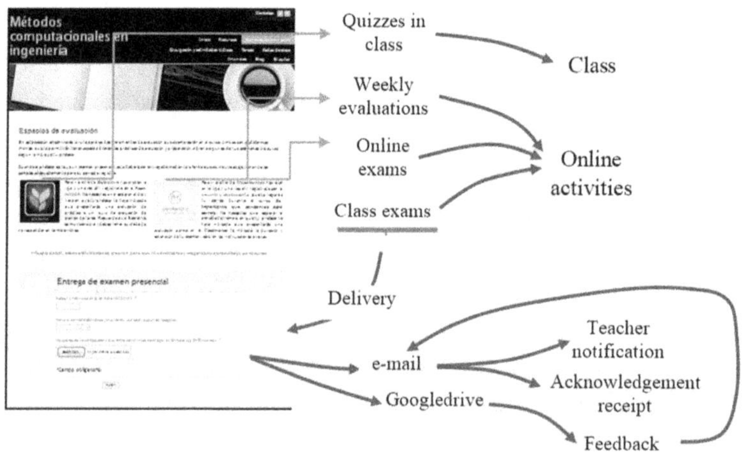

discriminate the student performance, then a wider horizon of flexibility in the course could be attempted in combination with a selected substitution of the class by the electronic activities included in the site.

Other Complementary Sections: Announcements, Ludic zone, Blog and Stats

Class announcements and scheduled activities are published in a specific section devoted to that. Homework due dates, activities for the class, evaluation dates and all kind of complementary activities are registered, recorded and scheduled in this section. The Blog is an interactive student section with trends, questions and answers related to the course topics, homework or other associated issues in the course. It is normally in the students' charge with occasional intervention by the teacher. The Ludic zone includes some activities and a comic, not necessarily related to the course: a science divulgation channel, review

Figure 7. Country accesses during the first six months of analytics and prospective accesses in a specific month compared and projected with the previous

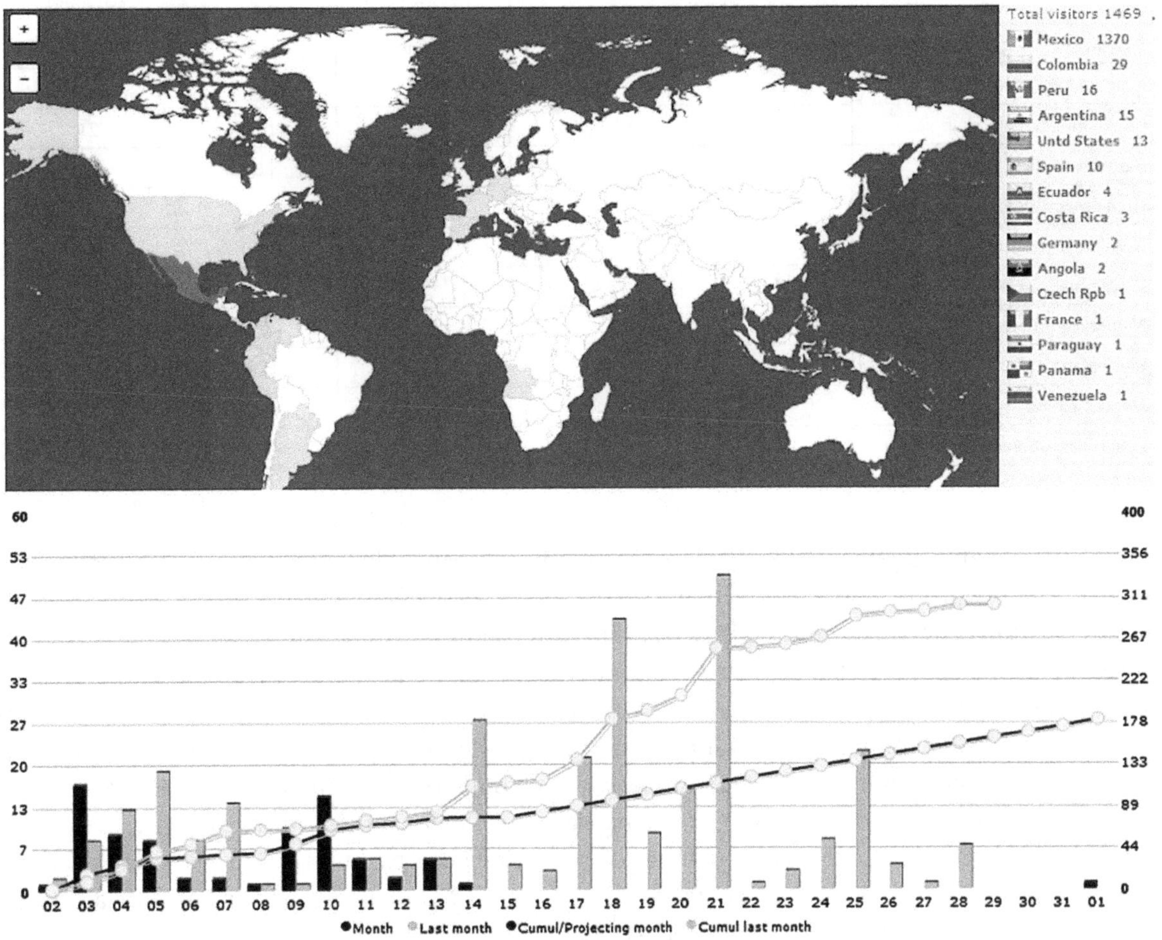

concepts games and video links to possible problems or applications. Usage stats in terms of access to each page, time of use and other analytics is included. These analytics record the non-formal access by the users (Figure 7), revealing the impact of open access resources on the site.

Data from Analytics: Formal and Informal Users

As was presented in Figure 7, several analytics show this site is being reached overseas by other users as a kind of complementary resource. Together, a Facebook group for Numerical Methods course is opened each semester for students enrolled and it works as an extension of the site's Blog. In addition, the YouTube channel is an open independent site (not expected to serve only the enrolled students) related informally with the course site from there. This site contains screencasts with short videos from Numerical Methods implementations. As for the course site, the YouTube channel (Delgado, 2012) analytics show almost 3,000 visits around world in the two last years, distributed as shown in Table 1.

Table 1. Analytics of the You Tube Numerical Methods channel in 2013-2014

Video	Visualizations		Visualization Time		Visualization Fraction
Statistical	Times	%	Minutes	%	%
Mathematica management	357	27%	385	18%	39%
Gauss-Jordan Method	258	19.0%	223	10.0%	21%
Bairstow Method	128	9.3%	224	11.0%	36%
High order differential equations	124	9.0%	247	12.0%	72%
Non-linear equations	124	9.0%	324	15.0%	84%
Least squares	102	7.4%	194	9.1%	63%
Numerical differentiation	83	6.0%	179	8.4%	60%
Interpolation	63	4.6%	145	6.8%	51%
Jacobi Method	52	3.8%	73	3.4%	30%
First order differential equations	48	3.5%	101	4.7%	51%

Data shows the main topics visited in terms of access (absolute and relative), time devoted (absolute and relative) and average visualization fraction (part of video effectively seen).

Access analytics for the site in the last year (2014) are shown in Table 2 in terms of percentages, although visits were about 2,000. Percentages are calculated with respect their own group, namely, formal or informal students (the geographically non-local users). Statistics reflect that Resources and Daily notes sections are valuable for both types of user. While, contrasting with the users in the YouTube channel, both resources appear useful and complementary to each other (because these screencasts are oriented to show and to explain the contents of Mathematica files). Inner statistics for formal users reflect the more common activities in the course in terms of evaluation policies, but still, the access to Resources and Daily notes sections shows a bigger value than expected because they are not mandatory activities, considering that formal users are enrolled in the face to face course.

These analytics cannot fully account for the impact on the learning process versus the face to face class. Statistics in the alternative materials in the course site reflect only some possible valuable use compared with the class activities. Clearly, in the current blended scheme, both activities become complementary but they do not appear exclusive. Note that most of these online materials and activities were introduced recently with the mobile learning approach in the course, complementing the scaffolding for the weaknesses in the traditional face to face course carried out in the computer lab (Delgado, 2013).

In the primary approach for the course redesign, the impact of technology activities on the main abilities being developed in the course includes math skills (Delgado, 2013). Instead, in the present work, mobile technology is being considered as a scaffolding element in the course and it is being considered as a possible substitute for the face to face instruction when it is represented as a differentiated delivery media (face to face, online based on computer and mobile) in terms of ubiquity. Currently, the author has no irrefutable evidence about the effectivity of complementary resources as an isolated scenario of instruction compared to face to face. Recent statistics show only (in terms of demand) that those resources are being an efficient complement to traditional instruction for the average student population. Nevertheless, in the previous report it was stated that the use of information systems in programming, delivering, work, discussion and usage of computational tools enhanced the learning concepts towards

Table 2. Usage analytics for the Numerical Methods site in 2014

Section	Access Analytics for Numerical Methods Course Site in 2014			
	Formal: 97%		Informal: 3%	
Statistical	Total Accesses	Access Percentage	Total Accesses	Access Percentage
Resources	981	51%	32	53%
Evaluation	250	13%	1	1%
Homework	346	18%	1	2%
Daily notes	173	9%	22	37%
Announcements	115	6%	0	0%
Blog	58	3%	2	4%

problem solving, promoting the extended abilities related to employability. Thus, abilities required to work successfully in the current version of the course are evolving into an ability to learn online.

Another important aspect which normally concerns teachers is the online evaluations. Are online evaluations representative of student performance? Exams have been a headache for generations. For some educative researchers, they represent months or years of learning settled by one sheet of paper, constructing or breaking any future career prospects. The classical usage, in which an exam is programmed at some date and hour, privileges some students with adequate conditions to present the exam under imposed conditions, stating a mandatory competition. Better options are based on systems allowing students to advance at their own pace, in accordance with the covering of the learning materials. Skill based learning is a trend in which evaluation is tailored for each student to recognize a combined group of skills more than specific knowledge.

Whilst at undergraduate level a lot of emphasis is put on grades as a prerequisite to carry on to the next course, job or a graduate program, employers think that grades do not matter as much as experience and placements in industry (Clark, 2014). Some schools have been pushing boundaries about evaluation measuring the learning beyond the content knowledge, in the direction of skills generated (Schwartz, 2014). Traditional evaluation is limited to measuring the mastery of contents, not the skilled use of them, supporting the fact that in-class exams are not sufficiently representative of the real goals that education should be pursuing.

Online Activities Evaluation vs. Global Performance

Activities in the course can be classified as online or as face to face. Exams in the classroom and other class work being graded normally requires teacher intervention which could not be present in a pure online course. Otherwise, homework, teamwork and online quizzes are activities not strictly requiring teacher intervention (although they could require tutoring and their good or bad evaluation could be conditioned to face to face or online instruction), so they will be classified mainly as pure online activities.

With the coming of technology and information access, teachers tend to undervalue the online activities, mainly due to various kinds of plagiarism. A similar situation arises for teamwork because the performance of some students could be shielded by their partners' work. As a consequence, it generates sometimes an overvalued grade for the face to face activities because teachers believe they have more controls over them than the online ones. This aspect does not generate the best practices in evaluation in terms of learning styles, time management and individual meaning. As an exercise to analyse the level of discrimination for the online activities versus the general performance through the whole graded activities, the grades for students in the last two years have been classified in these terms. Based on a one hundred scale for the grades, Figure 8 shows dots for each student evaluated in the course between 2013 and 2014. The graph shows a dot in a two dimensional scale based on the average of the online teamwork (problem based learning involving programming and complex applications) versus the average of online individual work (exercise based homework and online quizzes). The size of each dot corresponds to the average of partial evaluations before the final grade (which is 70%, mainly based on class exams rather than other activities). Finally, the colour shows the final promotion in the course (green for a final grade between 86-100/100 and yellow for 70-85/100, the promoted students; red for less than 70/100, the not promoted students). There are some quantitative clear outcomes (each affirmation is based on a hypothesis testing or on an ANOVA test with one or two factors at time, being proved among multiple factors in the inquiry), some of them not surprisingly:

Figure 8. Graph showing a synthesis of the student grades in two years comparing online individual work versus online teamwork

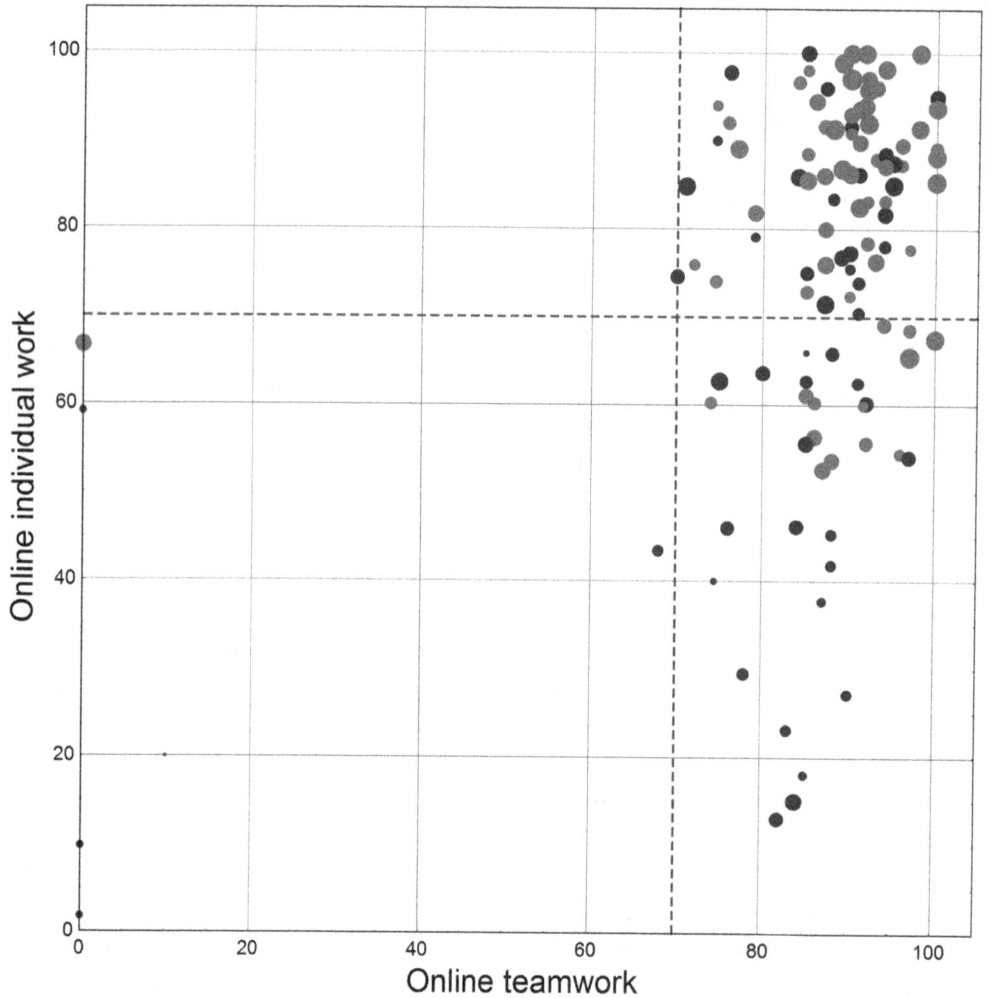

1. Online teamwork versus online individual work
    a. Online teamwork is better graded than individual work on average (better quality: finished and accomplished)
    b. Good online individual work is a good predictor of online teamwork
    c. Online individual work predicts meaningful good partial evaluations
2. Course promotion
    a. Better outcomes in partial evaluations (>70) are predictor of promotion
    b. Students promoted with distinction show best grades (>70) in both online works
    c. Students not promoted are related to poor individual work (<70)
    d. Promotion is associated with good online individual work (and teamwork too), while the opposite is not necessarily true, course failed is not meaningful to poor individual work
    e. Online individual work together with good partial evaluations (>70) predicts the promotion

These outcomes are not conclusive to assure that online individual and online teamwork are meaningful on the student performance, at least in the course being analysed. The grades suggest that for promoted students, good online individual work (normally under the pace of the student) is representative for the final promotion (mainly measured through individual and programmed in-class exams). In addition, for these students, good marks are obtained too in partial evaluations (mainly weighted with the online and in-class exams). In addition, outcomes show that outstanding students exhibit good results in the online teamwork activities, which are mostly based on contents not included explicitly in class, but instead based on research. Both aspects suggest the following observations:

1. Students obtaining good performance in the online individual homework normally demonstrate reliable consistent competencies in the online evaluations and similar outcomes could be expected for in-class exams.
2. As a consequence, it suggests that it is possible to design an alternative and flexible evaluation where students obtaining a potential good general performance in a face to face instruction could obtain similar outcomes under a faster, alternative and flexible instruction scheme too. It will be mainly based on online instruction.
3. Students not able to obtain good performance in the individual homework should be addressed to follow improved teaching methods based on blended schemes with alternative flexible schemes of evaluation than in-class exams.

Flexible alternative instruction could be centred in terms of each student constructing their own learning path, selecting their type of instruction for the whole course or for each theme. Inclusion in this scheme could result from the students' own perspective or from an initial skill evaluation. Similarly, alternative evaluation to demonstrate the individual performance, more than just in-class exams, should be included to warrant the acquisition of modular competencies (more than a global unique grade), possibly with redundant or extraordinary evaluations, only in necessary cases. Thus, each student selects their own approaches, preferred resources and choices among several kinds of learning activities and evaluations. Anyway, instruction should include challenging and motivational applied activities, as well as online evaluations and formative quizzes. In addition, it should state an underlying division in the student performance measurement in terms of each learning path draft.

SOLUTIONS AND RECOMMENDATIONS

STEM education is an acronym of Science, Technology, Engineering and Mathematics. This movement began in The U.S., but has spread to several countries of the world (Gonzalez and Kuenzi, 2012). This initiative includes educational activities (formal and informal) across all grade levels, from pre-school to post-graduate levels. Their perspective looks at emergent requirements in the Science and Engineering workforce market for the next few years as a boost to the current education data in those disciplines. They were low valued for decades and much actual technological development and economic growth is based on discoveries made two or more generations before those currently being educated ('70s & '80s). Inclusion of women, as an unrepresented group in Science, Engineering and Architecture, is an important goal in almost all countries having a STEM initiative. This is a big question for educators and industry, how to effectively prepare and integrate women and some other minorities into these disciplines?

The current perspective under STEM initiatives for the United States (NSF, 2013) is to construct flexible and effective instruction for most of the students in those areas. Experimentation of new instruction schemes based on Informational Technologies is being required and boosted. Thus, this is an important issue in a course such as Numerical Methods where mathematics, sciences, engineering and technology integration are converging.

Hybrid Schemes and Evaluation Flexibility

Competency-based learning is a trend in which evaluation is tailored for each student to recognize skills more than knowledge. Technology does not manage evaluation only, but manages material learning too. If the learning-evaluation process is inverted by measuring decisions when the student is learning, by analysing each decision and keystroke, applying some learning analytics criteria, then evaluation could run in parallel. This is a real continuous evaluation instead of exams applied apparently under the same conditions. Thus, if the course materials are offered online, the students' inputs could be analysed simultaneously by the computer software, so students can be assessed individually and then should identify when a given student has reached their full potential on some content (Barras, 2014).

Observations at the end of the last section suggested taking care of the two extremes in the potential performance spectrum in which students could be located. Both the outstanding students and those who require support or alternatives to demonstrate competency (more than the classical exam in the classroom), can benefit from a flexible scheme of instruction. Thus, the first group could finish some sections of instruction with limited effort to concentrate on other themes or courses, while the second group can be evaluated more efficiently under other schemes. Thus, the blended approach could be managed through flexible activities selected among face to face and online instruction. If it is precise, it should be done as a function of the student decisions, previous knowledge, necessities or interest. Figure 9 shows a proposed outline of a Numerical Methods course to be built as an evolution of the current blended strategy. The course contains several main themes (Figure 9a) which are taken by each student in an online or face to face version as a function of a previous evaluation or their own decision. Then, each module or theme is addressed following a menu of activities and evaluations to demonstrate competence (Figure 9b). Evaluations could be presented more than once and they should be evenly addressed, avoiding concentration on a few of them. Another possibility is to transform the current individual homework in an intelligent adaptive tutorial. This proposal should first be evaluated with selected students to get reliability and to be analysed as a differentiated scheme of instruction.

Figure 9. Outline for a hybrid version of the course, allowing flexibility in the selection of activities and evaluation. a) Path is variable for each student from theme to theme, and b) each theme states a group of selectable activities to demonstrate minimal specific competencies to be promoted by modules

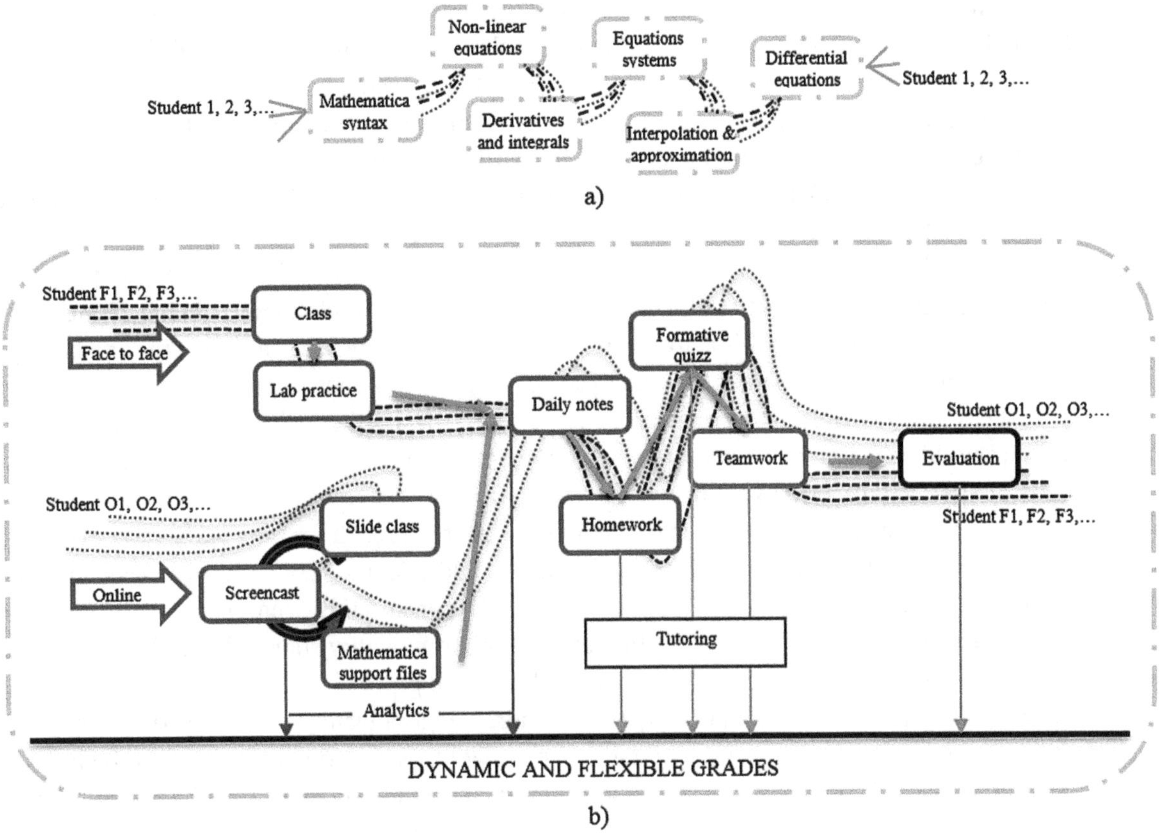

Dissemination and Faculty Working Together

The Numerical Methods course has evolved from a traditional approach into a course based on the use of technology and applied concepts. In this process, other teachers for the same course have considered changes in the current work. These changes were made as a personal perspective in terms of tools and applied problems being considered in the course. Anyway, teaching practices have changed in recent years becoming more accessible. In a world dominated by technology, at least in education, it is not possible to adopt only one trend. Instead we need to take several tracks to construct own initiatives in educative innovation. In the coming years, this course will evolve not only as a result of one teacher, but with the collaboration, sharing and openness of the whole faculty who teach it. The Physics and Math department has promoted a weekly seminar about the use of technology in the classroom, partially sparked off by the Numerical Methods course faculty and other parallel projects implementing and combining several technologies for Math and Physics courses. At the end of the day, it is the spirit that is most valuable in education.

FUTURE RESEARCH DIRECTIONS

Pau Kelly, founder of Wired Magazine, believes that the next 20 years in educational technology will be radical. So much that now is believed about technological advances, will be only a ghost in the near future (Shontell, 2014). Although great steps have taken place, there is uncertainty still. But now, there are a great number of proposals and experiments running. It is expected for the future to have a clearer panorama for educative technology and how it will mould more than break the educational structure in the university, in terms of effectiveness and flexibility. Still, education is based on the enthusiasm to improve teaching and to try elements in the environment supporting it (technological or not). Honest self-assessment, reflection and vision to leave practices giving good results and then improving again in each step. Thus, the worst enemy is to try think that there are universal and final trends stopping the daily teacher creativity (the same is true if it is believed there are not). Technology is being created with correct intent, but a group of students is a diverse universe in which many facts can converge and emerge, making it impossible to apply general solutions to solve all learning issues.

In the analysis developed, in future years the course should retain the collaboration among faculty. Collaboration should be maintained not only as in the previous years: selection of professional tools and construction of applied problems to be included in the course. Instead, renewed schemes of instruction and educational research should be run as a team to get a better approach for the course promotion and flexibility offered to students. In this initiative, faculty should adopt several roles: offering modular courses, offering lab workshops considering alternative tools, constructing scenarios of evaluation and defining alternative paths of instruction fulfilling some minimum quality standards. In these schemes, technology should be a basic support to their effective functioning.

The scheme being presented in Figure 9 should be addressed as a faculty project. Fortunately, there are trends in the institution for this experimentation. In a few years, several courses should be evolved to MOOC, SPOC, MOOR or alternative schemes. Numerical Methods courses have been evolving for some years, and faculty are now prepared with a strong training in new technologies based on the design of blended scenarios, online practice and construction of intelligent adaptive tutorials. In the coming years, faculty should change and share roles in teaching, more than just offering several copies of the course. This scenario requires a change in the way that each teacher sees their own teaching work and each student decides the way to learn under more self-directed strategies.

CONCLUSION

Anyone can neglect the spread of technology and the changes it is generating, in particular for education. Challenges in education as well as in professional life are becoming continuously deeper. On the way, those called digital natives normally are attributed with abilities that professionals do not always have, though they possess a deeper adherence to it because they are naturally advantaged towards experimentation. In the same way, digital immigrants are not handicapped in their capability to acquire those abilities. These groups have been adapting continuously to change, in the same way that new generations will with the future technology. Sometimes education is drastically divided in those terms, as two different species unable to agree, communicate and share knowledge between them. In some sense, this era is similar to others in terms of technology development.

Currently, technology is finding more effectively its role in education. Compared with ten years before technology, current technology is being well directed to reproduce and improve educative tasks that before were restricted to the classroom. But now, the easy access to information is speeding up the learning and the technology is broadening its learning approach. Faculty in the universities should note this trend and begin to design teaching strategies to reach social requirements. Technology is there to help, but without the correct direction to experiment with new ideas, few goals can be reached. In a new world where self-learning is a valuable skill, many students will require support to learn it. Thus, blended or hybrid schemes, in the sense depicted here, should be experimented with and improved for the new generations. These schemes should be prepared not only to teach through technology, instead, to consider teaching scenarios where theoretical concepts (as in Math) should be applied and combined with technology (as in programming) to solve complex problems (as those meet in Engineering or Science). This is the main focus in the coming years for the Numerical Methods course depicted here, a key link between theoretical and applied courses.

If we take a look around the world, we can find examples of simple but superb technology being created by skilled groups of professionals. MIT (2014) has marked some developments in technology, which supersede any other. But clearly this kind of challenge difficulty could be achieved with traditional instruction based mainly on raw knowledge and exams. Instead, they require integrated knowledge, practice, simulation and prototyping, as well as effective teamwork. For the current era, teachers should know the effervescent media of technology, settling some ideas about how new generations are perceiving the current world and how, on average, their personal histories have constructed their capacities to learn, work, understand and be interested in scientific and technological topics.

REFERENCES

Barras, C. (2014). Can technology kill off the exam? *Future*. Retrieved November 15, 2014, from http://www.bbc.com/future/story/20140328-can-tech-kill-off-the-exam

Blackboard. (2015). *Blackboard Webpage*. Retrieved February 25, 2015, from http://www.blackboard.com

Clark, N. (2014). Are grades at postgraduate level irrelevant? *The Guardian*. Retrieved January 5, 2014, from http://www.theguardian.com/education/mortarboard/2014/apr/01/students-grades-postgraduate-course

Classmarker. (2015). *Classmarker Webpage*. Retrieved March 25th, 2015, from http://www.classmarker.com

Conole, G., Oliver, M., Falconer, I., Littlejohn, A., & Harvey, J. (2007). Designing for learning. In G. Conole & M. Oliver (Eds.), *Contemporary perspectives in e-learning research: themes, methods and impact on practice, Open and Distance Learning Series*. London: RoutledgeFalmer.

Delgado, F. (2012). *Numerical Methods You Tube Channel*. Retrieved November 15, 2014, from https://www.youtube.com/playlist?list=PLFF477660D37081E4

Delgado, F. (2013). A numerical methods course based on b-learning: Integrated learning design and follow up. *International Journal of Mobile and Blended Learning*, *5*(1).

Delgado, F. (2014). Actualización docente en tecnologías educativas y aprendizaje móvil: Desarrollo de un programa institucional. *Revista de Formación e Innovación Educativa Universitaria*, *7*(4), 211–226.

Delgado, F., & Martínez, S. (2011). Cambios curriculares generados por el empleo de tecnología en la enseñanza de los métodos numéricos. In *Memorias de la XXXVIII conferencia de ANFEI*. Querétaro, México: ANFEI.

Delgado, F., & Santiago, R. (2014). Designing a site to embed and to interact with Wolfram Alpha widgets in Math and Sciences courses. In *Proceedings of Mobile Learning International Conference 2014* (pp. 266-270). Madrid, Spain: IADIS. Retrieved March 15, 2014, from http://aleesp.hypotheses.org/rg/files/2014/03/ML2014_proceed.pdf

Delgado, F., Santiago, R., & Quezada, L. (2015). Widget based learning to develop meaningful comprehension in math and science courses. In *Proceedings of INTED 2015*.

Gonzalez, H., & Kuenzi, J. (2012). *Science, Technology, Engineering, and Mathematics (STEM) Education: A Primer*. Congressional Research Service Report. Retrieved January 20, 2015, from http://fas.org/sgp/crs/misc/R42642.pdf

Goral, T. (2014). SPOC's may provide what MOOC's can't. *University Business*. Retrieved January 5, 2014, from http://www.universitybusiness.com/article/spocs-may-provide-what-moocs-can%E2%80%99t

Gray, L., Thomas, N., & Lewis, L. (2010). *Teachers' Use of Educational Technology in U.S.* Washington, DC: Department of Education. Retrieved January 25, 2014, from http://nces.ed.gov/pubs2010/2010040.pdf

Hobbes, R. (2013). *Hobbes' Internet Timeline*. Retrieved October 20, 2014 from http://www.zakon.org/robert/internet/timeline/

Hosler, A. (2014). *Massive Open Online Research: The MOOC Evolves into the MOOR*. Retrieved November 5, 2014, from http://www.emergingedtech.com/2014/01/massive-open-online-research-the-mooc-evolves-into-the-moor/EmergingEdTech

Internet World Stats. (2014). Usage and Population Statistics. *Internet World Stats*. Retrieved February 15, 2014, from http://www.internetworldstats.com/stats.htm

Jardines, F. (2009). Historical development of distance education. *Innovaciones de Negocios*, *6*(2), 225–236.

Johnson, L., Adams, S., Cummins, M., Estrada, V., Freeman, A., & Ludgate, H. (2013). *NMC Horizon Report: Higher Education* (2013th ed.). Austin, TX: The New Media Consortium.

Johnson, L., Adams, S., & Haywood, K. (2011). *The NMC Horizon Report: 2011 K-12 Edition*. Austin, TX: The New Media Consortium.

Johnson, L., Smith, R., Willis, H., Levine, A., & Haywood, K. (2011). *The 2011 Horizon Report*. Austin, TX: The New Media Consortium.

Laurillard, D. (2002). *Rethinking university teaching: a conversational framework for the effective use of learning technologies* (2nd ed.). London: RoutledgeFalmer. doi:10.4324/9780203304846

Laurillard, D. (2005). E-Learning in Higher Education. In P. Ashwin (Ed.), *Changing Higher Education: The Development of Learning and Teaching*. London: RoutledgeFalmer.

Laurillard, D. (2008). *Digital technologies and their role in achieving our ambitions for education*. London: Institute of Education, University of London.

MIT. (2014). 10 Breakthrough technologies. *MIT Technology Review*. Retrieved October 20, 2014, from http://www.technologyreview.com/lists/technologies/2014/

NSF. (2013). *Women, Minorities, and Persons with Disabilities in Science and Engineering: 2013*. NSF. Retrieved January 20, 2015, from http://www.nsf.gov/statistics/wmpd/2013/pdf/nsf13304 _full.pdf

Prensky, M. (2001). Digital Natives, Digital Immigrants. *Horizon*, *9*(5), 1–6. doi:10.1108/10748120110424816

Sadowski, J. (2014). The "Digital Native," a Profitable Myth. *The Baffler*. Retrieved January 15, 2014, from http://www.thebaffler.com/blog/the-digital-native-a-profitable-myth

Schwartz, K. (2014). More Progressive Ways to Measure Deeper Levels of Learning. *Mind/Shift*. Retrieved December 20, 2014, from http://blogs.kqed.org/mindshift/2014/04/more-progressive-ways-to-measure-deeper-level-of-learning/

Shontell, A. (2014). The Next 20 Years are going to make the last 20 look like we accomplished nothing in tech. *Business Insider*. Retrieved September 25, 2014, from http://www.businessinsider.com/the-future-of-technology-will-pale-the-previous-20-years-2014-6#ixzz3AkcyE1gs

Socrative. (2015). *Socrative Webpage*. Retrieved February 25, 2015, from http://www.socrative.com

Stocks, T. (2014). Can Video Help Learning *And Development*. Retrieved from http://www.business-2community.com/strategy/can-video-help-learning-development-0951021#!bpxhpA

Straumsheim, C. (2014). After #MassiveTeaching, questions about MOOC quality control. *Inside Higher Education*. Retrieved January 18, 2015, from https://www.insidehighered.com/news/2014/07/15/after-massiveteaching-questions-about-mooc-quality-control#ixzz37X1bxB6p

Weebly. (2015). *Weebly Webpage*. Retrieved March 23, 2015, from http://www.weebly.com

Wood, R., Baker, J., & Hopper, D. (2004). Hybrid structures: Faculty use and perception of web based courseware as a supplement to face-to-face instruction. *The Internet and Higher Education*, *7*(4), 281–297. doi:10.1016/j.iheduc.2004.09.002

KEY TERMS AND DEFINITIONS

Blended Learning: An educative face to face formal program in which student learns online some delivery of content on his own pace.

Digital Immigrants: People who was born before1980 and great part of their life has been adopting emergent digital technology.

Digital Natives: People who was born after 1980 and supposedly have natural abilities in the use of technology due to in the most of their life have been in contact with this technology.

Hybrid Learning: In the context of current work, an educative formal program in which each student decides how combine learning materials and activities, either face to face or online, but based on a permitted flexible design in which evaluation is reached with alternative ways and with partial control of time, warranting a minimum of performance and possibly differentiated learnings.

Learning Analytics: A trend which proposes track and analyse all student computer behaviour and decision based on predefined activities and criteria, with the objective to state continuous evaluation on each student and carry out assessment to individuals and to the global teaching and learning process.

Mobile Learning: An online process of learning which is based in great extent on materials and tools being delivered through mobile devices, whenever it is combined with face to face instruction or not.

MOOC: The acronym for Massive Open Online Course, a free online open modality for education, typically at undergraduate level, offered for a spread and large public.

Online Education: Education which is delivered through a digital device with a formal pretension to obtain skills and/or knowledge.

Chapter 12
Mobile Application Development by Students to Support Student Learning

Sonal Dekhane
Georgia Gwinnett College, USA

Mai Yin Tsoi
Georgia Gwinnett College, USA

Cynthia Johnson
Georgia Gwinnett College, USA

ABSTRACT

The ubiquity of mobile devices, their portability and their unique gesture recognition features make them an apt educational tool, especially in the STEM fields. The traditional pen and paper practice method can be transferred to a digital format and the content can be made available to students anywhere and at any time. In this chapter the authors discuss how mobile application development projects can be used to benefit students in multiple STEM disciplines at the same time by creating a symbiotic, business-like relationship between students enrolled in a software development course and students in other STEM courses (end-users). The authors will discuss the learning needs of students and the details of thoughtful mobile application design incorporating various learning theories. This chapter will also discuss various examples of educational mobile apps created, their usability testing results and their effect on student learning.

INTRODUCTION

In this chapter, we explore the use of electronic learning within three separate classes at an undergraduate liberal arts college. E-learning is the concept that a student's learning experience can be positively enhanced and augmented by harnessing today's various technological advances.

DOI: 10.4018/978-1-5225-0359-0.ch012

One of the roadblocks to implementing e-learning in a classroom is the lack of materials supporting e-learning. Instructors may have the desire to teach a concept using a website or mobile application, but are not able to locate anything that best fit the learners' needs. Or sometimes there are plenty of resources from which to choose, but the instructor is at a loss as to how to integrate them into the curriculum and the classroom environment. As we investigated the number of mobile apps available for content at the post-secondary level, we realized that there is very little that has been developed for learners in the undergraduate setting (Shuler, 2012). This was the impetus of our project.

The study described in this chapter was performed at Georgia Gwinnett College (GGC), a midsized liberal arts college offering only bachelor degrees in the state of Georgia in the United States. GGC is an open-access institution with small class sizes (less than 30) and a diverse student population. The population is diverse in both ethnicity and educational background. To investigate and facilitate the use of e-learning in the post-secondary classroom, collaboration was formed between a junior-level information technology software development course and two other courses (Chemistry and Introductory Programming) at our institution. The intent was to give a "real-life" project to the software development students and then provide the non-software development Science, Technology, Engineering and Mathematics (STEM) students a functioning learning tool; a mobile app that helps tutors to teach a certain concept. The project attempted to bring e-learning into the post-secondary classroom and also to extend the impact of e-learning across disciplines and into other classrooms. The primary objective of this project was to increase student engagement in a software development course and to help those students gain valuable work skills by working with an actual client and real end-users. The secondary objective was to provide a mobile app to the other courses to enhance the learning experience of those students as well.

The dearth of students who graduate in information technology (ITEC) and computer science (CS) across the U.S. is causing a real challenge for employers, as there are not enough qualified and trained programmers and developers to meet the ever-growing need as the Technology Age speeds along (Lutz, Naveda, & Valino, 2014). In other STEM fields, we also see a growing decline of graduates who are able to effectively contribute to the work force in these areas. Students usually struggle with content in STEM courses for a variety of reasons, such as: the large amount of content covered in a course, the way in which content builds on itself throughout the course and finally, the need to implement a multitude of problem-solving and critical thinking skills (Lutz, Naveda, & Valino, 2014). Students who do not understand the fundamental concepts at the beginning of a course tend to quickly fall behind as content builds on itself throughout the duration of the course, thus spelling potential failure for later topics and even subsequent courses. Hands-on learning and application of these key fundamental topics play an important role in a student's success in both these disciplines.

In this study, the ITEC students worked with faculty in chemistry and programming to develop mobile apps that address some of the course's fundamental concepts in a new way by helping students generalize and reuse the information. In it was postulated that in practicing these fundamentals repeatedly, using an e-learning based tool, students may better understand and remember the core concepts, but more importantly are provided the environment and "playground" in which to repeatedly practice applying their newly gained knowledge in a variety of different situations. In this chapter, we discuss both the impact of creating the apps upon the ITEC students, and the additional learning support provided to the chemistry and programming students using the new mobile apps. The focus of the study is primarily qualitative rather than quantitative. Specifically, we ask whether the development and use of these mobile apps provided the students with a motivating learning experience that added to their understanding of certain concepts.

BACKGROUND

Mobile Learning

A subset of e-learning that has gained a lot of recent attention is mobile learning. Defined as the use of mobile devices to aid in learning, Hwang and Tsai (2011) believe that mobile learning or m-learning consists of a "…formal and informal set of contextualized and situated learning activities that are mediated by mobile technologies." Key to this idea is the fact that these devices facilitate learning and are able to create new possibilities in learning for students.

Smartphone users topped the 1 billion mark in 2012 (eMarketer, 2014). Technology and learning have both become more individualized and user-centered over the past 30 years (Sharples, Corlett, & Westmancott, 2002), and as the cost of ownership decreases, students are finding themselves more able to have a smart device or tablet at their fingertips (Petersen & Divitini, 2005). Today, the mobile device is as integrated into how students interact with their environment as the commonplace automobile. Because of this dependence on the mobile device, educators have looked to it as a tool to impact the learning process.

"Taking down the classroom walls" by the mobile device is termed ubiquitous learning, learning that can happen at any location and at any time designated by the learner (Hwang, Tsai & Yang, 2008; Shih, Chu, Hwang & Kinshuk, 2010). Because ubiquitous learning can sometimes translate to more time-on-task and reinforcement, educators are looking to mobile learning as playing a major role in today's educational curriculum.

Past studies have emphasized that mobile learning should be used in addition to an existing learning environment. Indeed, as the learner interacts with all modes of information (textbook, lecture, online discussions etc.), it is thought that the synthesis of the information as the learner evaluates, understands, and applies the new content is what makes up the learning experience. Therefore, mobile devices should be investigated as a promising platform on which students can access information, but also perhaps obtain assistance in synthesizing all the sources of content.

The challenge, then, is "…adapting and appropriating the technology for learning in a way consistent with learning goals and principles" (Hoppe, 2004). Further, the mobile content must be developed in such a way as to appeal to a wide audience, a large range of abilities, and varied levels of technological know-how. It should be appropriate to the age level and interests of the target learning group and use valid and reliable teaching pedagogy. Therefore, mobile applications have focused on short, defined goals or tasks so as to be more effective in capturing the interest of users and to simplify the mobile device-user interaction.

There are some problems that educators encounter when using m-learning in their classrooms. Some specific domain areas are not covered in apps that deliver specific content or are not apparent in apps that aim to help students develop targeted skills. These specific domain areas can be a reason why educators sometimes find m-learning frustrating; m-learning sometimes can require extra teaching before students are able to utilize those apps. Another aspect is availability of appropriate apps. Besides elementary aged apps that teach reading, writing, and math, it appears that, for the most part, upper level content areas are not covered extensively in mobile apps so far. The most popular apps for post-secondary students are those that help with organizational skills, planning, and communication (Shuler, 2012). Very few actually help deliver the post-secondary curriculum.

Some of the other disadvantages of mobile learning include limitations by the device itself, such as limited battery life and capacity for large data files, multiple standards and operating systems, and risk of sudden obsolescence (Mehdipour & Zerehkafi, 2013). Other barriers include cost of ownership, supporting access to digital content inside and outside the classroom, content security, tracking of learning results and risk of distraction by other applications and uses of the mobile device.

Software Development Education

The decision to use software development students as our app developers was seen as an opportunity to integrate learning at multiple levels. Many of today's graduates in software development (sometimes called software engineering) are not well equipped to work effectively in today's workplace (Truong, 2014). It appears that very few graduates are afforded actual experiences working in a team environment, even though this is commonly how software is developed today. Coursework in colleges and universities does not always offer opportunities to hone communication skills and planning abilities even though most software is developed in a project format with multiple stakeholders and parties involved (Lutz, Naveda, & Valino, 2014). Employers cite widespread ignorance on the aspect of software testing, even though this vital stage of the process ensures that a quality product reaches the market. A study conducted at Microsoft Corporation indicated that skills in coding, design, persistence and problem solving were the strengths of recent software development graduates, while skills in communication, collaboration, technical aspects (in terms of using tools like version control system), cognition, and orientation were the graduates' most common limitations (Begel and Simon, 2008). The graduates in this study also had problems working with a pre-existing codebase that lacked documentation in quantity and quality.

In addition, students are expected to attain, along the way, "soft skills" that are fundamental to teamwork and company success. There are no formal opportunities for students to learn how to interact with others and how to manage interpersonal conflicts. Sometimes, it is within these interactions that students gain their first exposure to real-world tools such as version control software. Ideas such as estimation and tracking are rarely covered, but are extremely useful in the workplace.

The need for these seemingly ancillary concepts have actually been noted for years. The Computer Science Curricula 2013: Curriculum Guidelines for Undergraduate Degree Programs in Computer Science (CS2013), (ACM and IEEE, 2013) state that "Curricula must...include professional practice (for example, communication skills, teamwork, ethics) as components of the undergraduate experience. Computer science students must learn to integrate theory and practice, to recognize the importance of abstraction, and to appreciate the value of good engineering design."

Increasingly, computer science programs are integrating mobile learning into the curriculum in innovative ways. One way is to use the mobile device as the platform on which students learn and apply their newfound knowledge on software development. For example, Kurkovsky (2013) describes how he utilized the simpler format of game development on mobile devices to introduce programming concepts to Introductory and Intermediate Computer Science courses. He found that incorporating a device that is personal, familiar, and integral to the everyday life of his students helped increase motivation and engagement in his students.

LEARNING THEORIES

Primary Theories: Multiple Intelligence, Scaffolded Learning and Authentic Learning

With the advent of the Internet, there has been an increase in websites that offer other alternative ways to master material, with videos, simulations, and alternate models of visualizing the content. Indeed, educators have been turning to technology as a way to help learners succeed in areas that are more complex and difficult (Reiser, 2004). However, few of these modes of learning incorporate the Theory of Multiple Intelligences (Gardner, 1983), which states that there is not one way to measure intelligence and that humans learn through a variety of methods based on their strongest modes of intelligence. Educators utilize this theory in designing the curriculum in schools today, hoping to capture the interest of the greatest number of students by addressing a wide variety of learning strengths.

Other learning models have attempted to describe the process of learning. The behaviorism model from the 1960s looked at learning as a product-centered route, whereby students who learned could show their success through the products of their learning (Merriam and Caffarella 1991). Later on, learning focused on changes in behavior as a result of learning from experiences. Those behavior changes would then be observed in how a learner interacts with his/her reality (Maples & Webster, 1980). The cognitivist view of learning discussed by scientists like Jerome Bruner (1960) and David Ausubel (2000) describes learning as an internal mental process that includes insight, perception, and memory. The humanist orientation to learning (Maslow, 1943) and the Social/Situational orientation to learning (Lave, 1982) both view learning as learning to fulfill affective and social needs, as well as cognitive needs. This learning happens within a community.

Therefore, a mobile application that is able to accommodate the widest variety of learning styles would be the most useful and transferable of all apps currently available to the students. In addition, the level of mastery by each student must also be addressed in an effective application. Just as students learn via different modes of learning, students also learn at different paces depending on a variety of factors, such as intrinsic/extrinsic motivation, as suggested by Bandura (1981), and engagement. So the application must also incorporate the learning theory of "scaffolded learning", a theory introduced in the late 1950s by Jerome Bruner (Bruner, 2006). This theory states that people learn new material best when sufficient support is provided when first introduced to the material (Wood, Bruner, & Ross, 1976). As familiarity and mastery increases, the support is gradually removed to increase independence and incorporation of the new material into the existing set of content and skills possessed by the student. There are eight key aspects to scaffolded learning to help promote success in students. As summarized by Hogan & Pressley (1997), they are, in no particular order:

1. **Pre-engagement with the student and the curriculum:** The teacher considers curriculum goals and the students' needs to select appropriate tasks.
2. **Establish a shared goal:** The students may become more motivated and invested in the learning process when the teacher works with each student to plan instructional goals.
3. **Actively diagnose student needs and understandings:** The teacher must be knowledgeable of content and sensitive to the students (e.g., aware of the students' background knowledge and misconceptions) to determine if they are making progress.

4. **Provide tailored assistance:** This may include cueing or prompting, questioning, modeling, telling, or discussing. The teacher uses these as needed and adjusts them to meet the students' needs.
5. **Maintain pursuit of the goal:** The teacher can ask questions and request clarification as well as offer praise and encouragement to help students remain focused on their goals.
6. **Give feedback:** To help students learn to monitor their own progress, the teacher can summarize current progress and explicitly note behaviors that contribute to each student's success.
7. **Control for frustration and risk:** The teacher can create an environment in which the students feel free to take risks with learning by encouraging them to try alternatives.
8. **Assist internalization, independence, and generalization to other contexts:** This means that the teacher helps the students to be less dependent on the teacher's extrinsic signals to begin or complete a task and also provides the opportunity to practice the task in a variety of contexts.

In this study, students were not directly part of the key aspects 1-3. In choosing the topics for the apps and in deciding to commit to the app development project, instructors in the subject areas considered their curriculum goals and decided on the salient curriculum topics that needed additional support by examining assessment data from past students who have completed the course.

Learning by doing is considered the most effective way to learn by educators and also reported as motivational by students. As described in the white paper (Lombardi, 2007) by EDUCAUSE Learning Initiative, authentic learning focuses on real world problems and solutions and such learning environments are inherently multi-disciplinary. Students participating in authentic learning activities absorb the "portable skills" that newcomers to any discipline have difficulty acquiring on their own. Authentic learning provides learners with connections of content to the real world. It also provides learners with an opportunity to practice their skills and apply their knowledge in different scenarios, helping to make sustained attachment with the knowledge gained.

For software development students, the concept of practice goes one step further and involves the use of communication and leadership skills. The Software Development Life Cycle (SDLC) is a process consisting of different stages and planned activities involved in developing software. Interviewing clients to gather requirements, project planning, presenting to peers and clients and conducting usability testing with end-users are some of the real-world experiences that are invaluable for software development students. Usability testing is a technique used to evaluate a software product by testing it with representative users. It can be used to understand how real users use and experience the product. Combining technical knowledge of tools and languages with the actual experience of project management and client and end-user interaction can better prepare a software development graduate for the real world work environment. In the ever-changing world of technology, staying abreast of the advances in the field and learning new technology is also vital to a software development graduate. The ability to learn new languages and adopt new tools and environments is very important. Hence, providing an opportunity for students to update their skill set, learn a new set of tools and create a product using those tools is also essential to meet the learning needs of software development students. Maintaining an existing application, implementing someone else's design and testing someone else's code are common tasks in the real work environment. Understanding the learning needs of software development students, authentic learning theory should be adopted in the development of an educational mobile app.

Secondary Theories: Immediate Feedback, Motivation and Design for Learners with Special Needs

A secondary theory of immediate feedback (Gilmer, 1979), and the idea that the sooner the feedback is provided to a learner, the less permanent a misconception is rooted into knowledge, should also be incorporated into an app. This allows the learner to deconstruct and reform a new perception based on the feedback received. When feedback is received days or even weeks later, it is more difficult for the learner to then go back and uproot an erroneous idea, replacing it with a modified version that better approximates the correct way of thinking.

Finally, two other learning theories are important. The idea that people can be motivated by different factors, intrinsically and extrinsically, should be considered in how the apps "reward" their users. Bandura (1981) postulated that people could change thinking and behavior patterns if motivated to do so. The source of the motivation can come from within (intrinsic) or outside the person (extrinsic). Finally, learning theories dealing with learners with special needs, especially those with attention or visual processing issues should not be ignored. Minskoff and Allsopp (2002) suggest that by using colors, bold divisions, and clean designs, students are more apt to pay attention and to comprehend given information when the page has "sections" and definite boundaries.

DESIGNING AND DEVELOPING EDUCATIONAL MOBILE APPS: BLENDING LEARNING NEEDS AND LEARNING THEORIES

In this section, we discuss the concepts and issues that went into the design and development of the mobile apps. One of the first decisions that needs to be made when designing a mobile app is whether it should be native or web. While web apps can be made available to a larger audience without going through the app store's approval process, native apps work with the device's built-in features and perform fastest on the device. Because these apps go through the platform's market-place approval process, they are scrutinized for security. For developers, a standard set of tools is available for each platform that makes the native app development process easier (Xanthopoulos & Xinogalos, 2013).

The decision about which platform to use can also be reached by doing a quick study on the number of users and app downloads for various platforms. To reach the largest audience possible, at least the two major platforms, iOS and Android should be covered. In this study, we were limited by funding and could only implement one platform. Therefore, we chose iOS as the vehicle with which the students developed the apps. Had additional funding been available, we would have explored both platforms so that the apps would have had a larger target audience.

CASE STUDIES

As part of an interdisciplinary collaborative project, junior-level software development students created several chemistry and programming apps. Students worked in small teams with professors teaching these courses acting as their clients and content experts. The students enrolled in these professors' relevant classes acted as end-users, participating in usability testing, pre and post content-based testing, surveys and interviews. The development teams walked through all phases of the SDLC as they would on a real project. The TsoiChem App was created over a period of three semesters and students in the latter two semesters had an opportunity to build on and maintain existing code and documentation. Two apps to support learning needs of novice programming students were created from scratch by two different teams.

Development Methodology

A symbiotic relationship between software development and other STEM discipline students can be beneficial to both groups. While students at the receiving end can benefit from the development of an educational tool designed to aid in their learning, the software development students can benefit from authentic experiences by participating in every phase of the SDLC as discussed in the learning theories section. To incorporate the principles of authentic learning and to address the needs of improving communication skills and teamwork, our software development students participated as teams in all the phases of the SDLC.

- They interacted with the client (the educator) and the end-users (STEM students) to understand the domain for which they were developing and extracted requirements from the client. They participated in project planning, peer reviews (with their team mates and software development instructor) and walkthroughs (with team mates, instructor and client) at various stages of development. This gave them an authentic experience of interviewing a client and understanding the challenges of communicating with a non-technical client and diagnosing user needs, bringing in key aspects 2 and 3 of the scaffolded learning model: establishing shared goals and diagnosing student needs.
- They conducted prototype demonstrations for the client educator. This allowed the client to see what the end product would look like, even if it was not complete, and point out any changes or errors, if they existed. If there was any miscommunication between the client and the development teams, it could be easily identified at this stage and handled before more progress was made.
- They conducted usability testing with end-users, the students. This allowed the development teams to assess if the apps were usable. It also allowed them to identify any major design flaws, if they existed. Usability testing was done at the end of the semester. However, since the client was involved in the earlier stages of development, there were no major issues with the app design, as is discussed in the results section. Thus, the importance of clear and continuous communication with the client and end-users was experienced and learned by the software development students. Student comments related to this are discussed in the results section.

The actual development methodology used differed based on the application, timeline and the team itself. However, the experiences mentioned above were part of all team projects. These experiences were essential to provide the software development students with an authentic learning environment. The experiences mentioned here address the learning needs of the software development students as mentioned in the Background section ((Lutz, Naveda, & Valino, 2014)).

For a mobile app with a larger scope (TsoiChem), the design and development results of one team were passed down to another team in the following semesters, to provide the authentic experiences of maintaining pre-existing code as discussed by Begel and Simon (2008).

Apps for Organic Chemistry Students

The impetus for creating chemistry mobile apps was the low passing rate of students in Organic Chemistry across the country. Organic Chemistry tasks the student to synthesize several cognitive and problem-solving skills in order to solve problems. Many researchers have tried to implement interventions to address this issue, including active learning environments and changing the pacing/delivery of the curriculum (Bradley, Ulrich, Jones & Jones, 2002; Collard, Giradot, & Deutsch, 2002). Conceptual understanding has been noted as one of the factors that need to be addressed in improving student retention in Organic Chemistry (Szu, Nandagopal, Shavelson, Lopez, Penn, Scharberg, & Hill, 2011). We wanted to target this via mobile apps.

The first concept that we focused on was that of student recognition of organic functional groups. Many times, textbooks task students to memorize specific combinations of atoms because these groupings tend to react in specific ways. Unfortunately, the problem that students run into is that these groupings,

Figure 1. Splash screen for the TsoiChem app with the friendly avatar

Figure 2. Screenshot of a functional group problem in TsoiChem

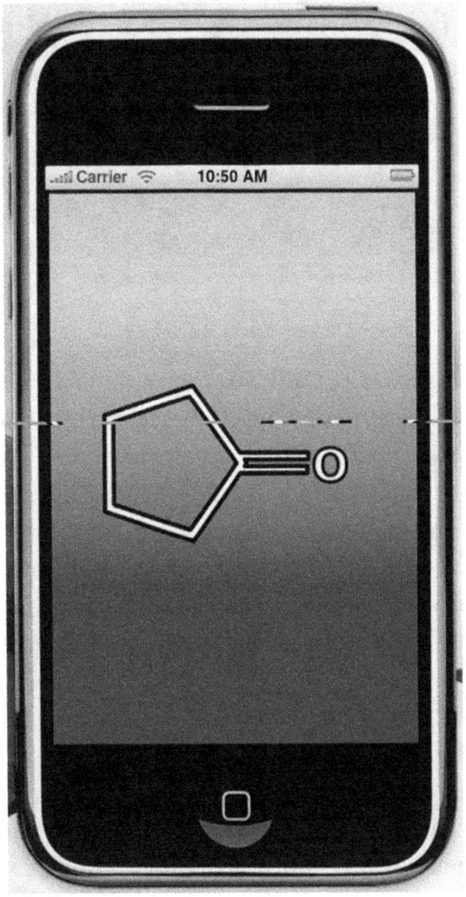

Figure 3. Screenshot of the partially highlighted functional group in a ketone functional group

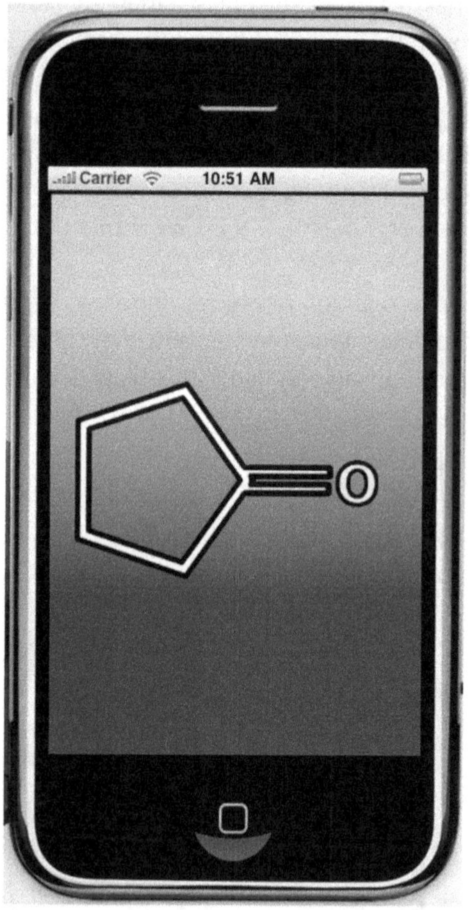

Figure 4. Screenshot of the correct answer screen

or "functional groups" can show up in different formats, which can cause students to be confused. Over a three-year period, our students' performance on this skill has averaged 64%.

The app TsoiChem was designed to address this issue. Instead of typically memorizing self-made flash cards with only one functional group format displayed, this app provided multiple opportunities for students to learn functional groups in a purposeful way. A cute avatar graphic was integrated throughout the app to give the students a sense of fun, ease and approachability – bringing in key aspect 5 of the scaffolded learning model – maintaining the student's pursuit of the learning goals. Bright colors, clear fonts, and a clean background were implemented to decrease distractions and aid focus for all types of learners – bringing in key aspect 7 of the scaffolded learning model – controlling for student frustration, along with addressing the needs of learners with special needs (figure 1).

By designing the application and selecting examples so as to address the common misconceptions that students have in learning this skill, the app was able to harness the touch-screen aspect of the mobile device in teaching functional group recognition. Each functional group was shown in several different formats, with multiple examples of each incorporating the theory of multiple intelligences. The app also included several examples where the functional groups were nested in a larger molecule, which addressed key aspect 8 of the scaffolded learning model – helping students generalize to new contexts. Moreover, the app does not let the student continue to the next step until all the parts of the functional group have been identified and touched, which addressed key aspect 6 – providing direct feedback to the user Students were able to self-correct themselves immediately, thus preventing misconceptions from becoming entrenched (figures 2-3).

There were also three "levels" of play: "Practice It", "Name It', and "Find It". At each level, the student was asked to complete more of the problem until the final level where they were tasked with locating the whole group and naming it as well. The amount of scaffolding provided decreased with each increasing level, addressing key aspects 4 and 8 in the scaffolded learning model – providing tailored assistance and assisting with independence. The app's scoring component, where students were rewarded for correct identifications of functional groups and were penalized points for incorrect answers, directly incorporates key aspect 5 – using praise to maintain pursuit of the goals and the theory of extrinsic motivation (figure 4).

Figure 5. Screen shot of the instructor graphic after one error

Figure 6. Screenshot of the instructor graphic after the user's errors caused an insinuated chemical explosion

In addition, the app provided an audio instructional feedback to help reinforce and motivate students to keep trying and to stay motivated throughout the app, as key aspect 4 of the scaffolded learning model posits that giving tailored assistance can really be beneficial to learning. This also supports the theory of designing for learners with special needs. To further engage students, humor was used when students made mistakes. A graphic of a scientist after an explosion in a beaker was shown when three errors were made. Many students commented that the graphic held a resemblance to the instructor and thought this was very funny, bringing in key aspect 5 – maintaining pursuit of the goals (figures 5 and 6).

Apps for Java Programming Novice Students

In addition to the mobile apps developed for Organic Chemistry students, two software development teams were tasked to develop apps to enhance the learning in an introductory Java programming course. Since this was a course that the developing students had already mastered, the topics covered by the app were left to the discretion of the student group. Professors currently teaching the course acted as "customers" for the developers. The customer provided sample quiz/test questions and practice problems and served as a sounding board for the group. In addition, the student developers were given the opportunity to do usability testing with an actual class of students taking introductory Java. While the customer professor provided a sanity check on the amount of material to be covered and checked the correctness of the material presented, the presentation and testing of the material was left to the developers. The reasoning was that the students had recently (within two years) taken the course and in most cases, were closer in age to the traditional student. Hopefully, this means they would be better equipped to design an app that appealed to the traditional student than a professor who took their first programming course 20-30 years ago. The two apps developed focus on variables/operators and for loops.

There is no shortage of either iPad or Android tablet based apps that promise to teach Java or other programming languages. This is not unlike the plethora of tutorials available on the Internet. However, most are designed for self-motivated learners and while they make successful tutorials, do not teach topics to the same depth as most college-level programming classes. The apps developed by the students were designed to only approach/enhance a single topic in Java rather than teach the language as a whole.

The Variables/Operators app was divided into three parts: learning, practice and quiz. The learning section provided the most scaffolding in the form of simulated flashcards where the user is presented with a question and then given an opportunity to see the answer. If more learning is needed, the app links to a YouTube video on the applicable topic, incorporating multiple intelligences theory. The practice section is designed as an example multiple-choice quiz with immediate feedback. Scaffolding is provided, but less than in the learning section. The immediate feedback allows the student to go back and try other options until they get the correct answer. This also helps in clearing up any misconceptions about the topic that the students may have. The quiz section is similar to the practice section with the exception that there are only 10 questions in the quiz, randomly selected from a database of questions, and results are not given until the end of the quiz. There is absolutely no scaffolding in this section and the students have to submit all answers before they receive any instructional feedback. Figure 7 shows an example of a learning screen with links to video and help.

The "for-loop" app focused on the for loop structure in the Java language. Developed at the same time as the variables app, it was only divided into two parts: learning and quiz. As seen in Figure 8, the learning portion was given as examples and text rather than links to videos as in the variable app. The text included visual examples depicting the correct syntax. It also included an interactive example where

Figure 7. Variable Apps Learning Mode

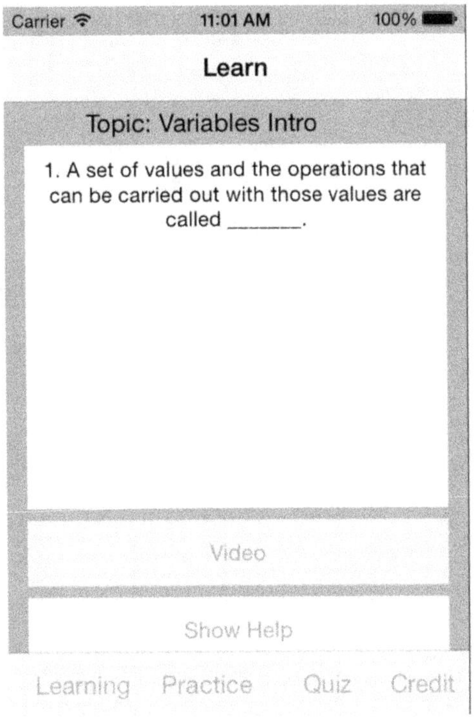

Figure 8. Screenshot of the "for-loop" Learning Mode

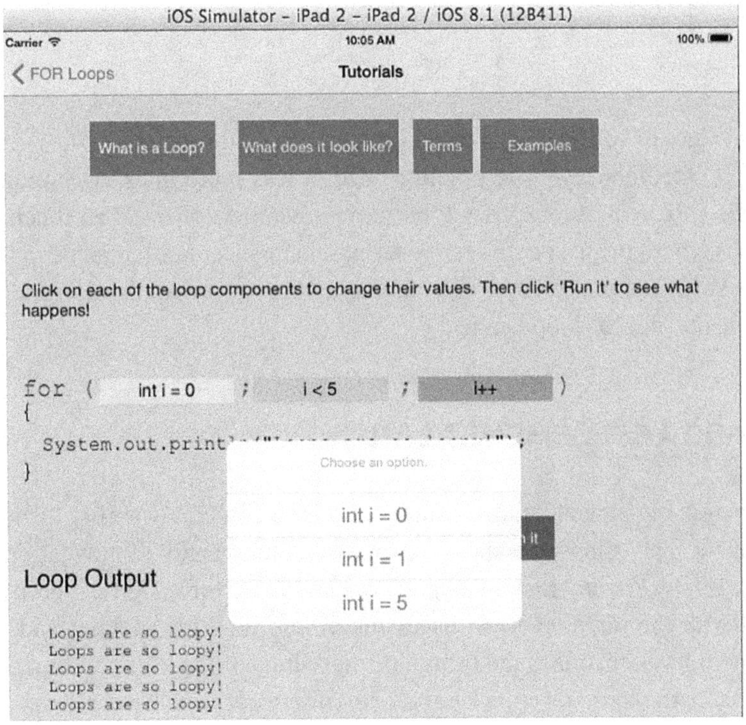

Figure 9. Screenshot of the "for-loop" Practice mode

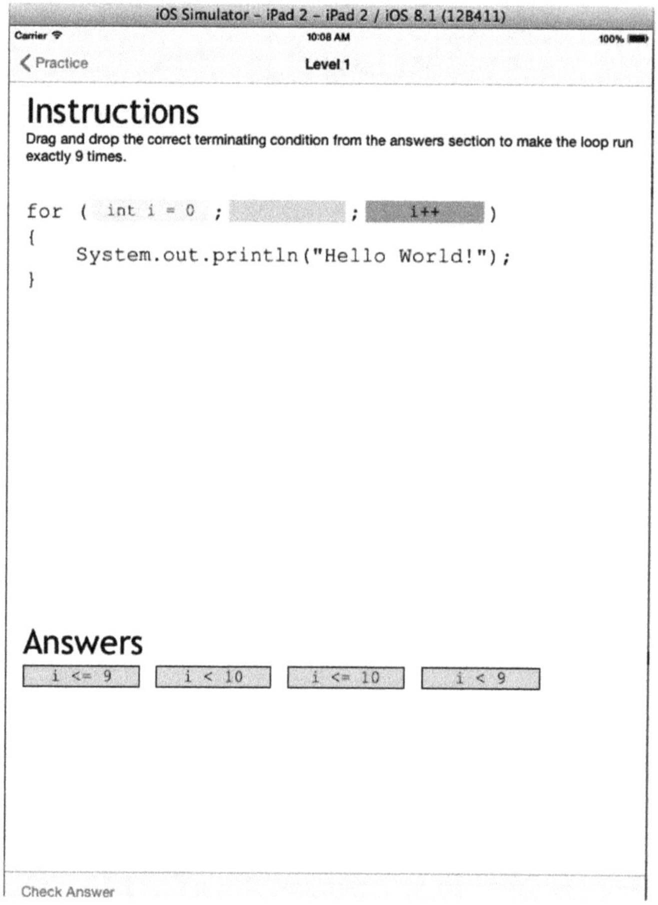

users could change values of various loop header components and run the loop to see the result, as shown in Figure 9 (multiple intelligences). The practice portion was more interactive than the variables app. Rather than multiple choice questions, they took more advantage of the iPad touch screen and created puzzle like questions where the user could move components in and out to achieve the desired for loop, as shown in Figure 9. This app also made effective and consistent use of color-coding to identify each of the three components of a for-loop header.

EVALUATION AND ASSESSMENT PLAN

One of the ways to evaluate an educational mobile app is to conduct usability testing of the app using the end-users as testers. This allows the app to be evaluated for its ease of navigation, user-friendliness, ease of use and also allows for any major design flaws and other errors to be identified. The evaluation plan was designed with the three levels of assessment described by Vavoula and Sharples (2008) in mind, where the micro level examines the individual activities of the user and looks at the accessibility and usability of the app, the meso level looks at the learning experience as a whole, and the macro level determines any long-term impact of the app on the user's learning journey. The end-user is the best tester possible to provide feedback on the usability of the app and hence they should be involved in the evaluation of the app. Depending on the schedule and resources available, usability testing should be conducted at least once during the development of the mobile app. Feedback about the usability of the app from the end-users should then be used to enhance the app and correct any errors that are identified.

The process that the software development students followed when conducting the usability testing was as follows:

1. End-users (Chemistry or Programming students) signed a consent form agreeing to participate in the study.
2. End-users completed a pre-survey that gathered information about their background, such as age, prior experience and extent of using mobile devices, mobile apps in general and educational mobile apps.
3. The purpose of the study was explained to end-users, but the app details were not described.
4. End-users were then given a list of tasks to achieve using the app. These tasks were some of the main features of the app such as answering a question in a specific mode, finding the score, etc.
5. While the end-users used the app to complete the tasks, the software development students observed their interaction with the app, without providing any assistance to the end-users. This allowed the software development students to identify if the end-users could figure out how to use the app on their own (not having seen it before), if they recovered from any mistakes they made while using the app or not, etc. The software development students made a note of such observations to decide if the app was usable enough or if it had some major design flaws.
6. Once the end-users completed their tasks, they participated in a survey rating their experience using the app. This survey included statements about the apps that the end-users could rate on a scale of 1-5, such as "This app was easy to use." The survey also included open-ended questions for the end-user to provide detailed feedback.
7. Using the note-taker's guide template obtained from usability.gov website, the observers (software development students) gave a rating on a scale of 0-2 for each task for each observer indicating if the task was not completed (0), or was completed with difficulty (1) or was easily completed (2). These results were tabulated and the average usability of each app was calculated on a scale of 0-2.

All usability testing followed this approach. There were slight differences among surveys and their rating scale used by different groups of software development students, however the methodology remained the same.

Usability testing results of each app varied. However, what the authors found was that user-friendly apps with good use of gesture recognition and animations received positive feedback from end-users. The following sections discuss the usability testing results of several educational mobile apps created for chemistry and novice programming students.

RESULTS

Usability Testing Results of TsoiChem

Usability testing of the TsoiChem app was conducted in Fall 2010 and Spring 2011, and the user feedback from one semester was used to enhance the app the following semester. 40 students enrolled in Organic Chemistry I course participated in the study during Fall 2010 and 94 during Spring 2011. 87.5% had used an iPhone/iPod in Fall 2010 and 70% in Spring 2011. Average usability of this app in Fall 2010 was 1.52, while that after the enhancement in Spring 2011 jumped to 1.9. As Figure 10 shows the survey results were more positive in Spring 2011 after enhancements, as compared to those in Fall 2010.

The standard deviation of end-user feedback as shown in Figure 11 is lower in Spring 2011 as compared to Fall 2010, indicating that the users received the enhancements made to the app favorably.

Figure 10. Results of the usability study: Average

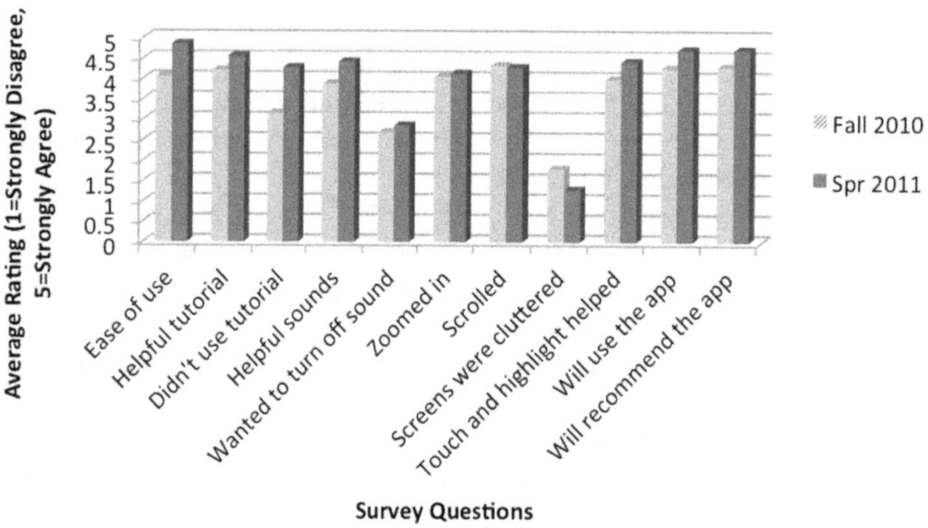

Figure 11. Results of the usability study: Standard Deviation

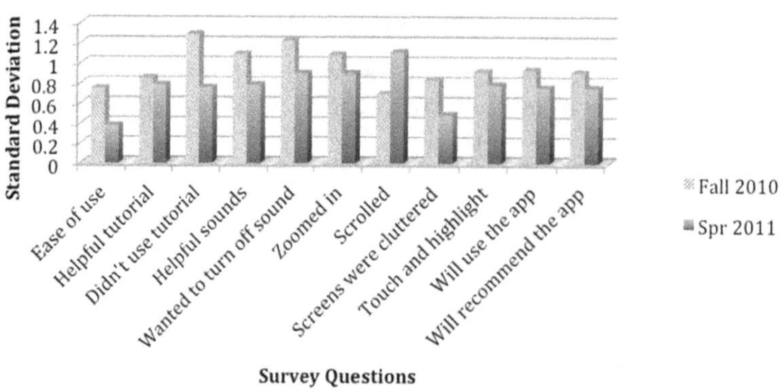

In the open ended section of the survey, most end-users described the TsoiChem app as interactive and easy to use and liked the touch and drag feature of the app along with the audio feedback, animations and the variety of modes available. Further details of this study can be found in Dekhane & Tsoi (2012).

Usability Testing Results of Java Apps

Usability testing of the "For Loops" app was conducted in Fall 2013. 20 end-users, who were students enrolled in Programming Fundamentals course tried a list of predefined tasks and then rated the app on a scale of 1-5 (1 being strongly disagree and 5 being strongly agree). By this time 100% of the participants had used a touchscreen device before. 67% of the participants were IT majors, while the remaining 33% were non-IT majors. 83% of the participants identified themselves as beginners in Java programming, while the remaining 17% identified themselves as intermediate level Java programmers.

The participant responses' average and standard deviation ratings were as shown in Table 1.

As can be seen from these results, the app was considered easy to use by the end users, while there was some variability of the app's visual appeal.

In the open ended section end-users commended the drag and drop feature of the app, it's interactivity, use of video tutorials for scaffolding and color coordination that was used in the app for identifying various components of a for loop. End-users reported this particular app to be a little dull, lacking vibrant color schemes and a theme. Intermediate users reported the questions to be easy. However, they also responded that they would recommend this app to novice students, which was the intended audience of the app. The usability of this app was 1.93.

Usability testing of the "Variables" app was also conducted in Fall 2013. 20 students enrolled in a different section of Programming Fundamentals course were the end users who participated in the usability testing. Once again all of the end users had used a touch screen device before. All of the participants had completed at least 2 IT classes before. 60% of the participants identified themselves as beginners in Java Programming and the remaining 40% as intermediate level. 71.43% of the participants liked the app.

Table 1. Average and Standard Deviation Ratings of For Loop App

App Related Statements	Average	Standard Deviation
Screens were easy to understand	4.5	0.71
Number of times asked for help	0	0
Ease of app use	4.06	0.79
Visual appeal of the app	2.93	1.06

Table 2. Average and Standard Ratings of Variables App

Statements	Average	Standard Deviation
Questions on the app were challenging	6.85	2.19
App was easy to navigate	8.57	1.13
User interface of the app was attractive	7.42	1.39
Would you recommend this app to others	7.57	2.43

End-users rated the app and its features on a scale of 1-10 (1 being strongly disagree, 10 being strongly agree). The average and standard deviation ratings were as shown in Table 2:

This app received very minor enhancement suggestions, such as changing the look of the buttons. All participants were able to complete all of the tasks and mentioned that the app was comprehensive and did not need any major changes or additions. However, as can be seen from the results, there was quite a bit of variability among participant ratings. This could have also been the result of using a much larger scale (1-10, instead of 1-5). Most end-users liked the use of swipe gesture on this app. The usability of this app was 1.93.

Learning for Students

To test the effectiveness of TsoiChem on students' understanding and identification of functional groups, the authors chose 20 structures that provided a good representation of all the functional groups in different formats and created a quiz. Students enrolled in an organic chemistry class were asked to voluntarily participate in the quiz after having an opportunity to use the app. The students also participated in a pre-quiz before using the app. Students received 2 points for a functional group if it was circled and named correctly. They received 1 point if it was circled correctly, but named incorrectly or only circled and 0 points if it was neither circled nor named correctly. A t-test analysis on the pre and post quiz scores indicated a significant difference between the pre and post scores ($p < 0.00001$).

In order to assess the amount of actual domain learning achieved by the Java Programming apps, short 5-10 question multiple choice quizzes were administered before and after the use of the learning app. The testing was done across two semesters. It is important to note that although the students assessed in both semesters were taught the same material by the same professor, the second semester classes were restricted to information technology majors only. During the first semester classes consisted of both majors and non-majors.

In the first semester, the average grades on the pre and post tests using both the variable app and for loop app improved by about 1% which can be considered statistically negligible. In the second semester, not only did both the pre and post test scores improved between 5 and 10% from the previous semester, but the amount of learning from the apps also showed statistically higher results. The variables app had an average score increase of 4.5 percentage points. The for-loop app had an average score increase of 9 percentage points.

More important than the increase in learning is the students' definite interest in using this kind of a learning tool. Almost all the students answering the post-survey agreed that the apps were a useful learning tool and thought we should continue to use them in future classes. Among the positive comments received were:

- "It helped me learn the vocabulary."
- "I was able to play around with the loops and find a better understanding of their components."
- "The practice section of the app was useful in demonstrating the variability of the code, and how different operators in the code would affect the outcome."

The negative comments included:

- "Repetitive".
- "Less glitches"

The negative comments also reflected the fact that the apps did have some minor bugs that need to be addressed before releasing these apps.

The primary benefit to the introductory Java students is that the apps provide a non-traditional and fun way to learn the material. Just taking the students out of the traditional classroom for one class period and exposing them to another perspective on the topic offered a refreshing change from routine. More important than the small increase in learning offered by the apps was the change in attitude of some of the students. Even the students who had no positive comments on the apps themselves agreed that it was a worthwhile endeavor and something worth repeating. In a novice student, the attitude is just as important as the aptitude in successfully passing the class and moving forward to learn more advanced programming skills. It is worth the effort of continuing to add and update the library of apps available to our students if it helps to increase student engagement in the material.

Learning was not limited to the students using the apps. The software development students developing the apps also learned from the experiences. These students were interviewed before and after the course to gauge their impressions of the project and their experiences throughout the app development cycle using a phenomenological paradigm. The interviews were semi-structured in that a set of general questions was posited to the students but as responses were recorded, the authors prompted for more detail or explanation on key points that arose. The responses were then transcribed and reviewed for salient themes relating to perceptions of usefulness, applicability, and authenticity (Punch, 1998). As a whole, almost all the interviewed students stated that going through the process of software development with a real client had definite benefits, to the point of wishing to continue the project even after the course ended. One student noted, "I was interested in maybe even extending it to like an internship or something like that." All students interviewed noted that the course seemed authentic in that the project lent a "real-world" aspect that they had not encountered in their previous undergraduate major courses. A student stated in his interview, "We actually got to work with a real client and we worked through each phase of SDLC instead of just reading about it. It's one thing if you just see it in a book and memorize it for a test. It's another thing when you actually are in the phase and get to see why it's important for the project."

Many times, students are given problems or in-class projects that are contrived by the instructor and limited by the curriculum in order to highly control the outcome and supposedly guide students into the content to be learned. However, the interviewees were both impressed and surprised that issues came up during the app development and deployment that they did not foresee at the start of the course. Many believed that as long as the requisite preparation and planning was completed at the beginning, then there would be no issues with completing the app. An interviewed student remarked, "You spend all night on it and come in the next day and figure out you didn't get most of the stuff (that was required) and (there's) a huge long list of "What if this happens?" It made me appreciate wanting to come up with a lot more stuff – made me want to think about it a lot more – but it took up a lot more time than I planned." By communicating on a weekly basis with the client, the students learned that what is envisioned by the client is not always congruent to what developers hear. The students went on to note that the constant communication helped bring their views back in line with what the client wanted, and this iterative process made them see that the software development process is time-consuming and somewhat complicated. Students were surprised that the types of problems that surfaced were of topics or issues that never occurred to them before; these issues were ones that were never discussed in the textbook (i.e. communication issues with team members, version control, etc.).

Several students expressed that going through this experience prepared them for the work force in that they have now had experience interviewing clients, gathering requirements, and navigating the delicate balance of interpreting client needs and merging them with technical limitations of the software and hardware involved. Also, they got a good dose of reality in working with others. They also realized that team dynamics and group work was fundamental to the success of the project and would translate to better work practices in their future jobs. One student explained, "If somebody is falling behind, you pick up part of their (job). I never had this experience in my other classes because I was the one responsible for turning things on time and didn't have a class of Chemistry students waiting on my work getting done."

Students had a greater appreciation of the soft skills and leadership needs of a team. Many of these students expressed that they were "good" at coding, when we interviewed them at the start of the course. However, in their post-interviews, less were as verbal about their own skills as they were about the necessity of teamwork and communication skills. Due to interpersonal conflicts and arguments among team members, the students were enlightened to the concept that a team is only as effective and productive as the weakest of the members. Thus, as the project went on, students were forced to collaborate and develop ways in which to work with each other. This was a positive outcome of the project, as these students really do not have many opportunities to hone these skills in college courses.

The majority of the students involved believed that they gained much needed experience with tools that today's employers might expect. Indeed, three of the students actually included the mobile app in their portfolio as they went on interviews for jobs. Along with this added experience, they also increased in their appreciation of the software development phases and documentation. One student complained, "I hate documentation. I think it is a waste of time. But I can see, after making TsoiChem, that if I don't document what I do and make it so that others can figure out what I'm coding, I can really cause a lot of problems down the road and even waste time because of it. I still don't like doing it, but I am now making sure I document everything I do. I have got to get better at it and I really need to start doing it all the time."

Another common theme that arose from the post-interviews was the appreciation of a "real-life" end user. It was noted in several pre-interviews that the students were aware of the benefits of having an actual client; the chemistry professor. They noted that the "pressure" of the chemistry students' future dependence on the app caused them to be more motivated, be more aware of quality assurance of the final product, increased their stress levels, made them more aware of time constraints and work flow, and amplified team dynamics, both positive and negative. Several noticed that no other school project had caused them this much distress or excitement, and it was probably because there were peers in the same school who were about to experience the product of their efforts, and that added an element of "reality" which made the project very real. As one student quietly whispered at the end of the interview, "It's your baby. It's your child. You put in your blood, sweat, and tears because others are going to use the app and their education depends on how good your app is. This is what it's like out there – and I really feel like I got a good taste of it doing this app."

Challenges

Significant challenges remain in the constantly evolving field of mobile application development. The development platform underwent minor changes every semester and major changes every year. This requires a serious time commitment to ensure that development tools and operating system are upgraded as required, while environments should be tested and curricula need to be updated as well. Sometimes these changes can happen in the middle of the semester. The authors were fortunate to have access to a dedicated lab, which could be upgraded as required. The minor changes were easy to handle, however the operating system upgrades were time consuming, but necessary for development and testing on devices. This constant upgrading and changes to technology introduced the students to real-time technical challenges that they could face in the work environment. Despite the many challenges, given the benefits of this project to students from various STEM disciplines, it is worth spending the time and effort required to carry on this project for multiple semesters.

Pre-planning for the project also involves significant time commitment by the involved faculty and should be taken into consideration. To simply show up to a software development class with a general, unspecified idea of what a needed app should look like or accomplish would be detrimental to the students' learning. Even though this partnership does try to mimic a real-life situation, there still needs to be parameters put into place to keep software development students from roaming too far off the learning journey and completely bypassing the intended learning objectives. As well, a clear, well-defined method of assessment is vital to the course. Without an understanding of how their efforts will be measured and graded, software development students are left with little information on what is valued and what is not in the software development cycle. Teamwork is only as effective as the lowest contributing team member; therefore, students need adequate training on how to work with each other, how to bring out the best in each other, and how to support continued growth in themselves. Although the software development students saw the inherent value of soft-skills and teamwork capabilities, there was not enough time or opportunities to mine these valuable lessons and reinforce their concepts.

CONCLUSION

Our findings showed that the benefits of authentic, integrated learning are immense for participating students. The junior level software development students developing the mobile apps showed the most benefit from this study. The relatively limited scope of mobile apps makes them a perfect candidate to provide an authentic software development experience to students. It is often difficult to find a "real-world" project that can be completed within the scope of a single semester course. The possibility that students can learn a completely new set of programming language and development tools to create a complete working product that can be used by end-users is a huge motivating factor and should be used to increase student engagement in the course. It is unclear whether developing mobile apps versus non-mobile apps played a role, but it is clear that the students felt that mobile apps were relevant to the real world.

For the introductory programming students, the quantifiable increase in learning was negligible, but both chemistry and programming students surveyed clearly saw the use of m-learning as relevant and more engaging than traditional learning methods. The apps provided an additional pathway to learning that was readily accepted by the students engaged. Since an engaged, interested student is one of the keys to student success, we found the use of the mobile apps was worth the time spent in class using them. By having upper level students develop the apps for use by freshman/sophomore level students, we doubled the benefit of this form of e-learning.

REFERENCES

ACM & IEEE. (2013). *Computer Science Curricula 2013: Curriculum Guidlines for Undergraduate Degree Programs in Computer Science*. IEEE and ACM.

Ausubel, D. P. (2000). *The Acquisition and Retention of Knowledge: A Cognitive View*. Springer. doi:10.1007/978-94-015-9454-7

Bandura, A., & Schunk, D. H. (1981). Cultivating competence, self-efficacy, and instrinsic interest through proximal self-motivation. *Journal of Personality and Social Psychology*, *41*(3), 586–598. doi:10.1037/0022-3514.41.3.586

Begel, A., & Simon, B. (2008). Strugges of new colege graduates in their first software development job. *SIGCSE Bulletin*, *40*(1), 226–230. doi:10.1145/1352322.1352218

Bradley, A. Z., Ulrich, S. M., Jones, M. Jr, & Maitland, S. M. (2002). Teaching the Sophomore Organic Course without a Lecture. Are You Crazy? *Journal of Chemical Education*, *79*(4), 514–519. doi:10.1021/ed079p514

Bruner, J. (1960). *The Process of Education*. Cambridge, MA: Harvard University Press.

Bruner, J. (2006). In Search Of Pedagogy: Vol. I. *The Selected Works of Jerome Bruner 1957-1978*. New York: Routledge.

Collard, D. M., Girardot, S. P., & Deutsch, H. M. (2002). From the Textbook to the Lecture: Improving Pre-Lecture Preparation in Organic Chemistry. *Journal of Chemical Education*, *79*(4), 520–523. doi:10.1021/ed079p520

Dekhane, S., & Tsoi, M. Y. (2012). Designing a Mobile Application for Conceptual Understanding: Integrating Learning Theory with Organic Chemistry Learning Needs. *International Journal of Mobile and Blended Learning*, *4*(3), 34–52. doi:10.4018/jmbl.2012070103

eMarketer. (2014, January 16). *Smartphone Users Worldwide Will Total 1.75 Billion in2014*. Retrieved from eMarketer: http://www.emarketer.com/Article/Smartphone-Users-Worldwide-Will-Total-175-Billion-2014/1010536

Gardner, H. (1983). *Frames of Mind: The Theory of Multiple Intelligences*. New York: Basic Books.

Gilmer, J. (1979). The effects of immediate feedback versus traditional no-feedback in a testing situation. *Proceedings of the 63rd Annual Meeting of the American Educational Research Association*.

Hogan, K., & Pressley, M. (1997). *Scaffoding student learning: Instructional approaches and issues*. Cambridge, MA: Brookline Books.

Hoppe, H. (2004). SMS-based discussions - Technology enhanced collaboration for a literature.*Proceedings of International Workshop on Wireless and Mobile Technologies in Education*.

Hwang, G. J., & Tsai, C. C. (2011). Research trends in mobile and ubiquitous learning: A review of publications in selected journals from 2001 to 2010. *British Journal of Educational Technology*, *42*(3), E65–E70. doi:10.1111/j.1467-8535.2011.01183.x

Hwang, G. J., Tsai, C. C., & Yang, S. J. (2008). Criteria, strategies and research issues of context-aware ubiquitous learning. *Journal of Educational Technology & Society, 11*(1), 81–91.

Kurkovsky, S. (2014). Mobile Game development: Improving student engagement and motivation in introductory computing courses. *Computer Science Education, 23*(2), 138–157. doi:10.1080/08993408.2013.777236

Lave, J. (1982). A comparative approach to educational forms and learning processes. *Anthropology & Education Quarterly, 13*(2), 181–187. doi:10.1525/aeq.1982.13.2.05x1832l

Lombardi, M. M. (2007). *Authentic Learning for the 21st Century: An overview. ELI Report No. 1.* Boulder, CO: EDUCAUSE Learning Initiative. Retrieved March 10, 2015, from http://www.educause.edu/ir/library/pdf/ELI3009.pdf

Lutz, M. J., Naveda, F., & Valino, J. R. (2014). Undergraduate Software Engineering. *Communications of the ACM, 57*(8), 52–28. doi:10.1145/2632361

Maples, M. F., & Webster, J. M. (1980). Thorndike's Connectionism. In G. M. Gazda & R. J. Corsini (Eds.), *Theories of Learning*. Itasca, IL: F.E. Peacock.

Maslow, A. H. (1943). The Theory of Human Motivation. *Psychological Review, 50*(4), 370–396. doi:10.1037/h0054346

Mehdipour, Y., & Zerehkafi, H. (2013). Mobile Learning for Education: Benefits and Challenges. *International Journal of Computational Engineering Research, 3*(6).

Merriam, S. B., & Caffarella, R. S. (1991). *Learning in Adulthood: A Comprehensive Guide*. San Francisco: Jossey-Bass Publishers.

Minskoff, E. H., & Allsopp, D. (2002). *Academic Success Strategies for Adolescents with Learning Disabilities/ADHD*. Baltimore, MD: Paul H Brookes Publishing.

Petersen, S. A., & Divitini, M. (2005). Language learning: from individual learners to communities. *Proceedings of the IEEE International Workshop on Wireless and Mobile Technologies in Education*. doi:10.1109/WMTE.2005.41

Punch, K. (1998). *Introduction to Social Research: Quantitative and Qualitative Approaches*. London: Sage Publications.

Reiser, B. J. (2004). Scaffolding Complex Learning: The Mechanisms of Structuring and Problematizing Student Work. *Journal of the Learning Sciences, 13*(3), 273–304. doi:10.1207/s15327809jls1303_2

Sharples, M., Corlett, D., & Westmancott, O. (2002). The design and implementation of a mobile learning resource. *Personal and Ubiquitous Computing, 6*(3), 220–234. doi:10.1007/s007790200021

Shih, J. L., Chu, H. C., & Hwang, G. J., & Kinshuk. (2010). An investigation on attitudes of students and teachers for participating in a context-aware ubiquitous learning activity. *British Journal of Educational Technology*.

Shuler, C. (2012). iLearn II; An Analysis of the Education Category of the iTunes App Store. *The Joan Ganz Cooney Center at Sesame Workshop*.

Szu, E., Nandagopal, K., Shavelson, R. J., Lopez, E. J., Penn, J. H., Scharberg, M., & Hill, G. W. (2011). Understanding Academic Performance in Organic Chemistry. *Journal of Chemical Education*, *88*(9), 1238–1242. doi:10.1021/ed900067m

Truong, D. (2014). How To Design a Mobile Application to Enhance Teaching and Learning? *International Journal of Interactive Mobile Technologies*, *9*(3), 4–11.

Vavoula, G. N., & Sharples, M. (2008). Challenges in Evaluating Mobile Learning. *Proceedings of Mlearn 2008.*

Wood, D. J., Bruner, J. S., & Ross, G. (1976). The role of tutoring in problem solving. *Journal of Child Psychiatry and Psychology*, *17*(2), 89–100. doi:10.1111/j.1469-7610.1976.tb00381.x PMID:932126

Xanthopoulos, S., & Xinogalos, S. (2013). A comparative analysis of cross-platform development approaches for mobile applications.*Proceedings of the 6th Balkan Conference in Informatics (BCI '13)* (pp. 213-220). New York, NY: ACM. doi:10.1145/2490257.2490292

Section 4
Literature Reviews

Chapter 13
A Trend Analysis of Mobile Learning

Serçin Karataş
Gazi University, Turkey

Nimet Özgül Ünsal Köse
Gazi University, Turkey

Onur Ceran
Gazi University, Turkey

Mustafa Kılıç
Gazi University, Turkey

Ülkü Ülker
Gazi University, Turkey

Gökçe Akçayır
Gazi University, Turkey

Ezgi Tosik Gün
Gazi University, Turkey

Zeynel Abidin Tok
Gazi University, Turkey

ABSTRACT

The aim of this study is to determine current tendencies regarding mobile learning in published research between 2010 and 2015. In this study, 221 articles collected from the Web of Science Core Collection database with SSCI index were examined by using the content analysis technique. In the analyses, eight criteria were used, namely; research technique, sampling size, sampling level, learning domain, topical domain, data collecting tool, data analysis method and mobile application development approach. The results suggest that the main tendencies under these categories were experimental method; sample sizes of 31-100 people; higher education students; humanities and social sciences domain; learner outcomes topical domain; mixed data collecting tools; mixed analysis methods; and native mobile application development approach.

INTRODUCTION

It is possible to say that we live the mobile age by looking at mobile phones, ATMs, mobile cars used as offices, timeless and placeless advertisements, and palmtop computer games (Sharples, 2006). The use of mobile technologies has become widespread as wireless communication technologies have improved (Chu, Hwang, Tsai & Tseng, 2010). Mobile technologies, which offer the opportunity to reach educational content at any place and time, have a great deal of potential in that they have the advantage

DOI: 10.4018/978-1-5225-0359-0.ch013

to make learning customized for the individual, and to eliminate time and place constraints (Attewell, 2005; Kukulska-Hulme, 2005; Mutlu, Yenigün, & Uslu, 2006; Motiwalla, 2007; Demsey, 2008; Wexler, Brown, Metcalf, Rogers, & Wagner, 2008; Wang, Wu, & Wang, 2009; Wagoner, Hover, & Ernest, 2011; Roberts, 2013; Ally & Tsinakos, 2014). Many higher education institutions favour mobile learning in order to ensure flexibility in education (Tsinakos & Ally, 2013). Likewise, learners prefer mobile devices to personal computers due to the flexibility they provide (Stockwell, 2010). When considered from this perspective, mobile technologies have become an up-to-date subject of study for educators and researchers (Kukulska-Hulme, & Traxler, 2007; Wu, Wu, Chen, Kao, Lin, & Huang, 2012).

Definition of Mobile Learning

As a result of these improvements, the definitions used for mobile learning (m-learning) have also varied over time. From the *technocentric point of view*, m-learning is learning which takes place through using any mobile device such as PDA, mobile phone, or iPod. For instance Quinn (2000) defined m-learning as electronic learning through mobile devices such as Palm, Windows CE machines, & digital cell phones, while Keegan (2005) defined it as enabling learning through PDA, smart and mobile phones. A relationship to e-learning perspective describes m-learning as an extension of e-learning; one group of researchers views m-learning as a *sub-branch of the study of e-learning* (Georgiev, Georgieva, & Smrikarov, 2004). Furthermore, another group states that the reason why m-learning and e-learning are different fields of study is due to a terminology difference (Parsons, 2007). In e-learning, terms such as multimedia and hyperlink are often used while; in m-learning, terms such as mobile, portable, personal are usually used. According to the *learner-centred perspective*, in mobile learning not only are devices used and life-long learning supported, but also it is a learning method, which is based on the mobility of learners (Sharples, 2000; Sharples, Corlett, & Westmancott, 2002; Winter, 2006).

To sum up, it is possible to say that the descriptions created in the early years about m-learning are technology focused, and stress the learning conducted through mobile devices and wireless technologies (Quinn, 2000; Traxler, 2005), whereas in later years, m-learning is defined as a learning method, in which learners can benefit from the learning advantages offered by mobile technologies without depending on a place or time, and in this definition learning is emphasized, instead of technology (Brown, 2010). In other words, m-learning is more concerned with the learner than technology, because the learner is mobile and is in the center of the act of learning, whereas the technology is only an enable for learning in any context. Mobile learning is a social event, which includes mobile people and enables the configuration of spontaneous learning contents, rather than a technological event (Vavoula and Sharples, 2009).

Technologies and Applications of m-Learning

The technologies used in mobile learning studies include devices such as tablet PCs, iPods, palm computers, PDAs, mobile phones, wireless infrastructures, laptops, and smart phones (Cobcroft, Towers, Smith, & Bruns, 2006; Corlett, Sharples, Bull, & Chan, 2005). The primary mobile devices people use for mobile learning are briefly explained in the following (Corbeil, & Valdés-Corbeil, 2007; Trinder, 2005; Woodill, 2011):

- **PDA-Personal Digital Assistants:** A system that includes calculation capacity, internet and network access, and is a productive mobile device in terms of its calendar, notebook, and address book.
- **Portable Media Players:** Devices that can play and record various digital media such as audio files, pictures, and videos (e.g. iPod).
- **MP3 Player:** Enable audio lessons and podcasts in MP3 format to be downloaded and played.
- **E-Book Reader:** For downloading text-based materials in various e–book formats.
- **USB Drive:** Used for storing and transferring data and serve only one purpose.
- **Laptop/Tablet PC:** The most integrative (containing more features) and functional system among the portable devices.
- **Ultra Mobile PC:** It contains most features of standard tablet PCs but is not as comprehensive.
- **Cell Phones:** They connect to special radio stations via radio signals. The radio stations work both as a provider and receiver, transferring received calls to mobile phones and sending back the returned calls to the wireless network.
- **Smart Phones:** A mobile device in which the features of a PDA and telephone are combined.
- **Tablet Devices:** Tablet devices may have touch and/or rotating screens.

The benefits of mobile technology usage in education come from mobile technology itself, because mobile technologies have the power of instant publishing, easily capturing and recording life events and distributed collaboration (Naismith, Lonsdale, Vavoula, & Sharples, 2006). According to Naismith, et al. (2006), mobile technologies impact learning due to the fact that they contribute to learner-centered learning environments, make meaningful connections among learners, resources and other people; promote learners to become investigators of their own environments; and help recall and collaborative reflection. M-learning can be adapted to various learning theories and approaches such as behaviorism, constructivism, situated learning, collaborative learning, informal learning, etc. (Naismith, et al. 2006).

Mobility enables learning and teaching outside the traditional classroom environment (Corbeil & Valdés-Corbeil, 2007). Considering the technologies of our age, it is seen that use of some devices such as PDAs and portable media players has declined. On the other hand, mobility has increased ever since devices such as laptops and smart phones have been equipped with wireless network features, which eliminated time and place limitations on access to information, and created a new dimension in education (Sharples, Arnedillo-Sánchez, Milrad, & Vavoula, 2009). Mobile learning provides learners with equal opportunities in terms of accessing learning resources (Ally & Prieto-Blázquez, 2014). Learners can reach updated and related learning resources via the Web and communicate with experts in their fields of study. They also get the chance to complete courses simultaneously while they work from their offices or any other place (Ally & Prieto-Blázquez, 2014). M-learning is beneficial for learners in that students can use mobile devices for; learning in their learning communities, authentic learning, context-aware learning, contingent learning, augmented reality learning, and individualized learning (Quinn, 2013; Traxler, 2010).

Historical Development of m-Learning

Individuals had previously been able to learn without time and place limitations thanks to distance education. Later, it became possible to access any content at any time and place as a result of technological developments, thus electronic learning (e-learning) came into being. Accessing any content has become convenient once internet technologies became wireless and devices became portable, enabling mobile learning.

Looking at the historical development of mobile learning; it first appeared in the 1970s with the Dynabook notion, and took place in schools for the first time through Microwriter (Psion Computer) palm devices. In the 1990s, m-learning applications of PDAs and tablet tools were used. Between the years of 2001 and 2003, the MOBILearn project was supported by the European Commission (Casey, 2009). MOBILearn was a development project, in which context-aware approaches to on-site learning, problem-based learning, and informal learning were examined through mobile technologies (MobiLearn, 2002).

Future of Mobile Learning

Researchers predict that technology-based learning will increase due to the newly developed devices or concepts, and furthermore believe that mobile and wireless communication technologies will play a major role throughout this process of educational revolution (Liu & Hwang, 2010). These rapid changes in the world will compel different stakeholders to work together towards developing new pedagogic models for new generation learners, who use mobile technologies (Brown, 2005).

Current mobile technologies could meet the increasing needs of learners in terms of accessing course contents. For that reason, mobile LMS use should be taken into consideration in higher education. The researches regarding Mobile LMS use are inadequate (Han & Han, 2014). Mobile LMS makes the co-operation and interaction of learners among themselves and the interaction between the learners and the instructors easier (Goh & Kinshuk, 2006).

Global virtual space will shrink as a result of mobile learning in future. Mobile technologies could be used by students from all over the world to produce information and share with each other (Ally & Prieto-Blázquez, 2014). The new generation of mobile learning will be freer of depending on time and place, moreover; there will be smart systems supporting learning everywhere and learners themselves will be mobile. Besides, new mobile technologies will be virtualized by virtual input and output skills (Ally & Prieto-Blázquez, 2014).

New mobile technologies, increase the attendance of students to indoor and outdoor activities through applications such as mobile augmented reality. Even though the results are similar to virtual reality, in augmented reality real-life contexts are simultaneously used with overlaid data (Martin, Diaz, Sancristobal, Gil, Castro, Peire, 2011).

Fascinating environments usually include games and virtual worlds. These environments enable learners to learn concepts more profoundly by making them connect with and drawn to the environment. Students today have difficulty in focusing on only one point in in-class courses (Sue, Maton, & Kervin, 2008). Many students spend long hours on activities such as watching TV, surfing on the internet, and playing online games on their computers or mobile phones (Prensky, 2001). This situation makes innovating teaching and learning methods necessary. It is possible to say that mobile learning through mobile technologies could support this innovation.

Researches on m-Learning

The interest in conducting research into mobile learning has become widespread enough to carry out significant projects in schools, workplaces, museums, cities, and rural areas (Sharples, 2006). Reynolds, Walker, & Speight (2010) have created a system for students to explore museum settings and collections through their mobile devices. Chiou, Tseng, Hwang, & Heller (2010) have developed an applicable navigation support system for students to guide while learning about museums, via mobile devices. Chu,

Hwang, & Tsai (2010) have organized nature science course as a ubiquitous learning activity, which could enable students to observe the characteristics of the plants around the school campus, through their mobile devices.

Many researchers have recently considered mobile technologies to be one of the most important research subjects in the field of education. Conducting learning activities through m-learning has become a strategic matter for many organizations nowadays (Ally & Prieto-Blázquez, 2014). 15 years ago, it was difficult to imagine the use of mobile technologies in learning, however, in today's world even libraries have been migrated to digital environments and information has become technologically usable (Ally & Prieto-Blázquez, 2014; Ally & Needham, 2010).

According to ECAR Mobile IT report, 53% of colleges and universities have activated at least one mobile assisted service, application or, website (Dobbin, Dahlstrom, Arroway, & Sheehan, 2011). Wu, Wu, Kao, Lin, & Huang (2012) have carried out a meta-analysis on mobile learning and pointed out that 86% of studies containing m-learning have positive learning outputs.

Literature expresses that m-learning is the future or an inseparable part of learning (Trifonova, 2003). Therefore, in this study, determining the current situation of the adoption of mobile learning in education, in order to see how this change is reflected in education, is planned as a step to highlight the expected situation. Within this direction, this trend study has been conducted in order to examine and analyze the current studies, and determine the tendencies in previous studies, and reveal the current situation.

It is believed that the concept of trend should be defined at this point. As mentioned by Visser & Gagnon (2005), trend is a term which refers to the statistically observable change or general orientation of a general movement. In order for an orientation or a tendency to become a trend, it must comprise a certain period of time, repeat itself during this period, be proven statistically, focus on a specific area, be based on data, and have a certain level of consistency. Within the international scope, trends enable enhancing the changes in education and having an international point of view, and also help nations to learn from one another. Trends report only the facts at the time the research is done, and they do not predict the future. Therefore, it is possible to say that trend studies are repeatable. Conducting more studies on less examined subjects and diversifying the subjects could be useful for eliminating that ambiguity. For that reason trend studies, which indicate where the studies in literature focus by using classifications such as what subjects are examined and what methods are employed, possess importance.

As mentioned before, up-to-dateness of the studies carried out in this field is of high importance due to the developing and increasing mobile technology use. This study examines the tendencies observed in 2001-2010, 2003-2008, 2003-2010 and 2008-2012 (Hwang & Tsai, 2011; Hung & Zhang, 2012; Wu, Wu, Huang, Chen, Kao, & Lin, 2012; Hwang & Wu; 2014). Table 1 sums up the categories used and articles examined by the researches about these trend analyses, which determined the general tendencies of m-learning themed studies covering different periods between 2001 and 2012, as well as their time intervals.

As Table 1 presents, the trend analyses include periodic intersecting studies, which are summarized in the following:

The results of the research of Hwang and Tsai (2011) indicated that the number of m-learning themed studies was the highest between the years of 2006 and 2010. The mostly studied subjects were motivation, perception and attitude at higher education level. Most studies were conducted in the field of engineering, language, art, and science. UK, and Taiwan were observed to have respectively contributed the most to this field.

Table 1. Demographic information about tendency studies on m-learning

Authors	Time Period	The Number of Articles	Criteria
Hwang & Tsai (2011)	2001-2010	154	* Topic * Sample Group * Learning Domain
Hung & Zhang (2012)	2003-2008	119	* Bibliometrics * Taxonomies * Across Time
Wu, Wu, Huang, Chen, Kao & Lin (2012)	2003-2010	164	* Purpose * Type of Mobile Devices * Representation of Academic Disciplines and Courses * Method * Outcomes * Highly Cited Articles
Hwang & Wu (2014)	2008-2012	214	* Learning Domain * Environments of M-learning * Type of Mobile Devices * Learning Achievement * Motivation * Interest * Sensing Technologies

The results of the study of Hung and Zhang (2012) suggested that the studies carried out in the field of m-learning, starting from 2003, had increased 4.5 times by 2008. Furthermore, the mostly studied subjects were effectiveness, evaluation, personalized systems, and studies on strategies and frameworks.

In the study conducted by Wu, et al. (2012) through a meta-analysis method, 164 researches published between 2003 and 2010 were investigated. Hwang & Tsai (2011) covered the same years in their trend researches as Hung & Zhang (2012) did, however, they deepened their study by adding objectives, method, and m-learning device types. The results indicated that the subjects mostly focused on designing the system and the system's efficiency. The frequently used experimental patterns were determined to be survey and experiment, while the frequently used mobile devices were found to be mobile phones and PDAs. It was concluded that the researches have usually left positive impressions.

The results of Hwang and Wu's (2014) study showed that the subject mostly focused on student's success. The studies conducted on success, motivation, and interest often had positive results. In the studies mostly examining language teaching, it was observed that smart phones and tablet PC's were frequently used.

Koszalka & Ntloedibe-Kuswani (2010) expressed the need for more studies on the use of mobile technology in education so as to increase the rate of m-learning in education. Moreover, it is a necessity that more extensive studies about m-learning be conducted, in order to develop m-learning applications in the 21st century (Ali & Irvine, 2009). For that reason, the aim of this study is to reveal the trends of the m-learning themed studies conducted in the last 5 years, in the light of the criteria mentioned above, and also to provide researchers with details, which could guide their future researches. This study is of importance in that it indicates how much this change is reflected on education, because determining the current status of the adoption of m-learning in education is a step to revealing the status it is expected to be. The answers to the questions below were sought throughout the process of this study:

Which,

1. Research Techniques
2. Sampling Sizes
3. Sampling Levels
4. Learning Field
5. Subject Area Classification
6. Data Collection Tools
7. Data Analysis Methods
8. Mobile Application Development Approach

were widely used in m-learning research?

Method

Trend studies must contain a certain period. For this reason, this research was limited to a 5-year period retrospectively from 2015. Since the study started in March 2015, it contains the studies in Web of Science Core Collection database and SSCI index from January 2010 to March 2015. As Hung and Zhang (2012) state, the fact that the Web of Science database is considered to be highly esteemed because it involves prestigious journals in scientific and social fields, was an important reason for us to choose it.

In the interview done with Donald P. Ely (Emeritus Professor in Instructional Design), he stressed that the subject field should be more customized in trend studies (Visser and Gagnon, 2005) by stating that:

Saying 'the use of educational technologies in distance education is increasing' is not quite practical in terms of trend. This is because educational technology is a broader field which contains different fields. However, saying that 'the increase in the use of computers in local schools' could reflect a trend.

Figure 1. Summary of article selection

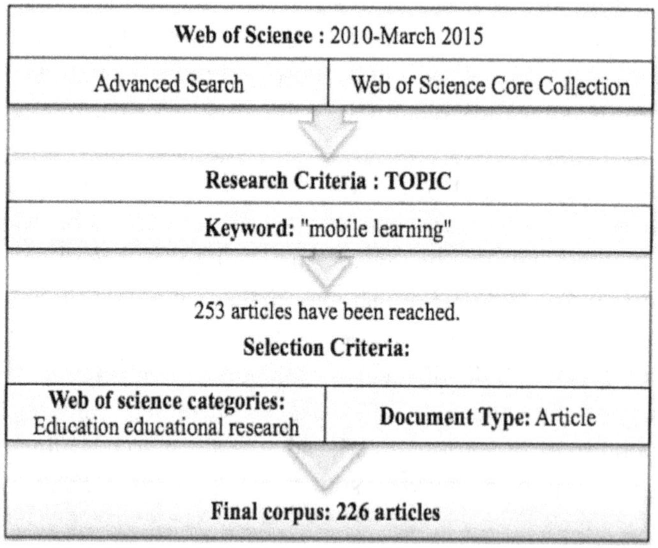

On this basis, the scope of the study was limited to *m-learning* and the keyword was selected as ''m-learning''. The search criterion was determined as *topic* because the database search properties did not offer the option to search with *keywords* or *abstract.* The articles to be examined within the scope of study were limited to the category of *education, educational research* in Web of Science Categories, and to *article* document type in the document type category.

The content analysis technique was employed in this study. Berelson (1952) defines content analysis as the objective, systematic, and quantitative description of a communication material. On the other hand, content analysis, whose coding codes are clear and has a large volume data, is a systematic and renewable research technique summarizing a few situations of the content category (Berelson, 1952; GAO, 1996; Krippendorf, 2004a). It is a useful technique for discovering and defining individual, corporate, in-group or social focuses of interest (Weber, 1990).

As seen in Figure 1, the researchers primarily decided on the search criteria and the database to be scanned. 253 articles were achieved as a result of the scanning done with the keyword ''m-learning''. The number of articles dropped to 226 due to the selection criteria applied on the scanned articles.

Krippendorff's Alpha Coefficient was examined in order to calculate the consistency among the coders. Krippendorff's Alpha (α), which presents the coefficient of multiple coders' reliability, also called the consistency coefficient among coders, was developed with the aim of measuring the reliability coefficient among observers, coders, evaluators, or measuring instruments. Krippendorff's Alpha (α) can be still widely used to reveal the level of reliability of the results achieved from the cases in which two or more methods are applied on the same datasets (Krippendorff, 2011). Four out of eight researchers worked as coders. The expert directing the study chose the researchers, who were personally, socially, economically or scientifically similar to each other, as coders.

One of the significant issues in content analysis is ensuring intercoder reliability. In order to evaluate intercoder reliability, a sampling representing 10-20% of the total is taken (Neuendorf, 2002). According to Krippendorf, so as to ensure multi coder reliability; coders should work independently from each other, however, use the same directions, work in the same groups; and use same units for coding in analysis (Krippendorf, 2004b). Therefore, in order to increase intercoder reliability, the meaning and content of the categories, sub-categories, and classification criteria used on coders are explained; furthermore, pilot studies were carried out on 23 articles representing approximately 10% of the total. The necessary changes were adjusted based on the results of the pilot study. The divergences coders had on expressing same article with different codes (using different methods, subject areas etc.) were decreased through the discussions on a randomly selected 23 articles, and expert support, afterwards; all of the articles were examined. The coders were provided with a coding form for the articles to be examined, and each coder classified articles based on the defined category. The data was transferred to SPSS 21 software and checked to find out if there were any data losses. The data was calculated via KALPHA macro, which is integrated into SPSS 21 software, and the value was determined as 0.7681.

As a result of the examinations of the articles, 5 articles were excluded from the examination by necessity. Since full texts of 2 articles were not found, 1 article which was written in Spanish (since all of the researchers only spoke English as their foreign language), the subject contents of 2 articles were irrelevant of m-learning, even though they had the keyword ''m-learning'', these articles were also excluded from the examination.

As Krippendorff (2004) states, content analysis must contain; a) what data to analyze, b) how to identify this data, c) the population from which the data is drawn, d) how the analyzed data is related to the content, e) what the extent of the analysis is, and f) what the aim of the conclusions is. The researchers created a content analysis classification made up of 8 criteria by adding two more questions; 'what method is used while analyzing the data' and 'what kind of mobile application is used since the use of mobile technology has increased in the recent years.'

Within the scope of this study, while creating each criterion (n=8) used in the examination of the articles, the researchers benefited from the relevant studies in the literature. Some new sub criteria were added to the classifications of the studies from which the researchers benefited, in accordance with the aims of the study. E.g. the researchers added a "Not Applicable" option to classifications due to the articles for which sampling use was not convenient and also "Not Found" option for the articles whose content did not state their sampling or "Mixed" option to this classification because there were more than one tool or method.

The research's criteria, sub categories, and the contents of these categories are presented in Table 11 in the Appendix.

Method classification, which is developed by Lee, Driscoll, & Nelson (2004), was employed for analyzing research methods used in the articles. In order to develop method classification, Lee et al. (2004) used classification categories, separately developed by Berge & Mrozowski (2001), Koble & Bunker (1997), Anglin & Morrison (2000), and Klein (2002).

For sampling size analysis, we used the sampling size classification used by Göktaş, Küçük, Aydemir, Telli, Arpacık, Yıldırım, & Reisoğlu (2012) in their research.

For sampling level analysis, the 'Learner Type' classification used by Drysdale, Graham, Spring, and Halverson (2013) in their research was employed; and the classification of 'Categories with how often addressed' was used so as to determine the subject areas of articles.

The learning area classification presented in Revised Field of Science And Technology (FoS) Classification In The Frascati Manual Report, which was prepared by Directorate For Science, Technology and Industry Committee for Scientific and Technological Policy (2007), was employed, in order to determine the learning area of the sampling which was investigated in the articles.

For the analyses of data collection tools and data analysis methods used in the articles; the data collection tool classification and data analysis method, developed by Beissel-Durant (2004) for the studies of ''Research Methods ESRC National Center'', were employed.

For the classification of mobile technology type used in the article, we applied the classification used in the study conducted by IBM (2012).

Findings and Results

The findings and results that were achieved from the examinations carried out within the scope of the study are presented in this section. Descriptions of all sub-categories which make up criteria are presented detailed in Table 11, in Appendix. The distribution of research methods of m-learning themed 221 articles based on the years are displayed in Table 2:

Table 2. indicates that the general tendency in research methods is experimental research (n=49) by 22.17%. Secondly the most widely used research methods were found to be developmental research (n=35) and survey research (n=35), and furthermore it was discovered that the general tendency had a share of 15.84%. The case study (n=22) had the least amount of tendency by its 9.95% share. Furthermore, since some of the articles stated their research methods as action research, diary research, and pilot study, 1.81% (n=4) of the researches were not found convenient for the research method criterion.

Table 2. The distribution of research methods used in the articles examined within the scope of the study

	2010		2011		2012		2013		2014		2015		TOTAL	
	n	%	n	%	n	%	n	%	n	%	n	%	n	%
Theoretical Inquiry	5	11.11	7	23.33	4	10.26	5	9.80	5	9.80	0	0	26	11.76
Experimental Research	7	15.56	10	33.33	10	25.64	15	29.41	7	13.73	0	0	49	22.17
Case Study	4	8.89	1	3.33	6	15.38	4	7.84	5	9.80	2	40	22	9.95
Evaluation Research	8	17.78	4	13.33	5	12.82	6	11.76	4	7.84	0	0	27	12.22
Developmental Research	13	28.89	5	16.67	4	10.26	6	11.76	7	13.73	0	0	35	15.84
Survey Research	3	6.67	1	3.33	5	12.82	10	19.61	14	27.45	2	40	35	15.84
Combination of Inquiries	2	4.44	1	3.33	5	12.82	5	9.80	9	17.65	1	20	23	10.41
Not Applicable	3	6.67	1	3.33	0	0	0	0	0	0	0	0	4	1.81
TOTAL	45	100	30	100	39	100	51	100	51	100	5	100	221	100

One of the criteria selected for the research is *sampling size* criterion. Table 3 displays the distribution of sampling size of m-learning themed 221 articles based on the years.

According to Table 3, the sampling size of the research was mostly 31-100 participants (n=79) 35.75%. It was also found out that the least preferred sampling size was 1-10 participants (n=11) 4.98%. The articles which were categorized in Not Applicable (n=24), referred to studies which the sample size were not convenient because of research methods like theoretical inquiry. The articles which were categorized in Not Found (n=13), referred to studies which had no specified sample size. To avoid duplicated statements, neither of these situations will be mentioned for the subsequent tables.

Sampling level criterion refers to the education level of the sampling in the articles that were examined. Table 4 displays the distribution of sampling levels of m-learning themed 221 articles based on the years:

Table 4. suggests that the general tendency of sampling levels, on which the researches were applied, was found as the individuals who have received education and training. Moreover, it was seen that the number of studies in which professionals participated was rather low. It was determined that in the groups, which were convenient to accept students, more studies were conducted with the higher educational level (n=99) with its percentage of 43.42% than K-12 level (n=81) with its percentage of 35.33%. The studies conducted with patterns, which did not contain samplings (n=25), consisted of 10.96%; while the research that did not determine what level its working group fell into (n=6), consisted of 2.63%. In 2011, the percentages of the studies applied in higher education and K-12 levels (n=13) equalled to 39.39%. When examined from the view point of yearly distribution, similar to general results, in each year, except for 2011, more studies were conducted in higher education level than K-12 or corporate levels.

Learning domain criterion displays what area of education the conducted studies belong to Table 5 indicates the distribution of learning domain of m-learning themed 221 articles based on the years:

Table 5. suggests that 221 researches fell into 3 learning domains, and interestingly their values were close to each other. Humanities (n=60) ranked as the first with 27.15%, social sciences (n=59) came as the second with 26.70% while natural sciences (n=55) ranked as the third most studied learning domain with 24.89%. The studies which reviewed the literature without having any learning domain (n=19) 8.60%, followed as medical and health (n=13) 5.88%, and engineering and technology (n=10) 4.52%. It was observed that the percentage of studies, which did not provide information about what

Table 3. The distribution of sampling size in the articles examined within the scope of the study

	2010		2011		2012		2013		2014		2015		TOTAL	
	n	%	n	%	n	%	n	%	n	%	n	%	n	%
1-10	2	4.44	1	3.33	1	2.56	2	3.92	5	9.80	0	0	11	4.98
11-30	3	6.67	4	13.33	3	7.69	6	11.76	5	9.80	1	20	22	9.95
31-100	21	46.67	9	30	13	33.33	20	39.22	15	29.41	1	20	79	35.75
101-300	6	13.33	4	13.33	14	35.90	11	21.57	6	11.76	1	20	42	19
301-1000	2	4.44	2	6.67	2	5.13	5	9.80	6	11.76	0	0	17	7.69
>1000	1	2.22	3	10	2	5.13	1	1.96	5	9.80	1	20	13	5.88
Not Applicable	5	11.11	7	23.33	4	10.26	4	7.84	4	7.84	0	0	24	10.86
Not Found	5	11.11	0	0	0	0	2	3.92	5	9.80	1	20	13	5.88
TOTAL	45	100	30	100	39	100	51	100	51	100	5	100	221	100

Table 4. The distribution of sampling levels used in the articles examined within the scope of the study

	2010		2011		2012		2013		2014		2015		TOTAL	
	n	%	n	%	n	%	n	%	n	%	n	%	n	%
K-12	22	44.90	13	39.39	11	28.21	16	31.37	17	33.33	2	40	81	35.53
Higher Education	20	40.82	13	39.39	20	51.28	24	47.06	20	39.22	2	40	99	43.42
Corporate	1	2.04	3	9.09	3	7.69	4	7.84	5	9.80	1	20	17	7.46
Not Applicable	4	8.16	4	12.12	5	12.82	6	11.76	6	11.76	0	0	25	10.96
Not Found	2	4.08	0	0	0	0	1	1.96	3	5.88	0	0	6	2.63
TOTAL	49	100	33	100	39	100	51	100	51	100	5	100	228	100

Table 5. The distribution of learning domains used in the articles examined within the scope of the study

	2010		2011		2012		2013		2014		2015		TOTAL	
	n	%	n	%	n	%	n	%	n	%	n	%	n	%
Natural Sciences	11	24.44	7	23.33	10	25.64	16	31.37	10	19.61	1	20	55	24.89
Engineering & Technology	1	2.22	1	3.33	0	0	2	3.92	6	11.76	0	0	10	4.52
Medical & Health	4	8.89	4	13.33	3	7.69	0	0	1	1.96	1	20	13	5.88
Agricultural Sciences	0	0	0	0	0	0	1	1.96	0	0	0	0	1	0.45
Social Sciences	10	22.22	4	13.33	21	53.85	11	21.57	12	23.53	1	20	59	26.70
Humanities	11	24.44	9	30	5	12.82	19	37.25	16	31.37	0	0	60	27.15
Not Applicable	8	17.78	5	16.67	0	0	1	1.96	5	9.80	0	0	19	8.60
Not Found	0	0	0	0	0	0	1	1.96	1	1.96	2	40	4	1.81
TOTAL	45	100	30	100	39	100	51	100	51	100	5	100	221	100

learning domain the study was conducted (n=4), equalled to 1.81%, moreover in the learning domain of agricultural sciences only 1 study was conducted. The reason why Humanities ranked as the first could be due to the fact that the education of m-learning and languages & literature are the areas which are mostly focused on. The studies conducted on the learning domains based on years were found to be parallel to general tendency.

The criterion the articles focused on were expressed through *topical domain*. The topical domain category explains the research topics of the learning domain in located in Table 5. Table 6 indicates the distribution of topical domains of m-learning themed 221 articles based on the years:

Table 6. suggests that the general tendency in topical domain is respectively as follows: first, learner outcomes (n=61) 27.48%, second, instructional design (n= 52), and third, disposition and technology (n=37) 16.67%.The least studied topical domains were demographic (n=1) 0.45% and other (n=6) 2.70% which includes benefits and challenges, access and availability, support system, time efficiency, nature, and international issues. Interaction (4.50%), with the value of n=10 in total, mostly indicates the interaction among the learners. The number of studies examining the interaction between the learners and educators is low. When the distribution values based on years were examined, it was observed that no changes took place in the most studied topical domains, but the ranking of topical domains based on how often they are studied might change. According to the yearly distribution values, subject values mostly consist of the areas including learner outcomes, instructional design, technology and disposition.

The tools that were used to collect the data in the examined articles were expressed through *data collecting tools.*Table 7 displays the distribution of data collecting tools of m-learning themed 221 articles based on the years:

Table 7. indicates that the general tendency in research methods is mixed (n= 98) by 44.34%. The mixed option shows that more than one data collecting tool was used. It was observed that in the studies in which mixed data collection tools were employed, at least two data collection tools in pairs such as survey-interview, questionnaire-observation, and observation-interview; or triple data collection tools such as questionnaire-interview-observation, and survey-interview-observation were used. Furthermore, it was also seen that in some studies, more varied data collection tools were used together. Particularly,

Table 6. The distribution of topical domain used in the articles examined within the scope of the study

	2010		2011		2012		2013		2014		2015		TOTAL	
	n	%	n	%	n	%	n	%	n	%	n	%	n	%
Learner Outcomes	16	35.56	2	6.45	6	15.38	15	29.41	20	39.22	2	40	61	27.48
Disposition	5	11.11	8	25.81	16	41.03	3	5.88	2	3.92	3	60	37	16.67
Instructional Design	15	33.33	8	25.81	9	23.08	10	19.61	10	19.61	0	0	52	23.42
Interaction	2	4.44	1	3.23	1	2.56	4	7.84	2	3.92	0	0	10	4.50
Comparison	0	0	3	9.68	3	7.69	1	1.96	4	7.84	0	0	11	4.95
Demographics	0	0	0	0	0	0	0	0	1	1.96	0	0	1	0.45
Technology	7	15.56	8	25.81	2	5.13	14	27.45	6	11.76	0	0	37	16.67
Professional Development	0	0	1	3.23	2	5.13	0	0	4	7.84	0	0	7	3.15
Other	0	0	0	0	0	0	4	7.84	2	3.92	0	0	6	2.70
TOTAL	45	100	31	100	39	100	51	100	51	100	5	100	222	100

Table 7. The distribution of data collecting tools used in the articles examined within the scope of the study

	2010		2011		2012		2013		2014		2015		TOTAL	
	n	%	n	%	n	%	n	%	n	%	n	%	n	%
Data Collection	0	0	4	13.33	6	15.38	1	1.96	2	3.92	0	0	13	5.88
Sampling	0	0	0	0	0	0	0	0	0	0	0	0	0	0
Self-Administrative Questioning	0	0	0	0	0	0	0	0	0	0	0	0	0	0
Interview	4	8.89	3	10	2	5.13	3	5.88	4	7.84	0	0	16	7.24
Questionnaire	10	22.22	9	30	12	30.77	12	23.53	9	17.65	2	40	54	24.43
Observation	1	2.22	0	0	0	0	0	0	0	0	0	0	1	0.45
Measurement	1	2.22	1	3.33	2	5.13	0	0	1	1.96	0	0	5	2.26
Use of Administrative Sources	0	0	0	0	0	0	0	0	0	0	0	0	0	0
Visual Methods	0	0	0	0	0	0	0	0	0	0	0	0	0	0
Advanced Technologies	1	2.22	1	3.33	2	5.13	0	0	4	7.84	0	0	8	3.62
Mixed	22	48.89	7	23.33	15	38.46	29	56.86	23	45.10	2	40	98	44.34
Not Applicable	5	11.11	4	13.33	0	0	5	9.80	8	15.69	0	0	22	9.95
Not Found	1	2.22	1	3.33	0	0	1	1.96	0	0	1	20	4	1.81
TOTAL	45	100	30	100	39	100	51	100	51	100	5	100	221	100

this situation was frequently encountered in the studies whose research method was evaluation, developmental, or experimental. The secondly most frequently used data collecting tool was found to be the questionnaire (n=54) 24.43%. The not applicable (n=22) option, which indicates the number of articles that were not convenient for a data collecting tool, increased up to 9.95%. The least frequently used data collecting tools were found to be observation (n=1) by 0.45% and measurement (n=5) by 2.26%. Sampling, self-administrative questioning, use of administrative sources, and visual methods were the data collecting tools that were never used. The percentage of not found (n=4), which included the studies that did not offer any information on what data collecting tools they used, equalled 1.81%.

Data analysis method was used to detect the data analysis of the researches or applications conducted in the examined studies. Table 8 displays the distribution of data analysis methods of m-learning themed 221 articles based on the years:

As seen in Table 8, the most frequently used data analysis method is the one in which both of the methods were employed by (n=73) 33.03% in the researches subject to examination. The shares of quantitative analysis techniques (n=65) 29.41%, and the approaches in which qualitative data analysis method was used (n=63)28.51% were determined to be approximate values. The number of publications, which were not convenient for the use of statistical data analysis methods in terms of their abstracts, only equalled to (n=6) 6.33%.

So as to indicate under which category the mobile application type used in the examined studies fell into, *mobile application development approach* criterion was used. Table 9 displays the distribution of mobile application development approach of m-learning themed 221 articles based on the years:

As Table 9 suggests, in the examined studies, the most frequently used one was native application by (n=107) 48.42%. The shares of the use of mixed approach mobile application (n=45) 20%, and the

Table 8. The distribution of analysis methods used in the articles examined within the scope of the study

	2010		2011		2012		2013		2014		2015		TOTAL	
	n	%	n	%	n	%	n	%	n	%	n	%	n	%
Qualitative	10	22.22	7	23.33	9	23.08	12	23.53	21	41.18	4	80	63	28.51
Quantitative	8	17.78	12	40	17	43.59	20	39.22	8	15.69	0	0	65	29.41
Mixed	21	46.67	6	20	13	33.33	19	37.25	14	27.45	0	0	73	33.03
Not Applicable	5	11.11	4	13.33	0	0	0	0	5	9.80	0	0	14	6.33
Not Found	1	2.22	1	3.33	0	0	0	0	3	5.88	1	20	6	2.71
TOTAL	45	100	30	100	39	100	51	10	51	100	5	100	221	100

Table 9. The distribution of mobile application development approach used in the articles examined within the scope of the study

	2010		2011		2012		2013		2014		2015		TOTAL	
	n	%	n	%	n	%	n	%	n	%	n	%	n	%
Web Application	6	13.33	7	23.33	13	33.33	9	17.65	4	7.84	2	40	41	18.55
Native Application	27	60	18	60	21	53.85	21	41.18	19	37.25	1	20	107	48.42
Mixed Approaches	7	15.56	1	3.33	2	5.13	18	35.29	17	33.33	0	0	45	20.36
Not Applicable	5	11.11	4	13.33	3	7.69	3	5.88	10	19.61	1	20	26	11.76
Not Found	0	0	0	0	0	0	0	0	1	1.96	1	20	2	0.90
TOTAL	45	100	30	100	39	100	51	100	51	100	5	100	221	100

use of web application (n=41) 18.55% are close to each other while the use of native applications is far from those. The Web application category contained the studies which were based on web browsers and included applications running on the Internet browser. Mixed approach category (n=45) contained both web application and native application together.

Table 10. displays the distribution of research methods, sampling size, sampling levels, learning domains, topical domains, data collecting tools, data analysis methods, and mobile application development approaches of m-learning themed 221 articles based on the years:

Table 10. suggests that in the published researches mostly, while in research method criterion (n=49), by 22.17% experimental research method was employed; in sampling size (n=79) by 35.57% (31-100) pitch was observed. Furthermore, in sampling level higher education weas mostly seen; in distribution of learning domain (n=60), by 27.15% humanities was studied. In the distribution of topical domain (n=61), by 27.48% learner outcomes were seen; in the distribution of data collecting tools (n=98), by 44.34% mixed data collecting tools were used; and in data analysis methods (n=73), by 33.03% at least two statistical analysis methods were used together. Moreover, it was observed that in mobile application development approach used in m-learning studies, native application ranked as the first (n=107) by 48.42%. In case there are some differences about distribution of the 8 investigated criteria according to the years, general results are not more interesting than the other studies results. Although there were some differences about distribution of the 8 investigated criteria by year, there was no remarkable diversity

Table 10. The M-learning Tendencies of January 2010 - March 2015

	2010	2011	2012	2013	2014	2015	General Trends
Research Methods	• Developmental Research	• Experimental Research	• Experimental Research	• Experimental Research	• Survey Research	• Case Study • Survey Research	• Experimental Research
	(n=13) 28.89%	(n=10) 33.33%	(n=10) 25.64%	(n=15) 29.41%	(n=14) 27.45%	(n=2) 40%	(n=49) 22.17%
Sampling Size	• 31-100	• 31-100	• 101-300	• 31-100	• 31-100	• 11-30 • 31-100 • 101-300 • >1000	• 31-100
	(n=21) 46.67%	(n=9) 30%	(n=14) 35.90%	(n=20) 39.22%	(n=15) 29.41%	(n=1) 20%	(n=79) 35.75%
Sampling Level	• K-12	• K-12 • Higher Education	• Higher Education	• Higher Education	• Higher Education	• K-12 • Higher Education	• Higher Education
	(n=22) 44.90%	(n=13) 39.39%	(n=20) 51.28%	(n=24) 47.06%	(n=20) 39.22%	(n=2) 40%	(n=99) 43.42%
Learning Domain	• Natural Science • Humanities	• Humanities	• Social Sciences	• Humanities	• Humanities	• Social Science • Medical& Health • Natural Science	• Humanities
	(n=11) 24.44%	(n=9) 30%	(n=21) 53.85%	(n=19) 37.25%	(n=16) 31.37%	(n=1) 20%	(n=60) 27.15%
Topical Domain	• Learner Outcomes	• Disposition • Instructional Design • Technology	• Disposition	• Learner Outcomes	• Learner Outcomes	• Disposition	• Learner Outcomes
	(n=16) 35.56%	(n=8) 25.81%	(n=16) 41.03%	(n=15) 29.41%	(n=20) 39.22%	(n=3) 60%	(n=61) 27.48%
Data Collection Tools	• Mixed	• Questionnaire	• Mixed	• Mixed	• Mixed	• Mixed • Questionnaire	• Mixed
	(n=22) 48.89%	(n=9) 30%	(n=15) 38.46%	(n=29) 56.86%	(n=23) 45.10%	(n=2) 40%	(n=98) 44.34%
Data Analysis Methods	• Mixed	• Quantitative	• Quantitative	• Quantitative	• Qualitative	• Qualitative	• Mixed
	(n=21) 46.67%	(n=12) 40%	(n=17) 43.59%	(n=20) 39.22%	(n=21) 41.18%	(n=4) 80%	(n=73) 33.03%
Development Approach for a Mobile Application	• Native Application	• Native Application	• Native Application	• Native Application	• Native Application	• Web Application	• Native Application
	(n=27) 60%	(n=18) 60%	(n=21) 53.85%	(n=21) 41.18%	(n=19) 37.25%	(n=2) 40%	(n=107) 48.42%

in overall results. Despite the increasing studies on m-learning, there was no different study to explain any various situation and gaps in the literature have been still remained.

CONCLUSION AND DISCUSSION

In this research, as a result of the evaluation of the m-learning studies based on their research method criterion, it was concluded that experimental research is the mostly used research method. This situation may result from extensively examining the impact of m-learning and m-learning environments on student's success and motivation, or students' attitudes towards m-learning or m-learning environment. It might have been assumed that the most convenient research method for measuring students' reactions was the studies containing experimental patterns. In the content analysis studies conducted by Kiliç-Çakmak, Çebi, Mihçi, Günbatar and Akçayir (2013), they found out that experimental research is the most frequently used research method. In trend studies about m-learning carried out by Soykan and Uzunboylu (2015), it was seen that more than half of the articles employed experimental research method. In that sense, the study coincides with the other researches in the literature. However, different studies reaching different conclusions also exist in literature. Wu, et al. (2012), found that, among the 164 studies, surveys were the primary research method, followed by experimental research methods.

As the number of the participants is examined in the reviewed articles, the most used sampling size is 31-100. The reason why the studies conducted with more people are less preferred might be due to the difficulty of reaching out to more people or finding groups with same characteristics. The same results were observed in the studies with similar content analysis in the literature (Alper & Gülbahar, 2009; Göktaş et al., 2012; Kiliç-Çakmak et al., 2013; Küçük, et al., 2013). However, when Bozkaya, Aydın, & Kumtepe (2012) examined the articles published between the years of 2008-2011, they realized that over 200 studies with samplings constituted the majority.

In this study it is observed that sampling level criterion mostly involves participants studying at higher education levels and secondly students studying at elementary schools are involved. The sampling sizes could be chosen from university students, since the studies are usually conducted by lecturers or instructors working at a university. It is possible to say that this case is preferred by researchers due to the convenience it provides to the researchers while conducting their studies. In the body of literature, there are studies which observed that undergraduate students were involved in these studies (Hawkridge, 1999; Latchem, 2006; Şimşek et al., 2008; Sert, 2010; Bozkaya et al., 2012; Kiliç-Çakmak et al., 2013; Soykan & Uzunboylu, 2015). Küçük et al. (2013) expressed that in their studies they mostly sampled prospective teachers and pointed out that the importance of the sampling level in the studies at higher education level. Soykan and Uzunboylu (2015) stated that in their trend studies about m-learning, the sampling levels of the articles were mostly at the higher education level. Hwang and Tsai (2011) found that from 2001 to 2010, research samples in higher education were selected most, followed by K-12 which has elementary school students and high school students. Only a few studies selected teachers, and working adult as the research sample. Wu, et al. (2012), found that the m-learning is most frequently used by higher education students, followed by elementary school students, adult learners, secondary (post-secondary) school students and disabled students, and that the number of mobile learners in many contexts increased sharply after 2009. We can think about the reason for adding participants who are from higher education is that they are using mobile technologies much according to their age range.

The distribution of learning domains in the articles examined within the scope of the study mostly consisted of humanities. Soykan and Uzunboylu (2015) concluded in their study that the most tendency was on the field of foreign language teaching. In that sense, it is clear that the study is in a linear relationship with that of Soykan and Uzunboylu (2015). The most prominent reason why m-learning is preferred in language teaching, which is one of the fields of Humanities learning domain, is because m-learning includes interactive applications (Gromik, 2012; Ağca and Özdemir, 2013; Sandberg, Maris, & Hoogendoorn, 2014). Hwang and Tsai (2011) found that most studies did not involve any learning domain, instead, they mainly focused on the investigation of motivations, perceptions and attitudes of students toward m-learning followed by engineering (including computers), language, art and science. Wu, et al. (2012), found that studies of m-learning for educational purposes focused most frequently on applications in the professions and applied sciences, followed by humanities, formal sciences, social sciences and natural sciences. In terms of sub-disciplines, languages and linguistics courses were the most common focus, followed by computer science, health sciences, environmental studies and forestry, physics, business, and journalism/media studies/communication. Chen, & Chung (2008) found in their study that participants who installed a mobile English dictionary application on PDAs have increased learning performance and interest. So we can reach the conclusion that mobile learning contributes on performance by increasing the interest.

The research indicates that based on the distribution of articles according to their topical domain; learner outcomes rank as the most studied topical domain while instructional design comes as the second. The reason why the tendency in the previously conducted studies accumulates on learner outcomes and instructional design, could be because of the rise of learner-centered applications and constructivist approach. It is seen as a possible result that in order to enable learners to learn on their own without depending on a place, within the scope of life-long learning, more attention be paid to subjects such as the impact of m-learning on teaching designs and learner outputs such as performance outcomes, student satisfaction, engagement, effectiveness, motivation & effort, independence in learning, and retention rates. The least studied topical domain were found as demographic and other subjects. According to the yearly distribution values, the most frequently studied topical domains are learning outcomes, instructional design, technology and disposition. The collected findings indicate that in accordance with the other studies in the body of literature, the mostly studied topical domain is learner outcomes while disposition topical domain comes as the second. Therefore, it is possible to say that in the field of m-learning scientific studies about topical domains such as learners' performance outcomes and satisfaction, entrepreneurship, activities, motivation, effort, and independency in learning were conducted.

Despite this study suggesting mixed data collection tool, which contains two of the data collecting tools that are significant factors for researchers for conducting a study in accordance with its purpose, is the most frequently used option; the body of literature proposes that the most frequently used data collecting tool is survey (Kiliç-Çakmak et al., 2013; Alper & Gülbahar, 2009; Göktaş et al., 2012). This study differs from the other studies in terms of its most frequently preferred data collecting tool type. This may result from the idea that simplex studies will not be able to satisfy the current needs, in the direction of improving technology and changing needs. As the number of criteria used in the studies increased, more various data was needed for examination and hence, the data collection tools used in only one study might have varied. Besides, it was observed in this study that survey is the secondly most preferred data collecting tool type. From this perspective, it is possible that in literature survey is usually preferred due to the fact that it is a data collecting tool which enables researchers to collect more data about subjects that with a specific sampling but different qualities in a short period of time.

The percentages of the other data collecting tools including interview, data collection, advanced technologies, sampling, self-administrative questioning, measurement, use of administrative sources, visual methods, and observation are not close to one another and are used less frequently compared to others.

Although there are studies with mixed data analysis method, which is the most frequently used statistical data analysis method consisting of both qualitative and quantitative studies, in the literature there are studies which suggest that content analysis studies of different fields mostly use quantitative method. (Şimşek et al., 2008; Alper & Gülbahar, 2009; Göktaşet al., 2012; Kiliç-Çakmak et al., 2013; Soykan & Uzunboylu, 2015). This shows parallelism with data collection tools. The more diversified data collection tools became, the more extensive the data analysis methods became in the study. Furthermore, it is possible to assert that the research method used in the study had an impact on this result. Since the research method contained both qualitative and quantitative characteristics, the data collection tools were adjusted in such a way that they could serve both purposes. This situation also affected data analysis method and it was found out that mixed data analysis method was intensely used.

When the tendencies of mobile application development approach were examined, it was observed that native mobile application development is mostly preferred while web application is less preferred in comparison. In the research carried out by IBM (2012), they stated that while setting out the mobile strategies the first thing many firms come across with is choosing their application development method, which has parameter such as budget, project deadline, and objectives. According to Godwin-Jones (2011), the phenomenal success Apple gained in 2010 after releasing their iPad tablet computers developed many native applications in this platform. He also mentioned that native application is more convenient for developing applications for special cases. Even though this follows a fluctuating between the dates when the analysis was conducted, it still explains why the native application rate is the highest. In his study comparing native applications to web applications, Jobe (2013) concluded that native applications are still the best choice for hardware-based studies. Considering the applications employed in the studies mostly used camera and GPS hardware, the fact that native applications have the highest rate is parallel to it.

Suggestions

Based on the results gained from the research, it is possible to propose new perspectives to the subject by expanding the scope of the m-learning themed studies and by examining them through more than eight different criteria. The scope of the study can be expanded by evaluating them together with the learning domains within the field of education, for the researchers conducting studies in this field. Making use of technology in an educational context can contribute to the development of individuals by emphasizing the importance of the contribution the m-learning makes to life-long learning. This study will light the way for m-learning related studies in future, since it reveals the trends of the m-learning themed studies between 2010 and 2015. The study could play a determinative role in directing researchers to work towards the missing points in literature.

The scope of the study is limited to the international studies which exist in the database of Web of Science and SSCI index, which was used throughout the study. It is possible to reach more studies through different databases, which could be useful for the study to collect more data. This situation will increase the internationality level of trend studies. Similarly, searching by different criteria such as *abstract, keyword*, etc. could enable evaluating the results in a more far-reaching manner, so that it could be possible to reach more documents whose contents are relevant. A broader period could be examined as a trend analysis study. By doing so, the level of consistency can be more clearly determined by defining the direction of tendency in the m-learning themed studies. The results of this study are limited to the classifications and criteria employed. Outside these classification and criteria, no generalizations should be made.

A meta-analysis of trend studies could be conducted by evaluating the trend analysis researches related to the subject. So that, it will be possible to determine the accuracy level of the tendencies acquired by evaluating the results of trend analysis within itself once again. It was observed that case study method is the least frequently used research method when the research method were investigated. Using case study method, which offers a more detailed examination, more studies could be conducted and different perspectives could be achieved.

Including the other sampling levels other than higher and elementary education to the researches could improve the scope of m-learning themed studies and give it a new direction. Groups such as occupational clusters, people working in in-service training, professionals could be also among the sampling levels. By this way, the studies will appeal to larger masses and more general results will be achieved in the trend analysis studies.

It was observed that the examined studies show tendency towards the learning domain of humanities, however, the number of studies conducted in the learning domain of agricultural and engineering technology was far lower. There could be more studies conducted in this learning domain and the effect the m-learning has on these learning fields could be investigates and furthermore; opinions on whether these domains might be used or to what degree they might be used, could be presented.

Concentrating on data collecting tools such as interview, data collection, advanced technologies, sampling, self-administrative questioning, measurement, use of administrative sources, visual methods, and observation in future studies could be useful for prospering and bringing a new perspective to the m-learning themed studies in distance education.

While tackling mobile application development approaches, the mixed approaches, in which native, and web application or both of them were used, were evaluated. Taking the programming languages, which were used while writing the application and the accessed smart phone hardware such as camera, GPS etc. as the subject matters of the new trend analysis studies could give direction to new studies in the field.

In the wake of researches examined within the scope of this study, it is suggested that applications, which could be used in m-learning environments be increased; and more language options be added to the studies which will be developed in future. By this way, it will be possible to provide everyone with equal opportunity to have life-long learning without classroom borders.

As mentioned above, m-learning was considered to be an extension of e-learning, however, as technology improved, it has proved itself to be an independent term. Albeit, it will be inevitable that this term will be defeated by time and new terminology will dominate in literature. Considering the characteristics of trend concept, repeating different terms such as virtual or augmented reality or ubiquitous learning in similar future studies could be significant for providing the researchers with new perspectives.

REFERENCES

Ağca, R. K., & Özdemir, S. (2013). Foreign language vocabulary learning with mobile technologies. *Procedia: Social and Behavioral Sciences*, *83*, 781–785. doi:10.1016/j.sbspro.2013.06.147

Ali, R., & Irvine, V. (2009). Current m-learning research: A review of key literature. In T. Bastiaens et al. (Eds.), *Proceedings of World Conference on E-Learning in Corporate, Government, Healthcare, and Higher Education 2009* (pp. 2353-2359). Chesapeake, VA: AACE.

Ally, M., & Needham, G. (Eds.). (2010). *M-Libraries: A virtual library in everyone's pocket*. London: Facet Publishing.

Ally, M., & Prieto-Blázquez, J. (2014). What is the future of m-learning in education? RUSC. *Universities and Knowledge Society Journal, 11*(1), 142–151. doi:10.7238/rusc.v11i1.2033

Ally, M., & Tsinakos, A. (Eds.). (2014). *Increasing access through m-learning*. Vancouver, BC: Commonwealth of Learning Press.

Anglin, G. J., & Morrison, G. R. (2000). An analysis of distance education research: Implications for the instructional technologist. *Quarterly Review of Distance Education, 1*, 189–194.

Attewell, J. (2005). Mobile technologies and learning. London: Learning and Skills Development Agency.

Berelson, B. (1952). *Content Analysis in Communication Research*. Glencoe, IL: Free Press.

Berge, Z. L., & Mrozowski, S. (2001). Review of research in distance education, 1990–1999. *American Journal of Distance Education, 15*(3), 5–19. doi:10.1080/08923640109527090

Bozkaya, M., Aydin, I. E., & Kumtepe, E. G. (2012). Research Trends and Issues in Educational Technology: A Content Analysis of TOJET (2008-2011). *Turkish Online Journal of Educational Technology-TOJET, 11*(2), 264–277.

Brown, J. (2010). *Can you hear me?* Retrieved from http://www.judybrown.com/docs/76100228.pdf

Brown, J. S. (2005). *New Learning Environments for the 21st Century*. Retrieved from www.johnseelybrown. com/newlearning.pdf

Casey, S. (2009). *The New literacies of m-learning*. (Unpublished master's thesis). Concordia University, Canada (UMR:67156).

Chen, C., & Chung, C. (2008). Personalized mobile English vocabulary learning system based on item response theory and learning memory cycle. *Computers & Education, 51*(2), 624–645. doi:10.1016/j.compedu.2007.06.011

Chiou, C.-K., Tseng, J. C. R., Hwang, G.-J., & Heller, S. (2010). An adaptive navigation support system for conducting context-aware ubiquitous learning in museums. *Computers & Education, 55*(2), 834–845. doi:10.1016/j.compedu.2010.03.015

Chu, H. C., Hwang, G. J., & Tsai, C. C. (2010). A knowledge engineering approach to developing mind tools for context-aware ubiquitous learning. *Computers & Education, 54*(1), 289–297. doi:10.1016/j.compedu.2009.08.023

Chu, H. C., Hwang, G. J., Tsai, C. C., & Tseng, J. C. R. (2010). A two-tier test approach to developing location-aware mobile learning systems for natural science courses. *Computers & Education, 55*(4), 1618–1627. doi:10.1016/j.compedu.2010.07.004

Cobcroft, R. S., Towers, S. J., Smith, J. E., & Bruns, A. (2006). *M-learning in review: Opportunities and challenges for learners, teachers, and institutions*. Retrieved from August 30, 2015, from: http://eprints.qut.edu.au/5399/1/5399.pdf

Corbeil, J. R., & Valdés-Corbeil, M. E. (2007). Are you ready for m-learning? *EDUCAUSE Quarterly*, *30*(2). Retrieved from http://er.educause.edu/articles/2007/1/are-you-ready-for-mobile-learning AreYouReadyforMobileLearning/157455

Corlett, D., Sharples, M., Bull, S., & Chan, T. (2005). Evaluation of an m-learning organiser for university students. *Journal of Computer Assisted Learning*, *21*(3), 162–170. doi:10.1111/j.1365-2729.2005.00124.x

Demsey, L. (2008). Always on: libraries in a world of permanent connectivity. In G. Needham & M. Ally (Eds.), *M-libraries: Libraries on the move to provide virtual access* (pp. xxv–lii). London: Facet Publishing.

Dobbin, G., Dahlstrom, E., Arroway, P., & Sheehan, M. (2011). *Mobile IT in higher education, 2011*. Educause Center for Applied Research, Research Report.

Drysdale, J. S., Graham, C. R., Spring, K. J., & Halverson, L. R. (2013). An analysis of research trends in dissertations and theses studying blended learning. *The Internet and Higher Education*, *17*, 90–100. doi:10.1016/j.iheduc.2012.11.003

Georgiev, T., Georgieva, E., & Smrikarov, A. (2004, June). M-learning-a New Stage of E-Learning. In *International Conference on Computer Systems and Technologies-CompSysTech* (Vol. 4, No. 28, pp. 1-4).

Godwin-Jones, R. (2011). Emerging technologies: Mobile apps for language learning. *Language Learning & Technology*, *15*(2), 2–11.

Goh, T., & Kinshuk, D. (2006). Getting ready for m-learning—adaptation perspective. *Journal of Educational Multimedia and Hypermedia, 15*(2), 175-198.Hassan, M., Hamdan, Z., & Al-Sadi, J. (2012). A new m-learning content design process. *International Journal of Academic Research*, *4*(1), 23–28.

Göktaş, Y., Küçük, S., Aydemir, M., Telli, E., Arpacık, Ö., Yıldırım, G., & Reisoğlu, İ. (2012). Türkiye'de eğitim teknolojileri araştırmalarındaki eğilimler: 2000-2009 dönemi makalelerinin içerik analizi. *Kuram, &. Uygulamada Eğitim Bilimleri.*, *12*(1), 177–199.

Gromik, A. N. (2012). Cell phone video recording feature as a language learning tool: A case study. *Computers & Education*, *58*(1), 223–230. doi:10.1016/j.compedu.2011.06.013

Gülbahar, Y., & Alper, A. (2009). A Content Analysis of the Studies in Instructional Technologies Area. *Ankara University Journal of Faculty of Educational Sciences*, *42*(2), 93–111.

Han, I., & Han, S. (2014). Adoption of the mobile campus in a cyber university. *The International Review of Research in Open and Distributed Learning*, *15*(6).

Hawkridge, D. (1999). Thirty years on, BJET! and educational technology comes of age. *British Journal of Educational Technology*, *30*(4), 293–304. doi:10.1111/1467-8535.00119

Hung, J. L., & Zhang, K. (2012). Examining m-learning trends 2003–2008: A categorical meta-trend analysis using text mining techniques. *Journal of Computing in Higher Education*, *24*(1), 1–17. doi:10.1007/s12528-011-9044-9

Hwang, G. J., & Tsai, C. C. (2011). Research trends in mobile and ubiquitous learning: A review of publications in selected journals from 2001 to 2010. *British Journal of Educational Technology*, *42*(4), E65–E70. doi:10.1111/j.1467-8535.2011.01183.x

Hwang, G. J., & Wu, P. H. (2014). Applications, impacts and trends of mobile technology-enhanced learning: A review of 2008–2012 publications in selected SSCI journals. *International Journal of M-learning and Organisation*, *8*(2), 83–95. doi:10.1504/IJMLO.2014.062346

Jobe, W. (2013). Native Apps vs. Mobile Web Apps. *iJIM, 7*(4), 27-32.

Keegan, D. (2005, October). The incorporation of m-learning into mainstream education and training. In *World Conference on M-learning* (p. 11).

Kenny, R. F., Park, C. L., Van Neste-Kenny, J. M. C., Burton, P., Park, C. L., & Qayyum, A. (2012). Using Self-Efficacy to Assess the Readiness of Nursing Educators and Students for M-learning. *International Review of Research in Open and Distance Learning*, *13*(3), 277–296.

Kiliç-Çakmak, E., Çebi, A., Mihçi, P., Günbatar, M. S., & Akçayir, M. (2013). A Content Analysis of Educational Technology Research in 2011. *Procedia: Social and Behavioral Sciences*, *106*, 74–83. doi:10.1016/j.sbspro.2013.12.010

Klein, J. D. (2002). Empirical research on performance improvement. *Performance Improvement Quarterly*, *15*(1), 99–110. doi:10.1111/j.1937-8327.2002.tb00243.x

Koble, M., & Bunker, E. (1997). Trends in research and practice: An examination of The American Journal of Distance Education 1987 to 1995. *American Journal of Distance Education*, *11*(2), 19–38. doi:10.1080/08923649709526959

Koszalka, T. A., & Ntloedibe-Kuswani, G. S. (2010). Literature on the safe and disruptive learning potential of mobile technologies. *Distance Education*, *31*(2), 139–157. doi:10.1080/01587919.2010.498082

Krippendorf, K. (2004b). Reliability in content analysis some common misconceptions and recommendations. *Human Communication Research*, *30*(3), 411–433.

Krippendorff, K. (2004). *Content analysis: An introduction to its methodology* (2nd ed.). Thousand Oaks, CA: Sage.

Krippendorff, K. (2011). *Computing Krippendorff's Alpha-Reliability*. Retrieved January 25, 2015 from http://repository.upenn.edu/asc_papers/43

Küçük, S., Aydemir, M., Yıldırım, G., Arpacık, O., & Göktaş, Y. (2013). Educational technology research trends in Turkey from 1990 to 2011. *Computers & Education*, *68*, 42–50. doi:10.1016/j.compedu.2013.04.016

Kukulska-Hulme, A. (2005). *M-learning: A handbook for educators and trainers*. Psychology Press.

Kukulska-Hulme, A., & Traxler, J. (2007). Learning design with mobile and wireless technologies. In H. Beetham & R. Sharpe (Eds.), *Rethinking pedagogy for the digital age: Designing and delivering e-learning* (pp. 180–192). London, UK: Routledge.

Latchem, C. (2006). Editorial: A content analysis of the British Journal of Educational Technology. *British Journal of Educational Technology, 37*(4), 503–511. doi:10.1111/j.1467-8535.2006.00635.x

Lee, Y., Driscoll, M. P., & Nelson, D. W. (2004). The Past, Present, and Future of Research in Distance Education: Results of a Content Analysis. *American Journal of Distance Education, 18*(4), 225–241. doi:10.1207/s15389286ajde1804_4

Liu, G. Z., & Hwang, G. J. (2010). A key step to understanding paradigm shifts in e-learning: Towards context-aware ubiquitous learning. *British Journal of Educational Technology, 41*(2), E1–E9. doi:10.1111/j.1467-8535.2009.00976.x

Martin, S., Diaz, G., Sancristobal, E., Gil, R., Castro, M., & Peire, J. (2011). New technology trends in education: Seven years of forecasts and convergence. *Computers & Education, 57*(3), 1893–1906. doi:10.1016/j.compedu.2011.04.003

Mobilearn. (2002). *Mobilearn projects website.* Retrieved from http://www.mobilearn.org/index.php

Motiwalla, L. F. (2007). M-learning: A framework and evaluation. *Computers & Education, 49*(3), 581–596. doi:10.1016/j.compedu.2005.10.011

Mutlu, M. E., Yenigün, H. U., & Uslu, N. (2006). Açıköğretimde Mobil Öğrenme: Açıköğretim E-Öğrenme Hizmetlerinden Mobil Bilişim Aygıtlarıyla Yararlanma Olanaklarının Değerlendirilmesi. *Bilgi Teknolojileri IV & Akademik Bilişim*, 9-11.

Naismith, L., Lonsdale, P., Vavoula, G., & Sharples, M. (2006). *Literature Review in Mobile Technologies and Learning. Report 11*. Futurelab.

Native, web or hybrid mobile-app development. (2012). Retrieved June 3, 2015, from ftp://public.dhe.ibm.com/software/pdf/mobile-enterprise/WSW14182USEN.pdf

Neuendorf, K. A. (2002). *The content analysis guidebook.* Thousand Oaks, CA: Sage Publications.

OECD. (2007). *Revised Field Of Science and Technology (Fos) Classification In The Frascati Manual.* Retrieved January 15, 2015, fromhttp://www.oecd.org/sti/inno/38235147.pdf

Parsons, D. (2007). M-learning. In D. Taniar (Ed.), Encyclopedia of Mobile Computing and Commerce, (pp. 525-527). Hershey, PA: Information Science Reference.

Prensky, M. (2001). Digital natives, digital Immigrants Part 1. *On the Horizon, 9*(5), 1–6. doi:10.1108/10748120110424816

Quinn, C. (2000). mLearning: mobile, wireless and in-your-pocket learning. *Line Zine Magazine.* Retrieved from http://www.linezine.com/2.1/features/cqmmwiyp.htm

Quinn, C. (2013). A future for M-Learning. In Z. L. Berge & L. Y. Muilenburg (Eds.), *Handbook of m-learning*. New York, NY: Routledge.

Reynolds, R., Walker, K., & Speight, C. (2010). Web-based museum trails on PDAs for university-level design students: Design and evaluation. *Computers & Education*, *55*(3), 994–1003. doi:10.1016/j.compedu.2010.04.010

Roberts, J. B. (2013). Accessibility in M-Learning: Ensuring equal access. In Z. L. Berge & L. Y. Muilenburg (Eds.), *Handbook of m-learning*. New York, NY: Routledge.

Sandberg, J., Maris, M., & Hoogendoorn, P. (2014). The added value of a gaming context and intelligent adaptation for an m-learning application for vocabulary learning. *Computers & Education*, *76*, 119–130. doi:10.1016/j.compedu.2014.03.006

Sert, G. (2010). *Öğretim teknolojileri eğitiminde yayınlanmış Türkiye adresli makalelerin içerik analizi.* (Unpublished master's thesis). Hacettepe Üniversitesi, Ankara, Türkiye.

Sharples, M. (2000). The design of personal mobile technologies for lifelong learning. *Computers & Education*, *34*(3-4), 177–193. doi:10.1016/S0360-1315(99)00044-5

Sharples, M. (2006). *Big issues in m-learning. Report of a workshop by the Kaleidoscope Network of Excellence M-learning Initiative*. Academic Press.

Sharples, M., Arnedillo-Sánchez, I., Milrad, M., & Vavoula, G. (2009). *M-learning* (pp. 233–249). Springer Netherlands.

Sharples, M., Corlett, D., & Westmancott, O. (2002). The design and implementation of an m-learning resource. *Personal and Ubiquitous Computing*, *6*(3), 220–234. doi:10.1007/s007790200021

Şimşek, A., Özdamar, N., Becit, G., Kılıçer, K., Akbulut, Y., & Yıldırım, Y. (2008). Türkiye'deki eğitim teknolojisi araştırmalarında güncel eğilimler. *Selçuk Üniversitesi Sosyal Bilimler Dergisi*, *19*, 439–458.

Soykan, E., & Uzunboylu, H. (2015). New trends on m-learning area: The review of published articles on m-learning in science direct database. *World Journal on Educational Technology*, *7*(1), 31–41. doi:10.18844/wjet.v7i1.22

Stockwell, G. (2010). Using mobile phones for vocabulary activities: Examining the effect of the platform. *Language Learning & Technology*, *14*(2), 95–110.

Sue, B., Maton, K., & Kervin, L. (2008). The "digital natives" debate: A critical review of the evidence. *British Journal of Educational Technology*, *39*(5), 775–786. doi:10.1111/j.1467-8535.2007.00793.x

Traxler, J. (2005). *Defining M-learning. IADIS International Conference M-learning*. Qawra, Malta

Traxler, J. (2010). Distance education and m-learning: Catching up, taking stock. *Distance Education*, *31*(2), 129–138. doi:10.1080/01587919.2010.503362

Trinder, J. J. (2005) Mobile technologies and systems. In M-learning: A Handbook for Educators and Trainers (pp. 7-24). London: Routledge.

Tsinakos, A., & Ally, M. (2013). *Global m-learning implementation and trends*. Beijing, China: CRTVU Press.

U.S. General Accounting Office. (1996). *Content Analysis: A Methodology for Structuring and Analyzing Written Material.* GAO/PEMD-10.3.1. Retrieved January 10, 2015, http://www.gao.gov/assets/80/76281.pdf

Vavoula, G., & Sharples, M. (2009). Meeting the Challenges in Evaluating M-learning: A 3-Level Evaluation Framework. *International Journal of Mobile and Blended Learning*, *1*(2), 54–75. doi:10.4018/jmbl.2009040104

Visser, L., & Gagnon, K. (2005). Defining "Trends" and "Issues" in Distance Education. Conversation with Donald Paul Ely. In Trends and Issues in Distance Education. International Perspectives.

Wagoner, T., Hoover, S., & Ernst, D. (2011). *CEHD iPad initiative*. Retrieved from August 30, 2015, from: http://www.cehd.umn.edu/Mobile/PDF/CEHDiPadReport.pdf

Wang, Y. S., Wu, M. C., & Wang, H. Y. (2009). Investigating the determinants and age and gender differences in the acceptance of m-learning. *British Journal of Educational Technology*, *40*(1), 92–118. doi:10.1111/j.1467-8535.2007.00809.x

Weber, R. P. (1990). *Basic Content Analysis* (2nd ed.). Newbury Park, CA.

Wexler, S., Brown, J., Metcalf, D., Rogers, D., & Wagner, E. (2008). *The e-learning Guild Report M-learning*. Retrieved from August 30, 2015, from: http://www.elearningguild.com/research/archives/index.cfm?id=132&action=viewonly

Winter, N. (2006). *What is m-learning. Big issues in m-learning*. Report of a workshop by the Kaleidoscope Network of Excellence M-learning Initiative.

Woodill, G. (2011). The M-learning Edge. McGraw Hill.

Wu, W. H., Jim Wu, Y. C., Chen, C. Y., Kao, H. Y., Lin, C. H., & Huang, S. H. (2012). Review of trends from m-learning studies: A meta-analysis. *Computers & Education*, *59*(2), 817–827. doi:10.1016/j.compedu.2012.03.016

Wu, W. H., Wu, Y. C. J., Chen, C. Y., Kao, H. Y., Lin, C. H., & Huang, S. H. (2012). Review of trends from m-learning studies: A meta-analysis. *Computers & Education*, *59*(2), 817–827. doi:10.1016/j.compedu.2012.03.016

APPENDIX

Table 11. The research's criteria, sub categories and contents of the categories

Analysis Criteria	Reference Work	Classification of Sub-Categories	Explanation of Categories
Research Method	Lee, Driscoll and Nelson (2004) (Lee and etc. (2004), with the aim of creating a research methods' classification, used the category classifications which were created by Berge and Mrozowski (2001), Koble and Bunker (1997), Anglin and Morrison (2000) and Klein (2002).)	• Theoretical Inquiry	• A theoretical review of literature and conceptual study for proposing new ideas in distance education
		• Experimental Research	• A study examining the effect of independent variable(s) on dependent variable(s)
		• Case study	• A study aimed at investigating a single individual, group, program, organization, qualitatively
		• Evaluation Research	• A study aimed at determining the impact of project, program, model or software
		• Developmental Research	• A study aimed at designing, developing and evaluating an existing or newly developed model, process, product or technique
		• Survey Research	• A study addressing the distribution and return of responses in a none experimental situation
		• Combination of Inquiries	• A study synthesizing two or more research methods
		• Not Applicable*	• A different method that is out of the classification
Sampling Size	Goktas et al., (2012) .	• 1-10	
		• 11-30	
		• 31-100	
		• 101-300	
		• 301-1000	
		• >1000	
		• Not Applicable*	• A different sample level that is out of the classification
		• Not Found*	• There is no sample or researchers did not explain the sample.
Sampling Level	Drysdale, Graham, Spring and Halverson (2013)	• K-12	• Participants are K-12 students (Elementary, middle, and high schools)
		• Higher Education	• Participants are higher education students (University, community college, and graduate studies.)
		• Corporate	• Participants are professionals (Businesses and military)
		• Not Applicable*	• A different sample level that is out of the classification
		• Not Found*	• There is no sample or researchers did not explain the sample.

continued on following page

Table 11. Continued

Analysis Criteria	Reference Work	Classification of Sub-Categories	Explanation of Categories
Learning Domain	Revised Field Of Science And Technology (Fos) Classification In The Frascati Manual	• Natural Sciences	• Mathematics • Computer and information sciences • Physical sciences • Chemical sciences • Earth and related environmental sciences • Biological sciences • Other natural sciences

		• Engineering& Technology	• Engineering
			  • Civil
			  • Electrical, electronic, information
			  • Mechanical
			  • Chemical
			  • Materials
			  • Medical
			  • Environmental
			• Environmental biotechnology
			• Industrial Biotechnology
			• Nano-technology
			• Other engineering and technologies

		• Medical& Health	• Basic medicine • Clinical medicine • Health sciences • Health biotechnology • Other medical sciences
		• Agricultural Sciences	• Agriculture, forestry, and fisheries • Animal and dairy science • Veterinary science • Agricultural biotechnology • Other agricultural sciences
		• Social Sciences	• Psychology • Economics and business • Educational sciences • Sociology • Law • Political Science • Social and economic geography • Media and communications • Other social sciences
		• Humanities	• History and archaeology • Languages and literature • Philosophy, ethics and religion • Art (arts, history of arts, performing arts, music) • Other humanities
		• Not Applicable*	• A different learning domain that is out of the classification
		• Not Found*	• Researchers did not explain the learning domain

continued on following page

Table 11. Continued

Analysis Criteria	Reference Work	Classification of Sub-Categories	Explanation of Categories
Topical Domain	Drysdale, Graham, Spring & Halverson (2013)	• Learner Outcomes	• Performance outcomes, student satisfaction, engagement, effectiveness, motivation and effort, independence in learning and retention rates.
		• Disposition	• Perceptions, attitudes, preferences, student expectations, and learning styles.
		• Instructional Design	• Models, strategies and best practices; design process; implementation; environment and course structure; and evaluation tools.
		• Interaction	• Student-to-instructor, general interaction, student-to-student, collaboration, community and social presence.
		• Comparison	• Blended &f ace-to-face and blended & online.
		• Demographics	• Student and faculty.
		• Technology	• Use& role, effect, type, implementation and familiarity.
		• Professional Development	• Professional development
		• Other	• Benefits & challenges, access & availability, support system, time efficiency, nature& role of blended learning, and international issues.
Data Collection Tools	Beissel &Durant (2004)	• Data Collection	• General
		• Sampling	• Sampling and survey designs, types of sampling (cluster sampling, multistage sampling etc.)
		• Self Administrative Questioning	• Question design, postal survey, mail survey, electronic mail survey, web-based survey, opinion polls,
		• Interview	• Question design, qualitative and quantitative, telephone, face-to face, Focus groups/ group interview, computerized, standardized and unstandardised, practice of interviewing, interviewers, procedure of interviewing, interviewer training, respondents, recording responses
		• Questionnaire	• Question design, self-administered questionnaire, mail questionnaire, questionnaire design, types of questions, question wording, structure of questionnaire, pre-test of questionnaire postal questionnaire, web-based questionnaire
		• Observation	• Field observation, field experimentation, participant observation, laboratory observation
		• Measurement	• Measurement of attitudes, behavior, ability etc.
		• Use of Administrative Sources	
		• Visual methods	
		• Advanced Technologies	• Computer-assisted data collection, grid technology, audio-and video, data mining, e- social science approaches to data collection
		• Mixed*	• Combinations of two or more data collection tools.
		• Not Applicable*	• Data collection tool is out of classification or not use full for the research method
		• Not Found*	• There is no data collection tool or researchers did not explain it

continued on following page

Table 11. Continued

Analysis Criteria	Reference Work	Classification of Sub-Categories	Explanation of Categories
Data Analysis Methods	Beissel&Durant (2004)	• Qualitative Approaches	• Qualitative Approaches (general) • Discourse Analysis • Interaction Analysis • Conversation Analysis • Content Analysis • Narrative Methods • Analysis of Composite Data • Corpus Analysis • Documentary Analysis • Biographical Methods/ Oral History • Grounded Theory • Ethnography • Phenomenology • Visual Methods • Thematic Analysis
		• Quantitative Approaches	• Quantitative Approaches (general) • Statistical Theory and Methods of Inference • Survey Data Analysis and Estimation • Micro data Methods • Regression Analysis • Multi level Modeling • Longitudinal Data Analysis • Event History Analysis • Spatial Data Analysis • Latent Variable Models • Structural Equation Models • Multivariate Analysis • Time Series Analysis • Data Mining • Non-Parametric Approaches
		• Mixed Approaches	• Mixed Methods Approaches (general) • Social Network Analysis • Combining Qualitative and Quantitative Approaches
		• Not Applicable*	• Data analysis method is out of classification or not use full for the research method
		• Not Found*	• There is no data analysis method or researchers did not explain it
Development Approach for a Mobile Application	IBM (2012)	• Web Application	• Browser-based applications
		• Native Application	• Native applications have binary executable files that are downloaded directly to the device and stored locally
		• Mixed Approaches*	• Using both of web and native applications
		• Not Applicable*	• Mobile technology type is out of classification
		• Not Found*	• There is no mobile technology type (if the research method is theoretical inquiry, usually there is no mobile technology type.)

* The categories, which are added by the researchers to the classification criteria.

Chapter 14
Mobile Learning in Science and Mathematics Teaching:
A Systematic Review

Rosiney Rocha Almeida
Universidade Cruzeiro do Sul, Brazil

Carlos Fernando Araújo Jr.
Universidade Cruzeiro do Sul, Brazil

ABSTRACT

This chapter, the authors analyze mobile learning literature addressing formal teaching situations in the field of science/mathematics education. The chapter describes a systematic review of relevant literature, investigating work published between 2005 and 2014. Based on the findings of this review, the results reveal that since 2009 interest in the academic community for research involving m-learning in science and mathematics teaching has intensified. An emphasis on the growing need for research involving m-learning at undergraduate levels is evident. The study notes positive reported results on the impact of m-learning in science and mathematics teaching, notably in assisting ease of learning, promoting student interest and collaborating with positive, motivating attitudes, encouraging meaningful commitment among students in learning activities.

INTRODUCTION

M-learning, a teaching modality which makes use of innovative technological tools, including hardware and software platforms, arose from the expansion of new possibilities for distance learning (Tarouco, Fabre, Konrath & Grando, 2004), as well as online teaching (e-learning.) M-learning also became possible due to the rise of interest and usage of the internet and widespread dissemination of information and communication technologies (ICTs) in computer networks. Hence, research in m-learning now seeks to understand how students' academic mobility, favored by advances in private and public technologies, can contribute to the process of knowledge acquisition, skills and experience in education (Sharples et al, 2014.) Faced with a still controversial and limiting context surrounding the definition of m-learning,

DOI: 10.4018/978-1-5225-0359-0.ch014

Parsons (2014) lists some myths and misconceptions surrounding this definition. For this study, the authors appropriated some of the m-learning characteristics identified and enumerated by Marçal et al (2005):

"...Provide access to educational content anywhere and at any time, according to web device connectivity; allow learners to personally shape their own knowledge; expand the internal and external boundaries of the classroom in a ubiquitous manner; provide means for the development of innovative teaching methods using contemporary computing dynamics and resources." (Marçal, Andrade & Rios, 2005, p.3.)

Mobile devices have led to changes in numerous segments of society. Embedded in our daily life, they reshape the way we live; communicate, relate, work and play. The spectrum of possibilities provided by mobile technologies, including portability, free access to information, flexibility and exchange of information, among others, lead us to question how society appropriates/incorporates these new features and how this affects socio-political, economic and, especially, learning spheres.

We must investigate specifically how these modern technological resources are inserted into teaching and learning subject study contents and specifically those related to Science and Mathematics. Studies like that of Chen (2009) point out that the teaching of Science and Mathematics is a continuous challenge to be overcome for many countries adopting many academic levels.

Drew (2011) also points to the lack of interest demonstrated by university students in scientific and technological careers, even in those who did well in math and sciences during High School. In this context, this chapter seeks to examine how mobile devices are used in formal teaching situations in the field of tutoring science and mathematics.

MOBILE TECHNOLOGIESS AND THE TEACHING-LEARNING PROCESS

Cook (2009) presents three stages of research on mobile learning, the first of which focuses on the devices involved; with widespread experimentation in PDAs, tablets, laptops and mobile phones, accentuating the advantages and disadvantages of using these devices in studies. The second stage concentrates on learning proposals *outside* of the classroom, including field trips, visits to museums, etc. The third stage of the research focuses on student mobility. This chapter focuses on student mobility, design/appropriation of learning spaces and lifelong learning experience.

Traxler (2009), while acknowledging progress thus far achieved, admits that this area still has far to go in technological as well as pedagogical terms. Mobile devices are mainly characterized by their mobility (wireless) and portability (small, lightweight devices) enabling the issuance, circulation and exchange of information during real time user movement.

The use of innovative strategies and tools assisting classroom or distance learning based on the expansion and accessibility of the internet enables new possibilities in the teaching and learning process. The educational process itself has also evolved with the advent of this rapidly advancing technology and its social, economic and cultural ramifications, since it constitutes an important link in inter-communication process mechanisms. In reference to this contemporary outline, technological developments in the expansion of online communication possibilities - now substantially mobile; anytime, anywhere and anyhow – enable learning through multiple dynamic interactive features in digital communication, bringing attention to the need to reformat different models and practices for educational purposes (Fedoce, 2010.)

As Metcalf (2006) suggested, communication functionality and increased data transfer speed favors the instructional value of mobile devices, thereby increasing learning opportunities. Pachler et al (2009) adds that this m-learning capacity and the everyday functionality of these technologies with the convergence of services and functions is a still recent advancement in this field. As a tool, it has been expanded, but at the same time reduced to a simple, single feature, a screen of omnipresent portable multifunctionality.

Mobile media presents very distinct characteristics, which require the accompaniment of content/ practices that take their technical, communicative potential and particularities into account. It becomes evident then, that numerous possibilities, such as educational games, augmented reality applications, electronic questionnaires, audiovisual recording of phenomena, collaborative content production etc. can all be incorporated into the teaching/ learning process. Mobile learning depends on the context, as well as the availability of cultural artifacts and researched materials and must prioritize real-time student activities.

Therefore, it is necessary to encourage activities that propose the search of knowledge in collaborative contexts and provide live interaction among participants while respecting student social context (Valentim, 2009). On the possibilities of using mobile devices in education, researchers recognize the need to create or adapt appropriate teaching practices for this new environment in which education and technology are interconnected. Graziola (2009) reaffirms some requirements for using this technology in education:

...differentiated pedagogical practices for a generation that behaves, thinks and learns differently; to create environments that allow students to continue learning when outside the institution/place of their formal education and/or in the context of mobility; open, flexible planning, which fundamentally respects the context of those involved in this process; creation of areas of reflection, namely through students' internalized knowledge; activities that valorize the particular sensations, subjectivities, impressions, desires and tastes of subjects involved in the educational process while checking to assure that didactic pedagogical issues in relation to technological issues prevail. (Graziola, 2009, pp. 9-10)

UNESCO (United Nations Educational, Scientific and Cultural Organization) launched a working paper series on mobile learning which seeks to improve the understanding of how mobile technologies can be used to improve access, equity and quality of worldwide education. These documents highlight the characteristics shared by successful initiatives in mobile learning and identify public policies favorable to its implementation, discussing likely impacts of mobile technology in future education. In the document "The Future of Mobile Learning", researchers argue that:

to the extent that it makes the compilation of information and acquisition of knowledge through mobile devices easier, analytical mobile learning can be used to boost the development of theories dealing with how people learn. (UNESCO, 2014 p. 50)

In addition, Reinhard et al (2007) explains that:

Mobile Learning is not a new idea - the possibility of learning anywhere and at any time has always been sought and enhanced with whatever available tools found at hand, such as books, notebooks and other mobile devices. (Reinhard, Saccol, Schlemmer, Barbosa & Kristoffersen, 2007, p.1.)

fValentim (2009) reaffirms that e-learning is a recently introduced learning model, and that one must keep in mind that it is solely an evolution of previous forms of learning. That being said, the adoption of technology adapted to the educational process does not indicate a distancing from classic educational methodologies. For Guerreiro (2013), the introduction of technology into the classroom has mobilized sweeping changes in how we teach and how we learn.

Educational Benefits of M-Learning

Adriani (2008) states that it is essential to understand the palpable educational benefits obtained with the use of modern teaching technologies. Portability and mobility are a few of the advantages identified in these devices by authors like Marçal et al (2005), such that these devices make access to educational content possible anywhere and at any time. Another asset of these devices, besides their physical mobility, is the temporal and contextual mobility (Schlemmer et al, 2007) innate to this technology and which can further the educational interests of all parties involved in the exchange of information; the party who accesses the information, as well as the information source. Both sides benefit with the possibility of updating accessed information in the context of the space and time in which it is accessed.

Parsons (2014) adds the possibility of ownership and streamlining of learning, conforming to the desire and subjectivity of students as one of the m-learning benefits. Recurring benefits should also be considered, indicated by various authors regarding the choice of these devices in education, mainly accessibility, collaboration and flexibility (Khaddage & Lattemann, 2009); the ability that youngsters demonstrate in working with this technology, the popularization of these benefits (Batista, 2011) and the prospect of practicing teamwork (Naismith, Lonsdale, Vavoula & Sharples, 2004; Keegan, 2007; Ally, 2009; Sharples et al, 2014).

In the opinion of Alexander (2004), the application of mobile technologies in education transforms it through the promotion of universal learning, enabling live intercommunication among teachers and students both inside and outside the classroom,

Klopfer et al (2002) identified five characteristics of mobile technologies that demonstrate unique potential for educational objectives: portability, interactivity, contextual sensitivity, connectivity and individuality.

Attewell (2008) points to numerous benefits related to m-learning which can help students, among which we highlight: a) improvement of literacy and calculation skills; b) teacher's awareness and student auto-recognition of their skills; c) the evolution and developing of individual and collaborative experiences; d) increasing self-esteem and self-confidence.

Among the many aforementioned advantages inherent to the use of mobile devices in education, interactivity deserves the greatest emphasis. To Adriani (2008) the characteristic of interactivity provided by these devices is a magnetically enticing and distracting driving agent, especially mesmerizing for children and adolescents. This interactivity needs to be studied, understood and interpreted, studied through the critical lens of theoretical references and teaching methodologies, in which the primary motivation of mobile learning use is not reduced exclusively to the use of the technology itself.

Challenges and Requirements of M-Learning

Despite the advantages presented, this fresh educational approach to education still faces difficulties and challenges in its widespread application.

To Traxler (2009), the expanding access to information at any time and place challenges the role of education, and in particular, formal education. The relationship between education, society and technology is increasingly dynamic. Authors such as Resnick (2002) and Vavoula & Sharples (2002) argue that this new generation of computer users needs innovative, modern learning methods, since they were born after digital technologies were readily available.

Some of the main recurring challenges to m-learning, such as the development and experience of effective teaching practices, have been discussed by researchers such as Naismith *et al* (2004); Chisholm (2005); Graziola (2009), Moura (2010) and Moran (2013). The understanding of learning processes using this technology, the endorsement of appropriate pedagogical platforms for its use and the acceptance/ utilization of postures in which its primary purpose is learning and not just using this technology as a source of entertainment have aroused the interest of researchers such as Roschelle (2003); Chisholm (2005), Adriani (2008); Franco (2010), Schelemmer (2007) and Batista (2011).

Fedoce and Squirra (2011) contribute to this debate pointing to the prerequisite time and commitment necessary for developing initiatives in mobile technology and in designing/projecting new methods of learning.

In summary, Naismith et al (2004, p.36) had predicted this when stating that, "whether they are welcome or not, the truth is that mobile technologies are going to enter classrooms in the pockets of children and young people." To this author, the challenge that arises for schools and educators is to include these technologies productively in their educational practices.

Roschelle (2003) argues that one of the challenges is the pedagogical application of these technologies, which are generally based on a complex view of technology and a simplistic view of social practices. Consequently, more research is imperative to discover richer pedagogical practices to adopt, arising from simple mobile technology. Chisholm (2005) concludes that more important and more complex than the technology itself are the pedagogical strategies and teaching methods developed and used with and for mobile technology and that these should be given the most attention. McInnes (2009) adds that the use of these devices will only translate into relevant teaching practices if they are properly designed and have a robust educational component. Parsons, Ryu & Cranshaw (2007) identify a set of requirements, principles and pertinent factors that should be considered in performing m-learning activities.

UNESCO (2013) states:

Lack of planning has led to a number of well-documented cases of failure in this approach to teaching that negatively affect the perception and interpretation of mobile learning among educators and policymakers. Such missteps represent a very evident danger to the future of mobile learning. This teaching modality will hardly flourish in formal settings if teachers do not adopt and adhere to the technology and pedagogy that supports it. (UNESCO, 2013, p. 47)

In this sense, the thoughts of Abreu (2012) are noteworthy, according to the author:

It is a difficult and laborious journey, but promises great satisfaction to those who learn to walk this path, being an avant-garde distinction in the art of mobile technology use in teaching / learning, along with the accompanying constructive and creative possibilities accompanying the consolidation of knowledge. (Abreu, 2012, p.84)

There is controversy as to the contribution of these resources for teaching and learning. Some authors enthusiastically support their educational potential while others summarily dismiss it as an unimportant and distracting novelty (Tremblay, 2010; Herrington et al, 2009; Lim et al, 2011). These authors warn of the risk of students making the mistake of neglecting the principal objective of e-learning and succumbing to the compelling superficial wonder of mobile technology, to their detriment.

There is, of course, academic activity that does not require the use of mobile technology for educational purposes in any way, whatsoever and even prohibits its use to ensure discipline (UNESCO, 2013). In view of this controversy relating to mobile phone use in the classroom, it is important at this point to bring to attention the rationality of Fedoce (2011) when he argues that these devices should not be banished from the school teaching / learning environment, but rather be a contributing factor, integrated into the student learning process. Nevertheless, Fonseca (2013) alerts us to the fact that technology can be inserted into the educational environment more as a 'marketing strategy' than as a learning strategy, and in this sense, may suppress and / or mask learning problems. For this reason it is necessary that schools adapt to this new social context and new forms of communication and education available through social networks and mobile technology (Göttsche, 2012), following and absorbing the changes in educational paradigms (Adriani, 2008; Fedoce & Squirra, 2011).

Another challenge to be met and overcome in m-learning is to ensure the veracity / authenticity of the results obtained in using these technologies, given that mobile devices provide clever new ways of illicitly obtaining answers to teachers' questions. This is in addition to the necessity of understanding the role of teachers and students in the processes of design, development and implementation of these new technologies (Adriani, 2008). Göttsche (2012) reiterates the frequent lack of a functional capacity of auto-criticism in children and the young on the use of information derived from the use of these technologies. Boulos et al (2006) highlight the disadvantage of the impossibility of guaranteeing the quality of information researched through these devices due to the absence of supervision in confirming the accuracy of information published freely on the web. Nevertheless, they speculate that these limitations are not unique to this area nor confined to information disseminated by mobile devices.

In addition to this, the digital environment calls attention to other concern. For example, how to keep students' attention in class? How to prevent the unmonitored exchange of messages and jokes among students in teaching situations? How to control plagiarism and the infamous *copy and paste* strategy? How to keep students' minds on the subject at hand and keep them from getting distracted by the many interesting things that students may accidently come upon while accessing subject material during class? Moreover, what happens if students investigate a class research topic on the net and formulate questions that professors cannot answer? Or even worse, correct the professor during class? (Ledesma, 2013). Educators must also deal with the technical limitations of mobile devices, such as mobile phone screen size and keyboards, besides difficulties arising from bad internet connections (Georgiev, Georgieva & Smrikarov, 2004; Koole, 2006; Keegan, 2007 and Ally, 2009), though these limitations may soon disappear as technology improves (Moura, 2010) and be absorbed by this new on-line generation (Fedoce & Squirra, 2011).

Moura (2010) also states that compared to computer resources, the use of mobile devices in education requires the critical analysis of the functionality, potential and limitations of these machines, weighing pros and cons and developing appropriate teaching methods for their best use.

SYSTEMATIC LITERATURE REVIEW

To present a working, state of the art environment regarding research on m-learning focused on science and mathematics teaching, a Systematic Literature Review was undertaken.

Review of Existing Literature on Mobile Learning

The first review in m-learning was conducted by Trifonova (2003), which discussed key points and basic essentials in the early era of m-learning. Its main focus was to understand how mobile devices could offer a better education experience. Based on a review of existing literature in papers published in international conference proceedings and journals in the 2002-2007 period, Petrova and Li (2009) proposed a classification framework for research in mobile learning and identified issues and trends in m-learning and similarly focused research in order to better understand the relationship when making a choice of an educational theory for mobile learning.

Fronberg et al (2009) offer a critical analysis of mobile learning projects published until the end of 2007. In this review, the authors used a mobile learning framework for assessing and categorizing 102 mobile learning projects using this criteria: Context, Tools Control, Communication, Theme and Goal. Wingkvist and Ericsson (2011) present a research study published in mobile learning. The authors investigated 114 articles from mLearn conferences in 2005, 2007 and 2008 and classified them according to two dimensions: research method and purpose of research.

Pereira and Rodrigues (2013) conducted an extensive review of literature on m-learning and presented a critical analysis of m-learning projects and related literature, with the aim of clarifying the question "What is possible in using recent m-learning technology?" Grund and Gil (2014) present research on the situation of Mobile Learning in Spain from a descriptive evaluation study of national scientific production from early 2009 to the end of 2013. In a later study, similar to that of Grund and Gil (2014), Hwang and Wu (2014) conducted a survey with a review of the publications in the period from 2008 to 2012 in seven well-recognized Assisted Learning Technology magazines. Despite this range of studies, no investigation or identification of work focusing on mobile-oriented learning activities in the teaching of science and mathematics was reported.

Figure 1. Stages of the Literature Review

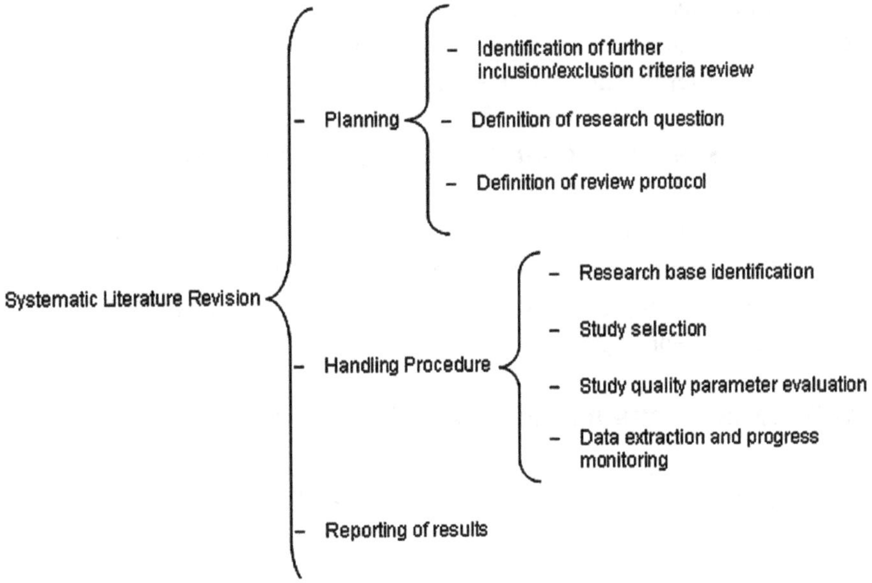

For this review, the authors' work was based on research by Kitchenham (2004), who conducted systematic reviews in medicine, and Barcelos (2014), who conducted a systematic review of literature on the relationship between computational thought and mathematics. Therefore, the literature review work was executed in three steps, namely: planning, handling and reporting of results, as shown in Figure 1.

Kitchenham (2004) explains that though the steps indicated above may seem sequential, it is important to recognize that several involve interaction with one another, as may be the case in the 'protocol setting definition' and 'procedure / handling of the review'. The review therefore proceeded observing this orientation.

Planning

Planning for the review was prompted by the observation and identification of the need for a review of the literature found on the topic, being that m-learning is a relatively recent field of study. Next, the following research question to be answered by the systematic literature review was defined:

How has mobile learning been used and evidenced in teaching science and mathematics in the literature?

The question to be answered through this literature review aims to include studies that address the use of mobile devices in formal science and mathematics teaching situations; the prioritized educational grade levels in these surveys, the subjects investigated and the main impacts resulting from this activity.

Finally, a 'review protocol' was defined, which is the last step in the planning and defines how the systematic review will be conducted. This protocol consists of the following steps:

Research Base Consultation Selection

For the purposes of this study, database research was undertaken (3 databases, shown below), which provided access to a large number of resources.

- SCIELO - Scientific Electronic Library Online.
- ACM Digital Library (DL) - Association of Computing Machinery.
- ERIC - Education Center of Information Resources.

Definition of String Search Research in Consulted Bases

String Search defines the consultation of research within consulted databanks. Since this is a recent research topic, the authors sought the most inclusive strategy possible to obtain the most comprehensive picture of the trends presented in this type of research through Boolean expression:

"mobile learning" or "m-learning"

Definition of Criteria for the Inclusion and Exclusion of Studies

Based on Barcelos (2014), from the results to be found in the data bases, the selection of studies to be included in the review was carried out in two steps, as shown in Table 1.

Table 1. Steps of study selection

Steps	
First	**Second**
1SI - immediate inclusion in the review	
1SE - Immediate exclusion in the review	
FA - indication of further analysis SV- later Inclusion in Review	2SI - later Inclusion in Review
	2SE - later Exclusion in Review

1SI = First Step Inclusion
1SE = First Step Exclusion
FA = Further Analysis Required
2SI = Second Step Inclusion
2SE = Second Step Exclusion

The first step focused on the title and summary analysis of each work. The work described as FA is selected for further review and analyzed in the second stage, which consists of reading the complete article or thesis / dissertation to indicate whether it is to be finally classified as 2SI or 2SE.

In both steps, the following inclusion and exclusion criteria are observed. Criteria definition was determined before search execution began on the chosen bases to avoid the *bias possibility factor* interfering in the chosen selection.

Inclusion Criteria

- The submitted work must be pertinent to the use of mobile devices in the field of formal education in science and mathematics.
- The paper should present some teaching experience with the use of mobile devices in teaching science and mathematics and include an experimental evaluation to provide partial or consolidated research results
- The work must be presented in Portuguese, English or Spanish for ease of correction:
- The time interval criterion for the accumulation of research throughout this stage was determined for the years 2005 – 2014. This is because the existence of available subject matter relating to this topic is very recent and therefore narrows down the search to this brief period.

Exclusion Criteria

- Work which is not directly related and pertaining to the use of mobile devices in teaching science and mathematics or when this criterion does apply in this respect, but does not apply to formal teaching situations in science and mathematics.
- The work proposes, discusses or defends the practice of m-learning, but does not suggest creative results (whether partial or total) in a practical application of the use of mobile devices in teaching science and mathematics.
- The work is presented in a language other than Portuguese, English or Spanish.

Table 2. Selected study statistics of the Systematic Review

Base	I	IS	E	ES	Total
SCIELO	3	0	18	3	24
ERIC	18	1	431	7	457
ACM DL	5	1	429	16	451
Total	26	2	878	26	932

- The work is done outside of the observed time limit investigation period criterion.
- Non-defined work research context. For example, the educational grade level or mobile device investigated content / discipline of the study is presented generically.
- Submission of duplicate articles.

Research Method Handling Criteria

The identification and selection of studies is executed by a manual search of the databases based on the string search and pre-defined period for the review. The criteria for the selection of previously described studies is also considered.

The research was done during June 2015, in the three databases listed above. The results of the search returned 932 studies. In Table 2, we present the study selection statistics, based on the two-step review. In this review, as shown in the table, 28 studies were chosen.

Captions:

I: Included in the first step from title and summary reading analysis.
IS: Included in the second step, from a full text reading analysis.
E: Excluded in the first step from title and summary analysis.
ES: Excluded in the second step from full text reading analysis.

As is shown in Table 2, the database identified as having a superior number of studies in this review was the ERIC database, which is understandable because this database focuses on research in education. Though a significant number of studies were found that orbit to a greater or lesser extent around m-learning, after analysis, many studies were found to be inadequate, judged by inclusion parameter criteria and were consequently excluded, e.g., the study not being related to science and mathematics teaching. Besides this, much of the research found in these student studies erroneously dealt with the individual student's focus of the perception/interpretation of *acceptance* of *student/teacher use of mobile devices* in teaching science and mathematics. This viewpoint completely misses the object of the study, which was to propose/conduct a study using these technological resources to realize a relevant proposed activity and further evaluate the results of this experience from said activity.

As instructed by Wohlin et al (2012) and Barcelos (2014), in more comprehensive cases of literature organization mapping, as in this research, the rigid definition of quality criteria does not become essential and can omit the inclusion of studies which may point to relevant tendencies in this field of research. Thus, for the purposes of this review, the authors chose not to define exclusion criteria based on a veracity quality check.

Figure 2. Study publication date/year

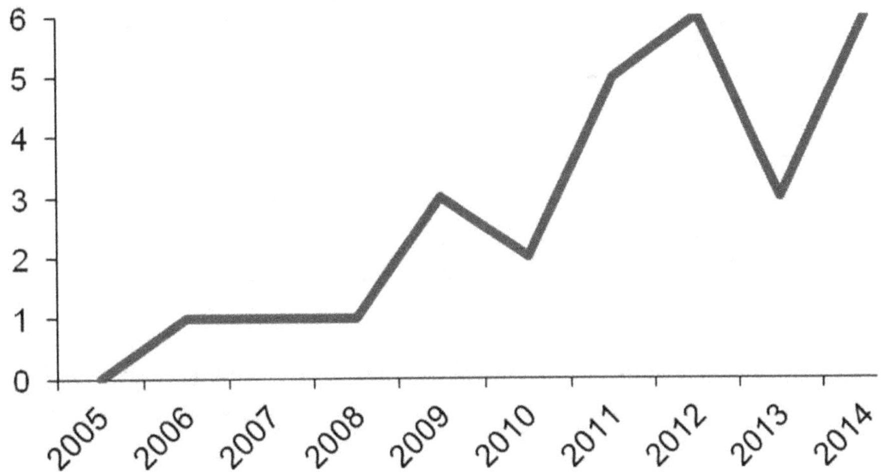

Based on the studies selected for review, data extraction, which allows analysis for reporting research findings was performed. The procedures laid down for the extraction of data was quantitative and qualitative.

Qualitative data was extracted based on the following analysis categories:

- The year of publication of the research
- The level of education addressed in research
- The subjects investigated in the research
- Main studied impacts presented

Reporting and Discussion of Results

Year of Publication

Figure 2. shows the number of selected publications, according to the criteria adopted in this study during 2005 and 2014. More frequent researcher interest from the year 2009 is seen in this field. This data is consistent with Higuchi (2011) when he says that constant insertion of mobile technologies into the teaching-learning process has favored the emergence of research with interest in investigating how these technologies (smartphones, tablets, etc.) can contribute to the m-learning process.

Level of Education in which the Studies were Conducted

In the research, it was sought to identify which levels of educational pedagogical activity using mobile devices had been proposed.

The results, presented in Table 3 show that the majority of the investigated studies are related to undergraduate level, with 11 studies focused on this level, though some studies were identified as being oriented toward middle school (08) and high school (07). This shows the necessity of focusing more m-learning studies on these levels, since mobile devices are commonly within reach of these adolescents.

Table 3. Level of education in which the studies were conducted

School Level	Number of Studies
Elementary School	3
Middle School	8
High School	7
Undergraduate	11
Graduate	1
Total	30

The authors clarify that the sum of the studies distributed to different educational levels (30) was superior to the number of papers surveyed (28), because some studies were applied to more than one grade level.

Subjects / Areas Investigated

As shown in Table 4, when analyzing which disciplines have been investigated in the area of science and mathematics teaching with mobile device use, researchers observed that of the 28 investigated studies, the majority of them (19) were directed towards science/biology education. This finding is important, being that these are subjects of a microscopic and dynamic nature, and that mobile technology can assist in facilitating study content comprehension of this nature.

No studies investigating physics teaching with the use of mobile devices in formal education were identified and only one study pertaining to chemistry teaching was identified. These results point to the necessity of intensifying research in m-learning involving such content.

Table 4. Area / content investigated in the research

Area/Discipline	No. Occurrences
Science/Biology	19
Mathematics	8
Physics	0
Chemistry	1
Total	28

Table 5. Impacts observed in research

	Positives	Negatives	Positives/Negatives
No. occurrences	26	0	2

Observed Impact Study Analysis

The following analysis refers to the types of impact evidenced with the use of mobile devices in the teaching of science and mathematics, presented in Table 5. Articles pointing to favorable outcomes in the educational use of mobile devices were considered as positive readings and the opposite as negative.

It was observed that nearly all the studies analyzed dealing with the use of mobile devices in formal teaching of science and mathematics indicated positive impacts, as identified in 26 of the 28 studies investigated. Only two of the studies pointed to negative, as well as positive aspects in m-learning activities. These results are consistent with the studies of Kumar et al (2011), indicating that the inclusion of these devices in the educational environment has been satisfactory for both students and teachers.

This data is also important in face of the current situation surrounding m-learning, as shown in findings "The Future of Mobile Learning: Implications for Planners and Policy Makers" (UNESCO, 2014). It states that though the benefits of mobile learning have been exploited by researchers for years, the educational community still seems skeptical about its potential benefit to the educational process.

The authors yet call attention to the fact that the most recurrent positive attitudes cited in these studies were shown to be superior in learning performance (12 studies - S004-ERIC, S005-ERIC, S006-ERIC, S007-ERIC, S009-ERIC, S014-ERIC, S015,ERIC, S017-ERIC, S019-ERIC, S004-ACM, S005-ACM, S006-ACM); followed by positive attitudes in participation and engagement in learning activities (04 studies - S013-ERIC ERIC S017-, S018- ERIC, S004-ACM), and motivation for learning (03 studies - S001-SCIELO, S008-ERIC, S016- ERIC).

Based on Caceffo et al (2011), the use of m-learning, when based on an active learning model, can offer many benefits, such as increased student interest, motivation and participation, stimulated by the practice of interactive lessons. To O'Nuallain and Brennan (2005), collaborative learning, which is based on mobile technology, helps the student learn because it boosts the retention of information as a result of increased motivation and understanding.

It was also observed that the negative aspects presented indicated technical problems, and were due to technical limitations such as screen size, etc. As discussed earlier, the difficulties found pertaining to these issues were already discussed in the studies of Georgiev et al (2004), Koole (2006), Keegan (2007) and Ally (2009). Naismith et al (2004) demonstrates the need for creating a solid technical infrastructure for learning using ICT. For UNESCO (2014), one of the main challenges of m-learning is to ensure that projects on mobile learning take into account the realities and limitations of the existing infrastructure. Fedoce and Squirra (2011) add that they can be absorbed by this new *e-generation*.

The technological fruit and skills of the 21st century require 21st century technological knowhow. General data gathered from this study shows a growing interest of researchers in this field of research, in parallel with the increasingly rapid adoption of mobile devices by students. The data also indicates the need to increase and amplify the range and number of research studies dealing with primary and secondary education. In addition, a deficit of studies reserved for teaching physics and chemistry was identified, pointing to the need for closer investigation of the teaching of these subjects using mobile device technology. The studies investigated demonstrated a majority of positive aspects experienced in using these resources in formal academic education, highlighting the improvement in learning performance; followed by demonstrated positive attitudes, enthusiastic participation and willing engagement in learning activities as well as an increased motivation for learning.

CONCLUSION

Based on this systematic review it is possible to infer that since 2009, a generalized academic community interest for research involving m-learning in formal education in science and mathematics teaching has been recurrent. The interest in this research and the need of intensifying studies in other levels of education is emphasized for undergraduate education levels. This review points to a greater involvement in research dealing with m-learning for other disciplines as well, such as physics and chemistry. Research shows mainly positive impacts on the use of these technologies in formal teaching of science and mathematics, mainly to facilitate, motivate and engage students in learning activities.

Both research and results point to the fact that, as suggested by Parsons (2014), researchers in m-learning must try to envision the future of m-learning, and try to further recognize and understand how mobile devices can be incorporated into the teaching environment, identify new methodologies that advocate their application in the teaching / learning process and its major implications for education. These reflections may help to understand how these devices can favorably modify the learning environment, stimulating and promoting active student participation in building knowledge.

ACKNOWLEDGMENT

The authors thank CAPES (Coordenação de Aperfeiçoamento de Pessoal de Nível Superior), Ministry of Education of Brazil, for their valuable support.

REFERENCES

Abreu, A. F. (2012). *Mobile learning: novos caminhos para o processo de ensino e aprendizagem.* Revista Discentis.

Adriani, R. L. S. B. (2008). Jogos e M-learning: do veículo de comunicação ao instrumento de ensino. 73 f. Dissertação (Mestrado em Comunicação e Semiótica). Pontifícia Universidade Católica de são Paulo, São Paulo.

Alexander, B. (2004). Going nomadic: Mobile learning in higher education. *EDUCAUSE Review*, *39*(5), 29–35.

Ally, M. (2009). *Mobile learning: Transforming the delivery of education and training*. Athabasca University Press.

Attewell, J. (2008). Towards sustainable large scale implementation of mobile learning: The mobile learning network (MoLeNET). In *The Bridge from text to context. Proceedings of the mLearn 2008 Conference* (pp. 28-35).

Barcelos, T. S. (2014). *Relação entre o pensamento computacional e a Matemática em atividades didáticas de construção de jogos digitais. Tese (Doutorado).* São Paulo: Universidade Cruzeiro do Sul.

Batista, S. C. F. (2011). *M - LEARNMAT: Modelo pedagógico para atividades m-learning em matemática. Tese (Doutorado).* Porto Alegre: Universidade Federal do Rio Grande do Sul.

Boulos, M. N., Maramba, I., & Wheeler, S. (2006). Wikis, blogs and podcasts: A new generation of Web-based tools for virtual collaborative clinical practice and education. *BMC Medical Education*, *6*(1), 41. doi:10.1186/1472-6920-6-41 PMID:16911779

Caceffo, R. E., da Rocha, H. V., & de Azevedo, R. J. (2011). Ferramenta de apoio para o aprendizado ativo usando dispositivos com caneta eletrônica. *Revista Brasileira de Informática na Educação*, *19*(02), 25.

Chen, X. (2009). *Students Who Study Science, Technology, Engineering, and Mathematics (STEM) in Postsecondary Education. Stats in Brief. NCES 2009-161*. National Center for Education Statistics.

Chisholm, L. A. (2005). *Micro-learning in the lifelong learning context*. Academic Press.

Cook, J. (2009). Phases of mobile learning. *Joint European summer school on technology enhanced learning*.

Drew, C. (2011). Why Science Majors Change Their Mind. *The New York Times*.

Fedoce, R. S. (2010). *A tecnologia móvel e os potenciais da comunicação na educação. Dissertação (Mestrado)*. São Bernardo do Campo, SP: Universidade Metodista de São Paulo.

Fedoce, R. S., & Squirra, S. C. (2011). A tecnologia móvel e os potenciais da comunicação na educação. *LOGOS, Comunicação & Universidade*, 267-278.

Fonseca, A. G. M. F. (2013). Aprendizagem, mobilidade e convergência: Mobile Learning com Celulares e *Smartphones*. *Revista Eletrônica do Programa de Pós-Graduação em Mídia e Cotidiano*, *2*(2), 265–283.

Franco, R. O. S. (2010). *Geolearning; Aplicativo móvel para apoio ao aprendizado de vocabulário de língua inglesa com conteúdo geolocalizado. Dissertação (Mestrado)*. Recife, PE: Universidade Federal de Pernambuco.

Frohberg, D., Göth, C., & Schwabe, G. (2009). Mobile learning projects–a critical analysis of the state of the art. *Journal of Computer Assisted Learning*, *25*(4), 307–331. doi:10.1111/j.1365-2729.2009.00315.x

Georgiev, T., Georgieva, E., & Smrikarov, A. (2004, June). M-learning-a New Stage of E-Learning. In *International Conference on Computer Systems and Technologies-CompSysTech* (Vol. 4, No. 28, pp. 1-4).

Göttsche, K. (2012). Tecnologias móveis: Uma mais valia em contextos educacionais? *Linhas*, *13*(2), 62–73. doi:10.5965/1984723813022012062

Graziola, P. G. G. (2009). Aprendizagem com mobilidade na perspectiva dialógica: Reflexões e possibilidades para práticas pedagógicas. *RENOTE*, *7*(3), 115–124.

Grund, F. B., & Gil, D. J. G. (2014). *Estado del Mobile Learning en España*. Educar em Revista.

Guerreiro, S. (2013). *A utilização das novas tecnologias enquanto estratégia facilitadora do processo de ensino-aprendizagem*. Academic Press.

Herrington, J., Herrington, A., Mantei, J., Olney, I. W., & Ferry, B. (2009). *New technologies, new pedagogies: Mobile learning in higher education*. Academic Press.

Higuchi, A. A. S. (2011). *Tecnologias móveis na educação: Um estudo de caso em uma escola da rede pública do estado de são Paulo. Dissertação (Mestrado)*. São Paulo, SP: Universidade Presbiteriana Mackenzie.

Hwang, G. J., & Wu, P. H. (2014). Applications, impacts and trends of mobile technology-enhanced learning: A review of 2008–2012 publications in selected SSCI journals. *International Journal of Mobile Learning and Organization*, *8*(2), 83–95. doi:10.1504/IJMLO.2014.062346

Keegan, D. (2007). *Mobile learning: a practical guide*. Academic Press.

Khaddage, F., & Lattemann, C. (2009). Towards an ad-hoc mobile social learning network using mobile phones. *Interactive Computer Aided Learning*, *12*, 374–380.

Kitchenham, B. (2004). Procedures for performing systematic reviews. Keele University.

Klopfer, E., Squire, K., & Jenkins, H. (2002). Environmental detectives: PDAs as a window into a virtual simulated world. In *Wireless and Mobile Technologies in Education, 2002. Proceedings. IEEE International Workshop on* (pp. 95-98). IEEE.

Koole, M. L. (2006, November). Practical issues in mobile education. In *Wireless, Mobile and Ubiquitous Technology in Education, 2006. WMUTE'06. Fourth IEEE International Workshop on* (pp. 142-146). IEEE. doi:10.1109/WMTE.2006.261363

Kumar, L. S., Jamatia, B., Aggarwal, A. K., & Kannan, S. (2011). Mobile Device Intervention for Student Support Services in Distance Education Context--FRAME Model Perspective. *European Journal of Open, Distance and E-Learning*.

Ledesma, F. (2013). *Mobile learning: Proibir ou integrar*. Retrieved from http://cefopna.edu.pt/revista/revista_09/es_02_09_fl. htm

Lim, T., Fadzil, M., & Mansor, N. (2011). Mobile learning via SMS at Open University Malaysia: Equitable, effective, and sustainable. *The International Review of Research in Open and Distributed Learning*, *12*(2), 122–137.

Marçal, E., Andrade, R., & Rios, R. (2005). Aprendizagem utilizando dispositivos móveis com sistemas de realidade virtual. *RENOTE, 3*(1).

McInnes, R. (2009). *Standards for Mobile Pedagogical Application Development: Literature Review*. Academic Press.

Metcalf, D. S., & De Marco, J. M. (2006). *mLearning: Mobile learning and performance in the palm of your hand*. Human Resource Development.

Moran, J. M. (2013). *Novas tecnologias e mediação pedagógica*. Papirus Editora.

Moura, A. M. C. (2010). *Apropriação do Telemóvel como Ferramenta de Mediação em Mobile Learning*. Braga, Portugal: Estudos de Caso em Contexto Educativo. Tese de Doutoramento. Universidade do Minho.

Naismith, L., Lonsdale, P., Vavoula, G., Sharples, M., & Series, N. F. (2004). *Literature review in mobile technologies and learning*. Futurelab Report 11, Retrieved from http://www2.futurelab.org.uk/resources/documents/lit_reviews/Mobile_Review.pdf

Naismith, L., Lonsdale, P., Vavoula, G. N., & Sharples, M. (2004). *Mobile technologies and learning*. futurelab.

O'Nuallain, C., & Brennan, A. (2005). How can one effectively assess students working in a collaborative mobile environment on an individual basis?. *Mobilelearning Anytime Everywhere*, 149.

Pachler, N., Bachmair, B., & Cook, J. (2009). *Mobile learning: structures, agency, practices*. Springer Science & Business Media.

Parsons, D. (2014). The future of mobile learning and implications for education and training. In M. Ally & A. Tsinakos (Eds.), *Increasing Access through Mobile Learning* (pp. 217–229). Commonwealth of Learning Press.

Parsons, D., Ryu, H., & Cranshaw, M. (2007). A design requirements framework for mobile learning environments. *Journal of Computers*, *2*(4), 1–8. doi:10.4304/jcp.2.4.1-8

Pereira, O. R., & Rodrigues, J. J. (2013). Survey and analysis of current mobile learning applications and technologies. *ACM Computing Surveys*, *46*(2), 27. doi:10.1145/2543581.2543594

Petrova, K., & Li, C. (2009). *Focus and setting in mobile learning research: A review of the literature*. Academic Press.

Reinhard, N., Saccol, A. Z., Schlemmer, E., Barbosa, J. L. V., & Kristoffersen, S. (2007). Aprendizagem com mobilidade no contexto organizacional. *UNISINOS*. Retrieved from http://www. joaomattar. com/7senaed/index. php

Resnick, M. (2002). *Rethinking learning in the digital age*. Retrieved from: https://llk.media.mit.edu/papers/mres-wef.pdf

Roschelle, J. (2003). Keynote paper: Unlocking the learning value of wireless mobile devices. *Journal of Computer Assisted Learning*, *19*(3), 260–272. doi:10.1046/j.0266-4909.2003.00028.x

Schlemmer, E., Saccol, A. Z., Barbosa, J., & Reinhard, N. (2007). *M-learning ou aprendizagem com mobilidade: casos no contexto brasileiro*. Academic Press.

Sharples, M., Adams, A., Ferguson, R., Gaved, M., McAndrew, P., Rienties, B., & Whitelock, D. (2014). *Innovating Pedagogy 2014*. Academic Press.

Tarouco, L. M., Fabre, M. C. J. M., Grando, A. R., & Konrath, M. L. (2004). Objetos de Aprendizagem para M-learning. In *Florianópolis: SUCESU-Congresso Nacional de Tecnologia da Informação e Comunicação*.

Traxler, J. (2009). Current state of mobile learning1. *Mobile learning*, 9.

Tremblay, E. (2010). Educating the Mobile Generation–using personal cell phones as audience response systems in post-secondary science teaching. *Journal of Computers in Mathematics and Science Teaching*, *29*(2), 217–227.

Trifonova, A. (2003). *Mobile learning-review of the literature*. Academic Press.

UNESCO. (2013). *The future of mobile learning: implications for planners and policy makers* (1st ed.). Paris, DC: Autor.

Valentim, H. D. (2009). *Para uma compreensão do Mobile Learning. Reflexão sobre a utilidade das tecnologias móveis na aprendizagem informal e para a construção de ambientes pessoais de aprendizagem. Dissertação de mestrado*. Lisboa, Portugal: Universidade Nova de Lisboa.

Vavoula, G. N., & Sharples, M. (2002). KLeOS: A personal, mobile, knowledge and learning organization system. In *Wireless and Mobile Technologies in Education, 2002. Proceedings. IEEE International Workshop on* (pp. 152-156). IEEE. doi:10.1109/WMTE.2002.1039239

Wingkvist, A., & Ericsson, M. (2011). A survey of research methods and purposes in mobile learning. *International Journal of Mobile and Blended Learning*, *3*(1), 1–17. doi:10.4018/jmbl.2011010101

Wohlin, C., Runeson, P., Höst, M., Ohlsson, M. C., Regnell, B., & Wesslén, A. (2012). *Experimentation in software engineering*. Springer Science & Business Media. doi:10.1007/978-3-642-29044-2

APPENDIX

Respondents files are presented in Tables 6, 7, and 8, as follows:

Table 6. Research Base SCIELO

Code	Reference
S001-SCIELO	García-Ruiz, M. A., Valdez-Velazquez, L. L., & Gómez-Sandoval, Z. (2012). Estudio de usabilidad de visualización molecular educativa en un teléfono inteligente. *Quim. Nova*, *35*(3), 648-653.
S002-SCIELO	Pérez Peregrino, L. E., Maritza Matallana, A., Rodríguez Pérez, F., Moreno Pinilla, A., & Herrera Cubides, J. (2011). TEACH-ME: implementation of mobile environments to the teach-learning process. *JISTEM-Journal of Information Systems and Technology Management*, *8*(1), 179-212.
S003-SCIELO	Tanaka, P. P., Hawrylyshyn, K. A., & Macario, A. (2012). Uso de tablet (ipad®) como ferramenta para ensino da anestesiologia em estágio de ortopedia. *Rev. Bras. Anestesiol*, *62*(2), 214-22.

Source: Research data, in 2015.

Table 7. Research Base ERIC

Code	Reference
S001-ERIC	Kalloo, V., & Mohan, P. (2012). Correlating Questionnaire Data with Actual Usage Data in a Mobile Learning Study for High School Mathematics.*Electronic Journal of e-Learning*, *10*(1), 76-89.
S002-ERIC	Kiger, D., Herro, D., & Prunty, D. (2012). Examining the influence of a mobile learning intervention on third grade math achievement. *Journal of Research on Technology in Education*, *45*(1), 61-82.
S003-ERIC	Hung, P. H., Hwang, G. J., Su, I., & Lin, I. H. (2012). A Concept-Map Integrated Dynamic Assessment System for Improving Ecology Observation Competences in Mobile Learning Activities. *Turkish Online Journal of Educational Technology-TOJET*, *11*(1), 10-19.
S004-ERIC	Pfeiffer, V. D., Gemballa, S., Jarodzka, H., Scheiter, K., & Gerjets, P. (2009). Situated Learning in the Mobile Age: Mobile Devices on a Field Trip to the Sea. *ALT-J: Research in Learning Technology*, *17*(3), 187-199.
S005-ERIC	Uzunboylu, H., Cavus, N., & Ercag, E. (2009). Using mobile learning to increase environmental awareness. *Computers & Education*, *52*(2), 381-389.
S006-ERIC	Teri, S., Acai, A., Griffith, D., Mahmoud, Q., Ma, D. W., & Newton, G. (2014). Student use and pedagogical impact of a mobile learning application.*Biochemistry and Molecular Biology Education*, *42*(2), 121-135.
S007-ERIC	DeWitt, D., Alias, N., & Siraj, S. (2014). The design and development of a Collaborative mLearning prototype for Malaysian secondary school science.*Educational Technology Research and Development*, *62*(4), 461-480.
S008-ERIC	Bragg, L. A. (2006). 'Hey, I'm Learning This'. *Australian Primary Mathematics Classroom*, *11*(4), 4.
S009-ERIC	Ahmed, S., & Parsons, D. (2013). Abductive science inquiry using mobile devices in the classroom. *Computers & Education*, *63*, 62-72.
S010-ERIC	White, T., & Martin, L. (2014). Mathematics and mobile learning. *TechTrends*,*58*(1), 64-70.
S011-ERIC	Vogel, B., Kurti, A., Milrad, M., Johansson, E., & Müller, M. (2014). Mobile inquiry learning in Sweden: Development insights on interoperability, extensibility and sustainability of the LETS GO software system. *Journal of Educational Technology & Society*, *17*(2), 43-57.
S012-ERIC	Roberts, N., & Vänskä, R. (2011). Challenging assumptions: Mobile learning for mathematics project in South Africa. *Distance Education*, *32*(2), 243-259.
S013-ERIC	Bruce-Low, S. S., Burnet, S., Arber, K., Price, D., Webster, L., & Stopforth, M. (2013). Interactive mobile learning: a pilot study of a new approach for sport science and medical undergraduate students. *Advances in physiology education*, *37*(4), 292-297.

continued on following page

Table 7. Continued

Code	Reference
S014-ERIC	Fozdar, B. I., & Kumar, L. S. (2007). Mobile Learning and Student Retention.*International Review of Research in Open and Distance Learning*, *8*(2), 1-18.
S015-ERIC	Looi, C. K., Zhang, B., Chen, W., Seow, P., Chia, G., Norris, C., & Soloway, E. (2011). 1: 1 mobile inquiry learning experience for primary science students: A study of learning effectiveness. *Journal of Computer Assisted Learning*, *27*(3), 269-287.
S016-ERIC	Bull, P., & McCormick, C. (2012). Mobile learning: Integrating text messaging into a community college pre-algebra course. *International Journal on E-Learning*, *11*(3), 233-245.
S017-ERIC	Liu, T. C., Peng, H., Wu, W. H., & Lin, M. S. (2009). The effects of mobile natural-science learning based on the 5E learning cycle: A case study.*Journal of Educational Technology & Society*, *12*(4), 344-358.
S018-ERIC	Mayfield, C. H., Ohara, P. T., & O'Sullivan, P. S. (2013). Perceptions of a mobile technology on learning strategies in the anatomy laboratory.*Anatomical sciences education*, *6*(2), 81-89.
S019-ERIC	Huang, Y. M., Lin, Y. T., & Cheng, S. C. (2010). Effectiveness of a mobile plant learning system in a science curriculum in Taiwanese elementary education. *Computers & Education*, *54*(1), 47-58.

Source: Research data, in 2015.

Table 08. Research Base ACM

Code	Reference
S001-ACM	Wyeth, P., & MacColl, I. (2010, June). Noising around: investigations in mobile learning. In *Proceedings of the 9th International Conference on Interaction Design and Children* (pp. 147-155). ACM.
S002-ACM	Miltchev, R., Tsavkov, E., & Milchev, G. (2011, June). Approaches to integration of the dichotomic tables in computer applications for the needs of electronic and mobile learning in the field of biological sciences. In*Proceedings of the 12th International Conference on Computer Systems and Technologies* (pp. 472-477). ACM.
S003-ACM	Wyeth, P., McEwan, M., Roe, P., & MacColl, I. (2011, November). Expressive interactions: tablet usability for young mobile learners. In*Proceedings of the 23rd Australian Computer-Human Interaction Conference*(pp. 311-314). ACM.
S004-ACM	Sánchez, J., & Salinas, A. (2008, October). Science problem solving learning through mobile gaming. In *Proceedings of the 12th international conference on Entertainment and media in the ubiquitous era* (pp. 49-53). ACM.
S005-ACM	Liu, T. C., Lin, Y. C., Tsai, M. J., & Paas, F. (2012). Split-attention and redundancy effects on mobile learning in physical environments. *Computers & Education*, *58*(1), 172-180.
S006-ACM	BjÁ¸rkli, K. (2014). The impact on learning outcomes in mathematics of mobile–enhanced, combined formative and summative assessment. *International Journal of Technology Enhanced Learning*, *6*(4), 343-360.

Source: Research data, in 2015.

Chapter 15
Taking Advantage of MOOCs in K-12 Education:
A Blended Approach

Samantha Briggs
Old Dominion University, USA

Helen Crompton
Old Dominion University, USA

ABSTRACT

Technological opportunities are opening new ways for educators to enhance K-12 instruction. While many educators are incorporating digital technologies into their teaching, there is evidence to show that K-12 educators have a lack of training, time, and resources to implement learner-centered digital instruction. Massive Open Online Courses (MOOCS) have gained a reputation for providing online learning in higher education and are now extending to K-12. The unique digital advantages as well as the rising number of students enrolled in schools has led to discussions about the potential of MOOCS for students in K-12 (ages 4-19 years) education. This chapter provides a review of the literature of early pioneering work of MOOCs in K-12 within North America. Early MOOC classes are explicated and positives and challenges discussed. It appears from the review that there is a place for MOOCs to be embedded into a blended K-12 environment to extend and enhance existing curricula.

INTRODUCTION

Learner-centered instructional environments facilitate the learning process and improve overall learning outcomes (Blumberg, 2008). However, researchers have found that instructors find difficulties designing courses based on learner-centered principles due to physical boundaries and time constraints (Brush, & Saye, 2000). With the emergence of Massive Open Online Courses (MOOCs), researchers (viz., Li, Zhang, Bonk, & Guo, 2015) posit that blending MOOCs into traditional instruction can support those learner-centered principles. Advocates, such as Thurne (2003), postulate that blended learning is a logical and natural evolution to a pedagogical framework. Thurne describes it as an elegant solution

DOI: 10.4018/978-1-5225-0359-0.ch015

to the challenges of tailoring instruction to meet the needs of individuals. It provides an opportunity to take advantage of both the technological advances provided by online learning and participation in traditional learning (Suprabha, & Subramonian, 2015). This chapter articulates the use of MOOCs in a K-12 blended learning environment in North America.

Massive Open Online Courses (MOOCs)

A MOOC is an online learning platform that delivers free education courses without enrollment restrictions. MOOCs have typically been designed for higher education courses. The rising number of students enrolled in K-12 education has led to discussions about the potential of MOOCs (Dermirci, 2014; Norris & Soloway 2012). The first MOOC was offered by the University of Manitoba, Canada in 2008 and had over 2000 students participate (Adair, Alman, Budzick, Grisham, Mancini, & Thackaberry, 2014; Dermirci, 2014; Liyanagunawardena, Adams, & Williams, 2013).

The nature of these courses contrasts with typical online courses because any person who wants to take the course can sign-up without any restrictions or having to be a registered student at a university. MOOCs originated from the philosophy of a few professors who felt knowledge should be free and accessible to anyone who wanted to learn (Bali, 2014; Dermirci, 2014; Johnston, 2013; Saadatmand, & Kumpulainen, 2014). Traditional higher education institutions may restrict the type of learner accepted based on economics, demographics, geography, prerequisites, or attendance limits (Liyanagunawardena et al., 2013; Saadatmand, & Kumpulainen, 2014). The companies and organizations that develop MOOCs break these barriers by providing the general public with free and equal access to high quality education (Holotescu, Grosseck, Cretu, Naaji, 2014; Liyanagunawardena et al., 2013).

Some diversity exists amongst pedagogical styles and credit options that developers of MOOCs offer (Bali, 2014; Pannoni, 2014). There are two distinct types, xMOOCs and cMOOCs. xMOOCs are a type developed by Coursera and edX that use a more traditional style of teaching with lectures, videos, and quizzes (Siemens, 2012). cMOOCs are centered on connectivity through technology platforms, such as digital social media tools, that allow students to communicate and share ideas (Adair et al., 2014; Dermirci, 2014; Ferdig, 2014; Horn, 2014; Liyanagunawardena et al., 2013). MOOCs started as a non-profit initiative by a few top universities but now for-profit companies like Coursera are becoming major developers. Even though the courses are free, most providers or developers of MOOCs offer options for college credit or a verified certificate for a small fee (Liyanagunawardena et al., 2013). The non-profit organization started by Harvard and MIT, called edX, offers the option to earn a verified certificate for a low fee or audit the course for free. Audited courses are still open to the public and free to anyone who wants to learn (Locke, 2013). Several states want to force universities to accept MOOCs as valid college credit in order to expedite baccalaureate degrees (Adair et al., 2014).

So how have discussions about MOOCs started to take precedence in K-12 education in the United States? The style of learning seen in today's students, known as the Net Generation, has changed and therefore educators are seeing the need to change their pedagogy to match those differences. The traditional K-12 classroom is outdated and is considered uninspiring or irrelevant by most students (Prensky, 2001). Today's students grew up in what we refer to as the Digital Age and have different expectations about learning. In addition, employers are seeking employees with the skill set necessary to thrive in the Digital Age. Many K-12 schools and colleges are failing to meet today's requirements because of inadequate resources and lack of teacher preparedness (Conley, 2010).

MOOCs are based on online learning tools like videos, games, and social media; tools that youth are both interested in and familiar with (Locke, 2013; Saadatmand & Kumpulainen, 2014). Students often prefer to use multiple resources, look-up information on their own, view information in short segments, and receive immediate feedback, and the model for delivery of knowledge in MOOCs often lends itself to this format (Saadatmand & Kumpulainen, 2014). There is noticeable potential for MOOCs to add enriched learning opportunities. Full establishment of these courses in K-12 education is gaining momentum by schools exploring and experimenting with different applications. Research to date, however limited, shows use of MOOCs in K-12 schools to have positive results (Pannoni, 2014; UK Department for Education, 2014). These studies show that K-12 students are already starting to use MOOCs on their own accord to learn about what interests them or prepare for tests and college. Learning online is something familiar to them and is a part of their daily learning outside the classroom. In order to engage students in the classroom, students not only need to have more control and responsibility for their learning but start using the tools they already use in daily life (Scherer, 2014).

The integration of MOOCs could solve many problems that exist in K-12 education. Technology integration is important to foster 21st century learning, but the cost of digital tools can be expensive. Free access to MOOCs could reduce the cost that districts spend on online educational resources. Online courses could also help reduce cost by lessening the need for paper in the classroom (Scherer, 2014). More importantly, these courses make it possible for students to have equal access to quality education; especially for disadvantaged lower income schools.

MOOCs in K-12 schools could help make up for teacher shortages and alleviate problems resulting from high student to teacher ratio (Locke, 2013). Students would be given the opportunity to work at their own pace and move ahead to more advanced curriculum that may not be regularly offered, including advanced placement (AP) courses and gifted programs. The learning environment delivered through MOOCs is enriched through student autonomy, rigor, relevance, and content mastery (Saadatmand & Kumpulainen, 2014). Currently, the most sought after use for MOOCs in the K-12 setting are advanced placement courses and college and test preparation.

A New Model for K-12 Learning

Pedagogies used in K-12 education are typically those designed for the industrial era and are not appropriate for this digital age (Mehta, 2013). Instruction is commonly dominated by memorization practices with pedagogies that do not help students understand the concepts (Fullan, & Langworthy, 2014; Ubuz & Üstün, 2004; Williams-Carling, 2009). It appears that educational leaders have become more focused on test grades than content mastery. In a positive move, philosophies and practice move toward learner-centered pedagogies and technology is providing affordances that are personalized and contextualized (Crompton, 2014).

To take advantage of these technologies, the blended approach may be considered by educators in K-12 settings. Blended learning refers to a mix of traditional face-to-face instruction and online components (Kassner, 2013). The blended approach has seen a massive growth in recent years and this trend is expected to rise (Staker, & Horn, 2012). The intention is not to swap one medium for another, but take the best of what each has to offer. In a survey involving 627 K-12 teachers, who were practicing the blended learning approach or had used the approach in the past, Werth, Werth, & Kellerer (2013) found teachers indicating student ability was either better or much better in their classes that used blended learning models. Similar results were also recorded for student engagement. Other positive results included better or much better teacher to teacher communication (61%), student-to-student and parent-to-teacher communication was the same or better (87%), teachers indicated that their ability to monitor student learning was better or much better (77.5), and classroom management was better or much better with the blended learning approach (64.1).

Empirical evidence such as this is becoming increasingly common in the past decade, supporting the argument that blended learning is more effective than face-to-face instruction or online instruction on its own (Means, Toyama, Murphy, Bakia, & Jones, 2010; Nagel, 2009). Online learning is being explored by educators for the unique affordances it can provide (Horn, 2014). The developers of MOOCs offer a new type of online learning that aligns to K-12 education. Research has been dedicated to online learning as a better method for K-12 students to learn (Barbour et al., 2006; Staker & Horn, 2012). Student-centered learning should be personalized, authentic, engaging, and have students critically thinking about the content (Humphries, & Ness, 2015).

It can be challenging for a K-12 teacher to provide student-centered learning in a traditional classroom setting (Brush, & Saye, 2000). "Increasingly, students are seen as the consumers of an educational service" (Adair et al., 2014). Many classroom settings are still teacher-centered with students as passive learners (Mueller, Knobloch, & Orvis, 2015). Online education could give students the opportunity to take a more active role in their learning. As a part of the online learning environment students can use social media outlets to communicate and collaborate with not only their peers but also students around the country and world (Ferdig, 2014). Learning through MOOCs requires teachers to take on the facilitator role, which is the first step to a student-centered learning environment (Adair et al., 2014). Furthermore, the flipped classroom could be another approach successful with MOOCs (Holotescu et al., 2014; Najafi, Evans & Federico, 2014).

Implementation of MOOCs in K-12 Education

Cost

Many companies providing online courses and curriculum are expensive (Horn, 2014). The high cost could eliminate online learning as an option for many schools. MOOCs can be used as a free option for online learning. Cost-effectiveness is crucial within educational institutions and requires a transparency among the various levels of governance which must consider a blend of online alternatives to traditional instruction (Montgomery, Hayward, Dunn, Carbonaro, & Amrhein, 2015). There are additional cost benefits, for example, cost of professional development is a necessary expense that schools could cut by using MOOCs to deliver such courses to teachers (Horn, 2014). A vast amount of professional development courses already exist for K-12 teachers. In addition, online courses also reduce the need for so much paper use in the classroom helping schools cut material cost which can be very expensive in many schools (Scherer, 2014).

Access and Equity

Disproportionate distribution of funding for public education has left the US facing a problem of educational inequality (Carneiro & Heckman 2003; DiPrete and Eirich, 2006). The mission of MOOCs specifically addresses the US national crisis by providing high quality education without any restrictions, including financial background. While there is a plethora of empirical and theoretical literature centered on cost to institutions, cost and equitable access to students using a blended approach with online access from home is lacking.

While students would have the same enrollment opportunities to free MOOC courses (Horn, 2014; Liyanagunawardena et al., 2013; Saadatmand & Kumpulainen, 2014), if K-12 students are required to work from home, they will need internet connectivity, a computer, and other peripherals, such as headsets and microphone to be able to access the learning materials. It should not be expected that students will have all these materials. If schools want students to conduct online learning at home, necessary measures should be implemented to that students have these resources. This may mean that students are loaned or even given these materials. If equipment is purchased by the school, this will obviously add to the overall cost and may not be as cheap to implement as MOOCs may initially appear.

Teacher Disparity

Schools are experiencing a shortage of teachers qualified to teach 21st Century skills. Math, science, and computer science are of major concern and highly qualified teachers in these areas of study are in high demand (Locke, 2013). MOOCs could serve as a means to compensate for the disparities (Locke, 2013). Professional development MOOCs could prepare teachers to become highly qualified in the needed areas. Computer science MOOCs could supplement teacher knowledge and sufficiently prepare students. Colleges more than support K-12 schools using MOOCs to learn challenging fields like math and science especially since many of the current MOOCs are led by top universities (Najafi et al., 2014).

Enriched Learning

While technology offers a great potential for K-12 education, many teachers lack the knowledge and skills to use technology effectively (Cheon, Lee, Crooks, & Song, 2012; Crow, Santos, LaBaron, McFaden, and Osborne, 2010). MOOCs can provide students with education from experts in content and in online learning pedagogies. This is a good example of how blended learning with MOOCs can be used to connect with experts at teaching online and experts in classroom learning.

Empirical evidence shows that MOOCs can be used to increase accessibility and personalization for students (Nelson, 2013). These elements are key for student engagement in learning (Ford, & Roby, 2013). Student motivation is increased when digital tools they are familiar with and already use outside of the class are also used inside class (Scherer, 2014). Student testimonies attest that MOOCs allow them to work at their own pace, focus on mastering the content, and learn valuable study skills (Atkeson, 2014; Bock & O'Dea, 2013; Holotescu et al., 2014). Online courses such as these help build student autonomy and teach them self-responsibility for their learning (Holotescu et al., 2014). Students who are actively involved in their learning progress are more likely to be successful (Ryan, & Deci, 2000). When learning is focused on mastery goal orientation instead of performance, student learning increases (Dickinson, 1995).

Furthermore, by lessening the time spent on lecturing and grading, the use of MOOCs would give teachers more time to facilitate learning and provide instant feedback. These courses are also great tools to provide data on learning progress (Bock & O'Dea, 2013). Having that data collected automatically could save teachers precious time. Having the data instantly would allow for faster teacher feedback.

MOOCs can increase learning through their connective design that brings students together with social media tools. For students to be able to communicate and collaborate with hundreds or even thousands of other students from all over the globe delivers an unmatched learning experience (Ferdig, 2014). Online discussion boards, blogs, and wikis support students who would never meet under normal circumstances to work together and share knowledge (Bock, & O'Dea, 2013; Holotescu et al., 2014; Saadatmand & Kumpulainen, 2014). Students would have the chance to work with higher-achieving peers, a strategy used in traditional classrooms, on a much broader scale (Bock, & O'Dea, 2013). The difference of having massive enrollment numbers in a MOOC can transform the learning experience and generate higher-level thinking that would not be possible without that technology (Ferdig, 2014). Research shows learning and interest are increased when students use social media tools to create and share knowledge. Social media tools also bring together students from all backgrounds and "eliminate geographic and economic barriers in education" (Saadatmand, & Kumpulainen, 2014).

AP Courses, Test and College Preparation

The benefits of using MOOCs for advance placement, test preparation, and college preparation are starting to be noticed (Bock, & O'Dea, 2013; Locke, 2013; Sandeen, 2013). "Online learning in general tends to be a really good fit for students that are self-motivated" (Bock, & O'Dea, 2013). Students who are actively involved in their learning progress are more likely to succeed when given autonomy (Holotescu et al., 2014; Ryan, & Deci, 2000). Several MOOCs are designed specifically as advanced placement and test preparation courses for K-12 students (Atkeson, 2014; Locke 2013). Many MOOCs are developed and led by professors from top universities, such as HarvardX, free MOOC courses provided by Harvard University. These courses can meet the challenging curriculum expected of AP courses. Scrutiny about the quality of AP courses has led some colleges to not even allocate credit for the course. However, if students were taking courses developed from universities like, MIT and Harvard, colleges would undoubtedly accept these credits (Horn, 2014). In one Florida case study, students who took online AP courses scored higher on the exams than the state average (Najafi et al., 2014).

The zero cost makes it possible for all schools to offer AP courses. Free quality education is vital for students of low-income schools that otherwise could not afford the materials and teachers to proctor AP courses (Najafi et al., 2014). In addition, many schools have a difficult time finding qualified teachers for advanced courses. Too many students are missing the opportunity for advanced placement due to these unfair constraints that are out of their control (Locke, 2013). Gifted students, who are often not challenged enough, could use MOOCs to explore and expand their interest (Horn, 2014; UK Department for Education, 2014).

A study by the National Center for Public Policy and Higher Education and the Southern Regional Education Board, showed 60 percent of first-year undergraduate students were not prepared academically for college level courses (Atkeson, 2014). Recently in 2014, edX created new courses specially tailored to prepare secondary students for college. The CEO of edX, Mr. Agarwal, announced "he hopes the new curriculum will give more high school students exposure to higher-level coursework, allowing them to enter college having already completed many of their first-year classes" (Atkeson, 2014). Several MOOCs are being created now to serve the necessity of college preparedness (Najafi et al., 2014). Students, regardless of access to MOOCs in school, are taking it upon themselves to prepare for college using these courses on their own accord (Locke, 2013).

Integration of MOOCs into K-12 education

Certain types of MOOCs are needed to best meet the needs of K-12 students (Najafi et al., 2014; Locke, 2013). Professors of leading universities are typically the developers and instructors of MOOCs. Although effective for undergraduate and advanced placement student populations, college professors would not be designing for the content required at the K-12 level. Furthermore, they would not be focusing pedagogical methods for a K-12 audience.

The main developers of MOOCs still use video recordings of lectures as the main delivery method (Horn, 2014; Johnston, 2013; Norris, & Soloway 2012). Traditional lectures are not engaging enough for primary and secondary students. Students also need more opportunities to apply the knowledge they have constructed. Using the MOOC model, educators could customize the courses specifically for K-12 students. The change is quickly taking place as courses are being created by subject experts in the primary and secondary field (Bock, & O'Dea, 2013; Locke, 2013; Pannoni, 2014). Customization directed at high school students was recently executed by edX (Atkeson, 2014; Najafi et al., 2014).

Research studies suggest it would be better to use MOOCs as a supplement in a blended setting of traditional face-to-face instruction and online learning (Atkeson, 2014; Locke, 2013; Pannoni, 2014). A blended learning framework could be practical for all grade levels (Locke, 2013). Research shows some students prefer face-to-face interaction with a teacher in order to get feedback (Najafi et al., 2014). Teacher presence also has shown to be beneficial for students' ability to follow the anticipated pace of the online course content (Najafi et al., 2014). Even Coursera, one of the largest MOOC companies, stated that MOOCs should serve to improve the traditional approach of learning not replace it completely. The company believes a blended classroom will better meet the needs of a diverse student population (Locke, 2013). "Test results showed that students who watched online lecture videos in addition to their face-to-face classroom scored higher than those who only relied on face-to-face classroom learning" (Najafi et al., 2014)

Like with any educational tool, the educator plays the decisive role in the tool's effectiveness. An iPad could be bought for every student but if the teacher doesn't know how to use the tool appropriately then that educational tool will not increase learning. Therefore, professional development for educators must be provided for effective integration. Fortunately, professional development MOOCs are offered for free.

MOOCs Concerns Addressed

There are several concerns surrounding the use of MOOCs in K-12 education. Will teachers be prepared or know how to use the courses? Will the online learning platform be able to serve primary and secondary students' needs given that these courses are mostly designed for college students? Is it easy for students to cheat? What about students who aren't motivated to learn on their own? Will students be prepared to take courses that are more challenging academically? Can schools afford enough computers or digital devices to deliver MOOCs? These concerns are addressed through the following research and practices. Educational leaders, principals, educators, and parents should work together to share their research and experience about using MOOCs. Continual contact and collaboration through digital social media tools would alleviate many of the present concerns (Ferdig, 2014; Saadatmand & Kumpulainen, 2014).

First, concerns of teacher readiness are addressed through professional development courses delivered through MOOCs. There are several free courses already available that instruct best practices for teaching in a blended setting (Horn, 2014). Unfortunately, many K-12 teachers lack understanding in how to effectively integrate technology into the classroom and technology is applied in a way that demonstrates a lack of breadth and depth (Groff, & Mouza, 2008; Levin, & Wadmany, 2008). Teachers must be competent using digital tools to evaluate and increase student learning. Teachers must also be able to model good digital citizenship in regards of safe, legal, and ethical use of digital information and technology (ISTE Standards for Teachers, 2008). These are all necessary skills in order to integrate MOOCs effectively. Teacher comfort level for using technology is consequential and tends to limit what they are willing to teach. Researchers suggest teachers should try completing a MOOC to familiarize themselves with the process. Teachers who have completed a MOOC state they are more likely to use the technology in their classroom (Holotescu et al., 2014; UK Department for Education, 2014). In addition, by participating in a MOOC teachers will know firsthand what works and what doesn't so they may adapt the course to best fit the needs of their students. Therefore, it is encouraged for teachers to take more than one given there are different styles of MOOCs available (Ferdig, 2014).

Additionally, teachers may be concerned about losing some control over what content students are exposed to (UK Department for Education, 2014). Using a blended model of traditional and online learning could moderate these concerns. Teachers could choose to supplement their curriculum with MOOCs where they see it enhancing their curriculum. MOOCs use open source content therefore teachers could use only certain parts of a course and adapt those parts in their classroom (Holotescu et al., 2014; Pannoni, 2014). MOOCs are designed to be non-committal courses so students could also take just select parts of courses that their teacher sees fit for a particular lesson (Ferdig, 2014). Schools should support teachers to build their own MOOCs or create their own videos for sharing content (UK Department for Education, 2014).

A blended learning environment also lessens the concern about cheating since teachers are there to oversee student work (Locke, 2013). Other critics of MOOCs propose the courses would only work for students who are self-motivated or have good study skills (Horn, 2014). This could be avoided in K-12 as a teacher facilitator would be overseeing the process and providing motivational tactics and necessary support (Holotescu et al., 2014). Research found that the vast communication and peer support unique to MOOCs can help unmotivated students who need extra support (Ferdig, 2014). Most critics of MOOCs complain about the low completion rate of approximately 10 percent. However, since K-12 students wouldn't be given the option of taking a MOOC commitment-free, as the general public is, the statistic is not relevant (Locke, 2013).

Student preparedness is a matter that should be determined by school counselors. Just as students should be supported now, there would need to be support from parents, teachers, and school staff for success in this new endeavor (Atkeson, 2014). Lastly, some schools fear that budgets will not cover the cost for computers and other technology needed to deliver MOOCs. However, President Obama set a policy for the federal government to help ensure that by 2017 every student will have a laptop, tablet, or smartphone and schools will have enough bandwidth to support every student being online (Scherer, 2014).

Current Integration

MOOCs are a recent phenomenon in higher education and are quickly becoming a reality in K-12 education (Norris, & Soloway 2012). As early as this year, the first MOOC designed for high school students was created (Atkeson, 2014; Horn, 2014). The number available for secondary students from edX is now 44 courses (Atkeson, 2014; https://www.edx.org/course). High school students account for 150,000 of the three million students enrolled at edX (Atkeson, 2014). Coursera does not yet offer courses solely for

K-12 students but they do have professional development courses for teachers. Another MOOC company known as, Instructure, has several courses now for teachers and students. Of particular interest to some teachers, the company has a course to help educators use the popular game of Minecraft for learning (Atkeson, 2014).

Now that more K-12 courses are available it will be easier to adopt them into school practice and policy. Regardless, many students are still using the courses on their own to help prepare for college and study for their AP and standardized exams. Unfortunately, authorizing credit for MOOCs it is not up to individual schools but state policy. Schools are still encouraging students to take these courses by offering non-academic credit (Atkeson, 2014).

EdX also realized the need to help schools provide quality computer science education. They designed a course able to be used by students independently or by teachers to deliver instruction (Najafi et al., 2014). In 2013, Florida allowed students to earn high school credit for completing a MOOC in any subject with an end-of-course exam. One Florida county is trying to develop their own MOOCs using game-based learning targeting high school students (Horn, 2014; McGrory, 2013).

Other cases include MOOCs designed for engineering career paths developed by Wendy Drexler, chief innovation officer at the International Society for Technology in Education

(Atkeson, 2014; Ferdig, 2014; Pannoni, 2014). In Pennsylvania, students were able to take edX courses as electives and reported the experience to be more challenging and made them feel better prepared for college (Atkeson, 2014). Even the Smithsonian Institute is using MOOCs to provide educational opportunities for students and teachers such as virtual field trips, interactive experiences, and teacher training (Horn, 2014).

CONCLUSION

In this chapter, a blended approach has been described that includes MOOCs in K-12 education. MOOCs have gained a name for providing online learning that is free and available to the masses without geographical or demographical restrictions (Liyanagunawardena et al., 2013; Saadatmand & Kumpulainen, 2014). MOOCs are available for a variety of subjects and the substantial student enrollment numbers and continually growth of the MOOC movement have forced educators to discuss how these courses could change the future of education (Dermirci, 2014; Norris & Soloway 2012).

This chapter highlights some of the benefits of using MOOCs within a K-12 blended approach, such as providing accessibility, personalization (Nelson, 2013), and self-autonomy, (Holotescu et al., 2014). The online component can enable students to learn valuable study skills and work at their own pace (Atkeson, 2014; Holotescu et al., 2014; Bock, & O'Dea, 2013). Many of these skills may be lacking from students who have only experienced a highly structured traditional schooling. Motivation is a crucial component driving a student to pay attention and be an active learner. Scherer (2014) described how the use of MOOCs increases student motivation and allow them to use digital tools they are familiar with and already use outside of the class to learn (Scherer, 2014).

Involving MOOCs in K-12 blended learning is still in its infancy. There are a great many questions that still remain unanswered. As Montgomery et al. (2015) described in the conclusions to their research, there is a great potential for the use of MOOCS to benefit all ages of students and it is important that these opportunities are explored.

REFERENCES

Adair, D., Alman, S. W., Budzick, D., Grisham, L. M., Mancini, M. E., & Thackaberry, A. S. (2014). Many shades of MOOCs. *Internet Learning Journal*, *3*(1), 53–72.

Atkeson, S. (2014, September 24). Harvard-MIT partnership unveils new MOOCs for K-12. *Education Week*, 8.

Bali, M. B. (2014). MOOC Pedagogy: Gleaning good practice from existing MOOCs. *Journal Of Online Learning & Teaching*, *10*(1), 44–56.

Barbour, M., Brown, R., Waters, L. H., Hoey, R., Hunt, J. L., Kennedy, K., Onsworth, C., Powell, A., Trimm, T., (2006). Online and blended learning: A survey of policy and practice from K-12 schools around the world. *International Association for K-12 Online Learning.*

Blumberg, P. (2008). *Developing learner-centered teachers: A practical guide for faculty*. San Francisco, CA: Jossey-Bass.

Bock, M., & O'Dea, V. (2013). Virtual educators critique the value of MOOCs for K-12. *Education Week*, *32*(20), 10.

Brush, T., & Saye, J. (2000). Implementation and evaluation of a student –centered learning unit: A case study. *Educational Technology Research and Development*, *48*(3), 79–100. doi:10.1007/BF02319859

Carneiro, P., & Heckman, J. (2003). Human Capital Policy. In Inequality in America: What Role for Human Capital Policies?. Cambridge, MA: MIT.

Cheon, J., Lee, S., Crooks, S. M., & Song, J. (2012). An investigation of mobile learning readiness in higher education based on the theory of planned behavior. *Computers & Education*, *59*(3), 1054–1064. doi:10.1016/j.compedu.2012.04.015

Conley, D. (2010). *College and career ready*. San Francisco, CA: Jossey-Bass. doi:10.1002/9781118269411

Crompton, H. (2014). A diachronic overview of mobile learning: A shift toward student-centered pedagogies. In M. Ali & A. Tsinakos (Eds.), Increasing access through mobile learning (pp. 7–16). British Columbia, Canada: Commonwealth of Learning Press. Available at http://www.col.org/PublicationDocuments/pub_Mobile%20Learning_web.pdf

Crow, R., Santos, I. M., LeBaron, J., McFadden, A. T., & Osborne, C. F. (2010). Switching gears: Moving from e-learning to m-learning. *Journal of Online Learning and Teaching*, *6*(1), 268–278.

Dermirci, N. D. (2014). What is Massive Open Online Courses (MOOCs) and what is promising us for learning? A review-evaluative article about MOOCs. (English). *Necatibey Faculty Of Education Electronic Journal Of Science & Mathematics Education*, *8*(1), 231–256. doi:10.12973/nefmed.2014.8.1.a10

Dickinson, L. (1995). Autonomy, self-direction and self-Access in language teaching and learning: The history of an idea. *System*, *23*(2), 165–174. doi:10.1016/0346-251X(95)00005-5

DiPrete, T. A., & Eirich, G. M. (2006). Cumulative Advantage as a Mechanism for Inequality: A Review of Theoretical and Empirical Developments. *Annual Review of Sociology*, *32*(1), 271–297. doi:10.1146/annurev.soc.32.061604.123127

Ferdig, R. E. (2014). Preparing for K-12 MOOCs. *Tech & Learning, 34*(6), 26–27.

Ford, V. B., & Roby, D. E. (2013). Why do high school students lack motivation in the classroom? *Global Education Journal, 2013*(2), 101-113.

Fullan, M., & Langworthy, M. (2014). *A rich seam: How new pedagogies find deep learning*. London: Pearson.

Groff, J., & Mouza, C. (2008). A framework for addressing challenges to classroom technology use. *AACE Journal, 16*(1), 21–46.

Holotescu, C., Grosseck, G., Cretu, V., Naaji, A. (2014). Integrating MOOCs in blended courses. *Elearning & Software for Education*, (1), 243-250. doi:.10.12753/2066-026X-14-034

Horn, M. B. (2014). MOOCs for high school. *Education Next, 14*(3), 82–83.

Humphries, J., & Ness, M. (2015). Beyond who, what, where, when, why, and how: Preparing students to generate questions in the age of Common Core Standards. *Journal of Research in Childhood Education, 29*(4), 551–564. doi:10.1080/02568543.2015.1073199

ISTE Standards for Teachers. (2008). *International Society for Technology in Education*. Retrieved from: http://www.iste.org/docs/pdfs/20-14_ISTE_Standards-T_PDF.pdf

Johnston, T. C. (2013). Massive Open Online Courses (MOOCs): Student perspective. *Society For Marketing Advances Proceedings, 25*, 179-180.

Kassner, L. (2013). Mix it up with blended learning in K-12 schools. *Metropolitan Educational Research Consortium*. Accessed 16th January 2016 from http://files.eric.ed.gov/fulltext/ED548381.pdf

Levin, T., & Wadmany, R. (2008). Teachers' views on factors affect ting effective integration of information technology in the classroom: Developmental scenery. *Journal of Technology and Teacher Education, 16*, 233–263.

Li, Y., Zhang, M., Bonk, C. J., & Guo, Y. (2015). Integrating MOOC and flipped classroom practice in a traditional undergraduate course: Students' experience and perceptions. *International Journal of Emerging Technologies in Learning, 10*(6), 4–10. doi:10.3991/ijet.v10i6.4708

Liyanagunawardena, T., Adams, A., & Williams, S. (2013). MOOCs: A systematic study of the published literature 2008-2012. *International Review of Research in Open and Distance Learning, 14*(3), 202–227. Retrieved from http://www.irrodl.org/index.php/irrodl/article/view/1455/2531

Locke, M. (2013). MOOC: Will these four letters change K-12? *Scholastic Administrator, 12*(6), 39–40.

McGrory, K. (2013). Florida K-12 system experimenting with MOOCs. *Tampa Bay Times*. Retrieved from: http://www.tampabay.com/blogs/gradebook/florida-k-12-system-experimenting-with-moocs/2151286

Means, B., Toyama, Y., Murphy, R., Bakia, M., & Jones, K. (2010, September). *Evaluation of evidence-based practices in online learning: A meta-analysis and review of online learning studies*. U.S. Department of Education. Retrieved from http://www2.ed.gov/rschstat/eval/tech/evidence-based-practices/finalreport.pdf

Mehta, J. (2013). *The Allure of Order: High Hopes, Dashed Expectations, and the Troubled Quest to Remake Schooling. Troubled Quest to Remake American Schooling*. New York: Oxford University Press.

Montgomery, A. P., Hayward, D. V., Dunn, W., Carbonaro, M., & Amrhein, C. G. (2015)... *Australasian Journal of Educational Technology*, *31*(6), 657–670.

Mueller, A. L., Knobloch, N. A., & Orvis, K. S. (2015). Exploring the effects of active learning on high school students' outcomes and teachers' perceptions of biotechnology and genetics instruction. *Journal of Agricultural Education*, *56*(2), 138–152. doi:10.5032/jae.2015.02138

Nagel, D. (2009, July). Meta-analysis: Is blended learning most effective? *THE Journal*. Retrieved 16th January 2016 from http://thejournal.com/articles/2009/07/01/meta-analysis-is-blended-learning-most-effective.aspx

Najafi, H., Evans, R., & Federico, C. (2014). MOOC integration into secondary school courses. *International Review of Research in Open and Distance Learning*, *15*(5), 306–322.

Nelson, R. F. (2013). *Scaling up: Using MOOCs to increase accessibility and personalization for students*. Academic Press.

Norris, C., & Soloway, E. (2012). Higher Ed has MOOCs, but K12 still needs to catch up. *District Administration*, *48*(9), 96.

Pannoni, A. (2014). *MOOCs a new tool for high school teachers*. Retrieved from: http://www.usnews.com/education/blogs/high-school-notes/2014/10/27/moocs-a-new-tool-for-high-school-teachers?int=a9e809

Prensky, M. (2001). Digital natives, digital immigrants. On the Horizon, 9(5).

Ryan, R. M., & Deci, E. L. (2000). Intrinsic and extrinsic motivations: Classic definitions and new directions. *Contemporary Educational Psychology*, *25*(1), 54–67. doi:10.1006/ceps.1999.1020 PMID:10620381

Saadatmand, M., & Kumpulainen, K. (2014). Participants' perceptions of learning and networking in connectivist MOOCs. *Journal Of Online Learning & Teaching*, *10*(1), 16–30.

Sandeen, C. (2013). The emerging "MOOC 3.0" era. *Change*, *45*(6), 34–39. doi:10.1080/00091383.2013.842103

Scherer, M. (2014). The paperless classroom is coming. *Time Magazine*. Retrieved from: http://time.com/3483905/the-paperless-classroom-is-coming/

Siemens, G. (2012). MOOCs are really a platform. *Elearnspace*. Retrieved from: http://www.elearnspace.org/blog/2012/07/25/moocs-are-really-a-platform/

Staker, H., & Horn, M. B. (2012). Classifying K-12 blended learning. *Innosight Institute*. Retrieved from: http://www.christenseninstitute.org/wp-content/uploads/2013/04/Classifying-K-12-blended-learning.pdf

Suprabha, K., & Subramonian, G. (2015). Blended learning approach for enhancing students learning experiences in a knowledge society. *Journal of Educational Technology*, *11*(4), 1–7.

Ubuz, B., & Üstün, I. (2004). Figural and conceptual aspects in defining and identifying polygons. *Eurasian Journal of Educational Research*, *16*, 15–26.

UK Department for Education. (2014). MOOCs: opportunities for their use in compulsory-age education. *E-Learning Update*, (46), 9-10.

Werth, E., Werth, L., & Kellerer, E. (2013). Transforming K-12 rural education through blended learning: Barriers and promising practices. *The International Association for K-12 Online Learning*. Accessed 16th January 2016 from http://files.eric.ed.gov/fulltext/ED561276.pdf

Williams-Carlin, R. (2009). *A comparative study of geometry curricula*. (Unpublished doctoral dissertation). Louisiana State University, Baton Rouge, LA.

Compilation of References

Abadines, A. (2012). *Part 2: The Future of Education: Flip it! The Filipino Teacher.* Retrieved April 4, 2012, from http://thefilipinoteacher.com/2012/04/04/part-2-the-future-of-education-flip-it/

Abas, Z. W., & Hussain, R. M. R. (2013). *Acceptance of tablet devices for learning and teaching in institutions of higher learning: The Malaysian perspective*. New York: Routledge.

Abdous, M., Camarena, M., & Facer, B. R. (2009). MALL technology: Use of academic podcasting in the foreign language classroom. *ReCALL, 20*(1), 1–20.

Abernathye, K., Abshire, S., Cummings, C., & Mason, D. (2012). *A Conversation with the Lamar Academic Technology Committee: An Overview of Technology-enabled 21st Century Learning Spaces.* Retrieved April 26,2012, from http://www.slideshare.net/dianemason/academic-computingpreso4-2612kaysheryldmccfinal

Abreu, A. F. (2012). *Mobile learning: novos caminhos para o processo de ensino e aprendizagem*. Revista Discentis.

Acedo, M. (2010). *Pros And Cons Of A Flipped Classroom.* Retrieved from http://www. teachthought. com/trends/10-pros-cons-flipped-classroom

ACM & IEEE. (2013). *Computer Science Curricula 2013: Curriculum Guidlines for Undergraduate Degree Programs in Computer Science*. IEEE and ACM.

Adair, D., Alman, S. W., Budzick, D., Grisham, L. M., Mancini, M. E., & Thackaberry, A. S. (2014). Many shades of MOOCs. *Internet Learning Journal, 3*(1), 53–72.

Adriani, R. L. S. B. (2008). Jogos e M-learning: do veículo de comunicação ao instrumento de ensino. 73 f. Dissertação (Mestrado em Comunicação e Semiótica). Pontifícia Universidade Católica de são Paulo, São Paulo.

Ağca, R. K., & Özdemir, S. (2013). Foreign language vocabulary learning with mobile technologies. *Procedia: Social and Behavioral Sciences, 83*, 781–785. doi:10.1016/j.sbspro.2013.06.147

Aghili, M., Palaniappan, A. K., Kamali, K., Aghabozorgi, S., & Sardareh, A. S. (2014). Unifying informal and formal learning environments: Educational use of social network sites through implementing community of inquiry framework. *International Journal of e-Education, e-Business, e- Management Learning, 4*(3), 191–196.

Ahmed, S., & Parsons, D. (2013). Abductive science inquiry using mobile devices in the classroom. *Computers & Education, 63*, 62–72. doi:10.1016/j.compedu.2012.11.017

Ajjan, H., & Hartshorne, R. (2008). Investigating faculty decisions to adopt Web 2.0 technologies: Theory and empirical tests. *The Internet and Higher Education, 11*(2), 71–80. doi:10.1016/j.iheduc.2008.05.002

Alebaikan, R., & Troudi, S. (2010a). Blended learning in Saudi universities: Challenges and perspectives. *Research in Learning Technology, 18*(1), 49–59. doi:10.1080/09687761003657614

Alebaikan, R., & Troudi, S. (2010b). Online discussion in blended courses at Saudi Universities. *Procedia: Social and Behavioral Sciences*, *2*(2), 507–514. doi:10.1016/j.sbspro.2010.03.054

Alexander, B. (2004). Going nomadic: Mobile learning in higher education. *EDUCAUSE Review*, *39*(5), 29–35.

Al-Hunaiyyan, A., & Al-Sharhan, S. (2009, November). The design of multimedia blended e-learning systems: Cultural considerations. In *3rd International Conference on Signals, Circuits and Systems* (pp. 1-5). IEEE.

Ali, R., & Irvine, V. (2009). Current m-learning research: A review of key literature. In T. Bastiaens et al. (Eds.), *Proceedings of World Conference on E-Learning in Corporate, Government, Healthcare, and Higher Education 2009* (pp. 2353-2359). Chesapeake, VA: AACE.

Al-Kathiri, F. (2014). Beyond the Classroom Walls: Edmodo in Saudi Secondary School EFL Instruction, Attitudes and Challenges. *English Language Teaching*, *8*(1), 189. doi:10.5539/elt.v8n1p189

Allan, B. (2007). *Blended learning: tools for teaching and training*. London: Facet.

Allen, I. E., & Seaman, J. (2008). *Staying the course: Online education in the United States*. The Sloan Consortium.

Ally, M. (2009). *Mobile learning: Transforming the delivery of education and training*. Athabasca University Press.

Ally, M. (Ed.). (2009). *Mobile learning: transforming the delivery of education and training*. UBC Press.

Ally, M., & Needham, G. (Eds.). (2010). *M-Libraries: A virtual library in everyone's pocket*. London: Facet Publishing.

Ally, M., & Prieto-Blázquez, J. (2014). What is the future of m-learning in education? RUSC. *Universities and Knowledge Society Journal*, *11*(1), 142–151. doi:10.7238/rusc.v11i1.2033

Ally, M., & Tsinakos, A. (Eds.). (2014). *Increasing access through m-learning*. Vancouver, BC: Commonwealth of Learning Press.

Alonso, F., Manrique, D., & Viñes, J. M. (2009). A moderate constructivist e-learning instructional model evaluated on computer specialists. *Computers & Education*, *53*(1), 57–65. doi:10.1016/j.compedu.2009.01.002

Al-Zoubi, A., Tahat, A., & Hasan, O. (2005). *Mobile virtual experimentation utilizing SMS*. Paper presented at the 4th IASTED International Conference on Communications, Internet, and Information Technology.

Ames, K. (2015). *Online Education - Reflecting on Practice*. Retrieved from https://onlineedreflections.wordpress.com/2013/02/05/reflecting-on-the-seven-principles/

Anglin, G. J., & Morrison, G. R. (2000). An analysis of distance education research: Implications for the instructional technologist. *Quarterly Review of Distance Education*, *1*, 189–194.

Aragon, S. R. (2010). *Creating social presence in online environments. In New Directions for Adult and Continuing Education* (pp. 57–68). San Francisco: Jossey Bass.

Arbaugh, J. B. (2000). Virtual classroom characteristics and student satisfaction with internet-based MBA courses. *Journal of Management Education*, *24*(1), 32–54. doi:10.1177/105256290002400104

Arbaugh, J. B. (2008). Does the community of inquiry framework predict outcomes in online MBA courses? *International Review of Research in Open and Distance Learning*, *9*(2), 1–6.

Arbaugh, J. B., Bangert, A., & Cleveland-Innes, M. (2010). Subject matter effects and the Community of Inquiry (CoI) framework: An exploratory study. *The Internet and Higher Education*, *13*(1), 37–44. doi:10.1016/j.iheduc.2009.10.006

Arbaugh, J. B., & Benbunan-Fich, R. (2007). The importance of participant interaction in online environments. *Decision Support Systems*, *43*(3), 853–865. doi:10.1016/j.dss.2006.12.013

Arbaugh, J., & Hwang, A. (2006). Does 'teaching presence' exist in online MBA courses? *The Internet and Higher Education*, *9*(1), 9–21. doi:10.1016/j.iheduc.2005.12.001

Ariffin, S. A., & Dyson, L. E. (2015). Culturally Appropriate Design of Mobile Learning Applications in the Malaysian Context. In *Cross-Cultural Design Applications in Mobile Interaction, Education, Health, Transport and Cultural Heritage* (pp. 3–14). Springer International Publishing. doi:10.1007/978-3-319-20934-0_1

Ariffin, S. A., Dyson, L. E., & Hoskins-McKenzie, D. (2012). Content is king: Malaysian industry experts' point of view on local content for mobile phones'. *Journal of Mobile Technologies. Knowledge in Society*, *2012*, 1–9.

Arimbuyutan, R. C., Kim, S., Song, J. G., & So, W. (2007). A Study on e-Learning for Philippines. *International Journal of Multimedia and Ubiquitous Engineering*, *4*(2), 49–53.

Armellini, A., & De Stefani, M. (2015). Social presence in the 21st century: An adjustment to the Community of Inquiry framework. *British Journal of Educational Technology*, n/a. doi:10.1111/bjet.12302

Ashton, D. (2009). *The professional in the age of the amateur: Higher education and journalism on-the-job*. Retrieved from http://theendofjournalism.wdfiles.com/local--files/danielashton/Daniel Ashton.doc

Atherton, T. (2002). Developing ideas with Multimedia in the Primary Classroom. In A. Loveless & B. Dore (Eds.), *ICT in the Primary Classroom*. Buckingham, UK: Open University Press.

Atkeson, S. (2014, September24). Harvard-MIT partnership unveils new MOOCs for K-12. *Education Week*, 8.

Atteberry, E. (2013). *Flipped classrooms may not have any impact on learning*. USA Today, Retrieved from http://www.usatoday. com/story/news/nationa/2013/10/22/flipped-classrooms-effectiveness/3148447

Attewell, J. (2005). Mobile technologies and learning. London: Learning and Skills Development Agency.

Attewell, J. (2008). Towards sustainable large scale implementation of mobile learning: The mobile learning network (MoLeNET). In *The Bridge from text to context.Proceedings of the mLearn 2008 Conference* (pp. 28-35).

Auerbach, C. F., & Silverstein, L. B. (2003). *Qualitative data: An introduction to coding and analysis*. New York: New York University Press.

Ausubel, D. P. (2000). *The Acquisition and Retention of Knowledge: A Cognitive View*. Springer. doi:10.1007/978-94-015-9454-7

Azucena, J. (2014). *Studying flipped. Manila Bulletin*. Retrieved October 8, 2014, from http://www.mb.com.ph/studying-flipped/

Bachmair, B., & Pachler, N. (2013). Appropriation and composition in a culture characterised by provisionality. In M. Böck & N. Pachler (Eds.), *Multimodality and Social Semiosis: Communication, Meaning-Making and Learning in the Work of Gunther Kress* (pp. 211–220). New York: Routledge.

Baepler, P., & Murdoch, C. J. (2010). Academic Analytics and Data Mining in Higher Education. *International Journal for the Scholarship of Teaching and Learning*, *4*(2), 1–9. doi:10.20429/ijsotl.2010.040217

Bali, M. B. (2014). MOOC Pedagogy: Gleaning good practice from existing MOOCs. *Journal Of Online Learning & Teaching*, *10*(1), 44–56.

Banaji, S., & Burn, A. (2007). Creativity through a rhetorical lens: Implications for schooling, literacy and media education. *Literacy*, *41*(2), 62–70. doi:10.1111/j.1467-9345.2007.00459.x

Banaji, S., Burn, A., & Buckingham, D. (2010). *The Rhetorics of Creativity: A literature review.* London: Creativity, Culture and Education.

Bandura, A., & Schunk, D. H. (1981). Cultivating competence, self-efficacy, and instrinsic interest through proximal self-motivation. *Journal of Personality and Social Psychology*, *41*(3), 586–598. doi:10.1037/0022-3514.41.3.586

Barak, M., & Ziv, S. (2013). Wandering: A Web-based platform for the creation of location-based interactive learning objects. *Computers & Education*, *62*, 159–170. doi:10.1016/j.compedu.2012.10.015

Barbour, M., Brown, R., Waters, L. H., Hoey, R., Hunt, J. L., Kennedy, K., Onsworth, C., Powell, A., Trimm, T., (2006). Online and blended learning: A survey of policy and practice from K-12 schools around the world. *International Association for K-12 Online Learning.*

Barbour, R. (2008). *Doing focus groups.* Sage Publications Ltd.

Barcelos, T. S. (2014). *Relação entre o pensamento computacional e a Matemática em atividades didáticas de construção de jogos digitais. Tese (Doutorado).* São Paulo: Universidade Cruzeiro do Sul.

Baron, P., & Corbin, L. (2012). Thinking like a lawyer/acting like a professional: Communities of practice as a means of challenging orthodox legal education. *The Law Teacher*, *46*(2), 100–119. doi:10.1080/03069400.2012.681176

Barras, C. (2014). Can technology kill off the exam? *Future*. Retrieved November 15, 2014, from http://www.bbc.com/future/story/20140328-can-tech-kill-off-the-exam

Bartle, E. K., Longnecker, N., & Pegrum, M. (2011). Collaboration, contextualisation and communication using new media: Introducing podcasting into an undergraduate chemistry class. *International Journal of Innovation in Science and Mathematics Education*, *19*(1), 16–28.

Bath, D., & Bourke, J. (2011). The blending of blended learning: An experiential approach to academic staff development. In G. Williams, P. Statham, N. Brown, & B. Cleland (Eds.), Changing Demands, Changing Directions. Proceedings ascilite Hobart 2011 (pp. 133–138). Available at http://www.ascilite.org.au/conferences/hobart11/downloads/papers/Bath-concise.pdf

Batista, S. C. F. (2011). *M - LEARNMAT: Modelo pedagógico para atividades m-learning em matemática. Tese (Doutorado).* Porto Alegre: Universidade Federal do Rio Grande do Sul.

Begel, A., & Simon, B. (2008). Strugges of new colege graduates in their first software development job. *SIGCSE Bulletin*, *40*(1), 226–230. doi:10.1145/1352322.1352218

Beghetto, R. A. (2005). Does Assessment Kill Student Creativity? *The Educational Forum*, *69*(3), 254–263. doi:10.1080/00131720508984694

Beghetto, R. A. (2007). Does creativity have a place in classroom discussion? Prospective teachers' response preferences. *Thinking Skills and Creativity*, *2*(1), 1–9. doi:10.1016/j.tsc.2006.09.002

Beghetto, R. A. (2013). Expect the unexpected: Teaching for creativity in the micromoments. In M. Gregerson, J. C. Kaufman, & H. Snyder (Eds.), *Teaching creatively and teaching creativity* (pp. 133–148). New York: Springer Science. doi:10.1007/978-1-4614-5185-3_10

Beghetto, R. A., & Kaufman, J. C. (2010). Broadening conceptions of creativity in the classroom. In R. A. Beghetto & J. C. Kaufman (Eds.), *Nurturing creativity in the classroom*. Cambridge, UK: Cambridge University Press. doi:10.1017/CBO9780511781629.010

Beichner, R. J., Saul, J. M., Abbott, D. S., Morse, J. J., Deardorff, D., Allain, R. J., ... & Risley, J. S. (2007). The student-centered activities for large enrollment undergraduate programs (SCALE-UP) project. *Research-Based Reform of University Physics, 1*(1), 2-39.

Bell, H., & Van Dillen, F. (2013, May). iPads supporting good teaching practice. *New Zealand Education Gazette, 92*(7), 2–3.

Benshoff, J. M., & Gibbons, M. M. (2011). Bringing Life to e-Learning: Incorporating a Synchronous Approach to Online Teachign in Counselor Education. *The Professional Counselor: Research and Practice, 1*(1), 21–28. doi:10.15241/jmb.1.1.21

Berelson, B. (1952). *Content Analysis in Communication Research*. Glencoe, IL: Free Press.

Berge, Z. L., & Mrozowski, S. (2001). Review of research in distance education, 1990–1999. *American Journal of Distance Education, 15*(3), 5–19. doi:10.1080/08923640109527090

Bergmann, J., & Sams, A. (2014). Flipped learning: Maximizing face time. *T+ D, 68*(2), 28-31.

Bergmann, J., & Sams, A. (2012). *Flip your classroom: Reach every student in every class every day*. International Society for Technology in Education.

Biggs, J. (2003). *Teaching for quality learning at university: What the student does* (2nd ed.). Maidenhead, UK: Open University Press.

Billings, D. (2000). A framework for assessing outcomes and practices in web-based courses in nursing. *The Journal of Nursing Education, 39*, 60–67. PMID:10688463

Bird, J. (2012). *The rabbit ate the grass! Exploring children's activities on digital technologies in an early childhood classroom.* (Unpublished doctoral dissertation). Australian Catholic University.

Bird, J., & Edwards, S. (2014). Children learning to use technologies through play: A digital play framework. *British Journal of Educational Technology*. doi:10.1111/bjet.12191

Bishop, J. L., & Verleger, M. A. (2013, June). The flipped classroom: A survey of the research. In *ASEE National Conference Proceedings*.

Bitzer, P., Lehmann, K., Hirdes, E. M., & Thillainathan, N. (2012). Managing the Masses - Developing an Educational Dashboard for Lecturers in Large-Scale Lectures. In *European Conference on Information Systems (ECIS)*.

Blackboard. (2015). *Blackboard Webpage*. Retrieved February 25, 2015, from http://www.blackboard.com

Blackboard.com. *(n.d.). Blackboard Analytics for Learn*. Washington, DC: Blackboard Inc.

Blackwell, C., Lauricella, A., & Wartella, E. (2014). Factors influencing digital technology use in early childhood education. *Computers & Education, 77*, 82–90. doi:10.1016/j.compedu.2014.04.013

Blaikie, N. (2009). *Designing social research*. Polity Press.

Blaiklock, K. (2008). A critique of the use of Learning Stories to assess the learning dispositions of young children. *New Zealand Research in Early Childhood Education, 11*, 77–87.

Bligh, D. A. (1998). *What's the Use of Lectures?* Eastbourne: Intellect books.

Blumberg, P. (2008). *Developing learner-centered teachers: A practical guide for faculty*. San Francisco, CA: Jossey-Bass.

Bock, M., & O'Dea, V. (2013). Virtual educators critique the value of MOOCs for K-12. *Education Week, 32*(20), 10.

Bolstad, R. (2004). *Role and potential of ICT in early childhood education: A review of New Zealand and international literature*. Wellington, New Zealand: Ministry of Education. Retrieved November 12, 2014, from http://www.educationcounts.govt.nz/publications/ict/4983

Bone, A. (2009). The twenty-first century law student. *The Law Teacher, 43*(3), 222–245. doi:10.1080/03069400903332496

Bonk, C. J., Kim, K.-J., & Zeng, T. (2004). *Future Directions of Blended Learning in Higher Education and Workplace Settings* (C. J. Bonk & C. Graham, Eds.). San Francisco, CA: Pfeiffer Publishing.

Boston, W., Diaz, S. R., Gibson, A. M., Ice, P., Richardson, J., & Swan, K. (2009). An exploration of the relationship between indicators of the Community of Inquiry framework and retention in online programs. *Journal of Asynchronous Learning Networks, 13*(3), 67–83.

Boulos, M. N., Maramba, I., & Wheeler, S. (2006). Wikis, blogs and podcasts: A new generation of Web-based tools for virtual collaborative clinical practice and education. *BMC Medical Education, 6*(1), 41. doi:10.1186/1472-6920-6-41 PMID:16911779

Boyd, D. M., & Ellison, N. B. (2007). Social network sites: Definition, history, and scholarship. *Journal of Computer-Mediated Communication, 13*(1), 210–230. doi:10.1111/j.1083-6101.2007.00393.x

Bozkaya, M., Aydin, I. E., & Kumtepe, E. G. (2012). Research Trends and Issues in Educational Technology: A Content Analysis of TOJET (2008-2011). *Turkish Online Journal of Educational Technology-TOJET, 11*(2), 264–277.

Bradley, A. Z., Ulrich, S. M., Jones, M. Jr, & Maitland, S. M. (2002). Teaching the Sophomore Organic Course without a Lecture. Are You Crazy? *Journal of Chemical Education, 79*(4), 514–519. doi:10.1021/ed079p514

Brady, K. P., Holcomb, L. B., & Smith, B. V. (2010). The use of alternative social networking sites in higher educational settings: A case study of the e-learning benefits of Ning in education. *Journal of Interactive Online Learning, 9*(2), 151–170.

Braun, V., & Clarke, V. (2006). Using thematic analysis in psychology. *Qualitative Research in Psychology, 3*(2), 77–101. doi:10.1191/1478088706qp063oa

Briggs, R. O. (2006). On theory-driven design and deployment of collaboration systems. *International Journal of Human-Computer Studies, 64*(7), 573–582. doi:10.1016/j.ijhcs.2006.02.003

Brooker, L., & Edwards, S. (Eds.). (2010). Engaging Play. Maidenhead, UK: Open University Press.

Brosius, F. (2008). *SPSS 16*: Hüthig Jehle Rehm.

Brown, J. (2010). *Can you hear me?* Retrieved from http://www.judybrown.com/docs/76100228.pdf

Brown, J. S. (2005). *New Learning Environments for the 21st Century*. Retrieved from www.johnseelybrown. com/newlearning.pdf

Brown, D. J., McHugh, D., Standen, P., Evett, L., Shopland, N., & Battersby, S. (2011). Designing location-based learning experiences for people with intellectual disabilities and additional sensory impairments. *Computers & Education, 56*(1), 11–20. doi:10.1016/j.compedu.2010.04.014

Brown, E. (Ed.). (2010). *Education in the wild: contextual and location-based mobile learning in action. A report from the STELLAR Alpine Rendez-Vous workshop series*. Learning Sciences Research Institute, University of Nottingham.

Brown, J. S., Collins, A., & Duguid, P. (1989). Situated cognition and the culture of learning. *Educational Researcher*, *18*(1), 32–41. doi:10.3102/0013189X018001032

Bruner, J. (1960). *The Process of Education*. Cambridge, MA: Harvard University Press.

Bruner, J. (1990). *Acts of Meaning*. Cambridge, MA: Harvard University.

Bruner, J. (2006). In Search Of Pedagogy: Vol. I. *The Selected Works of Jerome Bruner 1957-1978*. New York: Routledge.

Brush, T., & Saye, J. (2000). Implementation and evaluation of a student –centered learning unit: A case study. *Educational Technology Research and Development*, *48*(3), 79–100. doi:10.1007/BF02319859

Bryant, J., & Heath, R. (2000). *Human communication theory and research: concepts, contexts, and challenges*. Mahwah, NJ: Lawrence Erlbaum Associates.

Buchem, I., Jahnke, I., & Pachler, N. (Eds.). (2013). Guest editorial preface. In Special Issue on Mobile Learning and Creativity: Current Concepts and Studies. *International Journal of Mobile and Blended Learning*, *5*(3), 1–3.

Butrymowicz, S. (2012). *Promise of 'flipped classroom' may elude poorer school districts. The Hechinger Report.* Retrieved June 12, 2012, from http://hechingerreport.org/promise-of-the-flipped-classroom-eludes-poorer-school-districts/

Butt, A. (2014). Student views on the use of a flipped classroom approach: Evidence from Australia. *Business Education & Accreditation*, *6*(1), 33–43.

Caceffo, R. E., da Rocha, H. V., & de Azevedo, R. J. (2011). Ferramenta de apoio para o aprendizado ativo usando dispositivos com caneta eletrônica. *Revista Brasileira de Informática na Educação*, *19*(02), 25.

Cachia, R., Ferrari, A., Ala-Mutka, K., & Punie, Y. (2010). *Creative Learning and Innovative Teaching. Final Report on the Study on Creativity and Innovation in Education in the EU Member States*. Seville: EC JRC-IPTS.

Campbell, A. (2002). Research and the professional self. In O. McNamara (Ed.), *Becoming an evidence-based practitioner: A framework for teacher-researchers* (pp. 27–36). London: Routledge/Falmer.

Carneiro, P., & Heckman, J. (2003). Human Capital Policy. In Inequality in America: What Role for Human Capital Policies?. Cambridge, MA: MIT.

Carr, M., May, H., Podmore, V. N., Cubey, P., Hatherly, A., & Macartney, B. (2002). Learning and teaching stories: Action research on evaluation in early childhood in Aotearoa-New Zealand. *European Early Childhood Education Research Journal*, *10*(2), 115–125. doi:10.1080/13502930285208991

Carvalho, A. A., Aguiar, C., & Maciel, R. (2009). A taxonomy of podcasts and its application to higher education. In H. Damis & L. Creanor (Eds.), *"In dreams begins responsibility"—choice, evidence and change. The 16th Association for Learning Technology Conference (ALT-C 2009)*, (pp. 132–140). Oxford, UK: Association for Learning Technology.

Casey, S. (2009). *The New literacies of m-learning*. (Unpublished master's thesis). Concordia University, Canada (UMR:67156).

Cater, J. J. III, Michel, N., & Varela, O. E. (2012). Challenges of Online Learning in Management Education: An Empirical Study. *Journal of Applied Management and Entrepreneurship*, *17*(4), 76–96.

Chandra, V. (2004). *The impact of a blended web-based learning environment on the perceptions, attitudes and performance of boys and girls in junior science and senior physics*. Perth: Curtin University of Technology.

Chen, C., & Chung, C. (2008). Personalized mobile English vocabulary learning system based on item response theory and learning memory cycle. *Computers & Education*, *51*(2), 624–645. doi:10.1016/j.compedu.2007.06.011

Chen, C.-H., Liu, G.-Z., & Hwang, G.-J. (2015). Interaction between gaming and multistage guiding strategies on students' field trip mobile learning performance and motivation. *British Journal of Educational Technology*, n/a. doi:10.1111/bjet.12270

Chen, W., & Looi, C. K. (2007). Incorporating online discussion in face to face classroom learning: A new blended learning approach. *Australasian Journal of Educational Technology, 23*(3), 307–326.

Chen, X. (2009). *Students Who Study Science, Technology, Engineering, and Mathematics (STEM) in Postsecondary Education. Stats in Brief. NCES 2009-161*. National Center for Education Statistics.

Cheon, J., Lee, S., Crooks, S. M., & Song, J. (2012). An investigation of mobile learning readiness in higher education based on the theory of planned behavior. *Computers & Education, 59*(3), 1054–1064. doi:10.1016/j.compedu.2012.04.015

Chickering, A.W., & Ehrmann, S.C. (1996). Implementing the Seven Principles: Technology as Lever. *American Associates for Higher Education Bulletin, 49*(1-10), 3-6.

Chickering, A.W., & Gamson, Z.F. (1987). Seven Principles for Good Practice in Undergraduate Education. *American Associates for Higher Education Bulletin*, 3-7.

Chin, W. W. (1998). The partial least squares approach for structural equation modeling. In G. A. Marcoulides (Ed.), *Modern methods for business research* (p. 437). Mahwah, NJ: Lawrence Erlbaum Associates Publishers.

Chiou, C.-K., Tseng, J. C. R., Hwang, G.-J., & Heller, S. (2010). An adaptive navigation support system for conducting context-aware ubiquitous learning in museums. *Computers & Education, 55*(2), 834–845. doi:10.1016/j.compedu.2010.03.015

Chisholm, L. A. (2005). *Micro-learning in the lifelong learning context*. Academic Press.

Chu, H. C., Hwang, G. J., & Tsai, C. C. (2010). A knowledge engineering approach to developing mind tools for context-aware ubiquitous learning. *Computers & Education, 54*(1), 289–297. doi:10.1016/j.compedu.2009.08.023

Chu, H., Hwang, G., Tsai, C., & Tseng, J. C. R. (2010). A two-tier test approach to developing location-aware mobile learning systems for natural science courses. *Computers & Education, 55*(4), 1618–1627. doi:10.1016/j.compedu.2010.07.004

Clark, N. (2014). Are grades at postgraduate level irrelevant? *The Guardian*. Retrieved January 5, 2014, from http://www.theguardian.com/education/mortarboard/2014/apr/01/students-grades-postgraduate-course

Clark, R., Norman, B., & Besterfield-Sacre, M. (2014). Preliminary Experiences with "Flipping" a Facility Layout/Material. *Proceedings of the 2014 Industrial and Systems Engineering Research Conference.*

Classmarker. (2015). *Classmarker Webpage*. Retrieved March 25th, 2015, from http://www.classmarker.com

Claxton, G. (2000). A map of learning to learn. In B. Lucas & T. Graeny (Eds.), *Learning to Learn: Setting the Agenda for Schools in the 21st Century*. London: Campaign for Learning.

Clements, D. H., & Sarama, J. (2003). Strip Mining for Gold: Research and Policy in Educational Technology - A Response to "Fool's Gold". *Educational Technology Review, 11*(1), 7–69.

Coates, H. (2007). A model of online and general campus-based student engagement. *Assessment & Evaluation in Higher Education, 32*(2), 121–141. doi:10.1080/02602930600801878

Cobcroft, R. S., Towers, S. J., Smith, J. E., & Bruns, A. (2006). *M-learning in review: Opportunities and challenges for learners, teachers, and institutions*. Retrieved from August 30, 2015, from: http://eprints.qut.edu.au/5399/1/5399.pdf

Cochrane, T. (2011). Exploring the Pedagogical Affordances of Mobile Web 2.0. In L. Chao (Ed.), Open Source Mobile Learning: Mobile Linux Applications (pp. 196-215). Hershey, PA: Information Science Reference (an imprint of IGI Global).

Cochrane, T. (2012a). Critical success factors for transforming pedagogy with mobile Web 2.0. *British Journal of Educational Technology*. 10.1111/j.1467-8535.2012.01384.x

Cochrane, T., & Bateman, R. (2013). A mobile web 2.0 framework: Reconceptualizing teaching and learning. In M. Repetto & G. Trentin (Eds.), Using network and mobile technology to bridge formal and informal learning (10th ed.; pp. 57-92). Oxford, UK: Chandos Publishing.

Cochrane, T., & Withell, A. (2013). *Do 21st Century students dream of electric sheep? A mobile social media framework for creative pedagogies.* Paper presented at the Electric Dreams: Proceedings 30th ascilite Conference, Sydney, Australia.

Cochrane, T., Antonczak, L., Guinibert, M., & Mulrennan, D. (2014). *Developing a mobile social media framework for creative pedagogies.* Paper presented at the 10th International Conference on Mobile Learning, Madrid, Spain.

Cochrane, T., Antonczak, L., Guinibert, M., Withell, A., Mulrennan, D., & Mountfort, P. (2013). *Collaboration unplugged: Herding a flock of MOAs.* Paper presented at the Electric Dreams: 30th ascilite Conference, Sydney, Australia.

Cochrane, T., Black, B., Lee, M., Narayan, V., & Verswijvelen, M. (2012). Rethinking elearning support strategies. *International Journal for Academic Development,* 1-18.

Cochrane, T., Mulrennan, D., Sissons, H., Pamatatau, R., & Barnes, L. (2013). *Mobilizing Journalism Education.* Paper presented at the International Conference on Information Communication Technologies in Education (ICICTE 2013), Crete, Greece.

Cochrane, T., Munn, J., & Antonczak, L. (2013). *Design thinking for mlearning: Herding a flock of MOAs.* Paper presented at the 3rd Mobile Creativity and Innovation Symposium, Auckland, New Zealand.

Cochrane, T., Sissons, H., & Mulrennan, D. (2012). Journalism 2.0: Exploring the impact of Mobile and Social Media on Journalism Education. In I. A. Sánchez & P. Isaias (Eds.), *Proceedings of the IADIS International Conference on Mobile Learning 2012* (pp. 165-172). Berlin, Germany: IADIS International Association for Development of the Information Society.

Cochrane, T. (2010). Beyond the Yellow Brick Road: Mobile web 2.0 informing a new institutional elearning strategy. *ALT-J. Research in Learning Technology, 18*(3), 221–231.

Cochrane, T. (2012b). Secrets of mlearning failures: Confronting reality. In D. Hawkridge, S. Verjans, & G. Wilson (Eds.), *Proceedings of ALT-C 2012 - a confrontation with reality: The 19th international conference of the Association for Learning Technology* (pp. 123-134). ALT Association for Learning Technology.

Cochrane, T., & Bateman, R. (2010). Smartphones give you wings: Pedagogical affordances of mobile web 2.0. *Australasian Journal of Educational Technology, 26*(1), 1–14.

Cochrane, T., & Narayan, V. (2013). Redesigning professional development: Reconceptualising teaching using social learning technologies. *Research in Learning Technology, 21*(0), 1–19. doi:10.3402/rlt.v21i0.19226

Cochrane, T., Sissons, H., Mulrennan, D., & Pamatatau, R. (2013). Journalism 2.0: Exploring the impact of Mobile and Social Media on Journalism Education. *International Journal of Mobile and Blended Learning, 5*(2), 22–38. doi:10.4018/jmbl.2013040102

Cockbain, M., Blyth, C., Bovill, C., & Morss, K. (2009). Adopting a blended approach to learning: Experiences from radiography at Queen Margaret University, Edinburgh. *Science Direct, 15*, 242–246.

Collard, D. M., Girardot, S. P., & Deutsch, H. M. (2002). From the Textbook to the Lecture: Improving Pre-Lecture Preparation in Organic Chemistry. *Journal of Chemical Education*, *79*(4), 520–523. doi:10.1021/ed079p520

Conley, D. (2010). *College and career ready*. San Francisco, CA: Jossey-Bass. doi:10.1002/9781118269411

Connors, M., Bredeson, C., & Al-Shamoli, F. (2011). Distance Education Introductory Physics Labs: Online or In-Home? In J. G. Zubía & G. R. Alves (Eds.), *Using Remote Labs in Education* (pp. 309–321). Bilbao: University of Deusto.

Conole, G., & Dyke, M. (2004). What are the affordances of information and communication technologies? *ALT-J. Research in Learning Technology*, *12*(2), 113–124. doi:10.1080/0968776042000216183

Conole, G., Oliver, M., Falconer, I., Littlejohn, A., & Harvey, J. (2007). Designing for learning. In G. Conole & M. Oliver (Eds.), *Contemporary perspectives in e-learning research: themes, methods and impact on practice, Open and Distance Learning Series*. London: RoutledgeFalmer.

Conrad, D. (2005). Building and maintaining community in cohort based online learning. *Journal of Distance Education*, *20*(1), 1–20.

Cook, J. (2009). Phases of mobile learning. *Joint European summer school on technology enhanced learning*.

Coolican, H. (2014). *Research methods and statistics in psychology*. New York, NY: Psychology Press.

Corbeil, J. R., & Valdés-Corbeil, M. E. (2007). Are you ready for m-learning? *EDUCAUSE Quarterly*, *30*(2). Retrieved from http://er.educause.edu/articles/2007/1/are-you-ready-for-mobile-learning AreYouReadyforMobileLearning/157455

Cordes, C., & Miller, E. (2000). *Fool's gold: A critical look at computers in childhood*. College Park, MD: Alliance for Childhood.

Corlett, D., Sharples, M., Bull, S., & Chan, T. (2005). Evaluation of an m-learning organiser for university students. *Journal of Computer Assisted Learning*, *21*(3), 162–170. doi:10.1111/j.1365-2729.2005.00124.x

Corter, J. E., Nickerson, J. V., Esche, S. K., Hassapis, C., Im, S., & Ma, J. (2007). Constructing reality: A study of remote, hands-on, and simulated laboratories. *ACM Transactions on Computer-Human Interaction*, *14*(2), 7, es. doi:10.1145/1275511.1275513

Costa, R. (2005). Tele-Experimentação Móvel (Mobile Remote Experimentation)-Considerações sobre uma área emergente no ensino à distância. *Journal of Scientific Activity at ISEP, 15*.

Cottrell, D. M., & Robinson, R. A. (2003). Blended Learning in an Accounting Course. *Quarterly Review of Distance Education*, *4*(3), 261–269.

Craft, A. (2000). *Creativity across the primary curriculum: framing and developing practice*. London: Routledge.

Creswell, J. W. (2008). *Educational research: Planning, conducting, and evaluating quantitative and qualitative research*. Upper Saddle River, NJ: Pearson Prentice Hall.

Creswell, J. W. (2009). *Research design: Qualitative, quantitative, and mixed methods approaches*. Los Angeles, CA: Sage Publications.

Creswell, J., & Plano Clark, V. (2007). *Designing and Conducting Mixed Methods Research*. Thousand Oaks: Sage.

Crompton, H. (2014). A diachronic overview of mobile learning: A shift toward student-centered pedagogies. In M. Ali & A. Tsinakos (Eds.), Increasing access through mobile learning (pp. 7–16). British Columbia, Canada: Commonwealth of Learning Press. Available at http://www.col.org/PublicationDocuments/pub_Mobile%20Learning_web.pdf

Crouch, C. H., & Mazur, E. (2001). Peer instruction: Ten years of experience and results. *American Journal of Physics*, *69*(9), 970–977. doi:10.1119/1.1374249

Crow, R., Santos, I. M., LeBaron, J., McFadden, A. T., & Osborne, C. F. (2010). Switching gears: Moving from e-learning to m-learning. *Journal of Online Learning and Teaching*, *6*(1), 268–278.

Cruze, C. I. (2011). *TPACK: a framework for educators to implement technology in their classrooms*. Academic Press.

Csikszentmihalyi, M., & Sawyer, K. (1995). Creative insight: the social dimension of a solitary moment. In R. J. Sternberg & J. E. Davidson (Eds.), *The Nature of Insight* (pp. 329–363). Cambridge, MA: Bradford/MIT.

Dana, N. F., & Yendol-Silva, D. (2003). *The reflective educator's guide to classroom research: Learning to teach and teaching to learn through practitioner inquiry*. Thousand Oaks, CA: Sage.

Danvers, J. (2003). Towards a radical pedagogy: Provisional notes on learning and teaching in art & design. *International Journal of Art & Design Education*, *22*(1), 47–57. doi:10.1111/1468-5949.00338

Daspit, J. J., & D'Souza, D. E. (2012). Using the community of inquiry framework to introduce wiki environments in blended-learning pedagogies: Evidence from a business capstone course. *Academy of Management Learning & Education*, *11*(4), 666–683. doi:10.5465/amle.2010.0154

Davies, J., & Merchant, G. (2009). *Web 2.0 for Schools: Learning and Social Participation*. New York, NY: Peter Lang.

Davies, T. (1999). Taking Risks as a Feature of Creativity in the Teaching and Learning of Design and Technology. *The Journal of Design and Technology Education*, *4*(2), 101–108.

Davis, F. D. (1989). Perceived usefulness, perceived ease of use, and user acceptance of information technology. *Management Information Systems Quarterly*, *13*(3), 319–340. doi:10.2307/249008

Davis, N. E., & Roblyer, M. D. (2005). Preparing teachers for the "Schools that technology built" Evaluation of a program to Train teachers for virtual schooling. *Journal of Research on Technology in Education*, *37*(4), 399–409. doi:10.1080/15391523.2005.10782445

Dawabi, P., Dietz, L., Fernandez, A., & Wessner, M. (2003). Using PDAs to support face-to-face learning. In Computer Support for Collaborative Learning. Academic Press.

Dawley, L. (2009). Social network knowledge construction: Emerging virtual world pedagogy. *On the Horizon*, *17*(2), 109–121. doi:10.1108/10748120910965494

Day, J., & Foley, J. D. (2006). Evaluating a web lecture intervention in a human–computer interaction course. *Education. IEEE Transactions on*, *49*(4), 420–431.

de Clunie, G. T., Serrão, T., Monteiro Braz, J.-L., Serrão, T., Rangel, N., & Castillo, A. (2012). Developing an Android based learning application for mobile devices. In *Telematics and Information Systems (EATIS), 6th Euro American Conference on*. doi:10.1145/2261605.2261641

De Dios, A. C. (2014, June 2). *Philippine Basic Education "Pagputi Ng Uwak"*. Retrieved June 2, 2014, from http://philbasiceducation.blogspot.com/search/label/Classrooms#ixzz3ONET7KRR

De Lucia, A., Francese, R., Passero, I., & Tortora, G. (2012). A Collaborative Augmented Campus Based on Location-Aware Mobile Technology. *International Journal of Distance Education Technologies*, *10*(1), 55–73. doi:10.4018/jdet.2012010104

De Waard, I. (2014). *The Flipped Classroom. Presentation given at the Gulden Sporen College in Kortrijk, Belgium for one of their SOS sessions (pedagogical sessions). Educational researcher and practical implementer at Athabasca University and Open University of UK*. Retrieved November 20, 2014 from http://www.slideshare.net/ignatia/the-flipped-classroom-introduction-and-sources

Dekhane, S., & Tsoi, M. Y. (2012). Designing a Mobile Application for Conceptual Understanding: Integrating Learning Theory with Organic Chemistry Learning Needs. *International Journal of Mobile and Blended Learning*, *4*(3), 34–52. doi:10.4018/jmbl.2012070103

Delgado, F. (2012). *Numerical Methods You Tube Channel*. Retrieved November 15, 2014, from https://www.youtube.com/playlist?list=PLFF477660D37081E4

Delgado, F., & Santiago, R. (2014). Designing a site to embed and to interact with Wolfram Alpha widgets in Math and Sciences courses. In *Proceedings of Mobile Learning International Conference 2014* (pp. 266-270). Madrid, Spain: IADIS. Retrieved March 15, 2014, from http://aleesp.hypotheses.org/rg/files/2014/03/ML2014_proceed.pdf

Delgado, F. (2013). A numerical methods course based on b-learning: Integrated learning design and follow up. *International Journal of Mobile and Blended Learning*, *5*(1).

Delgado, F. (2014). Actualización docente en tecnologías educativas y aprendizaje móvil: Desarrollo de un programa institucional. *Revista de Formación e Innovación Educativa Universitaria*, *7*(4), 211–226.

Delgado, F., & Martínez, S. (2011). Cambios curriculares generados por el empleo de tecnología en la enseñanza de los métodos numéricos. In *Memorias de la XXXVIII conferencia de ANFEI*. Querétaro, México: ANFEI.

Delgado, F., Santiago, R., & Quezada, L. (2015). Widget based learning to develop meaningful comprehension in math and science courses. In *Proceedings of INTED 2015*.

Delialioğlu, Ö. (2012). Student engagement in blended learning environments with lecture-based and problem-based instructional approaches. *Journal of Educational Technology & Society*, *15*(3), 310–322.

Delialioglu, O., & Yildirim, Z. (2007). Students' perceptions on effective dimensions of interactive learning in a blended learning environment. *Journal of Educational Technology & Society*, *10*(2), 133–146. Retrieved from http://search.proquest.com/docview/61934868?accountid=16285

Demsey, L. (2008). Always on: libraries in a world of permanent connectivity. In G. Needham & M. Ally (Eds.), *M-libraries: Libraries on the move to provide virtual access* (pp. xxv–lii). London: Facet Publishing.

Dermirci, N. D. (2014). What is Massive Open Online Courses (MOOCs) and what is promising us for learning? A review-evaluative article about MOOCs. (English). *Necatibey Faculty Of Education Electronic Journal Of Science & Mathematics Education*, *8*(1), 231–256. doi:10.12973/nefmed.2014.8.1.a10

Derntl, M., & Motschnig-Pitrik, R. (2005). The role of structure, patterns and people in blended learning. *The Internet and Higher Education*, *8*(2), 111–130. doi:10.1016/j.iheduc.2005.03.002

Deslauriers, L., Schelew, E., & Wieman, C. (2011). Improved learning in a large-enrollment physics class. *Science*, *332*(6031), 862-864.

Dickinson, L. (1995). Autonomy, self-direction and self-Access in language teaching and learning: The history of an idea. *System*, *23*(2), 165–174. doi:10.1016/0346-251X(95)00005-5

Dill, E. M. (2012). *The Impact of Flip Teaching on Student Homework Completion, Behavior, Engagement and Proficiency*. University of New England.

DiPrete, T. A., & Eirich, G. M. (2006). Cumulative Advantage as a Mechanism for Inequality: A Review of Theoretical and Empirical Developments. *Annual Review of Sociology*, *32*(1), 271–297. doi:10.1146/annurev.soc.32.061604.123127

Dobbin, G., Dahlstrom, E., Arroway, P., & Sheehan, M. (2011). *Mobile IT in higher education, 2011*. Educause Center for Applied Research, Research Report.

Donohue, C. (2003). Technology in early childhood education. *Child Care Information Exchange*, *154*, 17–22.

Doyle, L., Brady, A., & Byrne, G. (2009). An overview of mixed methods research. *Journal of Research in Nursing*, *14*(2), 175–185. doi:10.1177/1744987108093962

Drew, C. (2011). Why Science Majors Change Their Mind. *The New York Times*.

Driscoll, M. (2002). Blended Learning: Let's get beyond the hype. *E-learning*, *1*(4).

Drysdale, J. S., Graham, C. R., Spring, K. J., & Halverson, L. R. (2013). An analysis of research trends in dissertations and theses studying blended learning. *The Internet and Higher Education*, *17*, 90–100. doi:10.1016/j.iheduc.2012.11.003

Dufresne, R. J., Gerace, W. J., Leonard, W. J., Mestre, J. P., & Wenk, L. (1996). Classtalk: A classroom communication system for active learning. *Journal of Computing in Higher Education*, *7*(2), 3–47. doi:10.1007/BF02948592

Duncan-Howell, J. A., & Lee, K.-T. (2007). *M-Learning–Innovations and Initiatives: Finding a place for mobile technologies within tertiary educational settings*. Auckland.

Dunlap, J. C., & Lowenthal, P. R. (2010). Investigating Twitter's ability to enhance social presence. In *Annual Meeting of the International Society for Technology in Education* (pp. 1-25). Denver, CO: Academic Press.

DuVall, J. B., Powell, M. R., Hodge, E., & Ellis, M. (2007). Text messaging to improve social presence in online learning. *EDUCAUSE Quarterly*, *30*(3), 24–28.

Dyson, L. E., Lawrence, E., Litchfield, A. J., & Zmijewska, A. (2008). M-fieldwork for information systems students. *Proceedings of the Forty-First Annual Hawaii International Conference on System Sciences (HICSS)* (pp. 1-10).

Dyson, L. E., & Litchfield, A. (2011). Advancing Collaboration between M-Learning Researchers and Practitioners through an Online Portal and Web 2.0 Technologies. *International Journal of Mobile and Blended Learning*, *3*(1), 64–72. doi:10.4018/jmbl.2011010104

Dyson, L. E., Litchfield, A., Raban, R., & Tyler, J. (2009). *Interactive classroom mLearning and the experiential transactions between students and lecturer*. Ascilite.

Eaton, J.S. (2001). Distance Learning: Academic and Political Challenges for Higher Education Accreditation. *Council for Higher Education Accreditation (CHEA), 1.*

Eco, U. (1964/1994). *Apocalypse Postponed* [Apocalittici e integrati]. R. Lumley (Ed.). Bloomington, In: Indiana UP.

Egan, G. (2014). *The Skilled Helper* (10th ed.). Belmont, CA: Brooks/Cole.

Eigen, M. (1983). A Note on the Structure of Freud's Theory of Creativity. *Psychoanalytic Review*, *70*(1), 41–45. PMID:6410431

El-Mowafy, A., Kuhn, M., & Snow, T. (2013). Blended learning in higher education: Current and future challenges in surveying education. *Issues in Educational Research*, *23*(2), 132–150.

eMarketer. (2014, January 16). *Smartphone Users Worldwide Will Total 1.75 Billion in2014*. Retrieved from eMarketer: http://www.emarketer.com/Article/Smartphone-Users-Worldwide-Will-Total-175-Billion-2014/1010536

Engeström, Y. (1996). Development as breaking away and opening up: A challenge to Vygotsky and Piaget. *Swiss Journal of Psychology, 55*(2/3), 126–132.

Engeström, Y. (1999). *Lernen durch Expansion.* Marburg: BdWi- Verlag.

Eom, S. B., Wen, H. J., & Ashill, N. (2006). The Determinants of Students' Perceived Learning Outcomes and Satisfaction in University Online Education: An Empirical Investigation*. *Decision Sciences Journal of Innovative Education, 4*(2), 215–235. doi:10.1111/j.1540-4609.2006.00114.x

Ernst & Young. (2012). *University of the future: A thousand year old industry on the cusp of profound change.* Retrieved from http://www.ey.com/Publication/vwLUAssets/ University_of_the_future/$FILE/University_of_the_future_2012.pdf

Ertmer, P. A. (2005). Teacher pedagogical beliefs: The final frontier in our quest for technology integration? *Educational Technology Research and Development, 53*(4), 25–39. doi:10.1007/BF02504683

Esperanza, P., & Toto, C. (2014, October). Does the Implementation of a Flipped Classroom Pedagogy changes the Students' Attitude towards Mathematics? In *World Conference on E-Learning in Corporate, Government, Healthcare, and Higher Education* (Vol. 2014, No. 1, pp. 552-557).

Estes, L. (2014). Flipped classrooms' get mixed reviews. *McClatchy Tribune Business News.* Retrieved April 08, 2014, from http://www.mcclatchydc.com/2012/06/12/152089/promise-of-flipped-classroom-may.html

Evans, C., & Gibbons, N. J. (2007). The interactivity effect in multimedia learning. *Computers &amp. Education, 49*(4), 1147–1160.

Fagan, T., & Coutts, T. (2012). *To iPad or not to iPad?* Christchurch, New Zealand: CORE Education. Retrieved November 3, 2014, from http://www.core-ed.org/thought-leadership/research/ipad-or-not-ipad

Fagen, A. P., Crouch, C. H., & Mazur, E. (2002). Peer instruction: Results from a range of classrooms. *The Physics Teacher, 40*, 206.

Farman, J. (2012). *Mobile Interface Theory.* New York: Routledge.

Fedoce, R. S., & Squirra, S. C. (2011). A tecnologia móvel e os potenciais da comunicação na educação. *LOGOS, Comunicação & Universidade*, 267-278.

Fedoce, R. S. (2010). *A tecnologia móvel e os potenciais da comunicação na educação. Dissertação (Mestrado).* São Bernardo do Campo, SP: Universidade Metodista de São Paulo.

Ferdig, R. E. (2007). Editorial: Examining social software in teacher education. *Journal of Technology and Teacher Education, 15*(1), 5–10.

Ferdig, R. E. (2014). Preparing for K-12 MOOCs. *Tech & Learning, 34*(6), 26–27.

Ferrari, A., Cachia, R., & Punie, Y. (2009). *Innovation and Creativity in Education and Training in the EU Member States: Fostering Creative Learning and Supporting Innovative Teaching. Literature review on Innovation and Creativity in E&T in the EU Member States (ICEAC).* Seville: EC JRC-IPTS.

Fidalgo, A. V. S., Rochandel, W., & Silva, J. B. d. (2013). *Remote experimentation using mobile devices.* Paper presented at the Engineering Education (CISPEE), 2013 1st International Conference of the Portuguese Society for.

Findlay-Thompson, S., & Mombourquette, P. (2013, July). Evaluation of a flipped classroom in an undergraduate business course. In *Global Conference on Business & Finance Proceedings* (Vol. 8, No. 2, p. 138). Institute for Business & Finance Research.

Fischer, G., Giaccardi, E., Eden, H., Sugimoto, M., & Ye, Y. (2005). Beyond binary choices: Integrating individual and social creativity. *International Journal of Human-Computer Studies*, *63*(4-5), 482–512. doi:10.1016/j.ijhcs.2005.04.014

Fischer, G., Scharff, E., & Ye, Y. (2004). Fostering Social Creativity by Increasing Social Capital. In M. Huysman & V. Wulf (Eds.), *Social Capital and Information Technology* (pp. 355–399). Cambridge, MA: MIT Press.

FitzGerald, E. (2012). Creating user-generated content for location-based learning: An authoring framework. *Journal of Computer Assisted Learning*, *28*(3), 195–207. doi:10.1111/j.1365-2729.2012.00481.x

Fitzgerg. (2012). *Flipping The Classroom Eduphoria*. Retrieved from http://www.slideshare.net/fitzgerg/flipping-the-classroom-mhs-tech-camp

FLN & Sophia Learning. (2014). *Growth in Flipped Learning: Transitioning the focus from teachers to students for educational success: Transitioning the focus from teachers to students for educational success*. Retrieved May, 2014, from http://flippedlearning.org/Page/95

FLN the Flipped Learning Network. (2014). Retrieved from http://flippedlearning.org/site/default.aspx?PageID=1

Fonseca, A. G. M. F. (2013). Aprendizagem, mobilidade e convergência: Mobile Learning com Celulares e *Smartphones*. *Revista Eletrônica do Programa de Pós-Graduação em Mídia e Cotidiano*, *2*(2), 265–283.

Ford, V. B., & Roby, D. E. (2013). Why do high school students lack motivation in the classroom? *Global Education Journal*, *2013*(2), 101-113.

Franco, R. O. S. (2010). *Geolearning; Aplicativo móvel para apoio ao aprendizado de vocabulário de língua inglesa com conteúdo geolocalizado. Dissertação (Mestrado)*. Recife, PE: Universidade Federal de Pernambuco.

Friesen, N., & Lowe, S. (2012). The questionable promise of social media for education connective learning and the commercial imperative. *Journal of Computer Assisted Learning*, *28*(3), 183–194. doi:10.1111/j.1365-2729.2011.00426.x

Frohberg, D., Göth, C., & Schwabe, G. (2009). Mobile learning projects–a critical analysis of the state of the art. *Journal of Computer Assisted Learning*, *25*(4), 307–331. doi:10.1111/j.1365-2729.2009.00315.x

Frydenberg, M. (2008). Principles and pedagogy: The two PS of podcasting in the information technology classroom. *Information Systems Education Journal*, *6*(6), 3–11.

Fullan, M., & Langworthy, M. (2014). *A rich seam: How new pedagogies find deep learning*. London: Pearson.

Fulton, K. (2012). Upside down and inside out: Flip Your Classroom to Improve Student Learning. *Learning and Leading with Technology*, *39*(8), 12–17.

Gagné, E. D., Yekovich, C. W., & Yekovich, F. R. (1993). *The cognitive psychology of school learning*. New York: HarperCollins College.

Gardner, H. (1983). *Frames of Mind: The Theory of Multiple Intelligences*. New York: Basic Books.

Garrett, N., Thoms, B., Soffer, M., & Ryan, T. (2007). *Extending the Elgg social networking system to enhance the campus conversation*. Pasadena, CA: Second Annual Design Research in Information Systems.

Garrison, D. R. & Anderson, T. (2003). *E-Learning in the 21st Century: A Framework for Research and Practice*. Routledge.

Garrison, D. R., & Akyol, Z. (2013). The community of inquiry theoretical framework. In Handbook of distance education (pp. 104-119). Routledge. doi:10.4324/9780203803738.ch7

Garrison, D. R., & Vaughan, N. D. (2011). *Blended Learning in Higher Education: Framework, Principles, and Guidelines*. Retrieved from http://jcu.eblib.com.au/patron/FullRecord.aspx?p=819029

Garrison, D. R., Anderson, T., & Archer, W. (2000). Critical inquiry in a text-based environment: Computer conferencing in higher education. *The Internet and Higher Education*, *2*(2), 87–105.

Garrison, D. R., Anderson, T., & Archer, W. (2001). Critical thinking, cognitive presence, and computer conferencing in distance education. *American Journal of Distance Education*, *15*(1), 7–23. doi:10.1080/08923640109527071

Garrison, D. R., Anderson, T., & Archer, W. (2010). The first decade of the community of inquiry framework: A retrospective. *The Internet and Higher Education*, *13*(1), 5–9. doi:10.1016/j.iheduc.2009.10.003

Garrison, D. R., & Arbaugh, J. B. (2007). Researching the community of inquiry framework: Review, issues, and future directions. *The Internet and Higher Education*, *10*(3), 157–172. doi:10.1016/j.iheduc.2007.04.001

Garrison, D. R., Cleveland-Innes, M., & Fung, T. S. (2010). Exploring causal relationships among teaching, cognitive and social presence: Student perceptions of the community of inquiry framework. *The Internet and Higher Education*, *13*(1), 31–36. doi:10.1016/j.iheduc.2009.10.002

Garrison, D., & Kanuka, H. (2004). Blended learning: Uncovering its transformative potential in higher education. *The Internet and Higher Education*, *7*(2), 95–105. doi:10.1016/j.iheduc.2004.02.001

Garrison, D., & Vaughan, N. (2008). *Blended learning in higher education: framework, principles, and guidelines*. San Francisco, CA: Jossey-Bass.

Gauntlett, D. (2011). *Making is connecting. The social meaning of creativity from DYI and knitting to YouTube and Web 2.0*. Cambridge, UK: Polity Press.

Gautier, P. A., & Klaauw, B. (2012). Selection in a field experiment with voluntary participation. *Journal of Applied Econometrics*, *27*(1), 63–84. doi:10.1002/jae.1184

Gawande Virendra. (2015). *Development of Blended Learning model based on the perceptions of students and faculty at Higher Education Institutes in Oman. CIDE*. Thailand: Assumption University.

Georgiev, T., Georgieva, E., & Smrikarov, A. (2004, June). M-learning-a New Stage of E-Learning. In *International Conference on Computer Systems and Technologies-CompSysTech* (Vol. 4, No. 28, pp. 1-4).

Gilmer, J. (1979). The effects of immediate feedback versus traditional no-feedback in a testing situation.*Proceedings of the 63rd Annual Meeting of the American Educational Research Association.*

Glahn, R., & Gen, R. (2002). Progenies in Education: The evolution of internet teaching. *Community College Journal of Research and Practice*, *26*(10), 777–785. doi:10.1080/10668920290104868

Godwin-Jones, R. (2011). Emerging technologies: Mobile apps for language learning. *Language Learning & Technology*, *15*(2), 2–11.

Goh, T., & Kinshuk, D. (2006). Getting ready for m-learning—adaptation perspective. *Journal of Educational Multimedia and Hypermedia, 15*(2), 175-198.Hassan, M., Hamdan, Z., & Al-Sadi, J. (2012). A new m-learning content design process. *International Journal of Academic Research*, *4*(1), 23–28.

Göktaş, Y., Küçük, S., Aydemir, M., Telli, E., Arpacık, Ö., Yıldırım, G., & Reisoğlu, İ. (2012). Türkiye'de eğitim teknolojileri araştırmalarındaki eğilimler: 2000-2009 dönemi makalelerinin içerik analizi. *Kuram, &. Uygulamada Eğitim Bilimleri.*, *12*(1), 177–199.

Gonzalez, H., & Kuenzi, J. (2012). *Science, Technology, Engineering, and Mathematics (STEM) Education: A Primer*. Congressional Research Service Report. Retrieved January 20, 2015, from http://fas.org/sgp/crs/misc/R42642.pdf

Goodman, N. (2013). Flipping the Classroom: Use online tools to provide content before the class ever meets—and still retain the value of interpersonal interaction. *Best practices- Training magazine*. Retrieved from https://trainingmag.com/content/best-practices-flipping-classroom

Goodwin, B., & Miller, K. (2013). Evidence on flipped classrooms is still coming in. *Educational Leadership*, *70*(6), 78–80.

Goral, T. (2014). SPOC's may provide what MOOC's can't. *University Business*. Retrieved January 5, 2014, from http://www.universitybusiness.com/article/spocs-may-provide-what-moocs-can%E2%80%99t

Gorsky, P., & Blau, I. (2009). Online Teaching Effectiveness: A Tale of Two Instructors. *International Review of Research in Open and Distance Learning*, *10*(3), 1–27.

Göttsche, K. (2012). Tecnologias móveis: Uma mais valia em contextos educacionais? *Linhas*, *13*(2), 62–73. doi:10.5965/1984723813022012062

Graham, C. (2004). Blended Learning Systems: Definiton, Current Trends, and Future Directions. In C. J. Bonk & C. Graham (Eds.), *Handbook of blended learning: Global perspectives, local designs*. San Francisco, CA: Pfeiffer Publishing.

Graham, C. (2006). Blended learning systems: definition, current trends, and future directions. In C. Bonk & C. Graham (Eds.), *Handbook of blended learning: global perspectives, local designs*. San Francisco, CA: Pfeiffer.

Grant, N. (2008, March). On the usage of social networking software technologies in distance learning education. In *Society for Information Technology & Teacher Education International Conference* (Vol. 2008, pp. 3755-3759). Association for the Advancement of Computing in Education (AACE).

Gray, L., Thomas, N., & Lewis, L. (2010). *Teachers' Use of Educational Technology in U.S.* Washington, DC: Department of Education. Retrieved January 25, 2014, from http://nces.ed.gov/pubs2010/2010040.pdf

Graziola, P. G. G. (2009). Aprendizagem com mobilidade na perspectiva dialógica: Reflexões e possibilidades para práticas pedagógicas. *RENOTE*, *7*(3), 115–124.

Greenbaum, T. L. (1998). *The handbook for focus group research*. Thousand Oaks, CA: Sage.

Greene, W. H., & Zhang, C. (2003). Econometric analysis. Prentice Hall.

Greene, J., Caracelli, V., & Graham, W. (1989). Toward a conceptual framework for mixed-method evaluation design. *Teaching and Teacher Education*, *26*, 1530–1543.

Groff, J., & Mouza, C. (2008). A framework for addressing challenges to classroom technology use. *AACE Journal*, *16*(1), 21–46.

Gromik, A. N. (2012). Cell phone video recording feature as a language learning tool: A case study. *Computers & Education*, *58*(1), 223–230. doi:10.1016/j.compedu.2011.06.013

Grund, F. B., & Gil, D. J. G. (2014). *Estado del Mobile Learning en España*. Educar em Revista.

Guerreiro, S. (2013). *A utilização das novas tecnologias enquanto estratégia facilitadora do processo de ensino-aprendizagem*. Academic Press.

Guilford, J. P. (1950). Creativity. *The American Psychologist*, *5*(9), 444–454. doi:10.1037/h0063487 PMID:14771441

Gülbahar, Y., & Alper, A. (2009). A Content Analysis of the Studies in Instructional Technologies Area. *Ankara University Journal of Faculty of Educational Sciences*, *42*(2), 93–111.

Gunawardena, C. N., & Zittle, F. J. (1997). Social presence as a predictor of satisfaction within a computer mediated conferencing environment. *American Journal of Distance Education*, *11*(3), 8–26. doi:10.1080/08923649709526970

Guro. (2013). Is the Philippines ready for the Flipped Classroom Model? *EdTech Guro Filipino Tech Educator in Ation.* Retrieved February 7, 2013, from https://edtechguro.wordpress.com/2013/02/07/is-the-philippines-ready-for-the-flipped-classroom-model/

Guskin, E. (2013). *Newspaper newsrooms suffer large staffing decreases.* Retrieved from http://www.pewresearch.org/fact-tank/2013/06/25/newspaper-newsrooms-suffer-large-staffing-decreases/

Haake, J. M., & Wessner, M. (2004). Kooperative Lernräume. In *CSCL-Kompendium. Lehr- und Handbuch zum computerunterstützten kooperativen Lernen* (pp. 109–117). München: Oldenbourg Wissenschaftsverlag GmbH. doi:10.1524/9783486593976

Hall, S. (2005). Introduction: Hidden in plain sight - journalism's critical issues. In A. Hall (Ed.), *Journalism: critical issues* (pp. 1–19). Berkshire, UK: Open University Press.

Hamdan, N., McKnight, P., McKnight, K., & Arfstrom, K. (2013, June). *A review of flipped learning.* Retrieved from Pearson - research & innovation network: http://researchnetwork.pearson.com/wpcontent/uploads/LitReview_flippedLearning1.pdf

Hamilton, J., & Tee, S. (2010). Smart utilization of tertiary instructional modes. *Computers & Education, 54*(4), 1036–1053. doi:10.1016/j.compedu.2009.10.009

Han, I., & Han, S. (2014). Adoption of the mobile campus in a cyber university. *The International Review of Research in Open and Distributed Learning, 15*(6).

Hansen, F. A., Kortbek, K. J., & Grønbæk, K. (2012). Mobile Urban Drama: Interactive storytelling in real world environments. *New Review of Hypermedia and Multimedia, 18*(1-2), 63–89. doi:10.1080/13614568.2012.617842

Harasim, L. K. (1989). *Online education: A new domain.* New York: Pergamon.

Hardless, C., Nilsson, M., & Nuldén, U. (2005). 'Copernicus' experiencing a failing project for reflection and learning. *Management Learning, 36*(2), 181–217. doi:10.1177/1350507605052557

Hargreaves, A., & Fullan, M. (2012). *Professional Capital: Transforming teaching in every school.* New York: Teachers College Press.

Harris, J., & Hofer, M. (2009). *Instructional planning activity types as vehicles for curriculum-based TPACK development.* Paper presented at the Society for Information Technology & Teacher Education International Conference.

Harwood, D., Bajovic, M., Woloshyn, V., Di Cesare, D. M., Lane, L., & Scott, K. (2015). Intersecting Spaces in Early Childhood Education: Inquiry-Based Pedagogy and Tablets. *The International Journal of Holistic Early Learning and Development, 1*, 53–67.

Hase, S., & Kenyon, C. (2001). From Andragogy to Heutagogy. *ultiBASE Articles,* 1-10. Retrieved from http://www.psy.gla.ac.uk/~steve/pr/Heutagogy.html

Hase, S., & Kenyon, C. (2007). Heutagogy: A child of complexity theory. *Complicity: an International Journal of Complexity and Education, 4*(1), 111–118.

Hawkridge, D. (1999). Thirty years on, BJET! and educational technology comes of age. *British Journal of Educational Technology, 30*(4), 293–304. doi:10.1111/1467-8535.00119

Haythornthwaite, C., Kazmer, M. M., & Robins, J. (2000). Community Development among Distance Learners: Temporal and Technological Dimensions. *Journal of Computer-Mediated Communication, 6*(1). doi:10.1111/j.1083-6101.2000.tb00114.x

Hedges, H. D. (2008). *Funds of knowledge in early childhood communities of inquiry*. (Unpublished doctoral dissertation). Massey University, Palmerston North, New Zealand.

Helou, A. M., & Rahim, N. Z. A. (2014). The Influence Of Social Networking Sites on Students' Academic Performance in Malaysia. *International Journal of Electronic Commerce*, *5*(2), 247–254. doi:10.7903/ijecs.1114

Henrie, C. R., Bodily, R., & Manwaring, K. C. (2015). Exploring Intensive Longitudinal Measures of Student Engagement in Blended Learning. *International Review of Research in Open and Distance Learning*, *16*(3).

Herreid, C. F., & Schiller, N. A. (2013). Case studies and the flipped classroom. *Journal of College Science Teaching*, *42*(5), 62–66.

Herrington, A., & Herrington, J. (2006). What is an Authentic Learning Environment? In A. Herrington & J. Herrington (Eds.), Authentic learning environments in higher education (pp. 1-13). Hershey, PA: Information Science Publishing. doi:10.4018/978-1-59140-594-8.ch001

Herrington, A., & Herrington, J. (2007). Authentic mobile learning in higher education. Australian Association for Research in Education (AARE) 2007 Conference (pp. 1-9).

Herrington, A., & Herrington, J. (2007). *Authentic mobile learning in higher education*. Paper presented at the AARE 2007 International Educational Research Conference. Retrieved from http://researchrepository.murdoch.edu.au/5413/

Herrington, J., Herrington, A., Mantei, J., Olney, I. W., & Ferry, B. (2009). *New technologies, new pedagogies: Mobile learning in higher education*. Academic Press.

Herrington, J., Mantei, J., Herrington, A., Olney, I., & Ferry, B. (2008). New technologies, new pedagogies: Mobile technologies and new ways of teaching and learning. In A. Farley & D. Holt (Eds.), *Proceedings of ASCILITE 2008* (pp. 419 - 427). ASCILITE.

Herrington, J., & Oliver, R. (2000). An instructional design framework for authentic learning environments. *Educational Technology Research and Development*, *48*(3), 23–48. doi:10.1007/BF02319856

Herrington, J., Reeves, T., & Oliver, R. (2009). *A guide to authentic e-learning*. London: Routledge.

Herrmann, T. (2009). Design heuristics for computer supported collaborative creativity. In *Proceedings of the 42nd Hawaii International Conference on System Sciences* (pp. 1-10). IEEE Computer Society.

Hevner, A. R., March, S. T., Park, J., & Ram, S. (2004). Design science in information systems research. *Management Information Systems Quarterly*, *28*(1), 75–105.

Higuchi, A. A. S. (2011). *Tecnologias móveis na educação: Um estudo de caso em uma escola da rede pública do estado de são Paulo. Dissertação (Mestrado)*. São Paulo, SP: Universidade Presbiteriana Mackenzie.

Hirst, M. (2011). *News 2.0: Can journalism survive the Internet?* Crows Nest, NSW, Australia: Allen & Unwin.

Hobbes, R. (2013). *Hobbes' Internet Timeline*. Retrieved October 20, 2014 from http://www.zakon.org/robert/internet/timeline/

Hobbs, R. (2013). Improvization and strategic risk-taking in informal learning with digital media literacy. *Learning, Media and Technology*, *38*(2), 182–197. doi:10.1080/17439884.2013.756517

Hockly, N. (2012). Mobile learning. *ELT Journal*, *67*(1), 80–84. doi:10.1093/elt/ccs064

Hoffman, E. S. (2013). Beyond the flipped classroom: Redesigning a research methods course for e3 instruction.[CIER]. *Contemporary Issues in Education Research*, *7*(1), 51–62. doi:10.19030/cier.v7i1.8312

Hofmann, J. (2006). Why blended learning hasn't (yet) fulfilled its promises: Answers to those questions that keep you up at night. In The handbook of blended learning: Global perspectives, local designs, (pp. 27-40). Academic Press.

Hogan, K., & Pressley, M. (1997). *Scaffoding student learning: Instructional approaches and issues*. Cambridge, MA: Brookline Books.

Holland, C., & Muilenburg, L. (2011). Supporting student collaboration: Edmodo in the classroom. In *Society for Information Technology & Teacher Education International Conference* (Vol. 1, pp. 3232-3236). Association for the Advancement of Computing in Education (AACE).

Holotescu, C., Grosseck, G., Cretu, V., Naaji, A. (2014). Integrating MOOCs in blended courses. *Elearning & Software for Education*, (1), 243-250. doi:.10.12753/2066-026X-14-034

Homer, B. D., Plass, J. L., & Blake, L. (2008). The effects of video on cognitive load and social presence in multimedia-learning. *Computers in Human Behavior*, *24*(3), 786–797. doi:10.1016/j.chb.2007.02.009

Hong, K., Lai, K., & Holton, D. (2003). Students' satisfaction and perceived learning with a web-based course. *Journal of Educational Technology & Society*, *6*(1), 116–124.

Hope, W., & Myllylahti. (2013). Fairfax in trouble. *Foreign Control Watchdog 132*. Retrieved from http://www.converge.org.nz/watchdog/32/index32.html

Hoppe, H. (2004). SMS-based discussions - Technology enhanced collaboration for a literature.*Proceedings of International Workshop on Wireless and Mobile Technologies in Education.*

Horn, M. (2013). The transformational potential of flipped classrooms. *Education Next*, *13*(3), 78–79.

Horn, M. B. (2014). MOOCs for high school. *Education Next*, *14*(3), 82–83.

Hosler, A. (2014). *Massive Open Online Research: The MOOC Evolves into the MOOR*. Retrieved November 5, 2014, from http://www.emergingedtech.com/2014/01/massive-open-online-research-the-mooc-evolves-into-the-moor/ EmergingEdTech

Howells, J. (2005). *The Management of Innovation and Technology: The Shaping of Technology and Institutions of the Market Economy*. London: Sage.

Hull, E., Jackson, K., & Dick, J. (2010). *Requirements engineering*. Springer Science & Business Media.

Humphries, J., & Ness, M. (2015). Beyond who, what, where, when, why, and how: Preparing students to generate questions in the age of Common Core Standards. *Journal of Research in Childhood Education*, *29*(4), 551–564. doi:10.1080/02568543.2015.1073199

Hung, J. L., & Zhang, K. (2012). Examining m-learning trends 2003–2008: A categorical meta-trend analysis using text mining techniques. *Journal of Computing in Higher Education*, *24*(1), 1–17. doi:10.1007/s12528-011-9044-9

Hussin, S., Manap, M. R., Amir, Z., & Krish, P. (2012). Mobile learning readiness among Malaysian students at higher learning institutes. *Asian Social Science*, *8*(12), 276–283. doi:10.5539/ass.v8n12p276

Hutcheson, G. D., & Sofroniou, N. (1999). *The Multivariate Social Scientist: Introductory Statistics Using Generalized Linear Models*. Sage Publication.

Hwang, G. J., & Chang, H. F. (2011). A formative assessment-based mobile learning approach to improving the learning attitudes and achievements of students. *Computers & Education*, *56*(4), 1023–1031. doi:10.1016/j.compedu.2010.12.002

Hwang, G. J., & Tsai, C. C. (2011). Research trends in mobile and ubiquitous learning: A review of publications in selected journals from 2001 to 2010. *British Journal of Educational Technology*, *42*(3), E65–E70. doi:10.1111/j.1467-8535.2011.01183.x

Hwang, G. J., Tsai, C. C., & Yang, S. J. (2008). Criteria, strategies and research issues of context-aware ubiquitous learning. *Journal of Educational Technology & Society*, *11*(1), 81–91.

Hwang, G. J., & Wu, P. H. (2014). Applications, impacts and trends of mobile technology-enhanced learning: A review of 2008–2012 publications in selected SSCI journals. *International Journal of M-learning and Organisation*, *8*(2), 83–95. doi:10.1504/IJMLO.2014.062346

Hyson, M. (2004). *The emotional development of young children: Building an emotion-centered curriculum*. Teachers College Press.

Iahad, N., Mirabolghasemi, M., & Huspi, S. H. (2012). A blended community of inquiry approach: The usage of social network as a support for course management system. In *International Conference on Computer & Information Science*, *1*, 180–183.

INEP/MEC. (2013a). *Censo da Educação Superior no Brasil 2013*. Author.

INEP/MEC. (2013b). *Censo Escolar Brasileiro 2013*. Retrieved from http://portal.inep.gov.br/basica-censo

Internet World Stats. (2014). Usage and Population Statistics. *Internet World Stats*. Retrieved February 15, 2014, from http://www.internetworldstats.com/stats.htm

Irby, J. (1992). *Creating the culture of journalism*. Paper presented at the the Annual Meeting of the Association for Education in Journalism and Mass Communication, Montreal, Quebec, Canada.

Ismail, I., Idrus, R. M., & Gunasegaran, T. (2010). Motivation, psychology and language effect on mobile learning in Universiti Sains Malaysia. *International Journal of Interactive Mobile Technologies*, *4*(4), 31–36.

ISTE Standards for Teachers. (2008). *International Society for Technology in Education*. Retrieved from: http://www.iste.org/docs/pdfs/20-14_ISTE_Standards-T_PDF.pdf

Ivey, A., Ivey, M., & Zalaquett, C. (2014). *Intentional interviewing and counselling. Facilitating client development in a multicultural society* (8th ed.). Belmont, CA: Brooks/Cole.

Jafari, A., McGee, P., & Carmean, C. (2006). Managing Courses Defining Learning: What Faculty, Students, and Administrators Want. *EDUCAUSE Review*, *41*(4), 50–70.

Jahnke, I. (2011). How to Foster Creativity in Technology Enhanced Learning. In B. White, I. King, & Ph. Tsang (Eds.), *Social Media Tools and Platforms in Learning Environments: Present and Future* (pp. 95–116). New York: Springer. doi:10.1007/978-3-642-20392-3_6

Janson, A., Ernst, S.-J., Lehmann, K., & Leimeister, J. M. (2014). Creating Awareness and Reflection in a Large-Scale IS Lecture–The Application of a Peer Assessment in a Flipped Classroom Scenario. In *4th Workshop on Awareness and Reflection in Technology-Enhanced Learning (ARTEL 2014) to be held in the context of EC-TEL 2014*.

Jardines, F. (2009). Historical development of distance education. *Innovaciones de Negocios*, *6*(2), 225–236.

Jaworski, B. (2006). Theory and practice in mathematics teaching development: Critical inquiry as a mode of learning in teaching. *Journal of Mathematics Teacher Education*, *9*(2), 187–211. doi:10.1007/s10857-005-1223-z

Jenkins, H., Clinton, K., Purushotma, R., Robinson, A., & Weigel, M. (2009). *Confronting the Challenges of Participatory Culture: Media Education for the 21st Century*. Chicago, IL: The John D. and Catherine T. MacArthur Foundation.

Jenkins, M., & Hanson, J. (2003). *A guide for senior managers. LTSN Generic Centre Elearningseries, No.1.* York: Learning and Teaching Support Network.

JISC. (2009). *Effective practice in a digital age: A guide to technology-enhanced learning and teaching.* Available at: http://www.jisc.ac.uk/media/documents/publications/effectivepracticedigitalage.pdf

Jobe, W. (2013). Native Apps vs. Mobile Web Apps. *iJIM, 7*(4), 27-32.

Johnson, L., Adams Becker, S., Estrada, V., & Freeman, A. (2014). NMC Horizon Report: Edição K12 (K-12/2014 ed.). Austin, TX: Estados Unidos.

Johnson, L., Becker, S. A., Cummis, M., & Estrada, V. (2014). 2014 NMC Technology Outlook for Brazilian Universities: A Horizon Project Regional Report. Academic Press.

Johnson, L., Adams, S., & Cummins, M. (2012). *The NMC horizon report: 2012 higher education edition.* Austin, TX: The New Media Consortium.

Johnson, L., Adams, S., Cummins, M., Estrada, V., Freeman, A., & Ludgate, H. (2013). *NMC Horizon Report: Higher Education* (2013th ed.). Austin, TX: The New Media Consortium.

Johnson, L., Adams, S., & Haywood, K. (2011). *The NMC Horizon Report: 2011 K-12 Edition.* Austin, TX: The New Media Consortium.

Johnson, L., Smith, R., Willis, H., Levine, A., & Haywood, K. (2011). *The 2011 Horizon Report.* Austin, TX: The New Media Consortium.

Johnston, T. C. (2013). Massive Open Online Courses (MOOCs): Student perspective. *Society For Marketing Advances Proceedings, 25*, 179-180.

Jonas, D., & Burns, B. (2010). The transition of e-learning. Changing the focus of educational delivery in children's pain management. *Nurse Educator, 10*(1), 1–7. doi:10.1016/j.nepr.2009.01.015 PMID:19264550

Jones, N. (2006). e-College Wales, a case study of blended learning. In Handbook of Blended Learning: Global Per spectives, Local Designs, (pp. 182–194). San Francisco, CA: Pfeiffer Publishing.

Jordan, J.D. (2015). *What's next: Blended Learning 2.0: The Hechinger Report.* Academic Press.

Jucevičienė, P., & Valinevičienė, G. (2015). A conceptual model of social networking in higher education. *Elektronika ir Elektrotechnika, 102*(6), 55-58.

Jung, M. L., Loria, K., Mostaghel, R., & Saha, P. (2008). E-Learning: Investigating University Student's Acceptance of Technology. *European Journal of Open, Distance and e-learning.* Available at: http://www.eurodl.org/materials/contrib/2008/Jung_Loria_Mostaghel_Saha.htm

Kampylis, P., & Berki, E. (2014). *Nurturing creative thinking.* International Bureau of Education (UNESCO) in collaboration with the International Academy of Education (IAE). Retrieved on 25th July 2015 from: http://www.ensembleproject.org/modules/wfdownloads/viewcat.php?cid=78

Kassner, L. (2013). Mix it up with blended learning in K-12 schools. *Metropolitan Educational Research Consortium.* Accessed 16th January 2016 from http://files.eric.ed.gov/fulltext/ED548381.pdf

Kato, C., Shiono, Y., Goto, T., & Tsuchida, K. (2011). Development of Online Counseling System and Usability Evaluation. *Journal of Emerging Technologies in Web Intelligence, 3*(2), 146–153. doi:10.4304/jetwi.3.2.146-153

Kearney, M., Schuck, S., Burden, K., & Aubusson, P. (2012). Viewing mobile learning from a pedagogical perspective. *Research in Learning Technology*, 20.

Keegan, D. (2007). *Mobile learning: a practical guide*. Academic Press.

Keegan, D. (2005, October). The incorporation of m-learning into mainstream education and training. In *World Conference on M-learning* (p. 11).

Kehinde, L. O., Ayodele, K. P., Akinwale, O. B., & Osasona, O. (2011). Remote Labs in Education. The Obafemi Awolowo University Experience. In J. G. Zubía & G. R. Alves (Eds.), *Using Remote Labs in Education: Two Little Ducks in Remote Experimentation* (pp. 81–111). Bilbao: University of Deusto.

Keil, M., & Johnson, R. D. (2002). Feedback channels: Using social presence theory to compare voice mail to e-mail. *Journal of Information Systems Education*, *13*(4), 295–302.

Keller, C., Hrastinski, S., & Carlsson, S. A. (2008) Students' Acceptance of E-Learning Environments: A Comparative Study In Sweden and Lithuania. In *Proceedings of the Fifteenth European Conference on Information Systems*, (pp. 395-406).

Kemper, E. A., Stringfield, S., & Teddlie, C. (2003). Mixed methods sampling strategies in social science research. In A. Tashakkori & C. Teddlie (Eds.), *Handbook of mixed methods in social and behavioral research*. Thousand Oaks, CA: Sage.

Kenny, R. F., Park, C. L., Van Neste-Kenny, J. M. C., Burton, P., Park, C. L., & Qayyum, A. (2012). Using Self-Efficacy to Assess the Readiness of Nursing Educators and Students for M-learning. *International Review of Research in Open and Distance Learning*, *13*(3), 277–296.

Kenwright, K. (2009). Clickers in the Classroom. *TechTrends*, *53*(1), 74–77. doi:10.1007/s11528-009-0240-7

Keppell, M., Suddaby, G., & Hard, N. (2011). *Good practice report: technology-enhanced learning and teaching*. Sydney, NSW: Australian Learning and Teaching Council.

Khaddage, F., & Lattemann, C. (2009). Towards an ad-hoc mobile social learning network using mobile phones. *Interactive Computer Aided Learning*, *12*, 374–380.

Khoo. E., Merry. R., Nguyen, N.H., Bennett. T., & MacMillan. N. (2013). Early childhood education teachers' iPad supported practices in young children's learning and exploration. *Computers in New Zealand Schools: Learning, Teaching, Technology*, *25*(1-3), 3-20.

Kiliç-Çakmak, E., Çebi, A., Mihçi, P., Günbatar, M. S., & Akçayir, M. (2013). A Content Analysis of Educational Technology Research in 2011. *Procedia: Social and Behavioral Sciences*, *106*, 74–83. doi:10.1016/j.sbspro.2013.12.010

Kim, K.-J., & Bonk, C. J. (2006). The Future of Online Teaching and Learning in Higher Education: The Survey Says. *EDUCAUSE Quarterly*, *29*(4), 22–30.

King, A. (1993). From sage on the stage to guide on the side. *College Teaching*, *41*(1), 30–35. doi:10.1080/87567555.1993.9926781

Kirschner, P., Sweller, J., & Clark, R. (2006). Why Minimal Guidance During Instruction Does Not Work: An Analysis of the Failure of Constructivist, Discovery, Problem-Based, Experiential, and Inquiry-Based Teaching. *Educational Psychologist*, *4*(2), 75–86. doi:10.1207/s15326985ep4102_1

Kitchenham, B. (2004). Procedures for performing systematic reviews. Keele University.

Klein, J. D. (2002). Empirical research on performance improvement. *Performance Improvement Quarterly*, *15*(1), 99–110. doi:10.1111/j.1937-8327.2002.tb00243.x

Klopfer, E., Squire, K., & Jenkins, H. (2002). Environmental detectives: PDAs as a window into a virtual simulated world. In *Wireless and Mobile Technologies in Education, 2002. Proceedings. IEEE International Workshop on* (pp. 95-98). IEEE.

Koble, M., & Bunker, E. (1997). Trends in research and practice: An examination of The American Journal of Distance Education 1987 to 1995. *American Journal of Distance Education, 11*(2), 19–38. doi:10.1080/08923649709526959

Koehler, M., & Mishra, P. (2009). What is technological pedagogical content knowledge (TPACK)? *Contemporary Issues in Technology & Teacher Education, 9*, 70.

Kofman, H. A., & Concari, S. B. (2011). Using Remote Lab for Physics Teaching. In J. G. Zubía & G. R. Alves (Eds.), *Using Remote Labs in Education: Two Little Ducks in Remote Experimentation* (pp. 293–309). Bilbao: University of Deusto.

Kolfschoten, G. L., & De Vreede, G. J. (2009). A design approach for collaboration processes: A multimethod design science study in collaboration engineering. *Journal of Management Information Systems, 26*(1), 225–256. doi:10.2753/MIS0742-1222260109

Koole, M. L. (2006, November). Practical issues in mobile education. In *Wireless, Mobile and Ubiquitous Technology in Education, 2006. WMUTE'06. Fourth IEEE International Workshop on* (pp. 142-146). IEEE. doi:10.1109/WMTE.2006.261363

Kopf, S., Haenselmann, T., Kiess, J., Guthier, B., & Effelsberg, W. (2011). Algorithms for video retargeting. *Multimedia Tools and Applications, 51*(2), 819–861. doi:10.1007/s11042-010-0717-6

Koszalka, T. A., & Ntloedibe-Kuswani, G. S. (2010). Literature on the safe and disruptive learning potential of mobile technologies. *Distance Education, 31*(2), 139–157. doi:10.1080/01587919.2010.498082

Kozan, K., & Richardson, J. C. (2014). Interrelationships between and among social, teaching, and cognitive presence. *The Internet and Higher Education, 21*, 68–73. doi:10.1016/j.iheduc.2013.10.007

Kreijns, K., Van Acker, F., Vermeulen, M., & Van Buuren, H. (2014). Community of Inquiry: Social presence revisited. *E-Learning and Digital Media, 11*(1), 5–18. doi:10.2304/elea.2014.11.1.5

Kress, G., & Pachler, N. (2007). Thinking about the 'm-' in mobile learning. In T. Hug (Ed.), *Didactics of Microlearning. Concepts, Discourse, and Examples* (pp. 139–154). Münster: Waxmann.

Krippendorff, K. (2011). *Computing Krippendorff's Alpha-Reliability*. Retrieved January 25, 2015 from http://repository.upenn.edu/asc_papers/43

Krippendorff, K. (2004). *Content analysis: An introduction to its methodology* (2nd ed.). Thousand Oaks, CA: Sage.

Krippendorf, K. (2004b). Reliability in content analysis some common misconceptions and recommendations. *Human Communication Research, 30*(3), 411–433.

Küçük, S., Aydemir, M., Yıldırım, G., Arpacık, O., & Göktaş, Y. (2013). Educational technology research trends in Turkey from 1990 to 2011. *Computers & Education, 68*, 42–50. doi:10.1016/j.compedu.2013.04.016

Kuh, G. D. (2001). Assessing What Really Matters to Student Learning Inside The National Survey of Student Engagement. *Change: The Magazine of Higher Learning, 33*(3), 10–17. doi:10.1080/00091380109601795

Kukulska-Hulme, A. (2005). *M-learning: A handbook for educators and trainers*. Psychology Press.

Kukulska-Hulme, A. (2010). Learning cultures on the move: Where are we heading? *Journal of Educational Technology & Society, 13*(4), 4–14.

Kukulska-Hulme, A. (2010). Mobile learning as a catalyst for change. *Open Learning: The Journal of Open and Distance Learning*, *25*(3), 181–185. doi:10.1080/02680513.2010.511945

Kukulska-Hulme, A., & Traxler, J. (2005). *Mobile learning a handbook for educators and trainers*. London: Routledge.

Kukulska-Hulme, A., & Traxler, J. (2007). Learning design with mobile and wireless technologies. In H. Beetham & R. Sharpe (Eds.), *Rethinking pedagogy for the digital age: Designing and delivering e-learning* (pp. 180–192). London, UK: Routledge.

Kukulska-Hulme, A., Traxler, J., & Pettit, J. (2007). Designed and user-generated activity in the mobile age. *Journal of Learning Design*, *2*(1), 52–65. doi:10.5204/jld.v2i1.28

Kumar, L. S., Jamatia, B., Aggarwal, A. K., & Kannan, S. (2011). Mobile Device Intervention for Student Support Services in Distance Education Context--FRAME Model Perspective. *European Journal of Open, Distance and E-Learning*.

Kumar, S., & Ritzhaupt, A. D. (2014). Adapting the Community of Inquiry survey for an online graduate program: Implications for online programs. *E-Learning and Digital Media*, *11*(1), 59–71. doi:10.2304/elea.2014.11.1.59

Kummervold, P. E., Gammon, D., Bergvik, S., Johnsen, J. A. K., Hasvold, T., & Rosenvinge, J. H. (2002). Social support in a wired world: Use of online mental health forums in Norway. *Nordic Journal of Psychiatry*, *56*(1), 59–65. doi:10.1080/08039480252803945 PMID:11869468

Kurkovsky, S. (2014). Mobile Game development: Improving student engagement and motivation in introductory computing courses. *Computer Science Education*, *23*(2), 138–157. doi:10.1080/08993408.2013.777236

Lachs, V. (2000). *Making Multimedia in the Classroom: A Practical Guide*. London: Routledge.

Lage, M. J., Platt, G. J., & Treglia, M. (2000). Inverting the classroom: A gateway to creating an inclusive learning environment. *The Journal of Economic Education*, *31*(1), 30–43. doi:10.1080/00220480009596759

Lambert, J. (2012). *Digital storytelling: capturing lives, creating community*. New York, NY: Routledge.

Landry, G. A. (2010). *Creating and validating an instrument to measure middle school mathematics teachers' technological pedagogical content knowledge*. TPACK.

Lape, N., Levy, R., & Yong, D. (2014). Can flipped classroom help students learn? We are trying to find out. *Slate*. Retrieved April 25, 2014, from http://www.slate.com/articles/technology/future_tense/2014/04/flipped_classrooms_can_they_help_students_learn.h

Latchem, C. (2006). Editorial: A content analysis of the British Journal of Educational Technology. *British Journal of Educational Technology*, *37*(4), 503–511. doi:10.1111/j.1467-8535.2006.00635.x

Laurillard, D. (2002). *Rethinking university teaching: a conversational framework for the effective use of learning technologies* (2nd ed.). London: RoutledgeFalmer. doi:10.4324/9780203304846

Laurillard, D. (2005). E-Learning in Higher Education. In P. Ashwin (Ed.), *Changing Higher Education: The Development of Learning and Teaching*. London: RoutledgeFalmer.

Laurillard, D. (2007). Pedagogcal forms of mobile learning: framing research questions. In N. Pachler (Ed.), *Mobile learning: towards a research agenda* (pp. 33–54). London: WLE Centre, Institute of Education.

Laurillard, D. (2008). *Digital technologies and their role in achieving our ambitions for education*. London: Institute of Education, University of London.

Lave, J. (1982). A comparative approach to educational forms and learning processes. *Anthropology & Education Quarterly, 13*(2), 181–187. doi:10.1525/aeq.1982.13.2.05x1832l

Lave, J., & Wenger, E. (1991). *Situated Learning: Legitimate peripheral participation.* Cambridge, UK: Cambridge University Press. doi:10.1017/CBO9780511815355

Lawlor, B., & Donnelly, R. (2010). Using podcasts to support communication skills development: A case study for content format preferences among postgraduate research students. *Computers & Education, 54*(4), 962–971. doi:10.1016/j.compedu.2009.09.031

Lazzari, M. (2009). Creative use of podcasting in higher education and its effect on competitive agency. *Computers & Education, 52*(1), 27–34. doi:10.1016/j.compedu.2008.06.002

Learning Analytics & Knowledge. (2011). *1st International Conference on Learning Analytics and Knowledge.*

Ledesma, F. (2013). *Mobile learning: Proibir ou integrar.* Retrieved from http://cefopna.edu.pt/revista/revista_09/es_02_09_fl. htm

Lee, M. J. W., McLoughlin, C., & Chan, A. (2008). Talk the talk: Learner-generated podcasts as catalysts for knowledge creation. *British Journal of Educational Technology, 39*(3), 501–521. doi:10.1111/j.1467-8535.2007.00746.x

Lee, M. J., & McLoughlin, C. (2010). *Beyond distance and time constraints: Applying social networking tools and Web 2.0 approaches to distance learning. In Emerging technologies in distance education* (pp. 61–87). Athabasca University Press.

Lee, Y., Driscoll, M. P., & Nelson, D. W. (2004). The Past, Present, and Future of Research in Distance Education: Results of a Content Analysis. *American Journal of Distance Education, 18*(4), 225–241. doi:10.1207/s15389286ajde1804_4

Lehmann, K., Oeste, S., Janson, A., Söllner, M., & Leimeister, J. M. (2015). Flipping the Classroom – IT-unterstützte Lerneraktivierung zur Verbesserung des Lernerfolges einer universitären Massenlehrveranstaltung. *HMD Praxis der Wirtschaftsinformatik*, 1-15.

Lehmann, K., & Leimeister, J.-M. (2015). Theory-driven Design of an IT-based Peer Assessment to Assess High Cognitive Levels of Educational Objectives in Large-scale Learning Services. In *23rd European Conference on Information Systems (ECIS 2015).*

Lehmann, K., Thillainathan, N., Bitzer, P., & Leimeister, J. M. (2012). Performance Dashboard für Dozenten in der universitären Lehre. In *Multikonferenz der Wirtschaftsinformatik.* Braunschweig, Deutschland: MKWI.

Leimeister, J. M. (2014). Collaboration Engineering. IT-gestützte Zusammenarbeitsprozesse systematisch entwickeln und durchführen. Heidelberg Springer DE.

Leimeister, J. M. (2012). *Dienstleistungsengineering und-management.* Berlin, Heidelberg: Springer DE. doi:10.1007/978-3-642-27983-6

Levin, T., & Wadmany, R. (2008). Teachers' views on factors affect ting effective integration of information technology in the classroom: Developmental scenery. *Journal of Technology and Teacher Education, 16*, 233–263.

Licona, M. M., & Gurung, B. (2011). Online Multicultural Education: Asynchronous Discussions in Online Multicultural Education. *Multicultural Education, 19*(1), 2–8.

Lim, K. Y. (2005). *Minister speech.* Retrieved June 7, 2011, from http://www.tmforum.org/WhyAttend/2719/home.html?linkid=31446

Lim, D. H., Morris, M. L., & Kupritz, V. W. (2014). Online vs. Blended Learning: Differences in Instructional Outcomes and Learner Satisfaction. *Journal of Asynchronous Learning Networks, 11*(2), 27–42.

Lim, T., Fadzil, M., & Mansor, N. (2011). Mobile learning via SMS at Open University Malaysia: Equitable, effective, and sustainable. *The International Review of Research in Open and Distributed Learning*, *12*(2), 122–137.

Littlejohn, A., & Pegler, C. (2007). *Preparing for blended e-learning*. New York: Routledge.

Liu, G. Z., & Hwang, G. J. (2010). A key step to understanding paradigm shifts in e-learning: Towards context-aware ubiquitous learning. *British Journal of Educational Technology*, *41*(2), E1–E9. doi:10.1111/j.1467-8535.2009.00976.x

Liu, T.-C., Liang, J.-K., Wang, H.-Y., Chan, T.-W., & Wei, L.-H. (2003). Embedding educlick in classroom to enhance interaction. In *International conference on computers in education (ICCE)*.

Livingstone, S., Mascheroni, G., Ólafsson, K., & Haddon, L. (2014). *Children's online risks and opportunities: comparative findings from EU Kids Online and Net Children Go Mobile. EU Kids Online*. London, UK: LSE.

Li, Y., Zhang, M., Bonk, C. J., & Guo, Y. (2015). Integrating MOOC and flipped classroom practice in a traditional undergraduate course: Students' experience and perceptions. *International Journal of Emerging Technologies in Learning*, *10*(6), 4–10. doi:10.3991/ijet.v10i6.4708

Liyanagunawardena, T., Adams, A., & Williams, S. (2013). MOOCs: A systematic study of the published literature 2008-2012. *International Review of Research in Open and Distance Learning*, *14*(3), 202–227. Retrieved from http://www.irrodl.org/index.php/irrodl/article/view/1455/2531

Locke, M. (2013). MOOC: Will these four letters change K-12? *Scholastic Administrator*, *12*(6), 39–40.

Lockyer, L., Heathcote, E., & Dawson, S. (2013, October01). (2103). Informing pedagogical action: Aligning learning analytics with learning design. *The American Behavioral Scientist*, *57*(10), 1439–1459. doi:10.1177/0002764213479367

Lombardi, M. M. (2007). *Authentic Learning for the 21st Century: An overview. ELI Report No. 1*. Boulder, CO: EDUCAUSE Learning Initiative. Retrieved March 10, 2015, from http://www.educause.edu/ir/library/pdf/ELI3009.pdf

Lombardo, V., & Damiano, R. (2012). Storytelling on mobile devices for cultural heritage. *New Review of Hypermedia and Multimedia*, *18*(1-2), 11–35. doi:10.1080/13614568.2012.617846

Looi, C. K., Seow, P., Zhang, B., So, H. J., Chen, W., & Wong, L. H. (2010). Leveraging mobile technology for sustainable seamless learning: A research agenda. *British Journal of Educational Technology*, *41*(2), 154–169. doi:10.1111/j.1467-8535.2008.00912.x

López-Pérez, M. V., Pérez-López, M. C., & Rodríguez-Ariza, L. (2011). Blended learning in higher education: Students' perceptions and their relation to outcomes. *Computers & Education*, *56*(3), 818–826. doi:10.1016/j.compedu.2010.10.023

Loughran, J. J. (2002). *Developing reflective practice: Learning about teaching and learning through modelling*. Routledge.

Loveless, A. (2002). *A Literature Review in Creativity, New Technologies and Learning: A Report for Futurelab*. Bristol: Futurelab.

Loveless, A. (2007). *Creativity, technology and learning – a review of recent literature. Report 4 – Update*. Bristol: Futurelab.

Lucas, B., Claxton, G., & Spencer, E. (2013). *Progression in Student Creativity in School: First Steps Towards New Forms of Formative Assessments*. OECD Education Working Papers, No. 86. OECD Publishing.

Luckin, R., Clark, W., Garnett, F., Whitworth, A., Akass, J., & Cook, J. (2010). Learner-Generated Contexts: A Framework to Support the Effective Use of Technology for Learning. In M. Lee & C. McLoughlin (Eds.), *Web 2.0-Based E-Learning: Applying Social Informatics for Tertiary Teaching* (pp. 70–84). Hershey, PA: IGI Global. doi:10.4018/978-1-60566-294-7.ch004

Lutz, M. J., Naveda, F., & Valino, J. R. (2014). Undergraduate Software Engineering. *Communications of the ACM*, *57*(8), 52–28. doi:10.1145/2632361

MacNaughton, G., Rolfe, S., & Siraj-Blatchford, I. (2010). *Doing early childhood research*. Berkshire, UK: McGraw-Hill Education.

Madden, M., Lenhart, A., Duggan, M., Cortesi, S., & Gasser, U. (2013). *Teens and Technology 2013*. Washington, DC: Pew Research Center's Internet & American Life Project.

Madeira, M. C.-A. (2010). *The development of pedagogical content knowledge in science teachers: New opportunities through technology-mediated reflection and peer-exchange*. University of Toronto.

Ma, J., Han, X., Yang, J., & Cheng, J. (2014). Examining the necessary condition for engagement in an online learning environment based on learning analytics approach: The role of the instructor. *The Internet and Higher Education*, *24*, 26–34. doi:10.1016/j.iheduc.2014.09.005

Ma, J., & Nickerson, J. V. (2006). Hands-on, simulated, and remote laboratories: A comparative literature review. *ACM Computing Surveys*, *38*(3), 7, es. doi:10.1145/1132960.1132961

Maples, M. F., & Webster, J. M. (1980). Thorndike's Connectionism. In G. M. Gazda & R. J. Corsini (Eds.), *Theories of Learning*. Itasca, IL: F.E. Peacock.

Marçal, E., Andrade, R., & Rios, R. (2005). Aprendizagem utilizando dispositivos móveis com sistemas de realidade virtual. *RENOTE, 3*(1).

Martin, S., Diaz, G., Sancristobal, E., Gil, R., Castro, M., & Peire, J. (2011). New technology trends in education: Seven years of forecasts and convergence. *Computers & Education*, *57*(3), 1893–1906. doi:10.1016/j.compedu.2011.04.003

Maslow, A. H. (1943). The Theory of Human Motivation. *Psychological Review*, *50*(4), 370–396. doi:10.1037/h0054346

Masrom, M. (2007). Technology Acceptance Model and E-learning.*12th International Conference on Education*. Sultan Hassanal Bolkiah Institute of Education.

Mastor, K. A., Jin, P., & Cooper, M. (2000). Malay culture and personality. *The American Behavioral Scientist*, *44*(1), 95–111. doi:10.1177/00027640021956116

Matheos, K., Daniel, K.B., & Mccalla, I.G. (2005). Dimensions for Blended Learning Technology: Learners' Perspectives. *Journal of Learning Design*, 56-74.

Matheos, K., Daniel, K., & McCalla, G. (2005). Dimensions for blended learning technology: Learners' perspectives. *Journal of Learning Design*, *1*(11), 56–76.

Mazur, A. D. (2013). *The Flipped Classroom: Theoretical Constructs*. Retrieved from https://ambermazur.files.wordpress.com/2013/06/flipped-classroom-theoretical-constructs-visual-with-explanation.pdf

Mazur, E. (1997). Peer Instruction: A User's Manual Series in Educational Innovation. Upper Saddle River, NJ: Prentice Hall. Retrieved from http://mazur.harvard.edu/publications.php?function=display&rowid=0

Mazur, E. (2009, November 12). *Confessions of a Converted Lecturer*. Retrieved from YouTube: http://www.youtube.com/watch?v=WwslBPj8GgI

McChesney, R., & Nichols, J. (2010). *The death and life of American journalism: The media revolution that will begin the world again*. New York: Nation Books.

McGarr, O. (2009). A review of podcasting in higher education: Its influence on the traditional lecture. *Australasian Journal of Educational Technology*, *25*(3), 309–321.

McGee, P., & Reis, A. (2012). Blended Course Design: A Synthesis of Best Practices. *Journal of Asynchronous Learning Networks*, *16*(4), 7–22.

McGiboney, M. (2009). Twitter's Tweet Smell Of Success. *Nielsen Online*. Retrieved from http://blog.nielsen.com/nielsenwire/online_mobile/twitters-tweet-smell-of-success/

McGreen, N., & Arnedillo Sanchez, I. (2005). Mobile Phones: Creative learning tools. In P. Isaias, C. Borg, P. Kommers, & P. Bonanno (Eds.), *Mobile learning 2005* (pp. 241–245). Malta: International Association for Development of the Information Society Press.

McGrory, K. (2013). Florida K-12 system experimenting with MOOCs. *Tampa Bay Times*. Retrieved from: http://www.tampabay.com/blogs/gradebook/florida-k-12-system-experimenting-with-moocs/2151286

McInnes, R. (2009). *Standards for Mobile Pedagogical Application Development: Literature Review*. Academic Press.

McKerlich, R., Riis, M., Anderson, T., & Eastman, B. (2011). Student Perceptions of Teaching Presence, Social Presence and Cognitive Presence in a Virtual World. *Journal of Online Learning and Teaching*, *7*(3), 324–336.

McLaughlan, R. G. (2007). Instructional strategies to educate for sustainability in technology assessment. *International Journal of Engineering Education*, *23*(2), 201–208.

McLaughlin, J. E., Griffin, L. M., Esserman, D. A., Davidson, C. A., Glatt, D. M., Roth, M. T., & Mumper, R. J. (2013). Pharmacy student engagement, performance, and perception in a flipped satellite classroom. *American Journal of Pharmaceutical Education*, *77*(9), 196. doi:10.5688/ajpe779196 PMID:24249858

McLaughlin, J. E., Roth, M. T., Glatt, D. M., Gharkholonarehe, N., Davidson, C. A., Griffin, L. M., & Mumper, R. J. (2014). The flipped classroom: A course redesign to foster learning and engagement in a health professions school. *Academic Medicine*, *89*(2), 236–243. doi:10.1097/ACM.0000000000000086 PMID:24270916

McLoughlin, C., & Lee, M. (2007). Social software and participatory learning: Pedagogical choices with technology affordances in the Web 2.0 era. In R. J. Atkinson, C. McBeath, S. K. A. Soong & C. Cheers (Eds.), Proceedings of Ascilite 2007, ICT: Providing Choices for Learners and Learning (pp. 664-675). ASCILITE.

McLoughlin, C., & Lee, M. (2010). Pedagogy 2.0: Critical challenges and responses to web 2.0 and social software in tertiary teaching. In M. Lee & C. McLoughlin (Eds.), *Web 2.0-Based E-Learning: Applying Social Informatics for Tertiary Teaching* (pp. 46–69). Hershey, PA: IGI Global. doi:10.4018/978-1-60566-294-7.ch003

McManis, L. D., & Gunnewig, S. B. (2012). Finding the education in educational technology with early learners. *Young Children*, *67*, 14–24.

Means, B., Toyama, Y., Murphy, R., Bakia, M., & Jones, K. (2010, September). *Evaluation of evidence-based practices in online learning: A meta-analysis and review of online learning studies.* U.S. Department of Education. Retrieved from http://www2.ed.gov/rschstat/eval/tech/evidence-based-practices/finalreport.pdf

Means, B., Toyama, Y., Murphy, R., & Baki, M. (2013). The Effectiveness of Online and Blended Learning: A Meta-Analysis of the Empirical Literature. *Teachers College Record*, *115*(3).

Mehdipour, Y., & Zerehkafi, H. (2013). Mobile Learning for Education: Benefits and Challenges. *International Journal of Computational Engineering Research*, *3*(6).

Mehta, J. (2013). *The Allure of Order: High Hopes, Dashed Expectations, and the Troubled Quest to Remake Schooling. Troubled Quest to Remake American Schooling*. New York: Oxford University Press.

Meishar-Tal, H., Kurtz, G., & Pieterse, E. (2012). Facebook groups as LMS: A case study. *The International Review of Research in Open and Distributed Learning*, *13*(4), 33–48.

Melero, J., Hernández-Leo, D., Sun, J., Santos, P., & Blat, J. (2015). How was the activity? A visualization support for a case of location-based learning design. *British Journal of Educational Technology*, *46*(2), 317–329. doi:10.1111/bjet.12238

Merriam, S. B., & Caffarella, R. S. (1991). *Learning in Adulthood: A Comprehensive Guide*. San Francisco: Jossey-Bass Publishers.

Mertens, D. M. (2010). *Research and evaluation in education and psychology*. Sage Publications Inc.

Metcalf, D. S., & De Marco, J. M. (2006). *mLearning: Mobile learning and performance in the palm of your hand*. Human Resource Development.

Meyer, K. A. (2004). Evaluating online discussions: Four different frames of analysis. *Journal of Asynchronous Learning Networks*, *8*(2), 101–114.

Meyer, R. (2013). The post-lecture classroom: How will students fare. *Atlantic (Boston, Mass.)*.

Ministry of Education. (1996). *Te Whāriki: Te Whāriki matauranga mo nga mokopuna o Aotearoa, Early childhood curriculum*. Wellington: Learning Media.

Minskoff, E. H., & Allsopp, D. (2002). *Academic Success Strategies for Adolescents with Learning Disabilities/ADHD*. Baltimore, MD: Paul H Brookes Publishing.

Mirabolghasemi, M., & Iahad, N. A. (2013). Social network intervention for the enhancement of students' social presence in relation to satisfaction and perceived learning. *International Journal of Web Based Communities*, *9*(2), 233–242. doi:10.1504/IJWBC.2013.053246

Mishra, P., & Koehler, M. (2006). Technological pedagogical content knowledge: A framework for teacher knowledge. *Teachers College Record*, *108*(6), 1054. doi:10.1111/j.1467-9620.2006.00684.x

Missildine, K., Fountain, R., Summers, L., & Gosselin, K. (2013). Flipping the classroom to improve student performance and satisfaction. *The Journal of Nursing Education*, *52*(10), 597–599. doi:10.3928/01484834-20130919-03 PMID:24044386

MIT. (2014). 10 Breakthrough technologies. *MIT Technology Review*. Retrieved October 20, 2014, from http://www.technologyreview.com/lists/technologies/2014/

Mobilearn. (2002). *Mobilearn projects website*. Retrieved from http://www.mobilearn.org/index.php

Moeller. (2015). Didactic & Socratic models of teaching and learning. *Moeller Humanities Book*. Retrieved from http://collegeenglishbooks.com/two-models-of-teaching-learning.html

Mok, H. N. (2014). Teaching tip: The flipped classroom. *Journal of Information Systems Education*, *25*(1), 7.

Montgomery, A. P., Hayward, D. V., Dunn, W., Carbonaro, M., & Amrhein, C. G. (2015).. . *Australasian Journal of Educational Technology*, *31*(6), 657–670.

Moore, A., Masterson, J. T., Christophel, D. M., & Shea, K. A. (1996). College Teacher Immediacy and Student Ratings of Instruction. *Communication Education*, *45*(1), 29–39. doi:10.1080/03634529609379030

Moore, M. G. (1989). Editorial: Three types of interaction. *American Journal of Distance Education*, *3*(2), 1–7. doi:10.1080/08923648909526659

Moran, J. M. (2013). *Novas tecnologias e mediação pedagógica*. Papirus Editora.

Morgan, N., & Saxton, J. (1991). *Teaching, questioning, and learning*. New York: Routledge.

Morrison, D. A. (2012). *How Learning Analytics Can Make Instructors More Effective in an Online Course*. Retrieved 20 July 2015, 2015, from https://onlinelearninginsights.wordpress.com/2012/10/15/how-learning-analytics-can-make-instructors-more-effective-in-the-online-course/

Motiwalla, L. F. (2007). M-learning: A framework and evaluation. *Computers & Education*, *49*(3), 581–596. doi:10.1016/j.compedu.2005.10.011

Moura, A. M. C. (2010). *Apropriação do Telemóvel como Ferramenta de Mediação em Mobile Learning*. Braga, Portugal: Estudos de Caso em Contexto Educativo. Tese de Doutoramento. Universidade do Minho.

Mueller, A. L., Knobloch, N. A., & Orvis, K. S. (2015). Exploring the effects of active learning on high school students' outcomes and teachers' perceptions of biotechnology and genetics instruction. *Journal of Agricultural Education*, *56*(2), 138–152. doi:10.5032/jae.2015.02138

Mutlu, M. E., Yenigün, H. U., & Uslu, N. (2006). Açıköğretimde Mobil Öğrenme: Açıköğretim E-Öğrenme Hizmetlerinden Mobil Bilişim Aygıtlarıyla Yararlanma Olanaklarının Değerlendirilmesi. *Bilgi Teknolojileri IV & Akademik Bilişim*, 9-11.

Myers, M. D., & Avison, D. (Eds.). (2002). *Qualitative research in information systems: a reader*. London: Sage Publications.

Nack, F., & Waern, A. (2012). Mobile digital interactive storytelling—a winding path. *New Review of Hypermedia and Multimedia*, *18*(1-2), 3–9. doi:10.1080/13614568.2011.641418

Nagel, D. (2009, July). Meta-analysis: Is blended learning most effective? *THE Journal*. Retrieved 16th January 2016 from http://thejournal.com/articles/2009/07/01/meta-analysis-is-blended-learning-most-effective.aspx

Naismith, L., Lonsdale, P., Vavoula, G. N., & Sharples, M. (2004). *Mobile technologies and learning*. futurelab.

Naismith, L., Lonsdale, P., Vavoula, G., Sharples, M., & Series, N. F. (2004). *Literature review in mobile technologies and learning*. Futurelab Report 11, Retrieved from http://www2.futurelab.org.uk/resources/documents/lit_reviews/Mobile_Review.pdf

Naismith, L., Lonsdale, P., Vavoula, G., & Sharples, M. (2006). *Literature Review in Mobile Technologies and Learning. Report 11*. Futurelab.

Najafi, H., Evans, R., & Federico, C. (2014). MOOC integration into secondary school courses. *International Review of Research in Open and Distance Learning*, *15*(5), 306–322.

Nanayakkara, C. (2007). *A model of user acceptance of learning management systems: A study within tertiary institutions in New Zealand*. Retrieved from http://citeseerx.ist.psu.edu/viewdoc/download

Native, web or hybrid mobile-app development. (2012). Retrieved June 3, 2015, from ftp://public.dhe.ibm.com/software/pdf/mobile-enterprise/WSW14182USEN.pdf

Nedic, Z., Machotka, J., & Nafalski, A. (2003). *Remote laboratories versus virtual and real laboratories*. Paper presented at the Frontiers in Education, 2003. FIE 2003 33rd Annual. doi:10.1109/FIE.2003.1263343

Nelson, R. F. (2013). *Scaling up: Using MOOCs to increase accessibility and personalization for students*. Academic Press.

Neuendorf, K. A. (2002). *The content analysis guidebook*. Thousand Oaks, CA: Sage Publications.

Ng, W., & Nicholas, H. (2013). A framework for sustainable mobile learning in schools. *British Journal of Educational Technology*, *44*(5), 695–715. doi:10.1111/j.1467-8535.2012.01359.x

Nielsen, J. (1994). Estimating the number of subjects needed for a thinking aloud test. *International Journal of Human-Computer Studies*, *41*(3), 385–397. doi:10.1006/ijhc.1994.1065

Norris, C., & Soloway, E. (2012). Higher Ed has MOOCs, but K12 still needs to catch up. *District Administration*, *48*(9), 96.

NSF. (2013). *Women, Minorities, and Persons with Disabilities in Science and Engineering: 2013*. NSF. Retrieved January 20, 2015, from http://www.nsf.gov/statistics/wmpd/2013/pdf/nsf13304 _full.pdf

Nunan, T., George, R., & McCausland, H. (2000). Rethinking the ways in which teaching and learning are supported: The Flexible Learning Centre at the University of South Australia. *Journal of Higher Education Policy and Management*, *22*(1), 85–98. doi:10.1080/713678130

Nunnally, J. C. (2010). *Psychometric Theory 3E*. New York: Tata McGraw-Hill Education.

Nuri, K., Cansu Cigdem, A., & Kursat, C. (2014). User Study of a New Smart Toy for Children's Storytelling. *Interactive Learning Environments*, *22*(5), 551–563. doi:10.1080/10494820.2012.682587

O'Nuallain, C., & Brennan, A. (2005). How can one effectively assess students working in a collaborative mobile environment on an individual basis?. *Mobilelearning Anytime Everywhere*, 149.

O'Quinn, L., & Correy, M. (2002). Factors that deter faculty from participating in distance education. *Online Journal of Distance Learning Administration*, *5*(4), 1–16.

Ocak, M. (2011). Why are faculty members not teaching blended courses? Insights from faculty members. *Computers & Education*, *56*(3), 689–699. doi:10.1016/j.compedu.2010.10.011

OECD. (2007). *Revised Field Of Science and Technology (Fos) Classification In The Frascati Manual*. Retrieved January 15, 2015, fromhttp://www.oecd.org/sti/inno/38235147.pdf

Oh, E., & Park, S. (2009). How are universities involved in blended instruction? *Journal of Educational Technology & Society*, *12*(3), 327–342.

Oliva, A., Mack, M. L., Shrestha, M., & Peeper, A. (2004). Identifying the perceptual dimensions of visual complexity of scenes. In *The 26th Annual Meeting of the Cogn. Sci. Soc.*

Oliver, B. (2005). Mobile blogging, 'Skyping' and podcasting: Targeting undergraduates' communication skills in transnational learning contexts. *Microlearning*, *107*(4), 587–600.

Oliver, M., & Trigwell, K. (2005). Can "Blended Learning" Be Redeemed? *E-learning*, *2*(1), 17. doi:10.2304/elea.2005.2.1.2

O'Neil, K., Kelly, T., & Bone, S. (2012, June). We turned learning on its ear: Flipping the developmental classroom. In *World Conference on Educational Multimedia, Hypermedia and Telecommunications* (Vol. 2012, No. 1, pp. 2752-2756).

Osguthorpe, R. T., & Graham, C. R. (2003). Blended Learning Environments: Definitions and Directions. *Quarterly Review of Distance Education*, *4*(3), 227–233.

Osguthorpe, R., & Graham, C. (2003). Blended learning environments: Definitions and directions. *The Quarterly Review of Distance Education*, *4*(3), 227–233.

Ostendorf, V. A. (1997). *Teaching and learning at a distance: What it takes to effectively design, deliver, and evaluation programs*. San Francisco: Jossey-Bass.

Özmen, B., & Atıcı, B. (2014). The Effects of Social Networking Sites in Distance Learning on Learners' Academic Achievements. *European Journal of Open. Distance and E-Learning*, *17*(2), 61–75.

Pachler, N., Bachmair, B., & Cook, J. (2010a). *Mobile learning: structures, agency, practices*. New York: Springer. doi:10.1007/978-1-4419-0585-7

Pachler, N., Cook, J., & Bachmair, B. (2010b). Appropriation of mobile cultural resources for learning. *International Journal of Mobile and Blended Learning*, *2*(1), 1–21. doi:10.4018/jmbl.2010010101

Paechter, M., & Maier, B. (2010). Online or face-to-face? Students' experiences and preferences in e-learning. *The Internet and Higher Education*, *13*(4), 292–297. doi:10.1016/j.iheduc.2010.09.004

Paladini, S., Silva, J. B. d., Alves, G. R., Fischer, B. R., & Alves, J. B. M. (2008). *Using remote lab networks to provide support to public secondary school education level*. Paper presented at the Computational Science and Engineering Workshops, 2008. CSEWORKSHOPS'08. 11th IEEE International Conference on. doi:10.1109/CSEW.2008.76

Pannoni, A. (2014). *MOOCs a new tool for high school teachers*. Retrieved from: http://www.usnews.com/education/blogs/high-school-notes/2014/10/27/moocs-a-new-tool-for-high-school-teachers?int=a9e809

Papacharissi, Z. (2011). *A networked self: identity, community and culture on social network sites*. New York, London: Routledge.

Parkinson, D., Greene, W., Kim, Y., & Marioni, J. (2003). Emerging themes of student satisfaction in a traditional course and a blended distance course. *TechTrends*, *47*(4), 22–28. doi:10.1007/BF02763508

Parsons, D. (2007). M-learning. In D. Taniar (Ed.), Encyclopedia of Mobile Computing and Commerce, (pp. 525-527). Hershey, PA: Information Science Reference.

Parsons, D. (2014). The future of mobile learning and implications for education and training. In M. Ally & A. Tsinakos (Eds.), *Increasing Access through Mobile Learning* (pp. 217–229). Commonwealth of Learning Press.

Parsons, D., Ryu, H., & Cranshaw, M. (2007). A design requirements framework for mobile learning environments. *Journal of Computers*, *2*(4), 1–8. doi:10.4304/jcp.2.4.1-8

Parthasarathy, B. (2008). *The ethnographic case study approach*. Technology & Social Change Group, University of Washington. Retrieved from http://www. globalimpactstudy. org/2008/07/the-ethnographic-case-study-approach/

Partridge, H., Ponting, D., & McCay, M. (2011). *Good Practice Report: Blended Learning*. Sydney, NSW: Australian Learning and Teaching Council.

Peffers, K., Tuunanen, T., Gengler, C. E., Rossi, M., Hui, W., & Virtanen, V. (2006). The design science research process: a model for producing and presenting information systems research. In *Proceedings of the First International Conference on Design Science Research in Information Systems and Technology (DESRIST 2006)*.

Pegrum, M., Bartle, E., & Longnecker, N. (2015). Can creative podcasting promote deep learning? The use of podcasting for learning content in an undergraduate science unit. *British Journal of Educational Technology*, *46*(1), 142–152. doi:10.1111/bjet.12133

Pereira, O. R., & Rodrigues, J. J. (2013). Survey and analysis of current mobile learning applications and technologies. *ACM Computing Surveys*, *46*(2), 27. doi:10.1145/2543581.2543594

Perry, B., & Edwards, M. (2005). Exemplary Online Educators: Creating a Community of Inquiry. *Turkish Online Journal of Distance Education*, *6*(2), 1–10.

Petersen, S. A., & Divitini, M. (2005). Language learning: from individual learners to communities.*Proceedings of the IEEE International Workshop on Wireless and Mobile Technologies in Education*. doi:10.1109/WMTE.2005.41

Peters, O. (2001). *Learning and Teaching in distance education*. London: Routledge Falmer.

Petrova, K., & Li, C. (2009). *Focus and setting in mobile learning research: A review of the literature*. Academic Press.

Petrovic, N., Jeremic, V., Cirovic, M., Radojicic, Z., & Milenkovic, N. (2014). Facebook versus Moodle in practice. *American Journal of Distance Education*, *28*(2), 117–125.

Petrović, N., Petrović, D., Jeremić, V., Milenković, N., & Ćirović, M. (2012). Possible educational use of Facebook in higher environmental education. In *International Conference on Information Communication Technologies in Education* (pp.355-362).

Pettit, J., & Kukulska-Hulme, A. (2007). Going with the grain: Mobile devices in practice. *Australasian Journal of Educational Technology*, *23*(1), 17–33.

Phillips, C. R., & Trainor, J. E. (2014). Millenial students and the flipped classroom. *ASBBS Proceedings*, *21*(1), 519-530.

Phillips, R. (2013). Computer giveaway designed to give freshmen running start. *The Record.* Retrieved June 18, 2013, from http://www.recordnet.com/article/20130618/A_NEWS/306180317

Picciano, A. G. (2002). Beyond student perceptions: Issues of interaction, presence, and performance in an online course. *Journal of Asynchronous Learning Networks*, *6*(1), 21–40.

Picciano, A. G. (2006). Blended learning: Implications for growth and access. *Journal of Asynchronous Learning Networks*, *10*(3), 95–102.

Picciano, A. G. (2009). Blending with purpose: The multimodal model. *Journal of the ResearchCenter for Educational Technology*, *5*(1), 4–14.

Pilemalm, S., & Timpka, T. (2008). Third generation participatory design in health informatics—Making user participation applicable to large-scale information system projects. *Journal of Biomedical Informatics*, *41*(2), 327–339. doi:10.1016/j.jbi.2007.09.004

Pilli, O. (2014). LMS Vs. SNS: Can Social Networking Sites Act as a Learning Management Systems. *American International Journal of Contemporary Research*, *4*(5), 90-97.

Pinnock, M. (2015). *Reaching More Learners: Using Flipped Classrooms.* Retrieved January 8, 2015, from http://www.slideshare.net/michelepinnock/flipped-learning-intro

Poerksen, B. (2011). *The creation of reality: A constructivist epistemology of journalism and journalism education* (A. Kock & W. Kock, Trans.). Exeter, UK: Imprint Academic.

Ponsford, D. (2014). Labour force survey suggests surge of 12,000 in number of UK journalists since 2009. *Press Gazette*. Retrieved from http://www.pressgazette.co.uk/print/content/labour-force-survey-suggests-surge-12000-number-uk-journalists-2009

Prensky, M. (2001). Digital natives, digital immigrants. On the Horizon, 9(5).

Prensky, M. (2001). Digital Natives, Digital Immigrants. *Horizon*, *9*(5), 1–6. doi:10.1108/10748120110424816

Prunuske, A. J., Batzli, J., Howell, E., & Miller, S. (2012). Using online lectures to make time for active learning. *Genetics*, *192*(1), 67–72. doi:10.1534/genetics.112.141754 PMID:22714412

Puentedura, R. (2006). *Transformation, Technology, and Education.* Retrieved 18 February, 2013, from http://hippasus.com/resources/tte/puentedura_tte.pdf

Punch, K. (1998). *Introduction to Social Research: Quantitative and Qualitative Approaches.* London: Sage Publications.

Quinn, C. (2000). mLearning: mobile, wireless and in-your-pocket learning. *Line Zine Magazine*. Retrieved from http://www.linezine.com/2.1/features/cqmmwiyp.htm

Quinn, C. (2013). A future for M-Learning. In Z. L. Berge & L. Y. Muilenburg (Eds.), *Handbook of m-learning*. New York, NY: Routledge.

Raaij, E. M. & Schepers, J. J. L. (2008). *The Acceptance and Use of a Virtual Learning Environment*. Academic Press.

Radich, J. (2013). Technology and interactive media as tools in early childhood programs serving children from birth through age 8. *Every Child*, *19*(4), 18–19.

Ranieri, M., & Bruni, I. (2013a). Mobile storytelling and informal education in a suburban area: A qualitative study on the potential of digital narratives for young second generation immigrants. *Learning, Media and Technology*, *38*(2), 217–235. doi:10.1080/17439884.2013.724073

Ranieri, M., & Bruni, I. (2013b). Racconto e mi racconto con il cellulare. In A. Parola, A. Rosa, & R. Giannatelli (Eds.), *Media, Linguaggi, Creatività* (pp. 143–165). Trento: Erickson.

Ratto, M., Shapiro, R. B., Truong, T. M., & Griswold, W. G. (2003). The ActiveClass project: Experiments in encouraging classroom participation. In Computer support for collaborative learning. doi:10.1007/978-94-017-0195-2_57

Reid, M., Burn, A., & Parker, D. (2002). *Evaluation Report of the Becta Digital Video Pilot Project*. Report prepared for Becta. Retrieved from www.becta.org.uk/research/reports/digitalvideo/index.html

Reinhard, N., Saccol, A. Z., Schlemmer, E., Barbosa, J. L. V., & Kristoffersen, S. (2007). Aprendizagem com mobilidade no contexto organizacional. *UNISINOS*. Retrieved from http://www. joaomattar. com/7senaed/index. php

Reinhardt, W., Sievers, M., Magenheim, J., Kundisch, D., Herrmann, P., & Beutner, M. (2012). "PINGO: peer instruction for very large groups. In *Seventh European Conference on Technology Enhanced Learning*. doi:10.1007/978-3-642-33263-0_51

Reiser, B. J. (2004). Scaffolding Complex Learning: The Mechanisms of Structuring and Problematizing Student Work. *Journal of the Learning Sciences*, *13*(3), 273–304. doi:10.1207/s15327809jls1303_2

Resnick, M. (2002). *Rethinking learning in the digital age*. Retrieved from: https://llk.media.mit.edu/papers/mres-wef.pdf

Resnick, M., Berg, R., & Eisenberg, M. (2000). Beyond Black Boxes: Bringing Transparency and Aesthetics Back to Scientific Investigation. *Journal of the Learning Sciences*, *9*(1), 7–30. doi:10.1207/s15327809jls0901_3

RExLab. (n.d.). *Experimentação Remota*. Retrieved from www.rexlab.ufsc.br

Reynolds, R., Walker, K., & Speight, C. (2010). Web-based museum trails on PDAs for university-level design students: Design and evaluation. *Computers & Education*, *55*(3), 994–1003. doi:10.1016/j.compedu.2010.04.010

Richard, G. (2005). *E-learning in tertiary education: Where do we stand?* Paris: Centre for Education Research and Innovation.

Richards, G. (2011). *Measuring Engagement: Learning Analytics in Online Learning*. Paper presented at the Electronic Kazan 2011, Russia.

Richardson, J. C., & Swan, K. (2003). Examing social presence in online courses in relation to students' perceived learning and satisfaction. *Journal of Asynchronous Learning*, *7*(1), 68–88.

Roberts, J. B. (2013). Accessibility in M-Learning: Ensuring equal access. In Z. L. Berge & L. Y. Muilenburg (Eds.), *Handbook of m-learning*. New York, NY: Routledge.

Robinson, C. C., & Hullinger, H. (2008). New Benchmarks in Higher Education: Student Engagement in Online Learning. *Journal of Education for Business*, *84*(2), 101–108. doi:10.3200/JOEB.84.2.101-109

Roblyer, M. D., & Ekhaml, L. (2000). How interactive are your distance courses? A rubric for assessing interaction in distance learning. *Online Journal of Distance Learning Administration, 3*(2). Retrieved from http://www. Westga. edu/~distance/roblyer32.html

Roehl, A., Reddy, S. L., & Shannon, G. J. (2013). The flipped classroom: An opportunity to engage millennial students through active learning strategies. *Journal of Family and Consumer Sciences, 105*(2), 44–49. doi:10.14307/JFCS105.2.12

Rogers, E. M. (2010). *Diffusion of innovations*. Simon and Schuster, Inc.

Roschelle, J., Patton, C., & Pea, R. (2002). To unlock the learning value of wireless mobile devices, understand coupling. In Wireless and Mobile Technologies in Education. doi:10.1109/WMTE.2002.1039214

Roschelle, J. (2003). Keynote paper: Unlocking the learning value of wireless mobile devices. *Journal of Computer Assisted Learning, 19*(3), 260–272. doi:10.1046/j.0266-4909.2003.00028.x

Roshan, S., & Roshan, W. (2012). *My View: It's never too late to begin flipping your classroom. CNN Schools of Though.* Retrieved August 24, 2012, from http://schoolsofthought.blogs.cnn.com/2012/08/24/my-view-its-never-too-late-to-begin-flipping-your-classroom/

Ross, B., & Gage, K. (2006). Global Perspectives on Blending Learning. In C. J. Bonk & C. Graham (Eds.), *The Handbook of Blended Learning: Global perspectives, local designs.*

Roth, W. M., McGinn, M. K., Woszczyna, C., & Boutonne, S. (1999). Differential participation during science conversations: The interaction of focal artifacts, social configurations, and physical arrangements. *Journal of the Learning Sciences, 8*(3-4), 293–347. doi:10.1080/10508406.1999.9672073

Rourke, L., Anderson, T., Garrison, D., & Archer, W. (1999). Assessing social presence in asynchronous text-based computer conferencing. *Journal of Distance Education, 14*(2), 50–71.

Rovai, A., & Jordan, H. (2004). Blended learning and a sense of community: A comparative analysis with traditional and fully on-line graduate courses. *International Review of Research in Open and Distance Learning, 5*(3), 197–211.

Rožac, J., Buendía, F., Ballester, J., Kos, A., & Pogačnik, M. (2012). Integration of learning management systems with social networking platforms. In *The Fourth International Conference on Mobile, Hybrid, and On-line Learning* (pp. 100-105). Athabasca University.

Rubin, B., Fernandes, R., & Avgerinou, M. D. (2013). The effects of technology on the Community of Inquiry and satisfaction with online courses. *The Internet and Higher Education, 17*, 48–57. doi:10.1016/j.iheduc.2012.09.006

Ruddick, K. W. (2012). *Improving chemical education from high school to college using a more hands-on approach.* Academic Press.

Rudd, T., Gifford, C., Morrison, J., & Facer, K. (2006). *What if...? Re-imagining Learning Spaces.* Bristol: Futurelab.

Runco, M. A. (2003). Education for Creative Potential. *Scandinavian Journal of Educational Research, 47*(3), 317–324. doi:10.1080/00313830308598

Runco, M. A. (2007). *Creativity: theories and themes: research, development, and practice.* Amsterdam, London: Elsevier Academic Press.

Rusbridger, A. (2012, 29 February). *The Guardian: a world of news at your fingertips* [The Guardian's Three Little Pigs open journalism ad campaign]. Weblog posted to http://www.guardian.co.uk/help/insideguardian/2012/feb/29/open-journalism-at-the-guardian

Ryan, R. M., & Deci, E. L. (2000). Intrinsic and extrinsic motivations: Classic definitions and new directions. *Contemporary Educational Psychology*, *25*(1), 54–67. doi:10.1006/ceps.1999.1020 PMID:10620381

Saadatmand, M., & Kumpulainen, K. (2014). Participants' perceptions of learning and networking in connectivist MOOCs. *Journal Of Online Learning & Teaching*, *10*(1), 16–30.

Sadowski, J. (2014). The "Digital Native," a Profitable Myth. *The Baffler*. Retrieved January 15, 2014, from http://www.thebaffler.com/blog/the-digital-native-a-profitable-myth

Saldaña, J. (2009). *The coding manual for qualitative researchers*. London: Sage.

Saltman, D. (2011, November-December). Flipping for Beginners: Inside the New classroom Craze. *Harvard Education Letter, 29*(6). Retrieved from http://hepg.org/hel/article/517

Sandberg, J., Maris, M., & Hoogendoorn, P. (2014). The added value of a gaming context and intelligent adaptation for an m-learning application for vocabulary learning. *Computers & Education*, *76*, 119–130. doi:10.1016/j.compedu.2014.03.006

Sandeen, C. (2013). The emerging "MOOC 3.0" era. *Change*, *45*(6), 34–39. doi:10.1080/00091383.2013.842103

Sarah, T. (2014). *The Flipped Classroom. Research Synthesis EDET.* Retrieved June 24, 2014, from http://www.slideshare.net/sarahdtolson/the-flipped-classroom-36263899

Saritepeci, M., & Çakir, H. (2015). The Effect of Blended Learning Environments on Student Motivation and Student Engagement: A Study on Social Studies Course. *Egitim ve Bilim, 40*(177).

Scheele, N., Wessels, A., & Effelsberg, W. (2004). Die interaktive Vorlesung in der Praxis. In DeLFI 2004: Die 2. e-Learning Fachtagung Informatik.

Scherer, M. (2014). The paperless classroom is coming. *Time Magazine*. Retrieved from: http://time.com/3483905/the-paperless-classroom-is-coming/

Schlemmer, E., Saccol, A. Z., Barbosa, J., & Reinhard, N. (2007). *M-learning ou aprendizagem com mobilidade: casos no contexto brasileiro*. Academic Press.

Schmidt, D. A., Baran, E., Thompson, A. D., Mishra, P., Koehler, M. J., & Shin, T. S. (2009). Technological pedagogical content knowledge (TPACK) the development and validation of an assessment instrument for preservice teachers. *Journal of Research on Technology in Education*, *42*(2), 123–149. doi:10.1080/15391523.2009.10782544

Schneider, B., Wallace, J. M., Blikstein, P., & Pea, R. (2013). Preparing for future learning with a tangible user interface: The case of neuroscience. *Learning Technologies. IEEE Transactions on*, *6*(2), 117–129.

Schon, D., Klinger, M., Kopf, S., & Effelsberg, W. (2012). MobileQuiz–A Lecture Survey Tool using Smartphones and QR Tags. *International Journal of Digital Information and Wireless Communications*, *2*(3), 231–244.

Schrum, L., & Berge, Z. L. (1997). Creating Student Interaction within the Educational Experience: A Challenge for Online Teachers. *Canadian Journal of Educational Communication*, *26*(3), 133–144.

Schwartz, K. (2014). More Progressive Ways to Measure Deeper Levels of Learning. *Mind/Shift*. Retrieved December 20, 2014, from http://blogs.kqed.org/mindshift/2014/04/more-progressive-ways-to-measure-deeper-level-of-learning/

Seipold, J., Pachler, N., Bachmair, B., & Döbeli Honegger, B. (2013). Mobile Learning – Strategies for planning and implementing learning with mobile devices in secondary school contexts. In N. Pachler & M. Leask (Eds.), *M-Learning. Learning to teach using ICT in the secondary school: a companion to school experience* (pp. 185–204). New York: Routledge.

Sells, J., Tan, A., Borgan, J., Dahlen, U., & Stupart, Y. (2012). Preparing International Counselor Educators Through Online Distance Learning. *International Journal for the Advancement of Counseling*, *34*(1), 39–54. doi:10.1007/s10447-011-9126-4

Selwyn, N. (2000). Creating aConnected'Community? Teachers' Use of an Electronic Discussion Group. *Teachers College Record*, *102*(4), 750–778. doi:10.1111/0161-4681.00076

Sert, G. (2010). *Öğretim teknolojileri eğitiminde yayınlanmış Türkiye adresli makalelerin içerik analizi.* (Unpublished master's thesis). Hacettepe Üniversitesi, Ankara, Türkiye.

Severson, A. (2004). *Faculty Support Required for the Implementation of a New Learning Management System.* (Master Thesis). Simon Fraser University, Canada.

Shamburg, C. (2009). *Student-powered podcasting: teaching for 21st-century literacy*. Eugene, OR: ISTE.

Sharp, C. (2004). Developing Young Children's Creativity: What can we learn from research? *Topic*, *32*, 5–12.

Sharples, M. (2006). *Big issues in m-learning. Report of a workshop by the Kaleidoscope Network of Excellence M-learning Initiative*. Academic Press.

Sharples, M., Adams, A., Ferguson, R., Gaved, M., McAndrew, P., Rienties, B., & Whitelock, D. (2014). *Innovating Pedagogy 2014*. Academic Press.

Sharples, M., Taylor, J., & Vavoula, G. (2005). Towards a theory of mobile learning. *Proceedings ofmLearn 2005, 1*(1), 1-9.

Sharples, M. (2000). The design of personal mobile technologies for lifelong learning. *Computers & Education*, *34*(3-4), 177–193. doi:10.1016/S0360-1315(99)00044-5

Sharples, M., Arnedillo-Sánchez, I., Milrad, M., & Vavoula, G. (2009). *M-learning* (pp. 233–249). Springer Netherlands.

Sharples, M., Corlett, D., & Westmancott, O. (2002). The design and implementation of a mobile learning resource. *Personal and Ubiquitous Computing*, *6*(3), 220–234. doi:10.1007/s007790200021

Shea, P., & Bidjerano, T. (2010). Learning presence: Towards a theory of self-efficacy, self-regulation, and the development of a communities of inquiry in online and blended learning environments. *Computers & Education*, *55*(4), 1721–1731. doi:10.1016/j.compedu.2010.07.017

Shea, P., Hayes, S., Uzuner-Smith, S., Gozza-Cohen, M., Vickers, J., & Bidjerano, T. (2014). Reconceptualizing the community of inquiry framework: An exploratory analysis. *The Internet and Higher Education*, *23*, 9–17. doi:10.1016/j.iheduc.2014.05.002

Shea, P., Hayes, S., Vickers, J., Gozza-Cohen, M., Uzuner, S., Mehta, R., & Rangan, P. (2010). A re-examination of the community of inquiry framework: Social network and content analysis. *The Internet and Higher Education*, *13*(1), 10–21. doi:10.1016/j.iheduc.2009.11.002

Shea, P., Li, C. S., & Pickett, A. (2006). A study of teaching presence and student sense of learning community in fully online and web-enhanced college courses. *The Internet and Higher Education*, *9*(3), 175–190. doi:10.1016/j.iheduc.2006.06.005

Shih, J. L., Chu, H. C., & Hwang, G. J., & Kinshuk. (2010). An investigation on attitudes of students and teachers for participating in a context-aware ubiquitous learning activity. *British Journal of Educational Technology*.

Shih, W. C., Tseng, S. S., Yang, C. C., Lin, C. Y., & Liang, T. (2012). A folksonomy-based guidance mechanism for context-aware ubiquitous learning: A case study of Chinese scenic poetry appreciation activities. *Journal of Educational Technology & Society*, *15*(1), 90–101.

Shontell, A. (2014). The Next 20 Years are going to make the last 20 look like we accomplished nothing in tech. *Business Insider*. Retrieved September 25, 2014, from http://www.businessinsider.com/the-future-of-technology-will-pale-the-previous-20-years-2014-6#ixzz3AkcyE1gs

Short, J., Williams, E., & Christie, B. (1976). *The social psychology of telecommunications*. London: John Wiley & Sons.

Shotsberger, P. G., & Vatter, R. (2001). Teaching and learning in the wireless classroom. *Computer*, *34*(3), 110–111. doi:10.1109/2.910902

Shroff, R., & Vogel, D. (2010). An Investigation on Individual Students' Perceptions of Interest Utilizing a Blended Learning Approach. *International Journal on E-Learning*, *9*(2), 279–294.

Shuler, C. (2012). iLearn II; An Analysis of the Education Category of the iTunes App Store. *The Joan Ganz Cooney Center at Sesame Workshop*.

Shulman, L. S. (1986). Those who understand: Knowledge growth in teaching. *Educational Researcher*, *15*(2), 4–14. doi:10.3102/0013189X015002004

Siau, K., Sheng, H., & Fui-Hoon Nah, F. (2006). Use of a classroom response system to enhance classroom interactivity. *IEEE Transactions on Education*, *49*(3), 398–403. doi:10.1109/TE.2006.879802

Siemens, G. (2012). MOOCs are really a platform. *Elearnspace*. Retrieved from: http://www.elearnspace.org/blog/2012/07/25/moocs-are-really-a-platform/

Silva, J. B. d. (2006). *A Utilização da experimentação remota como suporte para ambientes colaborativos de aprendizagem*. Academic Press.

Silva, M. S. (2012). jQuery Móbile–Desenvolva aplicações web para dispositivos móveis com HTML 5, CSS#, AJAX, jQuery e jQuery UI. São Paulo: Novatec.

Silva, J. B. (2013). Aplicação de dispositivos móveis e experimentação remota para o ensino de Física na Educação Básica. In R. Famelli & A. P. P. Nogueira (Eds.), *Educação no Século XXI - Mobilidade* (Vol. 5). São Paulo: Fundação Telefônica.

Silva, J. B. d., Rochadel, W., Schardosim, J. P., & Fidalgo, A. V. S. F. (2014). Adaptation Model of Mobile Remote Experimentation for Elementary Schools. Tecnologias del Aprendizaje. *IEEE Revista Iberoamericana de*, *9*(1), 28–32.

Silva, J., & Salinas, J. (2014). *Innovando con TIC en la formación inicial docente: aspectos teóricos y casos concretos*. Santiago.

Şimşek, A., Özdamar, N., Becit, G., Kılıçer, K., Akbulut, Y., & Yıldırım, Y. (2008). Türkiye'deki eğitim teknolojisi araştırmalarında güncel eğilimler. *Selçuk Üniversitesi Sosyal Bilimler Dergisi*, *19*, 439–458.

Sims, R. (2003). Promises of interactivity: Aligning learner perceptions and expectations with strategies for flexible and online learning. *Distance Education*, *24*(1), 87–103. doi:10.1080/01587910303050

Singh, H. (2003). Building effective blended learning programs. *Educational Technology*, *43*(6), 51–54.

Sinnappan, S., & Zutshi, S. (2011). Using Microblogging to facilitate Community of Inquiry: An Australian tertiary experience. In *ASCILITE-Australian Society for Computers in Learning in Tertiary Education Annual Conference* (Vol.1, pp. 1123-1135). Australasian Society for Computers in Learning in Tertiary Education.

SKMM Guidelines. (2012). *Content code*. SKMM, Malaysia. Retrieved January 2, 2013 from http://www.skmm.gov.my/Resources/Guidelines/Content-Code.aspx

Smith, B. (2001). Just give us the right answer. In H. Edwards, B. Smith, & G. Webb (Eds.), Lecturing: Case studies, experience and practice (pp. 123-129). London: Kogan.

Snell, Y. S., & Linda, S. (1999). Interactive lecturing: Strategies for increasing participation in large group presentations. *Medical Teacher*, *21*(1), 37–42. doi:10.1080/01421599980011

Socrative. (2015). *Socrative Webpage*. Retrieved February 25, 2015, from http://www.socrative.com

So, H. J., & Brush, T. A. (2008). Student perceptions of collaborative learning, social presence and satisfaction in a blended learning environment: Relationships and critical factors. *Computers & Education*, *51*(1), 318–336. doi:10.1016/j.compedu.2007.05.009

Sohaei, S., & Iahad, N. A. (2014). Review on Social Network Sites for Teaching and Learning. *Journal of Information Systems Research And Innovation*, *7*(1), 10–16.

Söllner, M., Hoffmann, A., Hoffmann, H., Wacker, A., & Leimeister, J. M. (2012). Understanding the Formation of Trust in IT Artifacts. In *The International Conference on Information Systems (ICIS)*.

Soykan, E., & Uzunboylu, H. (2015). New trends on m-learning area: The review of published articles on m-learning in science direct database. *World Journal on Educational Technology*, *7*(1), 31–41. doi:10.18844/wjet.v7i1.22

Spencer, K.M., Coutts, T., Fagan, T., & King, A. (2013). Connections, Diversity, Coherence: Three Vignettes Exploring Learning with iPads in Primary Schools. *Computers in New Zealand Schools: Learning, Teaching, Technology, 25*(1–3), 38–55

Stacey, E., & Gerbic, P. (2009). Introduction to blended learning practices. In E. Stacey & P. Gerbic (Eds.), *Effective blended learning practices: evidence-based in perspectives in ICT-facilitated education.* Hershey, PA: IGI Global. doi:10.4018/978-1-60566-296-1.ch001

Staker, H., & Horn, M. B. (2012). Classifying K-12 blended learning. *Innosight Institute*. Retrieved from: http://www.christenseninstitute.org/wp-content/uploads/2013/04/Classifying-K-12-blended-learning.pdf

Stanciu, A., Mihai, F., & Aleca, O. (2012). Social networking as an alternative environment for education. *Accounting and Management Information Systems*, *11*(1), 56–75.

Star, S. L. (2010). This Not a Boundary Object: Reflections on the Origin of a Concept. *Science, Technology & Human Values*, *35*(5), 601–617. doi:10.1177/0162243910377624

Stein, D. S., Wanstreet, C. E., Glazer, H. R., Engle, C. L., Harris, R. A., Johnston, S. M., & Trinko, L. A. (2007). Creating shared understanding through chats in a community of inquiry. *The Internet and Higher Education*, *10*(2), 103–115. doi:10.1016/j.iheduc.2007.02.002

Sternberg, R. J. (1999). *Handbook of creativity*. Cambridge, UK: Cambridge University Press.

Sternberg, R. J., Kaufman, J. C., & Pretz, J. E. (2002). *The creativity conundrum: A propulsion model of kinds of creative contributions*. Philadelphia: Psychology Press.

Sternberg, R. J., & Lubart, T. I. (1999). The concept of creativity: Prospects and paradigms. In R. J. Sternberg (Ed.), *Handbook of creativity* (pp. 3–15). Cambridge, UK: Cambridge University Press.

Stocks, T. (2014). Can Video Help Learning *And Development*. Retrieved from http://www.business2community.com/strategy/can-video-help-learning-development-0951021#!bpxhpA

Stockwell, G. (2010). Using mobile phones for vocabulary activities: Examining the effect of the platform. *Language Learning & Technology*, *14*(2), 95–110.

Stoilescu, D. (2011). *Technological Pedagogical Content Knowledge: Secondary School Mathematics Teachers' Use of Technology*. University of Toronto.

Stolle, M. (2014). *Teacher of the Month: Bamber Valley teacher 'flips' the classroom. Postbulletin.com.* Retrieved February 1, 2014, from http://m.postbulletin.com/news/local/teacher-of-the-month-bamber-valley-teacher-flips-the-classroom/article_ebff74ad-0aa7-5a45-aed6-1a581a95edb2.html?mode=jqm

Stone, B. B. (2012). Flip your classroom to increase active learning and student engagement. In *Proceedings from 28th Annual Conference on Distance Teaching & Learning.*

Straumsheim, C. (2013). Still in Favor of the Flip. *October, 30.*

Straumsheim, C. (2014). After #MassiveTeaching, questions about MOOC quality control. *Inside Higher Education.* Retrieved January 18, 2015, from https://www.insidehighered.com/news/2014/07/15/after-massiveteaching-questions-about-mooc-quality-control#ixzz37X1bxB6p

Su, B., Bonk, C. J., Magjuka, R. J., Liu, X., & Lee, S.-h. (2005). The importance of interaction in web-based education: A program-level case study of online MBA courses. *Journal of Interactive Learning Research, 4*(1), 1–19.

Su, C.-H., & Cheng, C.-H. (2015). A mobile gamification learning system for improving the learning motivation and achievements. *Journal of Computer Assisted Learning, 31*(3), 268–286. doi:10.1111/jcal.12088

Sue, B., Maton, K., & Kervin, L. (2008). The "digital natives" debate: A critical review of the evidence. *British Journal of Educational Technology, 39*(5), 775–786. doi:10.1111/j.1467-8535.2007.00793.x

Sunguro, S. (2013). *Flipped classroom.* Retrieved October 14, 2013, from http://www.slideshare.net/stephens2042/flipped-classroom-27164701

Suprabha, K., & Subramonian, G. (2015). Blended learning approach for enhancing students learning experiences in a knowledge society. *Journal of Educational Technology, 11*(4), 1–7.

Swan, K. (2003). Developing Social Presence in Online Discussions. In S. Naidu (Ed)., *Learning and Teaching with Technology: principles and practices* (pp. 147–164). London: Kogan Page.

Swan, K., & Ice, P. (2010). The Community of Inquiry Framework Ten Years Later: Introducation to the Special Issue. *The Internet and Higher Education, 13*(1-2), 1–4. doi:10.1016/j.iheduc.2009.11.003

Swan, K., Shea, P., Richardson, J., Ice, P., Garrison, D. R., Cleveland-Innes, M., & Arbaugh, J. B. (2008). Validating a measurement tool of presence in online communities of inquiry. *E-mentor, 2*(24), 1–12.

Swantz, M. L. (2008). Participatory Action Research as Practice. In P. Reason & H. Bradbury (Eds.), *The SAGE Handbook of Action Research: Participative Inquiry and Practice* (2nd ed.; pp. 31–48). London: SAGE Publications. doi:10.4135/9781848607934.n8

Szeto, E. (2015). Community of Inquiry as an instructional approach: What effects of teaching, social and cognitive presences are there in blended synchronous learning and teaching? *Computers & Education, 81*, 191–201. doi:10.1016/j.compedu.2014.10.015

Szu, E., Nandagopal, K., Shavelson, R. J., Lopez, E. J., Penn, J. H., Scharberg, M., & Hill, G. W. (2011). Understanding Academic Performance in Organic Chemistry. *Journal of Chemical Education, 88*(9), 1238–1242. doi:10.1021/ed900067m

Tabaque, M. (2014). *The Teacher of the 21st century. Is The Philippines Ready For Elearning Or Mlearning?* Retrieved from http://www.minreeva.com/blog

Tabaque, P. (2013). *The Benefits of Learning for Schools in the Philippines.* Retrieved February 3, 2013, from http://www.minreeva.com/blog/the-benefits-of-mlearning-for-schools-in-the-philippines

Tabata, Y., Yin, C., Ogata, H., & Yano, Y. (2010). An iPhone quiz system for learning foreign languages. In *Informatics in Control, Automation and Robotics (CAR), 2010 2nd International Asia Conference on*. doi:10.1109/CAR.2010.5456691

Talbert, R. (2012). Inverted Classroom. *Colleagues*, Article 7.

Tarouco, L. M., Fabre, M. C. J. M., Grando, A. R., & Konrath, M. L. (2004). Objetos de Aprendizagem para M-learning. In *Florianópolis: SUCESU-Congresso Nacional de Tecnologia da Informação e Comunicação.*

Taylor, C. W. (1988). Various Approaches to and Definitions of Creativity. In R. Sternberg (Ed.), *The Nature of Creativity: Contemporary Psychological Perspectives* (pp. 99–121). New York: Cambridge University Press.

Teleco. (2014). *Estatísticas de Celulares no Brasil.* Retrieved from http://www.teleco.com.br/ncel.asp

The New Media Consortium. (2014). *The New Media Consortium Horizon Report: 2014 Higher Education Edition.* The New Media Consortium.

Thoms, B., & Eryilmaz, E. (2014). How media choice affects learner interactions in distance learning classes. *Computers & Education, 75*, 112–126. doi:10.1016/j.compedu.2014.02.002

Thorne, S. L., & Payne, J. S. (2005). Evolutionary trajectories, internet-mediated expression, and language education. *CALICO Journal, 22*(3), 371–397.

Thurmond, V., & Wambach, K. (2004). Understanding interactions in distance education: A review of the literature. *International Journal of Instructional Technology and Distance Learning, 1*(1), 9–26.

Tickodri-Togboa, S. S., Mwikirize, C., Tumusiime, A. A., & Musasizi, P. I. (2011). iLabs: Revolutionizing Teaching and Learning at Makerere University. In J. G. Zubía & G. R. Alves (Eds.), Using Remote Labs in Education: Two Little Ducks in Remote Experimentation (pp. 112-153). Bilbão: University of Deusto.

Timperley, H. (2010). *Using evidence in the classroom for professional learning.* Paper presented at the Ontario Education Research Symposium, Toronto, Ontario, Canada.

Timperley, H. S., Wilson, A., Barrar, H., & Fung, I. (2007). Teacher Professional Learning and Development: Best evidence synthesis iteration. Wellington, New Zealand: Ministry of Education. Retrieved from http://www.educationcounts.govt.nz/publications/series/2515/15341

Tobias, S., & Duffy, T. (Eds.). (2009). *Constructivist instruction: Success or failure?* London: Routledge.

Tolmie, A. (2001). Examining learning in relation to the contexts of use of ICT. *Journal of Computer Assisted Learning, 17*(3), 235–241. doi:10.1046/j.0266-4909.2001.00178.x

Tongmak, M. (2013). Social Network System in Classroom: Antecedents of Edmodo© Adoption. *Journal of e-Learning &. Higher Education.* doi:10.5171/2013.657749

Traxler, J. (2009). Current state of mobile learning1. *Mobile learning*, 9.

Traxler, J. (2011). Introduction. In J. Traxler & J. Wishart (Eds.), Making mobile learning work: case studies of practice (pp. 4-12). Bristol, UK: ESCalate, University of Bristol, Graduate School of Education.

Traxler, J. (2005). *Defining M-learning. IADIS International Conference M-learning*. Qawra, Malta

Traxler, J. (2010). Distance education and m-learning: Catching up, taking stock. *Distance Education, 31*(2), 129–138. doi:10.1080/01587919.2010.503362

Tremblay, E. (2010). Educating the Mobile Generation–using personal cell phones as audience response systems in post-secondary science teaching. *Journal of Computers in Mathematics and Science Teaching, 29*(2), 217–227.

Trifonova, A. (2003). *Mobile learning-review of the literature*. Academic Press.

Trinder, J. J. (2005) Mobile technologies and systems. In M-learning: A Handbook for Educators and Trainers (pp. 7-24). London: Routledge.

Truong, D. (2014). How To Design a Mobile Application to Enhance Teaching and Learning? *International Journal of Interactive Mobile Technologies*, *9*(3), 4–11.

Tsinakos, A., & Ally, M. (2013). *Global m-learning implementation and trends*. Beijing, China: CRTVU Press.

Tucker, B. (2012). The Flipped Classroom: Online Instruction at Home Frees Class Time for Learning. *Education Next*, *12*(1), 82–83. http://educationnext.org/files/ednext_20121_BTucker.pdf Retrieved June 24, 2012

Twigg, C. (2003). Improving quality and reducing costs: New models for online learning. *EDUCAUSE Review*, *38*(5), 28–38.

U.S. General Accounting Office. (1996). *Content Analysis: A Methodology for Structuring and Analyzing Written Material*. GAO/PEMD-10.3.1. Retrieved January 10, 2015, http://www.gao.gov/assets/80/76281.pdf

Ubuz, B., & Üstün, I. (2004). Figural and conceptual aspects in defining and identifying polygons. *Eurasian Journal of Educational Research*, *16*, 15–26.

UK Department for Education. (2014). MOOCs: opportunities for their use in compulsory-age education. *E-Learning Update*, (46), 9-10.

Umrani-Khan & Iyer. (2009). *ELAM: A Model for Acceptance and use of e-learning by Teachers and Students*. International Conference on E-Learning, Toronto, Canada.

UNESCO. (2013). *The future of mobile learning: implications for planners and policy makers* (1st ed.). Paris, DC: Autor.

UNESCO. (2014). *Diretrizes de políticas para a aprendizagem móvel*. UNESCO.

Univeristy of Illinois. (2010). *Weaknesses of Online Learning*. Retrieved from http://www.ion.uillinois.edu/resources/tutorials/overview/weaknesses.asp

Uzunboylu, H., Cavus, N., & Ercag, E. (2009). Using mobile learning to increase environmental awareness. *Computers & Education*, *52*(2), 381–389. doi:10.1016/j.compedu.2008.09.008

Valentim, H. D. (2009). *Para uma compreensão do Mobile Learning. Reflexão sobre a utilidade das tecnologias móveis na aprendizagem informal e para a construção de ambientes pessoais de aprendizagem. Dissertação de mestrado*. Lisboa, Portugal: Universidade Nova de Lisboa.

Van Rijsselbergen, D., Poppe, C., Verwaest, M., Mannens, E., & Van de Walle, R. (2012). Semantic Mastering: Content adaptation in the creative drama production workflow. *Multimedia Tools and Applications*, *59*(1), 307–340. doi:10.1007/s11042-010-0710-0

VanDeGrift, T., Wolfman, S. A., Yasuhara, K., & Anderson, R. J. (2002). Promoting interaction in large classes with a computer-mediated feedback system. In Computer Science Department and Microsoft Research.

Vaughan, N. (2007). Perspectives on blended learning in higher education. *International Journal on E-Learning*, *6*(1), 81–94.

Vaughan, N. (2014). Student Engagement and Blended Learning: Making the Assessment Connection. *Education Sciences*, *4*(4), 247–264. doi:10.3390/educsci4040247

Vaughan, N., & Garrison, D. R. (2005). Creating cognitive presence in a blended faculty development community. *The Internet and Higher Education*, *8*(1), 1–12. doi:10.1016/j.iheduc.2004.11.001

Vavoula, G. N., & Sharples, M. (2002). KLeOS: A personal, mobile, knowledge and learning organization system. In *Wireless and Mobile Technologies in Education, 2002. Proceedings. IEEE International Workshop on* (pp. 152-156). IEEE. doi:10.1109/WMTE.2002.1039239

Vavoula, G. N., & Sharples, M. (2008). Challenges in Evaluating Mobile Learning. *Proceedings of Mlearn 2008*.

Vavoula, G. (2007). Learning Bridges: A role for mobile technologies in education. *Educational Technology Magazine*, *47*(3), 33–37.

Vavoula, G., & Sharples, M. (2009). Meeting the Challenges in Evaluating M-learning: A 3-Level Evaluation Framework. *International Journal of Mobile and Blended Learning*, *1*(2), 54–75. doi:10.4018/jmbl.2009040104

Vavoula, G., Sharples, M., Rudman, P., Meek, J., & Lonsdale, P. (2009). Myartspace: Design and evaluation of support for learning with multimedia phones between classrooms and museums. *Computers & Education*, *53*(2), 286–299. doi:10.1016/j.compedu.2009.02.007

Venkatesh, V., & Davis, F. D. (2000). A Theoretical Extension of the Technology Acceptance Model: Four Longitudinal Field Studies. *Management Science*, *45*(2), 186–204. doi:10.1287/mnsc.46.2.186.11926

Venkatesh, V., Morris, M. G., Davis, G. B., & Davis, F. D. (2003). User Acceptance of Information Technology: Toward a Unified View. *Management Information Systems Quarterly*, *27*(3), 425–478.

Verenikina, I., & Kervin, L. (2011). iPads, Digital Play and Pre-schoolers. *He Kupu*, *2*(5), 4–19.

Villalba, E. (2008). *On Creativity: Towards an Understanding of Creativity and Innovation*. Luxemburg: European Commission / JRC. Retrieved on 25th July, 2015 from: http://publications.jrc.ec.europa.eu/repository/bitstream/JRC48604/eur_on%20creativity_new_.pdf

Visser, L., & Gagnon, K. (2005). Defining "Trends" and "Issues" in Distance Education. Conversation with Donald Paul Ely. In Trends and Issues in Distance Education. International Perspectives.

Vygotsky, L. S. (1978). *Mind in Society*. Cambridge, MA: Harvard University Press.

Vygotsky, L. S. (1978). *Mind in society: The development of higher psychological processes* (M. Cole, V. John-Steiner, S. Scribner, & E. Souberman, Eds.). Cambridge, MA: Harvard University Press.

Vygotsky, L. S. (1986). *Thought and language* (A. Kozulin, Ed.). Cambridge, MA: MIT Press.

Wagoner, T., Hoover, S., & Ernst, D. (2011). *CEHD iPad initiative*. Retrieved from August 30, 2015, from: http://www.cehd.umn.edu/Mobile/PDF/CEHDiPadReport.pdf

Wang, M. C., Haertel, G. D., & Walberg, H. J. (1990). What influences learning? A content analysis of review literature. *The Journal of Educational Research*, *84*(1), 30–43. doi:10.1080/00220671.1990.10885988

Wang, Y. S., Wu, M. C., & Wang, H. Y. (2009). Investigating the determinants and age and gender differences in the acceptance of m-learning. *British Journal of Educational Technology*, *40*(1), 92–118. doi:10.1111/j.1467-8535.2007.00809.x

Wantz, R.A., Tromski, D.M., Mortsolf, C.J., Yoxtheimer, G., Brill, S., & Cole, A. (2003). Incorporating Distance Learning into Counselor Education Programs: A Research Study. *Cybercounseling and Cyberlearing: An Encore*, 327-344.

Weber, R. P. (1990). *Basic Content Analysis* (2nd ed.). Newbury Park, CA.

Weebly. (2015). *Weebly Webpage*. Retrieved March 23, 2015, from http://www.weebly.com

Wells, J. D., Valacich, J. S., & Hess, T. J. (2011). What signal are you sending? how website quality influences perceptions of product quality and purchase intentions. *Management Information Systems Quarterly*, *35*(2), 373–396.

Welsh, E. T., Wanberg, C. R., Brown, K. G., & Simmering, M. J. (2003). E-learning: Emerging uses, empirical results and future directions. *International Journal of Training and Development*, *7*(4), 245–258. doi:10.1046/j.1360-3736.2003.00184.x

Wenger, E., White, N., & Smith, J. (2009). *Digital Habitats: stewarding technology for communities*. Portland, OR: CPsquare.

Wenger, E. (1998). *Communities of Practice: Learning, Meaning, and Identity*. Cambridge, UK: Cambridge University Press. doi:10.1017/CBO9780511803932

Wenger, E., White, N., Smith, J., & Rowe, K. (2005). Technology for Communities. In L. Langelier (Ed.), *Working, Learning and Collaborating in a Network: Guide to the implementation and leadership of intentional communities of practice* (pp. 71–94). Quebec City: CEFIRO.

Werth, E., Werth, L., & Kellerer, E. (2013). Transforming K-12 rural education through blended learning: Barriers and promising practices. *The International Association for K-12 Online Learning*. Accessed 16th January 2016 from http://files.eric.ed.gov/fulltext/ED561276.pdf

Wessner, M., & Dawabi, P. (2004). Interaktionsunterstützung für die Präsenzlehre-Szenarien und Design. In DeLFI 2004: Die 2. e-Learning Fachtagung Informatik.

Wexler, S., Brown, J., Metcalf, D., Rogers, D., & Wagner, E. (2008). *The e-learning Guild Report M-learning*. Retrieved from August 30, 2015, from: http://www.elearningguild.com/research/archives/index.cfm?id=132&action=viewonly

Weyant, L. E. (2013). Designing online management education courses using the Community of Inquiry framework. *Journal of Instructional Pedagogies*, *12*(1), 1–14.

Wheeler, S., Yeomans, P., & Wheeler, D. (2008). The good, the bad and the wiki: Evaluating student-generated content for collaborative learning. *British Journal of Educational Technology*, *39*(6), 987–995. doi:10.1111/j.1467-8535.2007.00799.x

Whitmer, J., Fernandes, K., & Allen, W.R. (2012). Analytics in Progress: Technology Use, Student Characteristics, and Student Achievement. *Educause Review Online*, 1-7.

Williams-Carlin, R. (2009). *A comparative study of geometry curricula.* (Unpublished doctoral dissertation). Louisiana State University, Baton Rouge, LA.

Wilson, S. E. (2001). *The application of technology to counselor education: Video conferencing in distance supervision.* (Doctoral dissertation), University of Texas, San Antonio.

Wilson, S. G. (2013). The Flipped Class A Method to Address the Challenges of an Undergraduate Statistics Course. *Teaching of Psychology*.

Wingkvist, A., & Ericsson, M. (2011). A survey of research method and purposes in mobile learning. *International Journal of Mobile and Blended Learning*, *3*(1), 1–17. doi:10.4018/jmbl.2011010101

Winter, N. (2006). *What is m-learning. Big issues in m-learning*. Report of a workshop by the Kaleidoscope Network of Excellence M-learning Initiative.

Wise, A., Chang, J., Duffy, T., & del Valle, R. (2004). The effects of teacher social presence on student satisfaction, engagement, and learning. *Journal of Educational Computing Research*, *31*(3), 247–271. doi:10.2190/V0LB-1M37-RNR8-Y2U1

Withell, A., Cochrane, T., Reay, S., Gaziulusoy, I., & Inder, S. (2012). Augmenting the Design Thinking Studio. In M. Brown (Ed.), ASCILITE 2012: Future challenges, sustainable futures (pp. 1071-1081). Wellington, New Zealand: ASCILITE.

Wohlin, C., Runeson, P., Höst, M., Ohlsson, M. C., Regnell, B., & Wesslén, A. (2012). *Experimentation in software engineering*. Springer Science & Business Media. doi:10.1007/978-3-642-29044-2

Wood, D. J., Bruner, J. S., & Ross, G. (1976). The role of tutoring in problem solving. *Journal of Child Psychiatry and Psychology*, *17*(2), 89–100. doi:10.1111/j.1469-7610.1976.tb00381.x PMID:932126

Wood, E., & Attfield, J. (2005). Play, learning and the early childhood curriculum. *Sage (Atlanta, Ga.)*.

Woodill, G. (2011). The M-learning Edge. McGraw Hill.

Wood, R., Baker, J., & Hopper, D. (2004). Hybrid structures: Faculty use and perception of web based courseware as a supplement to face-to-face instruction. *The Internet and Higher Education*, *7*(4), 281–297. doi:10.1016/j.iheduc.2004.09.002

Wortmann, F., Hirdes, E. M., Lehmann, K., & Zülich, J. (2014). IT-Unterstützung von Blended Learning Konzepten. In K. Möller & W. Schultze (Eds.), *Produktivität von Dienstleistungen*. Wiesbaden: Springer Gabler.

Wu, H., Lee, S. W., Chang, H., & Liang, J. (2013). Current status, opportunities and challenges of augmented reality in education. *Computers & Education*, *62*, 41–49. doi:10.1016/j.compedu.2012.10.024

Wu, W. H., Wu, Y. C. J., Chen, C. Y., Kao, H. Y., Lin, C. H., & Huang, S. H. (2012). Review of trends from mobile learning studies: A meta-analysis. *Computers & Education*, *59*(2), 817–827. doi:10.1016/j.compedu.2012.03.016

Wyse, D., & Spendlove, D. (2007). Partners in Creativity: Action Research and Creative Partnerships. *Education 3-13, 35*(2), 181-191.

Xanthopoulos, S., & Xinogalos, S. (2013). A comparative analysis of cross-platform development approaches for mobile applications.*Proceedings of the 6th Balkan Conference in Informatics (BCI '13)* (pp. 213-220). New York, NY: ACM. doi:10.1145/2490257.2490292

Yarbro, J., Arfstrom, K. M., McKnight, K., & McKnight, P. (2014). *Extension of a review of flipped learning*. Academic Press.

Yigitbasioglu, O. M., & Velcu, O. (2011). A review of dashboards in performance management: Implications for design and research. *International Journal of Accounting Information Systems*.

Young, J. R. (2002). Hybrid Teaching Seeks to End the Divide Between Traditional and Online Instruction. *The Chronicle of Higher Education*, *48*(28), 2–3.

Yuen, H. K. A., & Ma, W. W. K. (2008). Exploring teacher acceptance of e-learning technology. *Asia-Pacific Journal of Teacher Education*, *36*(3), 229–243. doi:10.1080/13598660802232779

Zhao, Y., Lei, J., Yan, B., Lai, C., & Tan, H. S. (2005). What makes the difference? A practical analysis of research on the effectiveness of distance education. *Teachers College Record*, *107*(8), 1836–1884. doi:10.1111/j.1467-9620.2005.00544.x

Ziegler, M., & Bühner, M. (2009). *Statistik für Psychologen und Sozialwissenschaftler*. Pearson Deutschland GmbH.

Zubía, J. G., & Alves, G. R. (2011). *Using Remote Labs in Education: Two Little Ducks in Remote Experimentation* (Vol. 8). Bilbao: Universidad de Deusto.

About the Contributors

David Parsons is National Postgraduate Director for The Mind Lab by Unitec, Auckland, New Zealand. He holds a PhD in Information Technology and a Master's degree in Computer Science, and has wide experience in both academia and the IT industry. He is the founding editor in chief of the International Journal of Mobile and Blended Learning (IJMBL) and author of a number of texts on computer programming, web application development and mobile learning. He chaired the Conference on Mobile Learning Technologies and Applications in 2007 and was co-editor of 'Innovative Mobile Learning: Techniques and Technologies' (Information Science Reference, 2009). He also edits the annual compilations of papers from IJMBL that are published as a regular series of books. He is a member of the International Association for Mobile Learning and a professional member of the British Computer Society.

* * *

Gökçe Akçayır is a PhD student and research assistant at Gazi University, Faculty of Gazi Education, Department of Computer Education and Instructional Technology. Her research interests include educational use of social network sites, augmented reality, digital natives and blended learning.

Carlos Fernando Araújo Jr. is a Researcher and Professor of the Master's and Doctorate Program in Science and Mathematics Teaching, University Cruzeiro do Sul, and Provost of Distance Education (Cruzeiro do Sul Educacional). Degree in Physics, Master degree in Physics and PhD in Theoretical Physics from the Universidade Estadual Paulista Julio de Mesquita Filho, São Paulo, Brasil. In recent years he served, such as Course Coordinator, Rector Adviser, Director in theTechnology and Computing. Has experience in the information technology, with emphasis on Information and Communication Technologies Applied to Education, acting on the following topics: Technologies applied to education, virtual learning environments, distance education, teaching science and mathematics and computer education.

Shamsul Arrieya Ariffin received his PhD from University of Technology Sydney, Australia. He is a senior lecturer at Lecturer (Computing Department), Faculty of Arts, Computing and Industry Creative, Sultan Idris Education University (UPSI) in Tanjung Malim Perak Malaysia. He is also a Deputy Director at International Office Affairs at UPSI. He has 20 years working in various IT fields including international companies. He is also a member of the International Association for Mobile Learning. His current interest is on mLearning and user perspectives on software usages in education.

Heather Bell is an education facilitator working for Massey University's Centre for Educational Development. Her work centres around improving school practice in general with particular interests in supporting schools to develop their concepts of inclusive education and in helping schools take advantage of eLearning and smart technologies to effectively shift learning. Often these two areas merge as schools are now strongly encouraged to focus on shifting learning outcomes for their at risk students. Heather's Masters research investigated effective facilitation strategies for blended learning communities. She uses this in her work with schools to build collaborative learning communities using a blend of online and face to face environments to sustain and grow the professional learning and development opportunities. She is equally passionate about using the Web to create permeable classroom walls so students can collaborate with others; across the region, country and the world to generate new knowledge through interacting with authentic audiences. Heather encourages teachers to embrace the technology that is a common part of students' lives to make this happen.

Prosper M. Bernard is an internationally respected lecturer and consultant with extensive experience in International Business Education especially in Latin America and in China. He is the Chairman of the Board of Trustees of the University Consortium of the Americas and member of the Board of a few private and public organizations.. He is a Professor of Business at the University of Québec at Montreal. He served as Vice-rector and member of the Board of the same university. He is a Canadian and an American citizen. He has published many books and articles.

Simone Meister Sommer Bilessimo is a professor at the Federal University of Santa Catarina – Campus Araranguá and researcher at the Remote Experimentation Lab (RExLab).

Samantha Briggs is a graduate of Old Dominion University. She is now an online biology teacher for the Insight School of Washington (ISWA).

Isabella Bruni, PhD in Communication Studies, is a media educator and researcher. She is a member of the Italian Association for Media Education (MED), and part the editorial board of the journal 'Media Education: Studi, Ricerche, Buone Pratiche'. Her research interests are in mobile learning and digital storytelling through social media.

Margaret Anne Carter has a legacy of being a passionate educator who energizes and inspires those around her. Working as a behaviour change specialist, teacher, academic and author, Dr Carter has a proven track record of working with individuals and groups across the lifespan. The focus of this work is social and emotional wellbeing, identity, parenting, and early years education. Dr Carter has presented keynote addresses and conference papers specific to these areas in a variety of professional and community contexts. She is the author of several books, including Young Children's Social Behaviour and Take a Stand, Lend a Hand – Stop Bullying Now. Dr Carter has co-authored social skills texts including Set for Success and Friendly Fact, in addition to consultancy and parenting programs. Currently Dr Carter is employed as senior lecturer in the School of Education at James Cook University.

Onur Ceran is a PhD student at Gazi University, Faculty of Gazi Education, Department of Computer Education and Instructional Technology. He got his MSc degree at Information Systems from Middle East Technical University. He is a cybercrime investigator at Department of Cybercrime, Turkish National Police. His research area includes theories and applications of distance education, designing and developing content for e-learning, big data, mobile learning and programming, designing and developing context aware systems.

Thomas Cochrane is an Academic Advisor and Senior Lecturer in educational Technology at AUT University's Centre for Learning and Teaching (CfLAT). In 2011 he was awarded as an Ascilite Fellow http://www.ascilite.org.au/index.php?p=awards. Previously Thomas Cochrane was an Academic Advisor (elearning and Learning Technologies) with Unitec from 2004 to 2011. His research interests include mobile learning, web 2.0, and communities of practice. His PHD thesis was titled: "Mobilizing Learning: Transforming pedagogy with mobile web 2.0". Thomas has managed and implemented over 40 mobile learning projects, with a recent focus upon Android and iOS smartphones and the iPad as catalysts to enable student-generated content and student-generated learning contexts, bridging formal and informal learning environments. He has over 100 peer reviewed publications, receiving best paper awards at Ascilite 2009, ALT-C 2011, ALT-C 2012, and has been invited to keynote at several international educational technology conferences.

Marta Adriana da Silva Cristiano Doctoral student in Knowledge Engineering and Management, Master in Computer Science and certified in Inclusive Education. Currently she is researcher in the Reamote Experimentation Laboratory (RExLab) and she is professor for Information Systems at UNIBAVE. She has solid experience on computer science with focus on Information and Communication Technology, with, working mainly in the following topics: Remote Experimentation, Multimedia, Education & Information Technology, M-Learning, B-Learning, Mobile Computing, Web-System Development, and Open Software.

Helen Crompton (PhD) is an Assistant Professor of Instructional Technology at Old Dominion University, Virginia. Her research is focused on mobile learning and the effective integration of technology into K-12 education.

Filomena Dayagbil is dean of Teacher faculty for many years and a well known researcher.

Sonal Dekhane is an Associate Professor of Information Technology at Georgia Gwinnett College (GGC). At GGC, Dr. Dekhane contributed to curriculum development and led program assessment initiatives. She is committed to increasing student engagement in her classroom and experiments with a variety of pedagogical techniques to achieve this goal. She has successfully used mobile technology to enhance students' learning experience. Her work has been presented at national and international conferences and published in national and international journals such as International Journal of Mobile and Blended Learning. Dr. Dekhane is currently involved in projects focusing on increasing women's participation in computing at GGC.

Francisco Delgado is Professor of Physics and Mathematics in the Tecnologico de Monterrey, Campus Estado de Mexico. He is a Ph.D. in Physics (Quantum Mechanics). He was chairman of Mathematics Department and he is former dean of Engineering. Actually, he is a researcher full time professor of that institute in Quantum Information and Quantum Control. Nevertheless, his main field is Physics, he has been working in educative innovation for science: curricular integration, problem based learning and use of technology. He has been involved with mobile learning since 2009, leading a cross research project for all disciplines implementing mobile and blended learning in his institution. In the last years, he was awarded with two national prizes: Technological Development by QUANTUM (a software for quantum simulation) and with Research in Educative Innovation for the private universities in Mexico.

Giovanni Ferreira de Farias is a Specialist and advisor on distance education and Moodle certified service provider, Giovanni Farias has been providing pedagogical and technical services for government departments, universities, schools and corporations in Brazil since 1997. He is an electronic engineer who has a master degree in Computer Science (Brazil) and an EdD degree in Distance Education (Canada). Currently, he takes part of the research group on media and knowledge of the Federal University of Santa Catarina, in Brazil.

Silvio Serafim da Luz Filho is a Doctor in Education, Master in Business, graduated in Psychology, with certification in Educational Advising and School Psychology. Currently, he is Associate Professor at the Department of Psychology of the Federal University of Santa Catarina, professor in the Graduate Program in Knowledge Engineering and Management, coordinator of the Center for Research in Psychology for Traffic and Transportation, and the Research Group on Vocation, Occupation and Labor. He is member of the Brazilian Association for School and Educational Psychology and of the Brazilian Institute of Psychology Evaluation.

Virendra Gawande is currently the Head of IT Department and Assistant Professor at Sur College of Applied Sciences (Ministry of Higher Education) in Oman. He holds a PhD in Information Technology and a Master's degree in Computer Science. He has got an academic experience of 20 years and taught a range of courses in Computer Sciences and Information Technology. He has supervised several software projects along with Masters and Doctoral thesis. He has published several scientific research papers in International journals on the research topics related to Database, eLearning methodology and e-government. He is a Member of International Association of Engineers (IAENG), since March 2013. Member for the Middle East Association of Computer Science and Engineering (MEACSE) in Scientific Technical Committee and Editorial Board Member for the International Journal of Computing Academic Research (IJCAR), since November 15, 2013. He is also associated with 'The Research Council', Oman, as a peer reviewer since August 2014.

Donna Goldie is a committed educator who has worked with people from all ages and in various locations. Donna is engaged in postgraduate studies in the areas of blended learning and supporting student engagement which includes the ethics of blended learning. Donna supports students through their studies by creating and implementing innovative teaching and learning experiences that can be accessed through all modes of learning.

Ezgi Tosik Gün received her BSc. degree in Computer and Instructional Technology from Necmettin Erbakan University in 2012. She completed her master's degree in the same branch in Gazi University in 2014. Now she is research asistant in Gazi University she continues phD.

Noorminshah A. Iahad is a senior lecturer at the Information Systems Department, Faculty of Computing, Universiti Teknologi Malaysia (UTM). She received her PhD degree at the University of

Manchester, United Kingdom. Her research interests focus on Cloud Computing in education and for SMEs, Computer Supported Collaborative Learning, and usage of mobile applications for health monitoring and management. She has graduated seven PhD students and three Masters research students. She is a member of the Association for Information Systems and she is one of the Editors of the International Journal of Social Networking and Virtual Communities.

Cynthia Johnson, Assistant Professor of Information Technology at Georgia Gwinnett College, received her doctoral degree in Computer Engineering at the University of Central Florida in 2011. Research interests include machine learning, teamwork, and improving computer programming education. Prior to her academic career, she worked as a software engineer on various projects for over 20 years. Projects worked on include missile navigation and guidance systems, locomotive control systems and military training products. Companies worked for include Sonalyts, SRI, SAIC, Boeing and Martin-Marietta.

Serçin Karataş is an associate professor at the Department of Computer Education and Instructional Technologies of Gazi Faculty of Education at Gazi University, Ankara, Turkey. She undertook graduate studies at Gazi University Turkey (MA. Computer Education) and at Ankara University, Turkey (Ph.D. Educational Technology). Dr. Karataş has over 15 years' experience in focusing on the theoretical and applications of distance education. Her research area includes theories and applications of distance education, designing and developing story board and content for e-learning, instructional design, design principles for e-learning, interaction applications for e-learning, L(C)MS, creativity for (e)learning and gamification for education.

Mustafa Kılıç is a lecturer in Computer Programming Department at Ahi Evran University and PhD Student at CEIT, Gazi University. He is working on e-learning, near field communication, encryption algorithms and Internet programming languages. His Masters thesis is about project based learning and project management. He has several projects and a patent about visually impaired people and one of them was supported financially by Ministry of Science, Industry and Technology.

Cecily Knight is Manager, Blended Learning & Innovation at James Cook University, Australia. She is passionate about the power of education to change lives and communities and sees the opportunities the digital age brings to this agenda. She has a keen interest in technology-enhanced learning and has been developing curriculum and researching that field since completing her Masters degree in 1994 which focused on use of computers in the classroom. She is interested in researching the student experience as well as how to support academic staff to best utilize technologies for learning. She has participated in a number of research projects including three national OLT Projects on: Graduate Employability; Advancing Academic Professionalisation; and Exploring opportunities for whole of institution collaborations that enhance student learning. Prior to her current role, Assoc Prof Knight was Associate Dean Teaching & Learning at James Cook University.

Nimet Özgül Ünsal Köse is a lecturer at Ankara University. She lives in Ankara, Turkey. She is a PhD student at Gazi University Department of Computer Education and Instructional Technology. She was a Master degree at Gazi University Department of Office Management Training. Her fields of interests are Education(al) Technology, Computer Education, Interaction Type in the learning management system, and Performance-Based Education.

Katja Lehmann is a research associate at the Chair for Information Systems at the University of Kassel, Germany, since 2011. Her research focuses on technology-mediated learning, learning success measurement, peer assessment and skills development. She has been involved in several research and industry projects funded by the European Union, German ministries, and industry partners.

Jan Marco Leimeister joined University of St.Gallen, Switzerland, in 2012 as a professor at the Institute of Information Management (IWI-HSG), and is also the Chair for Information Systems at the University of Kassel, Germany, since 2008. His research focuses on Service Engineering, Collaboration Engineering, IT Innovations and Technology Management.

Kathryn MacCallum is a senior lecturer at the School of Computing at Eastern Institute of Technology, New Zealand. Her PhD addressed the adoption of mobile learning by students and teachers in the tertiary sector. Her PhD research proposed and tested a structural model of the adoption of mobile learning in tertiary education. Her current research focus encompasses the use of mobile technology throughout all sectors of education. She has a strong interest in the effective use and adoption of technology in education.

Marva Mirabolghasemi is currently a PhD candidate at the Department of Information Systems, Universiti Teknologi Malaysia. Her research interests are virtual learning environments, health 2.0, social networking, effectiveness and impact studies.

Danielle Mulrennan is a founding member of the Enhancing Journalism Education (EJE) community of practice. EJE is a community of research-focused practitioners who aim to aim to critically explore mobile social media applications in order to untangle the complex publishing and archiving of contemporary news mediascape to enable student-directed Heutagogy leading to enhanced learning outcomes of news and values, social media publishing and social capital. The COP was established in 2012 and combines the expertise of reflective practitioners, lecturers and industry stakeholders through international collaboration, cultural, textual, audio and visual media. See more at https://www.aut.ac.nz/profiles/communications/lecturers/danielle-danni-mulrennan.

Priscila Cadorin Nicolete graduated in Information & Communication Technologies (2014) and master´s degree in Information & Communication Technologies (2016) at Federal University of Santa Catarina (UFSC), Brazil. She has co-authored in papers about remote experimentation and mobile learning; has experience in information technology in education.

Paul Pagliano works across the three campuses of James Cook University: Townsville, Cairns, and Singapore. He is Vice President of the South Pacific Educators in Vision Impairment, an Associate Editor of the International Journal of Disability, Development and Education, and on the editorial boards of the British Journal of Visual Impairment, Australasian Journal of Special Education, Journal of Intellectual and Developmental Disability, Journal of the South Pacific Educators in Vision Impairment, Australian Journal of Dyslexia and Other Learning Disabilities.

Michel Plaisent is a full professor in the Business School of University of Québec in Montréal and former director of the MIS program. His Ph.D. in Information System is from the University of Québec in

Montréal in 1987; as is his M.Sc. in Project Management in 1980, his B.Sc. in Computer Sciences from University of Montréal in 1971, and his B.A. from Sainte-Marie Collège in 1968. He has held positions with governments and enterprises as system programmer, analyst and CIO from 1971 to 1987, before he joined the faculty of University of Québec in Montréal. He still works as consultant for many small business and non-profit organizations. His doctorial thesis (1987) was on the use of e-mail as a management tool. Dr. Plaisent's research focus is on the concept of cognitive ergonomics, learning problems and personal productivity tools for managers. He has published several books and articles.

Angeline Pogoy is well known as a conference guest speaker, researcher, pre-service teaching mentor in the teaching of mathematics, a professor and the research coordinator of the College of Teacher Education.

Maria Ranieri, PhD in Telematics and Information Society, is an Associate Professor of Educational Technology at the Department of Educational Sciences and Psychology, University of Florence. Her main research areas include theory and methodology relating to media and technology in education, as well as work around teachers' practice and students' learning. She is a member of the Italian Society of Pedagogy (SIPED) and of the executive council of the Italian Association for Media Education (MED).

Vernon Rive joined the School of Law at AUT University Law School as a Senior Lecturer in 2009 after over 14 years of private practice, latterly as a partner in the Environmental/Resource Management team at one of New Zealand's leading national commercial law firms Chapman Tripp. Since commencing a career as a full time legal academic in 2009, his research activities have focused on three key areas of interest: climate change law, international environmental law and New Zealand environmental law. These research interests align with his teaching commitments which include lecturing in International Environmental Law and Resource Management Law for 3rd and 4th year law students, as well as a compulsory course in Judicial Review to second-year law students. He is an associated scholar in the Brussels-based Refract research network on "Fragmentation and Complexity in Global Governance". From 2016, will be undertaking PhD research examining phenomena of international law fragmentation and regime interaction in the context of cross-border climate displacement through Melbourne Law School.

Juarez Bento da Silva has a degree in Business Administration from the Catholic University of Rio Grande do Sul (1991), specialization in Networking and Telecommunications at the UFSC(1999), Masters in Computer Science from UFSC (2002), PhD in Engineering and Knowledge Management, UFSC(2007). He is currently Associate Professor at UFSC and coordinator of the Remote Experimentation Lab (RExLab). He is currently Associate Professor at the Federal University of Santa Catarina (UFSC) and coordinator of the Remote Experimentation Lab (RExLab). He has experience in Computer Science with emphasis in Computer Systems and acting on the following topics: remote experimentation, 3D virtual worlds, embedded systems, computers in education and digital inclusion. The main interest in research and current focus is the development and use of new information and communication technologies in teaching and learning. Currently we seek to develop and implement new devices for remote testing and work on the integration of hardware (remote experiments) in 3D virtual worlds, using open-source project. This is the link to his resume for more details: http://lattes.cnpq.br/1594563006260546.

Helen Sissons spent 17 years as a journalist in the US and the UK, the last ten with the BBC as a television journalist and reporter. She is the author of Practical Journalism: How to write news. Before coming to New Zealand, Helen also spent six years teaching journalism at the University of Leeds, UK. Her current research interests are the future of journalism, ethnography and news production, ethnography and public relations practice and the relationship between public relations and journalism. See more at: https://www.aut.ac.nz/profiles/communications/curriculum-leader,-journalism/helen-sissons.

Matthias Söllner is a postdoctoral researcher at the Chair for Information Systems at the University of Kassel, Germany, as well as at the Institute of Information Management at University of St.Gallen (IWI-HSG), Switzerland. His research focuses on designing trustworthy and used IT innovations especially in the domains of higher education and vocational training.

Zeynel Abidin Tok, a native of Adana/Turkey, is passionate about teaching and about science. He received a Bachelor's in Computer education and instructional technology from Çukurova Üniversity, NY, a Master's in Education Management from the Dicle Üniversity, and is still a PhD student at Computer Education and Instructional Technology at Gazi University. Recently, information technology is committed to the Society. He also carries out the data processing manager for the Ministry of National Education. Interest programming, database management systems, is on instructional design, distance education and e-learning.

Mai Yin Tsoi, a native of California, is passionate about teaching and about science. She received a Bachelor's in Chemistry from Vassar College, NY, a Master's in Organic Chemistry from the Georgia Institute of Technology, and a Doctorate of Philosophy in Science Education from the University of Georgia. She has conducted oceanographic research at the Massachusetts Institute of Technology, Texas A&M University, and the U.S. Naval Research Center in San Diego, CA. Her published papers and current research center on the use of technology in science teaching and learning, with a focus on mobile learning and software development. Recently, her scholarly work helped garner the 2011 Excalibur Award from the Technology Association of Georgia and the 2012 Catalyst Award in Mobile Innovation from Blackboard. Dr. Tsoi was named Gwinnett County Teacher of the Year and was the 2nd finalist in the Georgia State Teacher of the Year competition, both in 2006. She taught high school Chemistry and Physics for 10 years in Washington D.C and Georgia. She has been teaching General and Organic Chemistry as an Associate Professor at Georgia Gwinnett College since 2007 and has thrived on the challenge of developing the first 4-year college in Gwinnett County, where she calls home.

Ülkü Ülker is a PhD student at Gazi University, Faculty of Gazi Education, Department of Computer Education and Instructional Technology. She is also research assistant at the same department, Faculty of Education, Dicle University. Her research area includes instructional Technologies, instructional design theory and practice, 21.st century skills, distance education, mobile learning, serious games, gamification, machine learning, artificial intelligence systems, adaptive systems and cryptology.

Index

I

J

K

L

M

N

O

P

R

S

T

U

V

W

Y

www.ingramcontent.com/pod-product-compliance
Lightning Source LLC
LaVergne TN
LVHW060408300926
849162LV00006B/1376